Domain-informed Machine Learning for Smart Manufacturing

Qiang Huang

Domain-informed Machine Learning for Smart Manufacturing

 Springer

Qiang Huang
Daniel Epstein Department of Industrial and
Systems Engineering
University of Southern California
Los Angeles, CA, USA

ISBN 978-3-031-91630-4   ISBN 978-3-031-91631-1   (eBook)
https://doi.org/10.1007/978-3-031-91631-1

This Springer imprint is published by the registered company Springer Nature Switzerland AG
The registered company address is: Gewerbestrasse 11, 6330 Cham, Switzerland

If disposing of this product, please recycle the paper.

To my parents:
Chengxian and Xiufang

To my family:
Quan, Amy, and William

Preface

Over the past two decades, technology breakthroughs in manufacturing and computing have paved the way for the Fourth Industrial Revolution. Personalized products such as 3D-printed medical implants are more accessible to the public. The fast beats of Artificial Intelligence (AI), Machine Learning, and the digital transformation of manufacturing can be felt in our daily life. The transition from the century-old mass production paradigm to the personalized manufacturing paradigm has increasingly become a reality for value differentiation. It might be one of the best times for smart manufacturing research.

Product quality control remains essential for both the economy of scale in mass production and the economy of scope in personalized manufacturing. Statistics has provided a scientific basis to infer the quality of the whole product population in mass production where the cost of learning from product samples diminishes with the scale. In personalized manufacturing, quality control faces the fundamental question of how to ensure quality of every product variety with one or a few units; more specifically, how to infer the quality of the whole product variety population based on a sample or a subset of product varieties? Does that mean that we have to break away from the existing statistical quality control framework?

This book aims to present domain-informed machine/statistical learning as a viable strategy to address the quality control challenge in personalized manufacturing. By integrating domain knowledge, machine learning models can remain robust in the presence of imperfect or limited data, providing interpretable prediction and generalizable insights. However, machine learning for manufacturing is often motivated by real-world specific problems with many engineering details. Substantial training or experience is usually needed to define and formulate domain-meaningful learning problems for individual cases, collect the right data economically, identify contextualized information representation, devise interpretable models, generalize the insights, and deploy validated models for applications. The buildup of experience through learning isolated cases and connecting dots could take a long period of time.

Using additive manufacturing and nanomanufacturing as exemplary domains, this book walks through the development of methodology from simple cases to general forms for contextualized information representation, domain-informed dimension reduction, fabrication-aware machine learning and transfer learning, and process-informed optimal compensation. Through case studies, the book illustrates how manufacturing domain knowledge in non-analytical forms is integrated into various aspects of machine learning for smart manufacturing quality.

The book is aimed at manufacturing and quality engineering researchers and engineers, AI and machine learning professionals with interest in smart manufacturing and digital twins, and graduate students specialized in advanced manufacturing, quality engineering, or machine learning for engineering applications. It is suited for a second-year graduate-level course on Machine Learning for Smart Manufacturing, requiring prerequisites in multivariate statistics or graduate-level machine learning knowledge. The knowledge requirements for the manufacturing domain are minimal. Additional learning and teaching materials, including data, codes, slides, and other resources, can be found at http://huanglab.usc.edu/book.

The author sincerely thanks his mentor, Prof. Jianjun Shi, for his wise suggestions, unfailing support, and continuous encouragement. The author is also thankful to all his coauthors Hui Wang, Xi Zhang, Lijuan Xu, Li Wang, Yuanqin Duanmu, Yuanxiang Wang, Cesar Ruiz, Nathan Decker, Weizhi Lin, Shekhar Bhansali, Tirthankar Dasgupta, Yong Chen, Chongwu Zhou, Arman Sabbaghi, Malancha Gupta, Noah Malmstadt, Raquel de S.B. Ferreira, and Nabil Anwer, whose work contributed to the book content. Lastly, the author extends deep gratitude to the US National Science Foundation for its decade-long support of research in nanomanufacturing and additive manufacturing, as well as to the US Office of Naval Research for its support of cyber-enabled additive manufacturing research from 2011 to 2014.

Lastly, there are numerous open research challenges in domain-informed machine learning for smart manufacturing. This book represents author's initial exploration of the development of quality control theories and methods for personalized manufacturing.

Los Angeles, CA, USA   Qiang Huang
October 2024

Contents

Acronyms

2D	Two Dimensional
3D	Three Dimensional
AI	Artificial Intelligence
AM	Additive Manufacturing
CAD	Computer-Aided Design
CDMV	Cross-Domain Model Building and Validation
DIML	Domain-Informed Machine Learning
DIKW	Data-Information-Knowledge-Wisdom
FMP	Finite Manufacturing Primitive
GMRF	Gaussian Markov Random Field
IoT	Internet of Things
IR	Impulse Response
LPBF	Laser Powder Bed Fusion
ML	Machine Learning
ML4AM	Machine Learning for Additive Manufacturing
NM	Nanomanufacturing
PM	Personalized Manufacturing
QC	Quality Control
SLA	Stereolithography
V/Q	Variety-to-Quantity Ratio
WAAM	Wire and Arc Additive Manufacturing

Chapter 1
Introduction to Smart Manufacturing and Its Quality Control

1.1 Concept and Goal of Smart Manufacturing

Smart manufacturing refers to the next-generation manufacturing featured by "fully-integrated, collaborative manufacturing systems that respond in real time to meet changing demands and conditions in the factory, in the supply network, and in customer needs" (NIST, National Institute of Standards and Technology). It is "the scaled integration of networked data with plantwide optimization; physical and sustainable production; and resilient, demand-driven supply chains" [54].

The concepts of smart manufacturing and Industry 4.0 originated in the mid-2000s from Smart Plant in the United States and Smart Factory in Germany [11, 15]. The notation of Industry 4.0 has been expanded to be a synonym of the Fourth Industry Revolution. These concepts were driven by the digital transformation in manufacturing and increased availability of data at all levels of factories, supply chains, and ecosystems. The technologies that have expedited the digital transformation include advanced simulation and computer-aided engineering, sensors, Internet of Things (IoT), cloud and edge computing, cyber-physical systems, artificial intelligence (AI), machine learning (ML), autonomous robotics, additive manufacturing (AM), high-speed communication, digital twins, human-machine interactions, etc.

By integrating the aforementioned technologies to collect data and digitize processes, smart manufacturing seeks to effectively utilize data to visualize real-time operation states, develop insights into operation behaviors, predict or replicate operation outcomes, and determine the best human or machine actions [54]. In a smart manufacturing system, technical and business processes, physical and cyber systems, smart products, and production systems are fully integrated and connected. Operation events will be predicted or detected and, if necessary, trigger proactive and collaborative actions in the factories and supply chains. Relieved from manual and repetitive tasks, upskilled workforce will focus on more complex and creative tasks to gain and build new manufacturing knowledge through intelligent

© The Author(s), under exclusive license to Springer Nature Switzerland AG 2025

Q. Huang, *Domain-informed Machine Learning for Smart Manufacturing*,

https://doi.org/10.1007/978-3-031-91631-1_1

human-machine interfaces. Ultimately, manufacturers can better adapt to changes in demand, reduce cost, and improve productivity, agility, quality, resilience, and sustainability.

1.2 Role of Data, Machine Learning, and Human in Smart Manufacturing

Visualization of operation states, prediction and analysis of operation behaviors, and optimal decision-making for actionable items rely critically on right data collection, contextualized representation, and domain-meaningful analytics. Scaled integration and utilization of data from factory operations to supply chains is therefore a key characteristics of smart manufacturing [54].

The utilization of data generally follows the DIKW (Data-Information-Knowledge-Wisdom) pyramid or hierarchy [2, 50, 67]. In the DIKW pyramid model (Fig. 1.6), data are unorganized and unprocessed facts of observations without interpretation. Information is processed, organized, or contextualized data with a meaningful form, description, or representation. Information answers to questions of what, who, when, where, or how many [2]. Knowledge answers why and know-to questions and transform information into insights, instructions, and actionable items to improve efficiency. Involving human judgment, wisdom is the ability to increase effectiveness. Ackoff argued that intelligent machines can generate information and knowledge, but not wisdom, which is the pursuit of ideas and values unique to human [2]. In future manufacturing, human can play a central role because greater synergies and interactions can be achieved through the integration of judgment-focused humans and prediction-focused intelligent machines [29].

1.3 Challenge of Machine Learning for Smart Manufacturing

Machine learning has been increasingly deployed in smart manufacturing to transform data into information and knowledge. The ML models, however, inevitably have uncertainties, which include aleatoric and epistemic uncertainties [28]. Aleatoric or statistical uncertainty refers to randomness, natural or inherent variation in a system that is irreducible. Epistemic uncertainty arises from the lack of knowledge or data, which can be reduced through better understanding or information.

A major challenge of ML for smart manufacturing is therefore how to reduce epistemic uncertainty because a lack of knowledge and right data can co-exist. The lack of knowledge may arise from:

- *In-complete first-principle knowledge*: Understanding of advanced manufacturing processes such as AM and nanomanufacturing requires multiphysics and multiscale simulation. Simulating the production of life-size products can be computational prohibitive and time-consuming. The incomplete first-principle understanding of production processes can cause uncertainty in surrogate model specification.
- *Insufficient causal understanding on what data to collect*: The causality between sensing data and manufacturing outcomes fundamentally determines the quality of data and ML models. The collected data may not be relevant to the fit of the purpose. Critical process variables may be missed in data collection and become unknown lurking variables in the causal inference.
- *Limited sensing capability*: Not all process variables can be measured due to sensing constraints. Being potentially critical to ML models, unmeasurable process variables also constitute lurking variables whose data is unavailable for causality analysis.

Other than the cost constraints on producing sufficient data for ML model training, the lack of right data can be attributed to a number of reasons:

- *High dimensionality of target space*: Thanks to the trend of mass customization and personalization, a smart manufacturing system can produce a large variety of products. The target space for prediction and control is high dimensional.
- *Heterogeneity*: Products of different varieties can be manufactured under different machine settings. Even under the same machine setting, covariates such as build location and orientation can introduce another layer of heterogeneity when fabricating the same product variety.
- *Low volume*: In smart manufacturing, production volume for each product variety can be significantly reduced, even down to single units. Data are not available for new product variety without trial production.
- *Missing process data*: The lack of causal understandings or sensing capability can result in missing critical process data.

Consequently, ML for smart manufacturing has to overcome the challenge of the lack of knowledge and data because the cost of generating sufficient data for common ML methods can be prohibitive.

1.4 Domain of Interest: Quality Control (QC) for Smart Personalized Manufacturing

While the concept of smart manufacturing is applicable to various manufacturing paradigms such as mass production or mass customization, the domain of interest of this book is to explore the quality control (QC) theory and methods for personalized manufacturing.

Modern manufacturing began around 1850 as craft manufacturing of cars for individual customers and later evolved into mass production owing to Henry Ford's invention of assembly lines in 1913. Mass production achieves economy of scale through dedicated production systems with limited product variety. In response to market segmentation and consumers' changing demand, mass customization emerged in the late 1980s to offer consumers a larger numbers of product options. Its economy of scope is achieved through reconfigurable production systems [8, 66]. Personalized manufacturing was introduced around 2000 as consumers became increasingly involved in the product design and realization for value differentiation [23, 39]. On-demand production systems and AM technologies further expedite the trend of cost-effective scaled production of a large variety of personalized products. It has been predicted that personalized manufacturing represents a paradigm shift for future manufacturing [23, 24, 38].

Product QC plays a vital role for cost-effective manufacturing. Its scheme is dictated by the manufacturing paradigm. As the dominant manufacturing paradigm in the past century, mass production has achieved the cost-effective fabrication of a large quantity of given product designs. Initially the basic question for QC was how to ensure the quality of every unit of the same design. Walter A. Shewhart in the 1924 addressed the fundamental challenge by introducing statistical science into QC [55]. A probability distribution perfectly describes the quality characteristics of all units. Statistical science ensures that a random sample of units is sufficient to infer the quality of the whole product population. Statistical QC is therefore an ideal fit for mass production where the cost of learning from product samples diminishes with the large production scale.

On the verge of a Fourth Industrial Revolution, rapid advances in computing and digital manufacturing technologies have increasingly transformed personalized manufacturing paradigm into a reality. Through AM, a large variety of products with complex geometries can be produced in extremely low volumes. The basic question that QC faces in personalized manufacturing is how to ensure the quality of every product variety or 3D design with one or a few units. Since one-off products often constitute complex heterogeneous data, statistical QC cannot be directly applied because units are from different populations. The fundamental QC challenge has shifted from large product quantity in mass production to large product variety in PM or one-off production. This challenge raises pivotal QC questions: *How to infer the quality of the whole product variety population based on a sample or a subset of product varieties?*

Although mass production also faces the product variety issue, a manufacturer typically manage a much smaller set of product varieties. Even without changing materials and machine conditions, the product variety space in personalized manufacturing has an infinite dimension. The complexity of making large product varieties in low volumes grows exponentially in comparison to managing a limited variety in large quantities. For one-off production, QC requires "doing it right the first time" not only for tested designs but also for a great number of untried new designs. Unlike mass production, it is simply not cost-effective in personalized manufacturing to improve quality through multiple iterations of making the same

Table 1.1 Comparison of quality control in mass production and personalized manufacturing (PM)

Categories	Mass production	PM
Quantity per product Q	Large Q or $Q \approx \infty$	Small Q or $Q = 1$
Product variety V	Small V	$V = \infty$
Variety-to-quantity (V/Q) ratio	$V/Q = 0$	$V/Q = \infty$
Data quantity	Sufficient data	Small or no data
QC objective	Variation reduction	*To be defined*
Analytical basis	Statistics	*To be developed*

product. An analogy to the curse of dimensionality in data science, the "*curse of variety*" demands new QC theories and methods.

Table 1.1 shows a comparison of quality control scenarios for mass production and for personalized manufacturing. Denote Q the quantity or volume of each product, and V the variety of product designs to be manufactured. If we introduce a product "variety-to-quantity" ratio or V/Q ratio as

$$\frac{V}{Q} = \begin{cases} 0, & \text{mass product;} \\ \infty, & \text{personalized manufacturing,} \end{cases} \tag{1.1}$$

then V/Q approaches to 0 in mass production but to infinity in personalized manufacturing.

This huge contrast in V/Q ratios has profound impact on methodology development for QC. In mass production, the definition for quality naturally focuses on the variation among the large quantity of products. The objective of QC therefore aims at variation reduction [56]. Thanks to sufficient sample data, statistics-based QC methods such as statistical process control, acceptance sampling, and design of experiments have been established for quality improvement. Mean and variance estimated from sample data are frequently utilized to characterize quality characteristics.

On the contrary, the near-infinity V/Q ratio in personalized manufacturing implies data heterogeneity and lack of sufficient sample for each product variety. Without enough sample data to build credible statistical distributions for quality characterization, the long-established concept of quality for mass production cannot be directly adopted. The aggregated data from large product variety can be highly heterogeneous due to variations in product designs, materials, processes, and process conditions. The independent and identically distributed assumption critical for Statistical QC can hardly be satisfied. Furthermore, the bigger challenge of data availability arises for new product variety that have not been made before.

The V/Q ratio can be utilized as a quantitative index to classify different manufacturing paradigms. Figure 1.1 shows different manufacturing paradigms on the product variety-versus-quantity (V-Q) diagram . Koren (2021) presented the trend of manufacturing in a product volume-versus-variety plot without a quantitative index like V/Q ratio [38]. As can be seen in Fig. 1.1, craft manufacturing and mass

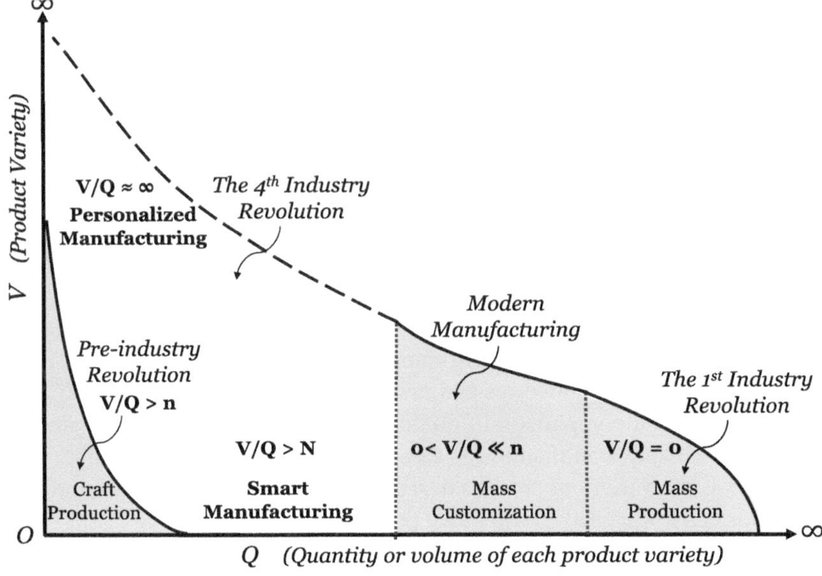

Fig. 1.1 Manufacturing paradigms on product variety-quantity (V-Q) diagram

production can be placed at the two ends of the V-Q diagram. Modern manufacturing including mass customization slightly increases V/Q ratio from zero. However, there is a still a large gap between craft manufacturing and mass customization on the V-Q diagram, where a larger number of product variety can be produced cost-effectively in smaller volumes. Featured by large V/Q ratios, smart manufacturing including personalized manufacturing will fill the gap in the era of the Fourth Industrial Revolution.

1.5 Product Geometric QC for Smart Additive Manufacturing (AM)

1.5.1 Brief Introduction of AM

Additive manufacturing (AM), or three-dimensional (3D) printing, refers to a new class of smart manufacturing technologies associated with the direct fabrication of physical products from computer-aided design (CAD) models. AM holds the promise of direct digital manufacturing [5, 7, 18, 22, 43] and is widely recognized as a disruptive technology, having the potential to fundamentally change the nature of future manufacturing [16]. In contrast to traditional material removal processes, AM builds products by adding material layer by layer without part-specific tooling and fixturing. Figure 1.2 illustrates the general procedure, where a 3D model of

Fig. 1.2 A 3D printing procedure

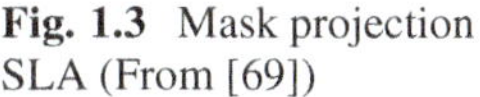

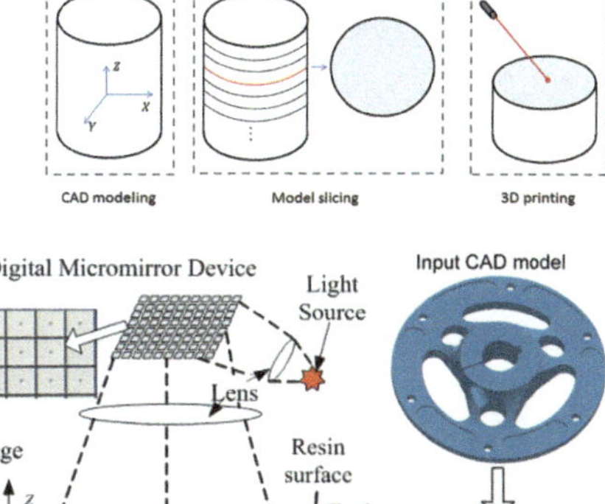

Fig. 1.3 Mask projection SLA (From [69])

the product is first built by CAD tools [69]. A specialized program slices the CAD model into cross sections and generates STL format files, which are sent to the AM machine to physically build each layer in a sequence.

Stereolithography (SLA) is the first commercialized technique for 3D printing patented in 1988 [30]. As illustrated in Fig. 1.3 [69], the SLA machine has liquid resin stored in a tank configured with a platform that can move vertically. During the printing process, the surface of the resin is exposed to light, which triggers the resin solidification. Control of light exposure area and intensity is through a digital micromirror device that receives commands from STL files for each layer. The platform in the tank moves down with the predefined thickness for printing the next layer when the previous layer is solidified.

Laser powder bed fusion (LPBF) is another widely known AM process with the ability to produce 3D lightweight metal parts with complex shapes and desired mechanical properties (Fig. 1.4). It uses laser to locally melt fine metal powder along pre-defined scanning paths layer by layer [19]. High-value-added products such as topologically optimized lightweight fuel nozzles and personalized orthopedic implants can be made for various industries (Fig. 1.4).

For large-size customized metal parts used in aerospace, aeronautics, and petroleum industries, wire and arc additive manufacturing (WAAM) makes use of traditional arc welding technology and robotic arms to deposit materials at much higher rates in larger build areas [34, 48] (Fig. 1.5). Unlike traditional manufacturing techniques for fabricating large-size metal parts, WAAM eliminates multiple energy-intensive manufacturing steps [40, 44, 46, 49]. It therefore can significantly simplify supply chains, shorten lead times, and reduce material waste and environmental impact.

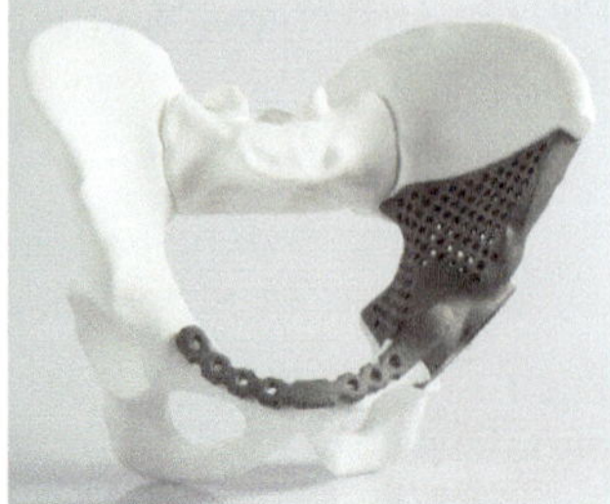

Fig. 1.4 Laser powder bed fusion (LPBF), GE fuel nozzle, and EOS hip implant

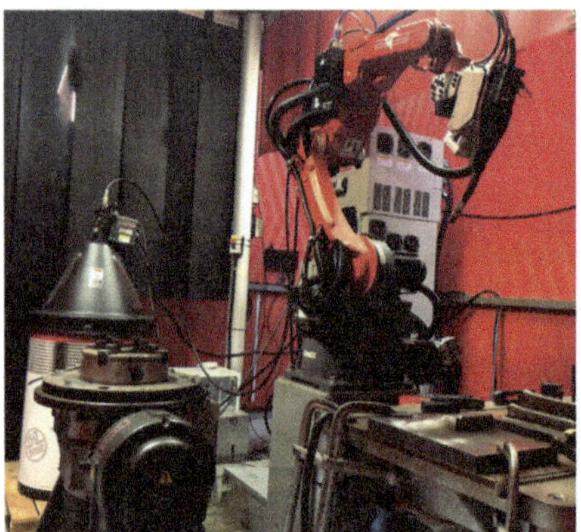

Fig. 1.5 Wire and arc additive manufacturing (WAAM) and a printed rocket propeller [21]

Despite its demonstrated capabilities and greater potentials, AM processes of various techniques often need iterative trials to optimize materials properties, minimize geometric deviations and distortion, improve dimensional accuracy, and reduce printing failures and defects, particularly for untried new product designs. In SLA processes, for example, material solidification is generally involved during layer formation, and this phase change from liquid to solid inevitably leads to shape shrinkage, a major root cause of geometric inaccuracy [64]. Quality issues in LPBF processes include dimensional and form errors, surface defects, porosity, residual stresses, cracks, and microstructural inhomogeneities and impurities [36]. In WAAM processes, layers are repeatedly subjected to a re-heating and cooling processes, and the generated heat stresses due to the multiple fusion, solidification, and phase change cycles are extremely nonlinear and transient. This results in significant residual stresses as well as large shape deviations and high surface roughness [31]. Effective quality control methodologies and tools are in need to materialize the potentials of cost-effective personalized manufacturing.

1.5.2   Problem Description for Product Geometric QC in Smart AM

Geometric and dimensional quality has long been essential to ensure product functionality and proper mechanical assemblies in complex engineering systems [57, 63]. As AM or 3D printing technologies advance rapidly with broader applications in aerospace, automotive, medical industries, and beyond, intelligent QC has become increasingly critical for cost-effective precision printing [9, 26, 27].

Yet AM poses unique challenges to QC [9]. Although it enables direct fabrication of products with complex geometries in a single production phase without tooling and fixturing, each phase actually consists of thousands of correlated steps forming and accumulating materials layer by layer. During printing, a multitude of factors such as materials, processing techniques, settings, and inter-layer interactions may cause printing defects. Physical modeling and simulation of complete layer-wise fabrication is still computationally prohibitive for timely prediction and QC operations [5, 37]. Multiple iterations or trial builds drive up printing costs significantly.

Quality verification of AM-built products is highly nontrivial as well. Established primarily for multistage subtractive manufacturing, standard quality verification methods often depend on features fabricated in the previous stages as datum references to verify the quality of those made later (e.g., distance or parallelism of plane feature relative to a datum). Since AM builds complex geometries in a layer-wise fashion, representation and verification of complex freeform surfaces demand AM-specific specification standards and tolerancing methods. This has been widely recognized by various international organizations and engineering communities as one of the top priorities to reduce the risk and cost of adopting rapid-growing AM technologies [3, 41, 47].

Unlike scale-driven manufacturing where operation expertise can be built up around a limited number of product families, AM aims at a versatile capability of making theoretically infinite variety of products. Learning from similar cases or past experience finds difficulty of defining similarity, particularly for low-volume AM with frequent design changes and heterogeneous AM processes. Current QC operations are heavily dependent on human expertise and intervention.

Lately, machine learning for AM (ML4AM) has emerged as a viable strategy to enhance 3D printing performance [12, 25–27, 35, 42, 51–53, 61, 65]. General-purpose machine learning and analysis of 3D shapes have been extensively studied in computer vision for shape analysis. It focuses on shape classification, matching, deformation (e.g., facial expression change), and correspondence [45, 62, 68]. But purely geometric analysis without consider engineering mechanisms of shape generation limits the scope of applying general-purpose ML techniques to manufacturing. For example, deviation patterns of 3D printed products not only vary with shapes but also object sizes owing to thousands of correlated steps of layer formation and accumulation [26, 27, 32]. Meanwhile, establishing fabrication-aware representation and learning of manifold-valued shape data have been open issues [4]. More progress of ML4AM has been made for specific tasks such as empirical and statistical modeling of AM processes [6, 59, 60, 70], sensing and inspection [17, 35, 53, 58], statistical shape analysis [10, 13, 14], and monitoring and detection [20].

Despite these advancements in ML4AM, the large amount of data required for model training and the lack of model interpretability and scalability are serious barriers for applying black-box learning methods to manufacturing. Product shape complexity, process complexity, and data heterogeneity in AM further complicate the QC efforts in AM. To achieve intelligent QC for precision 3D printing, four categories of ML4AM problems are defined, which are to be discussed in this book. Note that the problem definition below is far from being comprehensive.

Let input u to a 3D printing process be the designed shape of a 3D object. The set of design shapes is denoted as U. The output y represents the geometric quality or shape deviation of the actual printed product from its intended design u. The set of shape deviations of $u \in U$ is denoted as Y_U.

- **Offline learning problem (OLP)**: The offline learning objective is to establish functional mapping $f : U \to Y_U$, that is, to generate a model $f(u)$ to predict y_u by learning from a small set of training data $D_S = \{(u, y_u)|(u_i, y_{u_i}), i = 1, 2, .., n\}$.

  A product u to be predicted belongs to either source data D_S or a validation set D_V, i.e., $u \in D_S \cup D_V$ and $D_S \cap D_V = \varnothing$.
- **Offline control problem (OCP)**: The offline control objective is to solve the inverse problem by finding input adjustment δ_u for design $u \in D_S \cup D_V$ so that the modified design input $u + \delta(u)$ is expected to minimize the quality deviation $y_{u+\delta(u)}$ in reference to the intended design u.
- **Product qualification problem (PQP)**: The objective of product qualification is to define, measure, characterize, evaluate, and verify the quality of built products [1, 3, 41, 47, 71], specifically, the exact geometric deviation y_u of the built product from its intended design u. The qualification involves the pre-processing of the measurement/scanning data, automatic registration, and quantification of geometric deviations.
- **Generalization problem (GP)**: The generalization objective is to discover the structures of $f(u)$ (or $g(u, x, t)$) and $\delta(u)$ (or $\{\delta_u(t), \delta_x(t)\}$) and physical underpinnings so that solving learning and control problems for new product designs under new AM processes can be guided by principles. Transfer learning between source domain and target domain can also be achieved, where the domains can involve both product and process domains.

Other important categories of ML4AM problems that are not covered in this book include:

- **Real-time learning problem (RLP)**: The real-time learning objective is to establish functional mapping $g : U \times X \times t \to Y_U(t)$, that is, to generate a model $g(u, x, t)$ to predict product quality $y_u(t)$ based on sensing data $x(t)$ at time t. Real-time prediction is essential for process monitoring, change detection [9], and other digital twin applications.

- **Real-time control problem (RCP)**: The real-time control objective is to find real-time input adjustment $\{\delta_u(t), \delta_x(t)\}$ for design u so that the modified input $\{u(t+1) = u(t) + \delta_u(t),\ x(t+1) = x(t) + \delta_x(t)\}$ is expected to minimize the final quality deviation $y_u(\infty)$ in reference to the intended design u.
- **Machine qualification problem (MQP)**: The objective of machine qualification is to quantify the capability of a machine to realize the design u. MQP problem involves the machine capability definition quantification, test artifact design for machine capability qualification, test build analysis for machine capability evaluation, machine selection, etc.
- **Model verification, validation, and uncertainty quantification (VVUQ) problem**: The objective of VVUQ is to quantify the fidelity and reliability of computational models.

The descriptions of the aforementioned problems are relatively generic. As discussed previously, QC for AM faces issues of shape complexity, process complexity, data heterogeneity, and small data. These challenges impose domain-specific constraints for ML4AM. For example, a subset of problems can be defined for OLP, OCP, and GP in smart AM:

- **Input and output representation (OLP.1)**: A proper representation of (u, y_u) should enable small-sample learning. Training data D_S contains n printed products (e.g., $n < 10$) in different shapes and sizes. In general, the 3D shape u of AM-built products has infinite variety, and the shape space is manifold without natural linear structure. The output y_u is deviation of a product shape, not shape itself. The patterns of y_u depends not only on geometries of u but also on AM processes.
- **Model $f(u)$ generation and few-shot learning (OLP.2)**: Model $f(u)$ is expected to predict product quality of infinite variety by learning from small data D_S. This "1-to-∞" or few-shot learning problem prefers $f(u)$ being constructed with a limited number of building blocks, primitives, or basis functions for robustness and flexibility. Model interpretability is another key consideration to ensure better understanding of physical systems.
- **Optimal design compensation and optimality definition (OCP.1)**: Optimal criteria of minimizing shape deformation have to be defined with consideration of design specifications in AM. Optimal design compensation is to derive δ_u^* such that $y_{u+\delta_u^*}$ is minimized in reference to nominal design u for a given criterion.
- **Transfer function and system identification through transfer learning (GP.1)**: Deriving transfer function from $f(u)$ will provide compact description of AM processes and facilitate the understanding of physical process insights. Note that obtaining transfer functions independently for individual products will only result in a projected view of transfer functions in subspaces.
- **Principled design, learning, and improved quality control (GP.2)**: Combining transfer functions with block diagrams gives a powerful algebraic method to design, learn, and interpret AM processes.
- **Model adaptation and transfer learning with lurking variables (GP.3)**: Model $f(u)$ and D_S are associated with a specific process condition (materials,

process settings, printing methods, and machines). Let us denote the model $f(u)$ for the original source process with data D_S as $f^S(u)$. If the process condition is changed, model suitable for the new target process, denoted as $f^T(u)$, has to be established with new data set D_T. Obtaining $f^T(u)$ through transfer learning is to generate a new model from learning $f^S(u)$ and the combined data $D_S \cup D_T$. The model transfer/adaption may involve the change of model primitives or basis functions or model parameter changes for the same set of basis functions. Note that the actual differences between the source and target processes are unknown due to lurking or hidden process variables.

- **Design of source set D_S and target set D_T (PQP.1)**: Optimal design of source training set D_S is to minimize sample size n_S to obtain $f^S(u)$ with acceptable level of fidelity. Optimal design of target set D_T for transfer learning is to minimize n_T subject to $n_T \ll n_S$ and to obtain $f^T(u)$ from $D_S \cup D_T$ with acceptable level of fidelity.
- **Part-to-part (P2P) change detection and compensation**: Given D_S and a model $f(u)$ for an AM process, process changes shall be quickly detected either during or after printing a new or $(n+1)th$ product. If process change is detected, decision has to be made to either compensate changes or adjust process settings (including stopping).

1.6 Domain-Informed Machine Learning for Smart Manufacturing

Despite the ever-increasing sensing capability and amount of data collected in smart manufacturing, transforming data into interpretable information and knowledge faces challenges discussed in Sect. 1.3. Purely data-driven models not only require sufficient and potentially costly manufacturing data but also lack robustness for prediction and interpretability for manufacturing insights. Simulation based on the first principles can be prohibitively time-consuming and expensive for real-time manufacturing operations. Since seamless incorporation of data into simulation algorithms is challenging [33], integrating domain knowledge such as physical laws into statistical or machine learning approaches has long been a viable alternative when data is inadequate.

Domain-informed machine learning can be broadly defined as the problem-solving process that subject matter knowledge or physical understandings are leveraged to define and formulate machine learning problem, collect data through design, pre-process data for information representation, develop and train interpretable machine learning models, verify and validate models, and deploy the trained models with consideration of domain constraints and scalability requirements for real-world applications. By integrating subject matter knowledge, the domain-informed machine learning models can remain robust in the presence of imperfect or small data and provide interpretable prediction and generalizable insights.

One critical form of domain-informed machine learning is physics-informed machine learning, where the domain knowledge is presented in terms of partial differential equations, theoretical constraints, and boundary conditions. Various physics-informed neural networks have successfully solve both forward and inverse problems such as predicting material properties and discovering functional materials. Summarized in [33], physical knowledge can be embedded into learning models by introducing three types of biases:

- Observational bias: Observational data that reflect the underlying physical principles are intentionally collected or selected to cover the input domain. Often a large volume of data is needed to capture and reinforce the biases in data.
- Inductive bias: The learning model structure or architecture can be designed to incorporate prior assumptions so that the induction process implicitly satisfies the underlying physical laws.
- Learning bias: Rather than imposing hard structure constraints, soft constraints such as loss functions can be designed with great flexibility to choose models that approximately satisfy physical laws or constraints.

In manufacturing systems, domain knowledge can take analytical form like physical laws and mathematical models and, more often, non-analytical form like fabrication knowledge. Other than observational bias, inductive bias, and learning bias, domain knowledge can be integrated into machine learning through:

- Learning objective definition and problem formulation: The objective of machine learning problem and its formulation should be dictated by the problem domains such as manufacturing, not by specific machine learning techniques. Certainly, there may exist equivalence between different formulations [33].
- Information representation: Subject matter knowledge can assist in identifying the right form to represent data through feature representation, extraction, and dimension reduction.
- Data collection: Experiments can be designed to integrate domain knowledge, constraints, and information representation to collect data more economically and enhance observational bias more effectively.
- Verification, validation, and deployment: Verification, validation, and deployment of learning models need to consider domain constraints and scalability requirements for real-world applications.

1.7 Outline of the Book

This book intends to illustrate with examples how manufacturing domain knowledge in non-analytical forms is integrated into various aspects of machine learning for smart manufacturing quality. In the DIKW pyramid (Fig. 1.6), knowledge can inform the transformation from data to information and from information

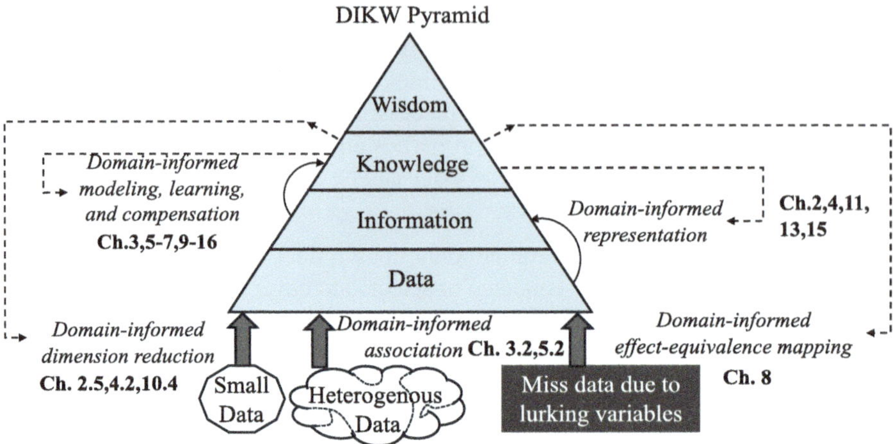

Fig. 1.6 Domain-informed machine learning for smart manufacturing quality

to knowledge, particularly in a small data environment with heterogeneity and unobservable or lurking variables.

The outline of the book is as follows:

- This chapter introduces the background and concept of smart manufacturing, challenges, and problems in domain-informed ML for smart manufacturing. Based on the author's published work, the rest of the fifteen chapters are divided into six parts.
- Part I introduces the fabrication-aware ML for additive manufacturing of 2D products, which covers domain-informed shape quality representation, dimension reduction, and small-sample learning and prediction of 2D geometric quality. It starts with simple geometries and generalizes the methodologies to 2D freeform geometries.
- Part II introduces the fabrication-aware ML for additive manufacturing of 3D products, which generalizes domain-informed shape quality representation, dimension reduction, and small-sample learning and prediction methods to 3D freeform geometries.
- Part III introduces process-informed optimal compensation strategies and algorithms for additive manufacturing. Applications in metal additive manufacturing and process monitoring are demonstrated.
- Part IV introduces domain-informed transfer learning and automated model generation for smart manufacturing. A small-sample transfer learning based on engineering effect equivalence is introduced to overcome the lack of data due to lurking variables. Principled design of Bayesian neural networks is also introduced to generate prediction models in additive manufacturing of 2D products.
- Part V introduces domain-informed machine/statistical learning for scale-up nanomanufacturing of nanostructures. It covers the domain-informed statistical

learning methods to understand the growth kinetics of nanoparticles, nanowires, and graphene.

- Part VI introduces domain-informed nanostructure characterization and defect detection with and without complete metrology data.

References

1. ASME Y14.46-2017 Product Definition for Additive Manufacturing. ASME (2017)
2. Ackoff, R.L.: From data to wisdom. J. Appl. Syst. Anal. **16**(1), 3–9 (1989)
3. Ameta, G., Lipman, R., Moylan, S., Witherell, P.: Investigating the role of geometric dimensioning and tolerancing in additive manufacturing. J. Mech. Design **137**(11) (2015). https://doi.org/10.1115/1.4031296
4. Bermano, A.H., Funkhouser, T., Rusinkiewicz, S.: State of the art in methods and representations for fabrication-aware design. In: Computer Graphics Forum, vol. 36, pp. 509–535. Wiley Online Library (2017)
5. Bourell, D.L., Leu, M.C., Rosen, D.W.: Roadmap for additive manufacturing: Identifying the future of freeform processing. Tech. rep., Sponsored by National Science Foundation and the Office of Naval Research (2009)
6. Campanelli, S., Cardano, G., Giannoccaro, R., Ludovico, A., Bohez, E.L.: Statistical analysis of the stereolithographic process to improve the accuracy. Comput.-Aided Design **39**(1), 80–86 (2007)
7. Campbell, T., Williams, C., Ivanova, O., Garrett, B.: Could 3d printing change the world? Technologies, potential, and implications of additive manufacturing (2011)
8. Chandler, A.D.: Scale and Scope. Harvard University Press, Harvard (1990)
9. Colosimo, B., Huang, Q., Dasgupta, T., Tsung, F.: Opportunities and challenges of quality engineering for additive manufacturing. J. Quality Technol. **50**(3), 233–252 (2018)
10. Colosimo, B., pacella, M., Semeraro, Q.: Statistical process control for geometric specifications: on the monitoring of roundness profiles. J. Quality Technol. **40**(1), 1–18 (2008)
11. Davis, J.: The 15th anniversary of smart manufacturing (2022). https://www.cesmii.org/the-15th-anniversary-of-smart-manufacturing/
12. de Souza Borges Ferreira, R., Sabbaghi, A., Huang, Q.: Automated geometric shape deviation modeling for additive manufacturing systems via Bayesian neural networks. IEEE Trans. Autom. Sci. Eng. **17**(2), 584–598 (2020). https://doi.org/10.1109/TASE.2019.2936821
13. del Castillo, E., Colosimo, B.M.: Statistical shape analysis of experiments for manufacturing processes. Technometrics **53**(1), 1–15 (2011)
14. del Castillo, E., Colosimo, B.M., Tajbakhsh, S.: Geodesic Gaussian processes for the reconstruction of a 3D free-form surface. Technometrics **57**(1), 87–99 (2015)
15. DIN VDE: German Standardization Roadmap Industrie 4.0 – Version 4. DIN e. V. (2020). https://www.din.de/resource/blob/65354/1bed7e8d800cd4712d7d1786584a7a3a/roadmap-i4-0-e-data.pdf
16. Economist: Manufacturing and innovation- A third industrial revolution (Special Report) (2012)
17. Everton, S.K., Hirsch, M., Stravroulakis, P., Leach, R.K., Clare, A.T.: Review of in-situ process monitoring and in-situ metrology for metal additive manufacturing. Mater. Design **95**, 431–445 (2016)
18. Gibson, I., Rosen, D., Stucker, B.: Additive Manufacturing Technologies: Rapid Prototyping to Direct Digital Manufacturing. Springer, Berlin (2009)
19. Gibson, I., Rosen, D.W., Stucker, B., et al.: Additive Manufacturing Technologies, vol. 238. Springer, Berlin (2010)

20. Grasso, M., Demir, A., Previtali, B., Colosimo, B.: In situ monitoring of selective laser melting of zinc powder via infrared imaging of the process plume. Robot Comput.–Integrated Manuf. **49**, 229–239 (2018)
21. He, T., Yu, S., Shi, Y., Dai, Y.: High-accuracy and high-performance waam propeller manufacture by cylindrical surface slicing method. Int. J. Adv. Manuf. Technol. **105**(11), 4773–4782 (2019)
22. Hilton, P., Jacobs, P.: Rapid Tooling: Technologies and Industrial Applications. CRC, Boca Raton (2000)
23. Hu, S.J., Ko, J., Weyand, L., ElMaraghy, H.A., Lien, T.K., Koren, Y., Bley, H., Chryssolouris, G., Nasr, N., Shpitalni, M.: Assembly system design and operations for product variety. CIRP Ann. **60**(2), 715–733 (2011)
24. Huang, Q.: Machine learning for quality control in additive manufacturing. In: Manufacturing in the Era of 4th Industrial Revolution: A World Scientific Reference Volume 1: Recent Advances in Additive Manufacturing, pp. 231–278. World Scientific, Singapore (2020)
25. Huang, Q., Nouri, H., Xu, K., Chen, Y., Sosina, S., Dasgupta, T.: Statistical predictive modeling and compensation of geometric deviations of three-dimensional printed products. ASME Trans. J. Manuf. Sci. Eng. **136**(6), 061008 (2014)
26. Huang, Q., Wang, Y., Lyu, M., Lin, W.: Shape deviation generator (SDG)—a convolution framework for learning and predicting 3d printing shape accuracy. IEEE Trans. Autom. Sci. Eng. **17**(3), 1486–1500 (2020)
27. Huang, Q., Zhang, J., Sabbaghi, A., Dasgupta, T.: Optimal offline compensation of shape shrinkage for 3d printing processes. IIE Trans. Quality Reliab. **47**(5), 431–441 (2015)
28. Hüllermeier, E., Waegeman, W.: Aleatoric and epistemic uncertainty in machine learning: an introduction to concepts and methods. Mach. Learn. **110**(3), 457–506 (2021)
29. Hwang, J.S.: Future manufacturing: Bracing for and embracing the postpandemic era. Bridge **51**(1), 7–13 (2021)
30. Jacobs, P.: Rapid Prototyping & Manufacturing: Fundamentals of Stereolithography. SME (1992)
31. Jafari, D., Wits, W.W., Vaneker, T.H., Demir, A.G., Previtali, B., Geurts, B.J., Gibson, I.: Pulsed mode selective laser melting of porous structures: Structural and thermophysical characterization. Addit. Manuf. **35**, 101263 (2020). https://doi.org/https://doi.org/10.1016/j.addma.2020.101263
32. Jin, Y., Qin, S., Huang, Q.: Offline predictive control of out-of-plane geometric errors for additive manufacturing. ASME Trans. Manuf. Sci. Eng. **138**(12), 121005 (2016)
33. Karniadakis, G.E., Kevrekidis, I.G., Lu, L., Perdikaris, P., Wang, S., Yang, L.: Physics-informed machine learning. Nat. Rev. Phys. **3**(6), 422–440 (2021)
34. Karunakaran, K., Suryakumar, S., Pushpa, V., Akula, S.: Low cost integration of additive and subtractive processes for hybrid layered manufacturing. Robot. Comput.-Integrated Manuf. **26**(5), 490–499 (2010)
35. Khanzadeh, M., Rao, P., Jafari-Marandi, R., Smith, B.K., Tschopp, M.A., Bian, L.: Quantifying geometric accuracy with unsupervised machine learning: using self-organizing map on fused filament fabrication additive manufacturing parts. J. Manuf. Sci. Eng. **140**(3), 031011
36. King, W.E., Anderson, A.T., Ferencz, R.M., Hodge, N.E., Kamath, C., Khairallah, S.A., Rubenchik, A.M.: Laser powder bed fusion additive manufacturing of metals; physics, computational, and materials challenges. Appl. Phys. Rev. **2**(4), 041304 (2015)
37. King, W.E., Anderson, A.T., Ferencz, R.M., Hodge, N.E., Kamath, C., Khairallah, S.A., Rubenchik, A.M.: Laser powder bed fusion additive manufacturing of metals; physics, computational, and materials challenges. Appl. Phys. Rev. **2**(4) (2015)
38. Koren, Y.: The local factory of the future for producing individualized products. Bridge **51**(1), 100 (2021)
39. Kumar, A.: From mass customization to mass personalization: a strategic transformation. Int. J. Flexible Manuf. Syst. **19**, 533–547 (2007)
40. Kunovjanek, M., Knofius, N., Reiner, G.: Additive manufacturing and supply chains a systematic review. Prod. Plan. Control **0**(0), 1–21 (2020)

41. Leach, R.K., Bourell, D., Carmignato, S., Donmez, A., Senin, N., Dewulf, W.: Geometrical metrology for metal additive manufacturing. CIRP Ann. **68**(2), 677–700 (2019). https://doi.org/10.1016/j.cirp.2019.05.004
42. Luan, H., Huang, Q.: Prescriptive modeling and compensation of in-plane shape deformation for 3-d printed freeform products. IEEE Trans. Autom. Sci. Eng. **14**(1), 73–82 (2017)
43. Melchels, F., Feijen, J., Grijpma, D.: A review on stereolithography and its applications in biomedical engineering. Biomaterials **31**(24), 6121–6130 (2010)
44. Mohd Yusuf, S., Cutler, S., Gao, N.: The impact of metal additive manufacturing on the aerospace industry. Metals **9**(12), 1286 (2019)
45. Montagnat, J., Delingette, H., Ayache, N.: A review of deformable surfaces: topology, geometry and deformation. Image Vis. Comput. **19**(14), 1023–1040 (2001)
46. Monteiro, H., Carmona-Aparicio, G., Lei, I., Despeisse, M.: Energy and material efficiency strategies enabled by metal additive manufacturing–a review for the aeronautic and aerospace sectors. Energy Rep. **8**, 298–305 (2022)
47. Morse, E., Dantan, J.Y., Anwer, N., Söderberg, R., Moroni, G., Qureshi, A., Jiang, X., Mathieu, L.: Tolerancing: Managing uncertainty from conceptual design to final product. CIRP Ann. **67**(2), 695–717 (2018). https://doi.org/10.1016/j.cirp.2018.05.009
48. Mughal, M., Fawad, H., Mufti, R.: Three-dimensional finite-element modelling of deformation in weld-based rapid prototyping. Proc. Inst. Mech. Eng. C: J. Mech. Eng. Sci. **220**(6), 875–885 (2006)
49. Narasimharaju, S.R., Zeng, W., See, T.L., Zhu, Z., Scott, P., Jiang, X., Lou, S.: A comprehensive review on laser powder bed fusion of steels: Processing, microstructure, defects and control methods, mechanical properties, current challenges and future trends. J. Manuf. Process. **75**, 375–414 (2022)
50. Rowley, J.: The wisdom hierarchy: representations of the DIKW hierarchy. J. Inform. Sci. **33**(2), 163–180 (2007)
51. Sabbaghi, A., Huang, Q.: Model transfer across additive manufacturing processes via mean effect equivalence of lurking variables. Ann. Appl. Stat. **12**(4), 2409–2429 (2018)
52. Sabbaghi, A., Huang, Q., Dasgupta, T.: Bayesian model building from small samples of disparate data for capturing in-plane deviation in additive manufacturing. Technometrics **60**(4), 532–544 (2018). https://doi.org/10.1080/00401706.2017.1391715
53. Samie Tootooni, M., Dsouza, A., Donovan, R., Rao, P.K., Kong, Z.J., Borgesen, P.: Classifying the dimensional variation in additive manufactured parts from laser-scanned three-dimensional point cloud data using machine learning approaches. J. Manuf. Sci. Eng. **139**(9), 091005 (2017)
54. National Academies of Sciences, E., Medicine, et al.: Options for a National Plan for Smart Manufacturing (2023)
55. Shewhart, W.A.: Economic quality control of manufactured product 1. Bell Syst. Tech. J. **9**(2), 364–389 (1930)
56. Shewhart, W.A., Deming, W.E.: Statistical Method from the Viewpoint of Quality Control. Courier Corporation (1986)
57. Srinivasan, V.: A geometrical product specification language based on a classification of symmetry groups. Comput. Aided Design **31**(1), 659–668 (1999)
58. Tapia, G., Elwany, A.: A review on process monitoring and control in metal-based additive manufacturing. ASME Trans. J. Manuf. Sci. Eng. **136**(6), 060801–060810 (2014)
59. Tong, K., Joshi, S., Lehtihet, E.: Error compensation for fused deposition modeling (FDM) machine by correcting slice files. Rapid Prototyping J. **14**(1), 4–14 (2008)
60. Tong, K., Lehtihet, E., Joshi, S.: Parametric error modeling and software error compensation for rapid prototyping. Rapid Prototyping J. **9**(5), 301–313 (2003)
61. Tsung, F., Zhang, K., Cheng, L., Song, Z.: Statistical transfer learning: a review and some extensions to statistical process control. Quality Eng. **30**(1), 115–128 (2018)
62. Van Kaick, O., Zhang, H., Hamarneh, G., Cohen-Or, D.: A survey on shape correspondence. In: Computer Graphics Forum, vol. 30, pp. 1681–1707. Wiley Online Library (2011)
63. Walker, R.K., Srinivasan, V.: Creation and evolution of the asme y14. 5.1 m standard. Manuf. Rev. **7**(1), 16–23 (1994)

64. Wang, W., Cheah, C., Fuh, J., Lu, L.: Influence of process parameters on stereolithography part shrinkage. Mater. Design **17**(4), 205–213 (1996)
65. Wang, Y., Ruiz, C., Huang, Q.: Learning and predicting shape deviations of smooth and non-smooth 3d geometries through mathematical decomposition of additive manufacturing. IEEE Trans. Autom. Sci. Eng. (2022). https://doi.org/10.1109/TASE.2022.3174228
66. Womack, J.P.: Mass customization: the new frontier in business competition. MIT Sloan Manag. Rev. **34**(3), 121 (1993)
67. Zeleny, M.: Management support systems: Towards integrated knowledge management. Hum. Syst. Manag. **7**(1), 59–70 (1987)
68. Zhang, D., Lu, G.: Review of shape representation and description techniques. Pattern Recogn. **37**(1), 1–19 (2004)
69. Zhou, C., Chen, Y.: Additive manufacturing based on optimized mask video projection for improved accuracy and resolution. J. Manuf. Process. **14**, 107–118 (2012)
70. Zhou, J., Herscovici, D., Chen, C.: Parametric process optimization to improve the accuracy of rapid prototyped stereolithography parts. Int. J. Mach. Tools Manuf. **40**(3), 363–379 (2000)
71. Zhu, Z., Anwer, N., Huang, Q., Mathieu, L.: Machine learning in tolerancing for additive manufacturing. CIRP Ann. **67**(1) (2018). https://doi.org/10.1016/j.cirp.2018.04.119

Fabrication-Aware Machine Learning for Additive Manufacturing: Two-Dimensional Shape Quality Representation, Learning, and Prediction

Chapter 2
Representation of Two-Dimensional (2D) Geometric Shape Quality

2.1 Point Cloud Data for AM Quality Control

Product geometric quality is critical to product functionality and proper fit [31]. In AM, the post-production measurement of product accuracy is often conducted through 3D scanning using structured light scanning or laser scanning [38, 76]. A cloud of points will be generated by the 3D scanner to digitally replicate the scanned object or manufactured part. Embedded in the Euclidean space, each point has its set of Cartesian coordinates measuring a point on the exterior surface or boundary of an object. Depending on the sensing and scanning techniques, each point may also contain RGB color data or intensity information. Point clouds have been widely used in many engineering and medical disciplines such as computer-aided design, reverse engineering, metrology and quality inspection, architecture modeling, geographic information systems, augmented reality, and medical imaging.

Point cloud data is a typical form of manifold data, where points on the exterior profile or surface of a product form a topological space that locally resembles the Euclidean space. Point cloud data analysis is a large field of study involving processing and interpreting point cloud data to extract meaningful information and knowledge about the shape, structure, and properties of objects or environments represented by the point cloud. The analysis includes point cloud registration, surface reconstruction, segmentation and localization, object detection, classification, feature extraction, and qualification [34, 76, 79, 92]. Figure 2.1 illustrates an example of a raw point cloud generated by laser-scanning a printed test plate with multiple parts. As can be seen, the raw point clouds need pre-processing such as cleaning and registration before assessing the printing quality.

Q. Huang, *Domain-informed Machine Learning for Smart Manufacturing*,
https://doi.org/10.1007/978-3-031-91631-1_2

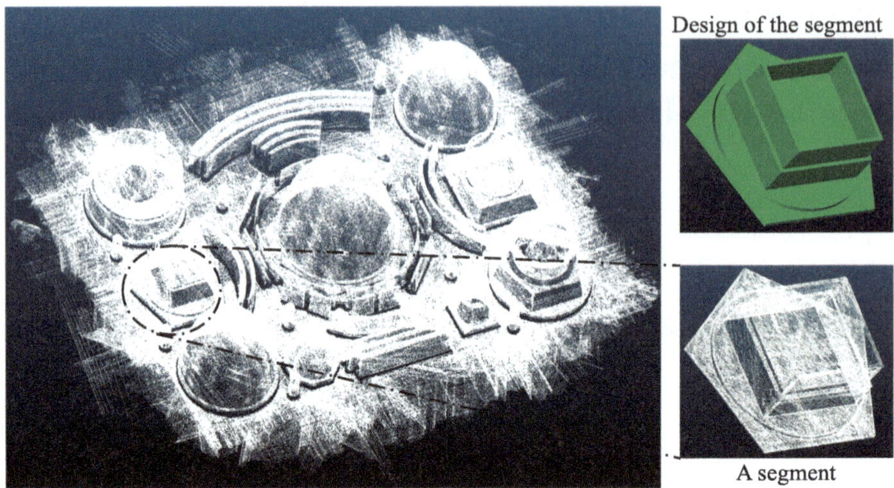

Fig. 2.1 Raw point cloud scanned from a test plate, a segment of the test plate, and its design

2.2   Basics of Feature Engineering: From Data to Information

In machine learning, models can be constructed either from raw data or feature, which is information extracted from that data. Feature by definition means a distinctive attribute, property, or characteristic of a data object. Although "feature," "variable," "attribute," "characteristic," and "parameter" are often used as synonyms, feature tends to be of broad scope and widely used in the context of machine learning. A feature can be an individual measurable characteristics in the raw data (e.g., weight or strength), a category encoder of an attribute (e.g., "0" for passing inspection and "1" for fail), a numerical transformation of an original variable (like scaling or logarithm), a linear or nonlinear combination of a set of original variables (e.g., through principal component analysis), or a constructed parameter through domain knowledge (e.g., a dimensionless Reynolds number in fluid dynamics).

Feature engineering is the art and science of transforming raw data into domain-meaningful representations that facilitate analysis, modeling, and decision-making. It is a crucial step in preparing data for machine learning tasks [8, 24, 25]. The engineered features can improve model performance by focusing on relevant patterns in raw data. Better data representation schemes can be identified for analysis and interpretation. Dimensionality reduction can be achieved to reduce complexity and computational costs. Feature engineering also allows domain knowledge to be integrated into the modeling process by selecting and constructing relevant features that align with the learning task. Problematic raw data can be pre-processed to handle missing values, outliers, and other imperfections before modeling.

In the past three decades, feature engineering in machine learning has been quickly expanded from feature selection [15, 57, 63] into a much broader area including feature extraction, feature transformation, feature construction, and feature engineering for various types of data and applications [25]. In the context of smart manufacturing, feature engineering becomes instrumental in various aspects of data-driven operations. By converting raw data into structured features, manufacturers can gain a deeper understanding of their processes, identify inefficiencies, predict equipment failures, optimize production, and ultimately enhance product quality. Not intended to be comprehensive, this section briefly reviews some useful feature engineering approaches for manufacturing applications.

2.2.1   Feature Selection

Feature selection involves identifying the most relevant subset of features from the original feature set so that the feature space is optimally reduced according to a certain criterion [15, 25]. Feature selection can date back to the variable selection for multiple regression [27, 44]. It regained attention in the 1990s when the number of variables or features grow significantly larger in data mining applications [15, 57]. In the era of big data, novel feature selection methods tend to be tailored for the specific nature and structures of data [62].

Feature selection methods can be classified as supervised, semi-supervised, and unsupervised methods, depending on whether the target variable or response is utilized in the selection process [25, 59, 88]. Depending on whether a machining learning model/algorithm is considered, feature selection methods have been categorized as wrapper, filter, and embedded strategies [59, 62],

- *Wrapper*: Wrapper method searches features by evaluating the predictive performance of a predefined learning algorithm/model. The search procedure involves subset generation, subset evaluation, and stopping decision. A candidate feature subset will be generated in each iteration and compared to the current best choice according to a specified evaluation criterion. The best feature subset will be updated until the stopping criterion is satisfied.

  The subsets can be generated by randomly selecting a fixed number of features, sequentially adding features, or iteratively selecting and removing features. For example, forward selection starts with an empty feature set and incrementally adds the feature that results in the best performance improvement at each step. Backward elimination begins with the full feature set and iteratively removes the least important feature in terms of performance. Recursive feature elimination starts with all features and iteratively removes the least important one, based on a ranking criterion, until the desired number of features is reached. Genetic algorithms search for the best subset of features by iteratively evolving a population of feature subsets.

  For each generated subset of features, the predefined machine learning model is trained on the corresponding training data and then evaluated on a validation or testing dataset. The evaluation metric could be prediction accuracy, mean squared error, or any other relevant metrics depending on the nature of the problem.

Wrapper methods capture feature interactions and their effects on the specific model being used. However, when the feature space is large, searching an optimal feature subset can be intractable [57] and computationally expensive. They may suffer from overfitting if the dataset is small or the model is complex.

Therefore, different search strategies such as sequential search [36], hill-climbing search, best-first search [6, 57], and branch-and-bound search [74] are proposed to yield a local optimum learning performance.

- Filter: Independent of specific machine learning models, filter method selects relevant features based on their characteristics. It first ranks features and then selects a subset of the top-ranked features for use in building machine learning models. Filter methods are generally faster and less computationally intensive compared to wrapper methods, making them suitable for large datasets.

  Filter method can rank features one by one or multiple features at the same time. Individual feature/variable evaluation or ranking has the benefits of simplicity and computational efficiency. But it is unable to capture the feature interactions and avoid redundancy caused by high correlations among certain features. Additionally, their success heavily depends on the choice of the ranking measure and the predefined threshold. The selected features may not be optimal for the target learning algorithms or models. When evaluating multiple features simultaneously, both the relevance and redundancy of a feature subgroup will be assessed. The search of the optimal feature subgroup can be NP-hard. Heuristic search is therefore often adopted to provide a sub-optimal solution [111].

  Feature ranking metrics include correlation [36] and mutual information [30, 36, 101] between a feature (or a feature subgroup) and the target/response variable. Another ranking metric is the discriminative ability of a feature or a feature subgroup that efficiently separate training samples into distinct classes or clusters [56, 94, 108]. In unsupervised feature selection, ranking criteria can be the ability of selected features to preserve original data manifold structure [17, 39, 112] or to recover the original data [28, 69].

- *Embedded method*: Embedded method incorporates feature selection directly into the model learning process. This integration allows feature selection interacting with the learning model and avoiding iterative feature evaluation.

  Embedded methods penalize or reward features based on their impact on model performance. For example, LASSO (least absolute shrinkage and selection operator) regression adds a L_1 regularization term to the penalty function [96]. This effectively performs feature selection as it shrinks less important features to zero. Decision tree-based algorithms like random forest compute feature importance scores during their training process. Features that contribute the most to reducing impurity are considered more important. Similar to random forest, gradient boosting algorithms also compute feature importance scores based on how features contribute to reducing the prediction error. As a combination of L_1 (LASSO) and L_2 (Ridge) regularization, elastic net balances between feature selection and feature grouping.

  Embedded methods offer several advantages, including the ability to handle complex interactions between features and the model, as well as efficient feature selection within the training process itself. They also mitigate the risk of overfitting to noisy or irrelevant features by prioritizing those that contribute the most to the model's predictive performance.

However, embedded methods are limited to the characteristics and capabilities of the chosen model. Therefore, selecting an appropriate model architecture and hyperparameters is crucial to achieving effective feature selection.

2.2.2 Feature Extraction

Rather than selecting features out of the existing variables or attributes, feature extraction aims to extract a completely new set of features. It involves converting existing features into new representations through functional mappings. In literature, the extracted feature space can also be referred to as lower-dimensional space through dimensionality reduction [99], embedded space through embedding [41, 78], encoded space through encoding [33], subspace through subspace learning [65, 103], submanifold through manifold learning [29], or representation space through representation learning [8, 32]. Common feature extraction methods include:

- Principal component analysis (PCA): PCA extracts new features or major principal components through linear combinations of original features [45]. New features represent the directions of largest variations in the data.
- Multidimensional scaling: Multidimensional scaling (MDS) finds a lower-dimensional representation of data while preserving the pairwise metric distance or similarity of data samples [21]. The similarity measure can be the Euclidean distance or other distance metrics. For example, isometric feature mapping (Isomap) preserves geodesic distance between two data samples [95].
- Kernel principal component analysis: PCA is limited to finding a linear subspace to extract features [85]. Kernel PCA uses kernel method to map data in a higher-dimensional space and then conduct dimensionality reduction. Furthermore, PCA, MDS, Isomap, locally linear embedding [78], Laplacian Eigenmap [7], and maximum variance unfolding [104] can all be viewed as kernel PCA with different kernels [37, 90]. Notably, locally linear embedding and Laplacian Eigenmap [7] find low-dimensional representation of the dataset that preserves local neighborhood information. Adjacency or similarity among data is captured by deterministic weights.
- Stochastic neighbor embedding: Stochastic neighbor embedding [41] preserves the distribution property of original data samples when extracting features in the lower-dimension space. The distribution of extracted features will have minimal Kullback-Leibler divergence from the original data distribution.
- Autoencoders and artificial neural networks (ANNs): An encoding function in ANNs transforms input data and learns an efficient representation (encoding) of the data, typically for dimensionality reduction and data compression [42].

2.3 Representation of Product Shape Quality

Representation of product geometric quality is related to both product representation itself and engineering context. Product representation describes the characteristic attributes of topological, geometrical, and context-dependent engineering information about the part [86]. As one aspect of product quality, geometric quality measures the discrepancy between the actual product geometry u_A and its design counterpart u_D. The geometric deviation of an manufactured product from its nominal design can be modeled by a set of discrepancy functions [13]:

$$y_u = d(R(\mathbb{P}_{u_D}), R(\mathbb{P}_{u_A})) \tag{2.1}$$

where $\mathbb{P}_u$ are designated regions of interest or physical portions of the product u that are of quality significance. The quality characteristics of regions $\mathbb{P}_u$ are defined or extracted through characterization function $R(\cdot)$. The discrepancy function $d(\cdot)$ measures the deviation of the quality characteristics from their target values.

A designated region $\mathbb{P}_u$ with context-dependent semantic meaning is commonly referred to as geometric feature in engineering design [87]. The frequently encountered geometric features such as cylinder, hole, and protrusion constitute a feature library for solid modelling and 3D object representation [91]. Characterization function $R(\cdot)$ usually extracts dimensional or form characteristics of geometric features [84, 87], for example, diameter or cylindricity. The quality of geometric features is assessed by comparing the discrepancy $d(R(\mathbb{P}_{u_D}), R(\mathbb{P}_{u_A}))$ to the geometric dimensioning and tolerancing (GD&T) requirements, as illustrated in Fig. 2.2 (top row). The feature-based representation and extraction approaches lack the descriptive capability for complex freeform geometries such as the dental models in Fig. 2.2 or customized medical implants [9]. An infinite number of geometric features or feature groups are required to capture shape complexity.

To accommodate complex geometries, surface patches have been utilized for region $\mathbb{P}_u$ selection and representation [84, 110]. Patches are typically identified in design stage on a case-by-case basis. Rather than using simple dimensional or form characteristics for geometric features, characteristics function $R(\cdot)$ has to model patch profile quality. Surface patches can take non-parametric functional forms such as non-uniform rational B-splines (NURBS)[110] and kernels [109], parametric functions [100], or predefined patch templates with specific geometries [35]. The patch quality can be assessed by checking whether $R(\mathcal{P}_u)$ falls within the specified envelope [67]. Despite its flexibility of representing 3D objects, it can be tedious and labor-intensive to specify and qualify surface patches for products with frequent design changes, for example, in one-off 3D printing.

Essentially, the representation of product geometric quality is determined by the selection of regions of interest $\mathbb{P}_u$ and characterization of the deviation patterns in the selected regions (i.e., $R(\mathbb{P}_u)$). The representation scheme should consider the challenges of lack of knowledge and right data discussed in Sect. 1.3, specifically, the high dimensionality of product variety space, the heterogeneity caused by

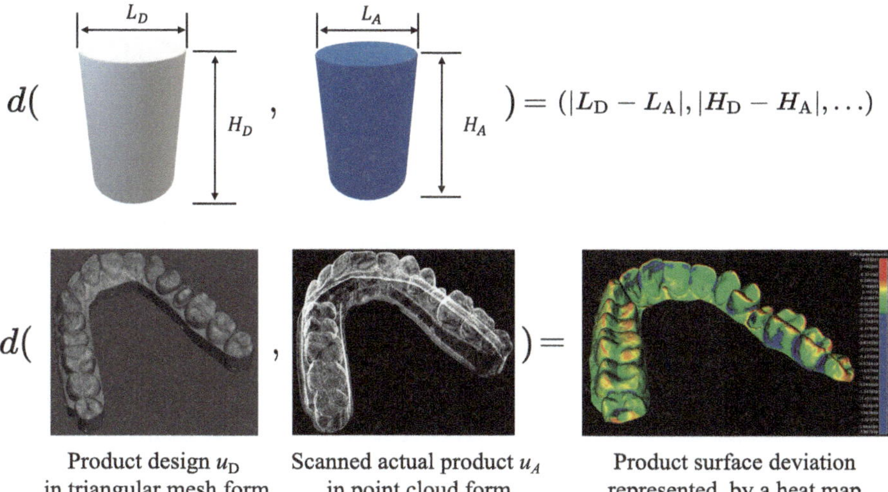

$$d\left(\;\raisebox{-0.5em}{\scriptsize H_D}\;,\;\raisebox{-0.5em}{\scriptsize H_A}\;\right) = (|L_D - L_A|, |H_D - H_A|, \ldots)$$

$$d\left(\;,\;\right) =$$

Product design u_D	Scanned actual product u_A	Product surface deviation
in triangular mesh form	in point cloud form	represented by a heat map

Fig. 2.2 Representation of product geometric quality using (1) dimensional characteristics for a simple geometric feature (top row) and (2) a heat map for a complex shape (bottom row): color values denoting the point-to-point distances between the design and the actual product

process covariates, small data due to low production volumes, and effects of lurking process variables. Representation of product geometric quality dictates quality control activities. A shape-dependent representation would likely lead to a shape-dependent accuracy control approach. Under a smart personalized manufacturing environment, a generic selection of regions of interest $\mathbb{P}_u$ and characterization of the deviation patterns $R(\mathbb{P}_u)$ of freeform geometries is desirable for quality control in AM. It will facilitate modeling and learning from limited sample data and, equally important, inferences on the prediction and compensation of untested shapes.

For the simplicity of illustration, Sect. 2.4 first discusses the representation of geometric quality of individual two-dimensional (2D) objects. Section 2.5 introduces the representation scheme for freeform 2D objects for the purpose of small-data machine learning of product quality. The representation of 3D geometric quality will be introduced in Chap. 4.

2.4   Representation of Shape Quality for Individual 2D Objects in AM

In AM, each layer to be printed can be approximated as a 2D object. In addition, the geometric deviation of a 3D object can be decomposed into in-plane ($x - y$ plane) and out-of-plane (z direction) shape deviations, respectively. Therefore, a general formulation of geometric quality for 2D objects and the in-plane shape deviation is introduced first.

Suppose the boundary deviations of individual 2D geometries are of interest, that is, $\mathbb{P}_u = u$. A set of points on the boundary of the built 2D object u_A is scanned and measured. Characterization of $R(\mathbb{P}_{u_A})$ can be conducted either in the Cartesian coordinate system using coordinates (x, y) or in the polar coordinate system (PCS) using (r, θ). The PCS representation is preferred as it greatly facilitates modeling and design for quality control in AM [50].

Assume that the center of the product is well-defined. The boundary of a 2D shape can be represented by a function $r_0(\theta)$ denoting the *nominal radius* at angle θ. For example, the function $r_0(\theta) = r_0$ for all θ defines a circle, and $r_0(\theta) = 2(1 - \sin\theta\cos\theta)$ and $r_0(\theta) = 2\big(1 - \sin(0.5\theta)\big)$ for $0 \le \theta \le 2\pi$ define the shapes in the left and right panels of Fig. 2.3, respectively. The *actual radius* at angle θ is a function of θ and $r_0(\theta)$ and is expressed as $r(\theta, r_0(\theta))$ (i.e., $R(\mathbb{P}_{u_A})$). The difference between the actual and nominal radius at an angle θ is essentially what defines the geometric deviation at θ. Therefore, the discrepancy function $d(R(\mathbb{P}_{u_D}), R(\mathbb{P}_{u_A}))$ of an AM-built 2D product can be conveniently represented as

$$y_u = d(R(\mathbb{P}_{u_D}), R(\mathbb{P}_{u_A})) = \Delta r(\theta, r_0(\theta)) = r(\theta, r_0(\theta)) - r_0(\theta). \qquad (2.2)$$

Like other manufacturing processes, the center of the product normally coincides with the origin of the part coordinate system defined by CAD software. For the convenience of building and measuring parts, we can also choose the center to be the origin of the machine or inspection coordinate system.

The representation of geometric quality in the Cartesian coordinate system has been previously studied in the literature [97, 98]. It faces a practical issue of correctly identifying shape deviation. As shown in Fig. 2.4, for a given nominal point $A(x, y)$, its final position A' is hard to be identified after the build. A practical solution is to fix the x or y coordinate and study the deviation of the other coordinate (Δx or Δy in Fig. 2.4). Another method is to study deviation components along x and y directions separately [97, 98]. But the apparent correlation among deviation components cannot be captured, potentially leading to prediction error.

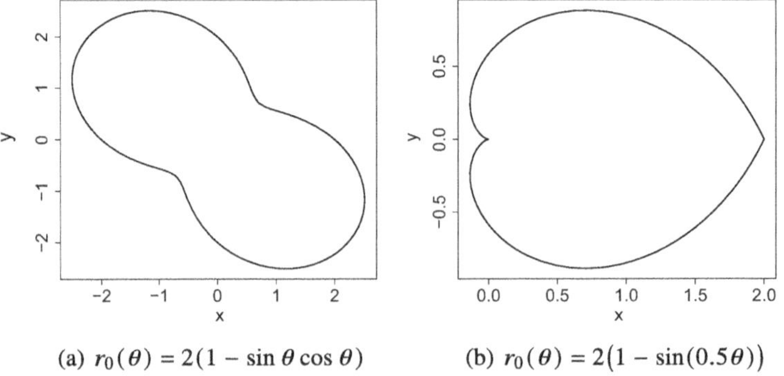

(a) $r_0(\theta) = 2(1 - \sin\theta\cos\theta)$ (b) $r_0(\theta) = 2\big(1 - \sin(0.5\theta)\big)$

Fig. 2.3 Shapes generated by different functional forms $r_0(\theta)$ [50]

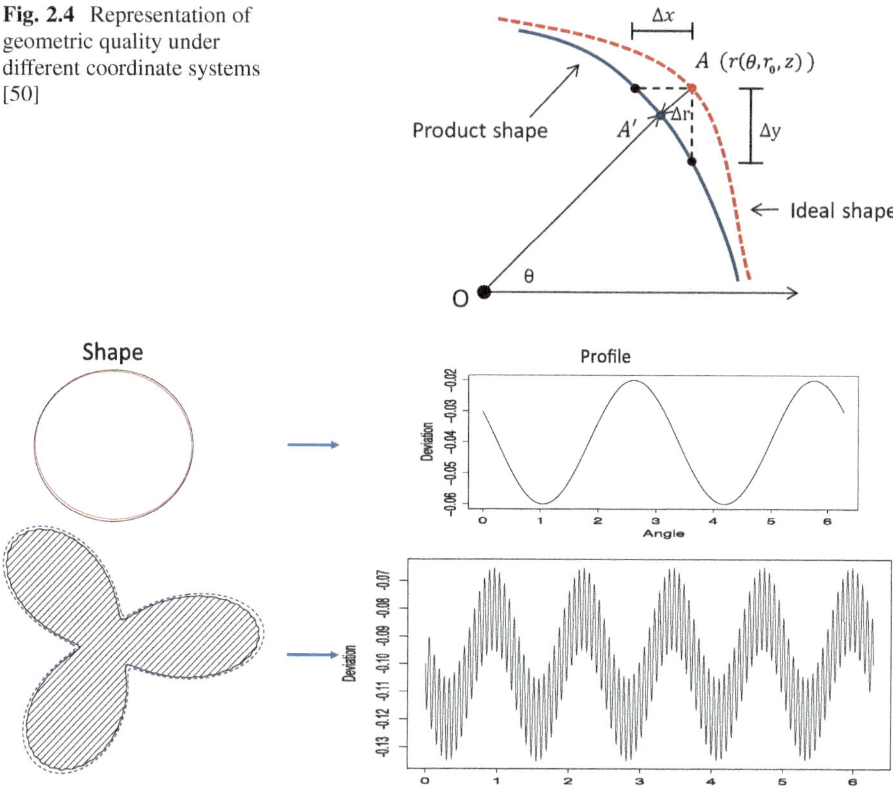

Fig. 2.4 Representation of geometric quality under different coordinate systems [50]

Fig. 2.5 Transform shape deviations to deviation profiles in the PCS [46]

In contrast, the definition of radius deviation in Eq. (2.2) naturally captures the deviation along radial direction and is convenient for visualizing deviation patterns. As illustrated in Fig. 2.5, once the shape deformation $\Delta r(\theta, r_0(\theta))$ are presented in the PCS as deviation profiles, modeling and analysis of geometric part errors are greatly alleviated from the original geometric complexity. The essence of this representation is to transform in-plane geometric errors into a functional profile defined on the interval $[0, 2\pi]$. This representation decouples the geometric shape complexity from the deformation modeling, and a generic formulation of shape deformation can thus be achieved.

For product shapes like those in Fig. 2.5, we could locate the origin of the PCS in such a way that for any given angle θ, there is only one unique point with radius $r(\theta)$ on the product boundary. The origin usually coincides with build center and measurement origin. The transformed shape deformation in the PCS will be a *single* continuous profile.

For more complex shapes, however, there could exist more than one point on the product boundary for a given angle θ (Fig. 2.6a). We then have to establish

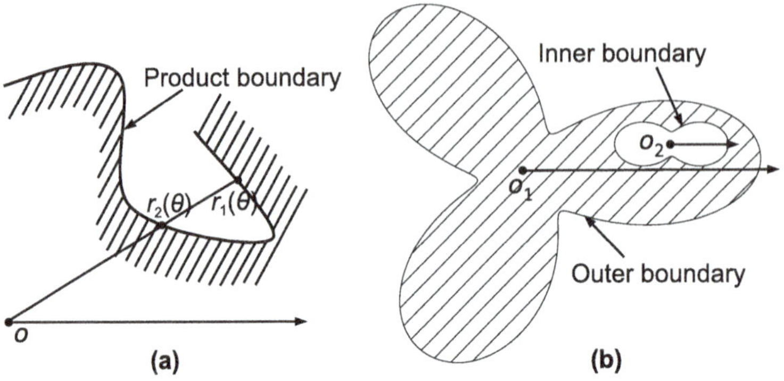

Fig. 2.6 Multiple profiles and PCSs [46]

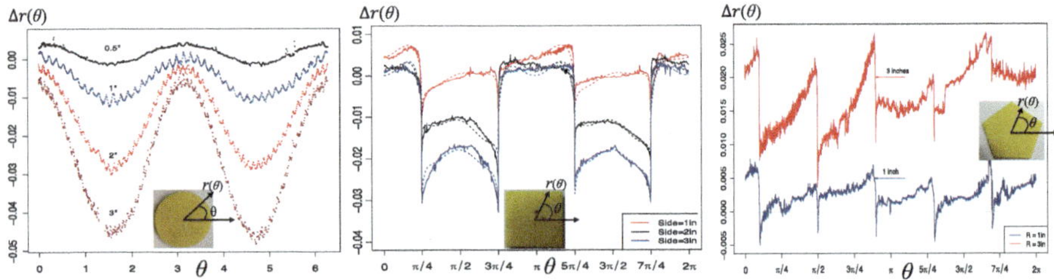

Fig. 2.7 Deviation profiles of disks [50] and polygon plates [48]

multiple PCSs in which each profile is uniquely defined. A special case shown
in Fig. 2.6b is a product with outer and inner boundaries defined in two PCSs,
and the transformation of shape deformation will yield multiple deviation profiles.
For complicated concave shapes, segmentation may be needed to generate multiple
deviation profiles as well. We should point out that the analysis herein will still
apply, except for the complication arising from the inference between deviation
profiles [80].

Figure 2.7 shows the deviation profiles of 2D plates printed in a mask image
projection stereolithography process, which include four flat disks with different
sizes [50], three square plates, and two pentagon plates of different sizes [48].
Clearly, representation of y_u in the PCS identifies systematic shape deviation
patterns.

2.5 Fabrication-Aware Representation of Shape Quality for 2D Freeform Objects in AM

Since AM is capable of manufacturing a large variety of products in extremely
low volumes, representation of product quality faces the challenge of describing

heterogeneous (manifold) data for the high-dimensional product variety space. The product variety refers to the wide range of 3D geometries or shapes, and product quality therefore concerns the deviation of complex geometries from their intended 3D designs. The measurement/training data is point-cloud data scanned from built products of finite varieties.

Although mass production also faces product variety issue, a manufacturer typically manage a much smaller set of product varieties through group technology, that is, grouping a finite amount of product features (e.g., holes) and their process techniques (e.g., drilling). Product quality is measured by a few characteristics associated with features (e.g., diameter and cylindricity).

Grouping complex freeform geometries in AM, however, is ineffective because it will eventually end up with a huge feature space. In CAD and computer graphics [2, 3, 10, 14, 19, 64, 89, 105], mesh and shape primitives (e.g., circular sector, line segment, and patch) are utilized as building blocks to reduce the dimension of the product variety space. However, geometric objects and their characteristics (e.g., diameter or length) cannot fully predict quality. As we discovered in AM [48, 50], the deviation pattern of the same geometric object varies not only with shape but also with its size due to the fabrication mechanism and materials shrinkage behaviors (Fig. 2.8).

Therefore, the geometry-centric dimension reduction strategy has limited applicability in AM because the deviation pattern of a product cannot solely be predicted by its geometry even when the materials and process condition are fixed. There has been a growing emphasis on fabrication-aware representation, verification, and learning methods for complex freeform products [1, 5, 11, 61, 72, 113]. But there is a lack of domain-informed dimension reduction and learning methodology to support the optimal selection a subset of product varieties for quality inference. As a result, numerous test artifacts or benchmarks have been developed to evaluate and compare the performance of AM processes [22, 51, 58, 60, 68, 73, 77].

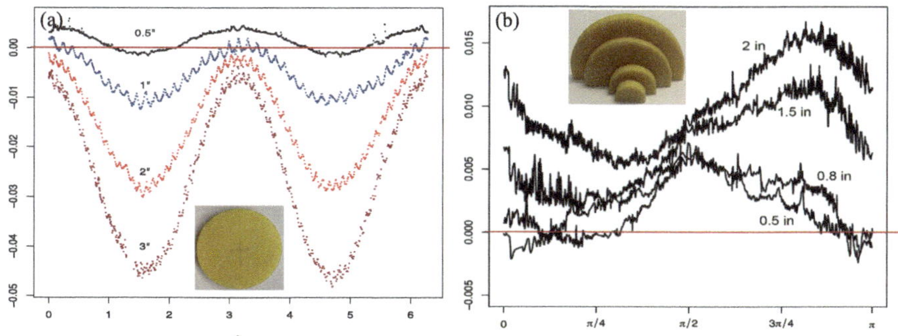

Fig. 2.8 Given the same materials and printing condition, (**a**) the deviation of a circular sector varies with the size and location of a printed disk. The smallest disk expands, while larger disks shrink. (**b**) Disks printed vertically through layer stack-up have more complicated deviation patterns

Domain-informed dimension reduction and machine learning in AM relies on fabrication-aware input and output representation (**OLP.1** problem laid out in Sect. 1.5). Representation of design input u involves mathematical description of 3D shapes. Depending on applications, various methods have been developed to describe 3D objects, for example, parametric surfaces, solids, or constructive solid geometry representations [2, 3, 10, 14, 89], finite element meshes [43, 55, 71, 75] for multi-physics analysis and slices in STL format and landmarks [4, 18, 26, 54], point clouds [12, 49, 50, 70, 83, 93, 102, 106, 107], or meshes [12, 93] for shape registration, inspection, and distortion control.

Representation of shape quality y_u involves the description and measure of the deviations between a build product and its nominal design. Generally there are two strategies to represent output quality y_u:

- *Extracting characteristic features or descriptors* from a shape and measuring their deviations: The common features or descriptors include geometric and dimensional measures related to angular or distance metrics (e.g., parallelism, Euclidean, Hausdorff, or geodesic distances) [5]; differential measures such as curvatures and surface roughness [84]; and integral descriptors such as areas and volumes (e.g., volume shrinkage factor [40].
- *Describing surface deviations/deformation* for complex freeform shapes: A finite number of descriptors is mathematically inadequate to represent complex shapes obtained by topological optimization in AM. Complete description of local surface deviations everywhere along product boundaries not only provides a comprehensive representation solution but also enables full control access to any region on a product surface [46, 49, 50]. Note that this strategy also facilitates comprehensive feature extraction and evaluation afterward.

Though each representation method is powerful in its own right, little work has concurrently considered the proper representation of design input u and quality output y_u for the purpose of small-sample learning, prediction, and control [47]. Some limited attempts have been made toward this direction [20, 23, 48–50, 52, 53, 66, 81, 82, 102]. Using input u, materials, and process information, finite element representation and modeling in theory is able to predict quality y_u with small sample data for parameter tuning and model validation. But it is computationally costly [16, 55] and has been difficult to generalize the knowledge to new part geometries without extensive rounds of new simulations or new test builds.

Huang [47] proposed the fabrication-aware dimension reduction and finite manufacturing primitive (FMP) method to address the large variety challenge in AM. The premise of dimension reduction is that a freeform product built in AM can be viewed as an assembly of a finite set of basic manufacturing objects called finite manufacturing primitives (FMPs). The product quality can be represented by a finite set of functions, where each function represents the quality model of one type of FMPs. By operating in a low-dimensional space spanned by FMPs, the FMP method will enable learning from a small sample of 3D product varieties to infer and control the quality of new 3D product varieties. The small-sample machine learning of product-variety problem will be transformed into large-sample learning of manufacturing-primitive problem.

The justification of using FMPs can be validated by the following three principles in AM:

P.1 **Dimension reduction principle**: Through a thinking or real printing experiment illustrated by Fig. 2.9, the quality of segment A will not be affected by the design choice among segments B_1, B_2, or B_3 when they are far apart enough. Quality analysis of the infinite variety of manufactured products can be reduced to the analysis of a basic set of manufacturing objects called finite manufacturing primitives (FMPs).

P.2 **Discretization principle**: To print a 3D object, the design input to an AM machine is discretized into a finite number of layers, and each layer is discretized through piecewise approximation. The discretization and piecewise approximation principle is self-evident in AM (Fig. 2.10), and it applies to the FMPs or the segmentation of a built product.

P.3 **Nonlinear addition principle**: The layer stack-up to form 3D objects or FMPs is nonlinear due to complex process physics and materials phase changes.

Principle P.1 implies that the computation and analysis can be conducted in the low-dimensional space of manufacturing primitives. Principles P.2 and P.3 suggest that modeling of manufacturing primitives should consider the discretized design input and nonlinear addition in manufacturing (i.e., segmenting each 2D layer into 2D FMPs and nonlinearly stacking up 2D FMPs into 3D FMPs).

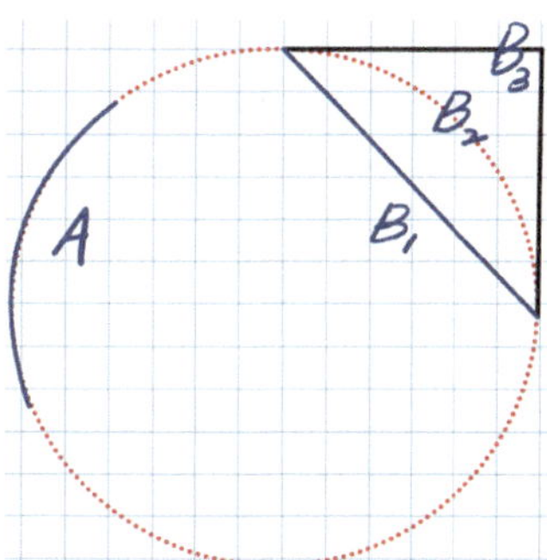

Fig. 2.9 Segments far apart are not correlated in AM

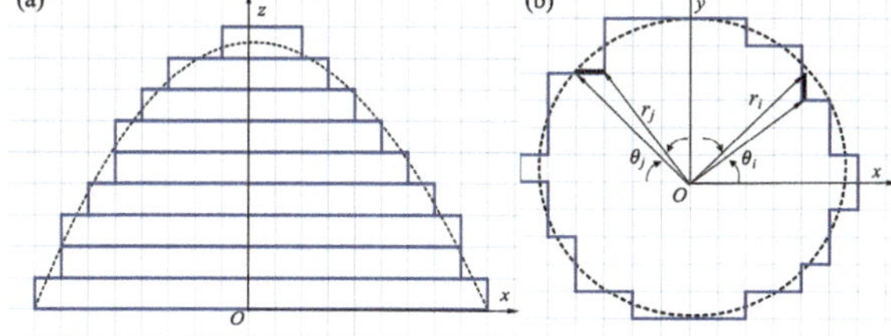

Fig. 2.10 Discretized input in AM: (**a**) discretized layers through slicing 3D shapes along z direction; (**b**) piecewise linear approximation of layer boundaries in the $x - y$ build plane [47]

Fabrication-Aware Input Representation Based on Constructive Shape Primitives

To enable engineering-informed small-sample learning, an input representation method based on the shape primitive concept and a fabrication-aware formulation were developed in [47]. Essentially, the small-sample learning of printed products is transformed into a large-sample learning of printed shape primitives.

Geometric (shape) primitives have been developed in computer graphics and CAD systems to construct 3D-shape objects [10]. Geometric primitives are simple shapes such as sphere, cube, cylinder, or surface patches. In 2D computer graphics, primitives include segments of straight lines, circles, and more complicated curves. The true input u to the AM processes, however, is not the smooth shape defined analytically, for example, a dome in dashed line illustrated in Fig. 2.10a but discretized or sliced layers (Fig. 2.10a) and piecewise linear approximation of layer boundaries (Fig. 2.10b). Accuracy of the approximation is not only determined by computational modeling but also by materials and processes. For example, resolution along the build direction (z) is determined by the layer thickness in Fig. 2.10a, while the resolution in the build plane $(x - y$ plane in Fig. 2.10b) or the minimum feature size is limited by materials properties and technologies such as the laser spot size or the pixel size of a digital light projector.

The input representation therefore entails the *representation of individual layers and layer stack-up*. Borrowing geometric primitive concepts from computer graphics, we propose to use line segments, circular sectors, and corners as primitive shapes to construct individual layer boundaries or 2D shapes (Fig. 2.11). Notice that the three 2D shape primitives have curvatures of zero, constant, and infinity, respectively. Comparing to line segments on the grids in Fig. 2.10b, the choice of primitives will reduce computational load of approximation and assist to interpret the deviation patterns. We therefore state the following accepted truth as the basis for the subsequent derivation:

Definition 2.1 (Primitive Manufacturing Input) A 3D object built in AM is based on primitives manufacturing inputs, that is, sliced layers to be stacked up and individual layers with boundaries composed by 2D-shape primitives.

Essentially, the design input to AM machines is the sliced and discretized model (e.g., in STL format defined by unit normal and vertices), as opposed to smooth

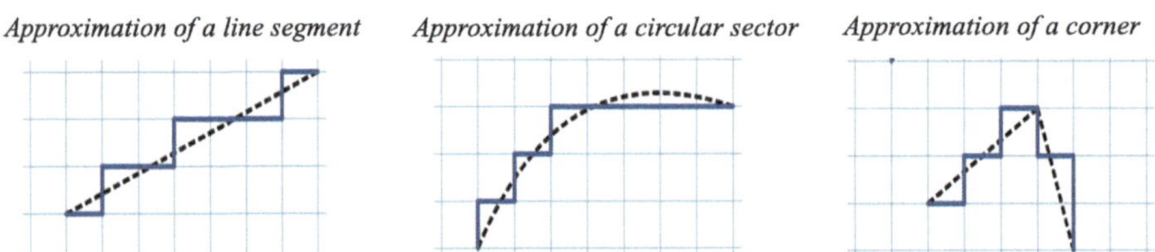

Fig. 2.11 2D geometric shape primitives for layer boundary approximation [47]

CAD models. Also note that even though the sliced layers can be geometrically stacked up in design input, the final print outcome is nonlinear due to physical layer interactions and accumulation. Considering layer stack-up in input/output representation is therefore critical to capture underlying physical mechanisms.

To conveniently represent shape deformation [46, 50], we represent layers or 2D shapes in the polar coordinate system (PCS). The layer boundary is represented as piecewise segments through a circular approximation with selective cornering strategy developed in [66]:

$$u(\theta) = \sum_{i=0}^{I} \mathbf{1}_{\theta \in [\theta_i, \theta_{i+1})} \, r_i(\theta) \tag{2.3}$$

where in the interval $[\theta_i, \theta_{i+1})$, $r_i(\theta)$ is either a line segment, circular sector, or a corner segment. Each segment consists of a set of points describing the primitive.

In essence, the definitions of primitive manufacturing input and shape primitives intend to establish a fabrication-aware formulation that enables understanding of the deviations of shape primitives. As such, deviation prediction of a freeform shape product can be achieved by predicting the deviation of the shape constructed by shape primitives. A freeform 2D shape can be constructed by the three basic types of 2D-shape primitives.

Output Representation Based on Transformation of Point-Cloud Data
The fabrication-aware output representation adopts the expression in Eq. 2.2, that is, $y_u = \Delta r(\theta, r_0(\theta)) = r(\theta, r_0(\theta)) - r_0(\theta)$, for individual layers or 2D shapes.

Difference between u and Nominal Design Note that the output y_u is the difference between the actual printed product $r(\cdot)$ and its intended design $r_0(\cdot)$, not the input u to the fabrication process. The fabrication input u is only a close approximation of design $r_0(\cdot)$ in the digital manufacturing. This is true for the 3D cases as well.

References

1. ASME Y14.46-2017 Product Definition for Additive Manufacturing. ASME (2017)
2. Achlioptas, P., Diamanti, O., Mitliagkas, I., Guibas, L.: Learning representations and generative models for 3D point clouds. In: Dy, J., Krause, A. (eds.) Proceedings of the 35th International Conference on Machine Learning. Proceedings of Machine Learning Research, vol. 80, pp. 40–49. PMLR, Stockholmsmässan, Stockholm (2018)
3. Alliez, P., Cohen-Steiner, D., Yvinec, M., Desbrun, M.: Variational tetrahedral meshing. In: ACM SIGGRAPH 2005 Papers, SIGGRAPH '05, pp. 617–625. Association for Computing Machinery, New York (2005). https://doi.org/10.1145/1186822.1073238
4. Alshraideh, H., Del Castillo, E.: Statistical performance of tests for factor effects on the shape of objects with application in manufacturing. IIE Trans. **45**(2), 121–131 (2013)

5. Ameta, G., Lipman, R., Moylan, S., Witherell, P.: Investigating the role of geometric dimensioning and tolerancing in additive manufacturing. J. Mech. Design **137**(11) (2015). https://doi.org/10.1115/1.4031296
6. Arai, H., Maung, C., Xu, K., Schweitzer, H.: Unsupervised feature selection by heuristic search with provable bounds on suboptimality. In: Proceedings of the AAAI Conference on Artificial Intelligence, vol. 30 (2016)
7. Belkin, M., Niyogi, P.: Laplacian eigenmaps for dimensionality reduction and data representation. Neural Comput. **15**(6), 1373–1396 (2003)
8. Bengio, Y., Courville, A., Vincent, P.: Representation learning: a review and new perspectives. IEEE Trans. Pattern Anal. Mach. Intell. **35**(8), 1798–1828 (2013)
9. Bermano, A.H., Funkhouser, T., Rusinkiewicz, S.: State of the art in methods and representations for fabrication-aware design. Comput. Graph. Forum **36**(2), 509–535 (2017). https://doi.org/10.1111/cgf.13146
10. Bermano, A.H., Funkhouser, T., Rusinkiewicz, S.: State of the art in methods and representations for fabrication-aware design. In: Computer Graphics Forum, vol. 36, pp. 509–535. Wiley Online Library (2017)
11. Bermano, A.H., Funkhouser, T., Rusinkiewicz, S.: State of the art in methods and representations for fabrication-aware design. In: Eurographics State of the Art Reports (2017)
12. Besl, P.J., McKay, N.D.: A method for registration of 3-d shapes. IEEE Trans. Pattern Anal. Mach. Intell. **14**(2), 239–256 (1992). https://doi.org/10.1109/34.121791
13. Biasotti, S., Cerri, A., Bronstein, A., Bronstein, M.: Recent trends, applications, and perspectives in 3D shape similarity assessment. Comput. Graph. Forum **35**(6), 87–119 (2016). https://doi.org/10.1111/cgf.12734
14. Bloomenthal, J.: Polygonization of implicit surfaces. Comput. Aided Geom. Design **5**(4), 341–355 (1988). https://doi.org/https://doi.org/10.1016/0167-8396(88)90013-1
15. Blum, A.L., Langley, P.: Selection of relevant features and examples in machine learning. Artif. Intell. **97**(1-2), 245–271 (1997)
16. Bugeda Miguel Cervera, G., Lombera, G.: Numerical prediction of temperature and density distributions in selective laser sintering processes. Rapid Prototyping J. **5**(1), 21–26
17. Cai, D., Zhang, C., He, X.: Unsupervised feature selection for multi-cluster data. In: Proceedings of the 16th ACM SIGKDD International Conference on Knowledge Discovery and Data Mining, pp. 333–342 (2010)
18. del Castillo, E., Colosimo, B.M.: Statistical shape analysis of experiments for manufacturing processes. Technometrics **53**(1), 1–15 (2011)
19. Chen, Z., Zhang, H.: Learning implicit fields for generative shape modeling. In: Proceedings of the IEEE Conference on Computer Vision and Pattern Recognition, pp. 5939–5948 (2019)
20. Cheng, L., Wang, A., Tsung, F.: A prediction and compensation scheme for in-plane shape deviation of additive manufacturing with information on process parameters. IISE Trans. **50**(5), 394–406 (2018)
21. Cox, T.F., Cox, M.A.: Multidimensional Scaling. CRC Press, Boca Raton (2000)
22. de Pastre, M.A., Tagne, S.C.T., Anwer, N.: Test artefacts for additive manufacturing: a design methodology review. CIRP J. Manuf. Sci. Technol. **31**, 14–24 (2020)
23. de Souza Borges Ferreira, R., Sabbaghi, A., Huang, Q.: Automated geometric shape deviation modeling for additive manufacturing systems via bayesian neural networks. IEEE Trans. Autom. Sci. Eng. (2019). https://doi.org/10.1109/TASE.2019.2936821
24. Domingos, P.: A few useful things to know about machine learning. Commun. ACM **55**(10), 78–87 (2012)
25. Dong, G., Liu, H.: Feature Engineering for Machine Learning and Data Analytics. CRC Press, Boca Raton (2018)
26. Dryden, I.L., Mardia, K.V.: Statistical Shape Analysis: With Applications in R, 2nd edn. John Wiley & Sons, London (2016)
27. Efroymson, M.: Multiple Regression Analysis. Mathematical Methods for Digital Computers. In: Ralston, A., Wilf, HS (eds.) (1960)

28. Farahat, A.K., Ghodsi, A., Kamel, M.S.: An efficient greedy method for unsupervised feature selection. In: 2011 IEEE 11th International Conference on Data Mining, pp. 161–170. IEEE, Piscataway (2011)
29. Fefferman, C., Mitter, S., Narayanan, H.: Testing the manifold hypothesis. J. Am. Math. Soc. **29**(4), 983–1049 (2016)
30. Gao, S., Ver Steeg, G., Galstyan, A.: Variational information maximization for feature selection. In: Advances in Neural Information Processing Systems, vol. 29 (2016)
31. Ge, Q., Chen, B., Smith, P., Menq, C.: Tolerance specification and comparative analysis for computer-integrated dimensional inspection. Int. J. Prod. Res. **30**(9), 2173–2197 (1992)
32. Ghojogh, B., Samad, M.N., Mashhadi, S.A., Kapoor, T., Ali, W., Karray, F., Crowley, M.: Feature selection and feature extraction in pattern analysis: a literature review (2019). arXiv preprint arXiv:1905.02845
33. Goodfellow, I.: Deep learning (2016)
34. Guo, Y., Wang, H., Hu, Q., Liu, H., Liu, L., Bennamoun, M.: Deep learning for 3d point clouds: a survey. IEEE Trans. Pattern Anal. Mach. Intell. **43**(12), 4338–4364 (2020)
35. Gupta, R.K., Gurumoorthy, B.: Automatic extraction of free-form surface features (FFSFs). Comput.-Aided Design **44**(2), 99–112 (2012)
36. Guyon, I., Elisseeff, A.: An introduction to variable and feature selection. J. Mach. Learn. Res. **3**(Mar), 1157–1182 (2003)
37. Ham, J., Lee, D.D., Mika, S., Schölkopf, B.: A kernel view of the dimensionality reduction of manifolds. In: Proceedings of the Twenty-First International Conference on Machine Learning, p. 47 (2004)
38. Harding, K.: Handbook of Optical Dimensional Metrology. CRC Press, Boca Raton (2013)
39. He, X., Cai, D., Niyogi, P.: Laplacian score for feature selection. In: Advances in Neural Information Processing Systems, vol. 18 (2005)
40. Hilton, P., Jacobs, P.: Rapid Tooling: Technologies and Industrial Applications. CRC Press, Boca Raton (2000)
41. Hinton, G.E., Roweis, S.: Stochastic neighbor embedding. In: Advances in Neural Information Processing Systems, vol. 15 (2002)
42. Hinton, G.E., Salakhutdinov, R.R.: Reducing the dimensionality of data with neural networks. Science **313**(5786), 504–507 (2006)
43. Ho-Le, K.: Finite element mesh generation methods: a review and classification. Comput.-Aided Design **20**(1), 27–38 (1988). https://doi.org/https://doi.org/10.1016/0010-4485(88)90138-8
44. Hocking, R.R.: A biometrics invited paper. the analysis and selection of variables in linear regression. Biometrics **32**, 1–49 (1976)
45. Hotelling, H.: Analysis of a complex of statistical variables into principal components. J. Educ. Psychol. **24**(6), 417 (1933)
46. Huang, Q.: An analytical foundation for optimal compensation of three-dimensional shape deformation in additive manufacturing. ASME Trans. J. Manuf. Sci. Eng. **138**(6), 061010 (2016)
47. Huang, Q.: An impulse response formulation for small-sample learning and control of additive manufacturing quality. IISE Trans. (2022). https://doi.org/10.1080/24725854.2022.2113186
48. Huang, Q., Nouri, H., Xu, K., Chen, Y., Sosina, S., Dasgupta, T.: Statistical predictive modeling and compensation of geometric deviations of three-dimensional printed products. ASME J. Manuf. Sci. Eng. **136**(6), 061008 (2014)
49. Huang, Q., Wang, Y., Lyu, M., Lin, W.: Shape deviation generator (SDG)—a convolution framework for learning and predicting 3d printing shape accuracy. IEEE Trans. Autom. Sci. Eng. **17**(3), 1486–1500 (2020)
50. Huang, Q., Zhang, J., Sabbaghi, A., Dasgupta, T.: Optimal offline compensation of shape shrinkage for 3d printing processes. IIE Trans. Quality Reliab. **47**(5), 431–441 (2015)
51. Jacobs, P.: Rapid Prototyping & Manufacturing, chap. Accuracy, pp. 287–315. Society of Manufacturing Engineers (1992)

52. Jin, Y., Pierson, H., Liao, H.: Scale and pose-invariant feature quality inspection for freeform geometries in additive manufacturing. J. Manuf. Sci. Eng. **141**(12), 1–15 (2019). https://doi.org/10.1115/1.4045174
53. Jin, Y., Qin, S., Huang, Q.: Offline predictive control of out-of-plane geometric errors for additive manufacturing. ASME Trans. Manuf. Sci. Eng. **138**(12), 121005 (2016)
54. Khanzadeh, M., Rao, P., Jafari-Marandi, R., Smith, B.K., Tschopp, M.A., Bian, L.: Quantifying geometric accuracy with unsupervised machine learning: using self-organizing map on fused filament fabrication additive manufacturing parts. ASME Trans. J. Manuf. Sci. Eng. **140**(3), 031011 (2017)
55. King, W., Anderson, A.T., Ferencz, R.M., Hodge, N.E., Kamath, C., Khairallah, S.A.: Overview of modelling and simulation of metal powder bed fusion process at Lawrence Livermore National Laboratory. Mater. Sci. Technol. **31**(8), 957–968 (2015)
56. Kira, K., Rendell, L.A.: A practical approach to feature selection. In: Machine Learning Proceedings 1992, pp. 249–256. Elsevier, Amsterdam (1992)
57. Kohavi, R., John, G.H.: Wrappers for feature subset selection. Artif. Intell. **97**(1-2), 273–324 (1997)
58. Kruth, J.P.: Material incress manufacturing by rapid prototyping techniques. CIRP Ann. **40**(2), 603–614 (1991)
59. Kuhn, M.: Applied predictive modeling (2013)
60. Leach, R., Bourell, D., Carmignato, S., Donmez, A., Senin, N., Dewulf, W.: Geometrical metrology for metal additive manufacturing. CIRP Ann. **68**(2), 677–700 (2019)
61. Leach, R.K., Bourell, D., Carmignato, S., Donmez, A., Senin, N., Dewulf, W.: Geometrical metrology for metal additive manufacturing. CIRP Ann. **68**(2), 677–700 (2019). https://doi.org/10.1016/j.cirp.2019.05.004
62. Li, J., Cheng, K., Wang, S., Morstatter, F., Trevino, R.P., Tang, J., Liu, H.: Feature selection: a data perspective. ACM Comput. Surv. **50**(6), 1–45 (2017)
63. Liu, H., Motoda, H.: Feature Extraction, Construction and Selection: A Data Mining Perspective, vol. 453. Springer, Berlin (1998)
64. Lorensen, W.E., Cline, H.E.: Marching cubes: A high resolution 3d surface construction algorithm. SIGGRAPH Comput. Graph. **21**(4), 163–169 (1987). https://doi.org/10.1145/37402.37422
65. Lu, H., Plataniotis, K.N., Venetsanopoulos, A.N.: A survey of multilinear subspace learning for tensor data. Pattern Recogn. **44**(7), 1540–1551 (2011)
66. Luan, H., Huang, Q.: Prescriptive modeling and compensation of in-plane shape deformation for 3-d printed freeform products. IEEE Trans. Autom. Sci. Eng. **14**(1), 73–82 (2017)
67. Luo, C., Franciosa, P., Ceglarek, D., Ni, Z., Jia, F.: A Novel geometric tolerance modeling inspired by parametric space envelope. IEEE Trans. Autom. Sci. Eng. **15**(3), 1386–1398 (2018). https://doi.org/10.1109/TASE.2018.2793920
68. Mahesh, M., Wong, Y., Fuh, J., Loh, H.: Benchmarking for comparative evaluation of RP systems and processes. Rapid Prototyping J. **10**(2), 123–135 (2004)
69. Masaeli, M., Yan, Y., Cui, Y., Fung, G., Dy, J.G.: Convex principal feature selection. In: Proceedings of the 2010 SIAM International Conference on Data Mining, pp. 619–628. SIAM (2010)
70. McConaha, M., Anand, S.: Additive Manufacturing Distortion Compensation Based on Scan Data of Built Geometry. J. Manuf. Sci. Eng. **142**(6), 061001 (2020). https://doi.org/10.1115/1.4046505
71. Mori, K., Osakada, K., Takaoka, S.: Simplified three-dimensional simulation of non-isothermal filling in metal injection moulding by the finite element method. Eng. Comput. **13**(2), 111–121 (1996)
72. Morse, E., Dantan, J.Y., Anwer, N., Söderberg, R., Moroni, G., Qureshi, A., Jiang, X., Mathieu, L.: Tolerancing: Managing uncertainty from conceptual design to final product. CIRP Ann. **67**(2), 695–717 (2018). https://doi.org/10.1016/j.cirp.2018.05.009
73. Moylan, S., Slotwinski, J., Cooke, A., Jurrens, K., Donmez, M.A., et al.: Proposal for a standardized test artifact for additive manufacturing machines and processes. In: Proceedings

of the 2012 Annual International Solid Freeform Fabrication Symposium, pp. 6–8. Austin (2012)

74. Narendra, Fukunaga: A branch and bound algorithm for feature subset selection. IEEE Trans. Comput. **100**(9), 917–922 (1977)

75. Pal, D., Patil, N., Zeng, K., Stucker, B.: An integrated approach to additive manufacturing simulations using physics based, coupled multiscale process modeling. ASME Trans. J. Manuf. Sci. Eng. **136**(6), 061022 (2014)

76. Qi, C.R., Su, H., Mo, K., Guibas, L.J.: Pointnet: Deep learning on point sets for 3d classification and segmentation. In: Proceedings of the IEEE Conference on Computer Vision and Pattern Recognition, pp. 652–660 (2017)

77. Rebaioli, L., Fassi, I.: A review on benchmark artifacts for evaluating the geometrical performance of additive manufacturing processes. Int. J. Adv. Manuf. Technol. **93**(5), 2571–2598 (2017)

78. Roweis, S.T., Saul, L.K.: Nonlinear dimensionality reduction by locally linear embedding. Science **290**(5500), 2323–2326 (2000)

79. Rusinkiewicz, S., Levoy, M.: Efficient variants of the icp algorithm. In: Proceedings Third International Conference on 3-D Digital Imaging and Modeling, pp. 145–152. IEEE, Piscataway (2001)

80. Sabbaghi, A., Dasgupta, T., Huang, Q., Zhang, J.: Inference for deformation and interference in 3d printing. Ann. Appl. Stat. **8**(3), 1395–1415 (2014)

81. Sabbaghi, A., Huang, Q.: Model transfer across additive manufacturing processes via mean effect equivalence of lurking variables. Ann. Appl. Stat. **12**(4), 2409–2429 (2018)

82. Sabbaghi, A., Huang, Q., Dasgupta, T.: Bayesian model building from small samples of disparate data for capturing in-plane deviation in additive manufacturing. Technometrics (2018). https://doi.org/10.1080/00401706.2017.1391715

83. Samie Tootooni, M., Dsouza, A., Donovan, R., Rao, P.K., Kong, Z.J., Borgesen, P.: Classifying the dimensional variation in additive manufactured parts from laser-scanned three-dimensional point cloud data using machine learning approaches. J. Manuf. Sci. Eng. **139**(9), 091005 (2017)

84. Savio, E., De Chiffre, L., Schmitt, R.: Metrology of freeform shaped parts. CIRP Ann. **56**(2), 810–835 (2007). https://doi.org/10.1016/j.cirp.2007.10.008

85. Schölkopf, B., Smola, A., Müller, K.R.: Kernel principal component analysis. In: International Conference on Artificial Neural Networks, pp. 583–588. Springer, Berlin (1997)

86. Shah, J.J.: Feature transformations between application-specific feature spaces. Comput.-Aided Eng. J. **5**(6), 247–255 (1988)

87. Shah, J.J., Mäntylä, M.: Parametric and Feature-Based CAD/CAM: Concepts, Techniques, and Applications. John Wiley & Sons, London (1995)

88. Sheikhpour, R., Sarram, M.A., Gharaghani, S., Chahooki, M.A.Z.: A survey on semi-supervised feature selection methods. Pattern Recogn. **64**, 141–158 (2017)

89. Snyder, J.: Generative Modeling for Computer Graphics and CAD. Academic Press, Boston (1992)

90. Strange, H., Zwiggelaar, R.: Open problems in spectral dimensionality reduction (2014)

91. Sunil, V., Pande, S.: Automatic recognition of features from freeform surface CAD models. Comput.-Aided Design **40**(4), 502–517 (2008). https://doi.org/10.1016/j.cad.2008.01.006

92. Tam, G.K., Cheng, Z.Q., Lai, Y.K., Langbein, F.C., Liu, Y., Marshall, D., Martin, R.R., Sun, X.F., Rosin, P.L.: Registration of 3d point clouds and meshes: A survey from rigid to nonrigid. IEEE Trans. Vis. Comput. Graph. **19**(7), 1199–1217 (2012)

93. Tam, G.K., Cheng, Z.Q., Lai, Y.K., Langbein, F.C., Liu, Y., Marshall, D., Martin, R.R., Sun, X.F., Rosin, P.L.: Registration of 3d point clouds and meshes: a survey from rigid to Nonrigid. IEEE Trans. Vis. Comput. Graph. **19**(7), 1199–1217 (2013). https://doi.org/10.1109/TVCG.2012.310

94. Tang, J., Hu, X., Gao, H., Liu, H.: Discriminant analysis for unsupervised feature selection. In: Proceedings of the 2014 SIAM International Conference on Data Mining, pp. 938–946. SIAM (2014)

95. Tenenbaum, J.B., Silva, V.d., Langford, J.C.: A global geometric framework for nonlinear dimensionality reduction. Science **290**(5500), 2319–2323 (2000)
96. Tibshirani, R.: Regression shrinkage and selection via the lasso. J. Roy. Stat. Soc. B (Methodological) **58**(1), 267–288 (1996)
97. Tong, K., Amine Lehtihet, E., Joshi, S.: Parametric error modeling and software error compensation for rapid prototyping. Rapid Prototyping J. **9**(5), 301–313 (2003)
98. Tong, K., Joshi, S., Amine Lehtihet, E.: Error compensation for fused deposition modeling (FDM) machine by correcting slice files. Rapid Prototyping J. **14**(1), 4–14 (2008)
99. Van Der Maaten, L., Postma, E.O., Van Den Herik, H.J., et al.: Dimensionality reduction: A comparative review. J. Mach. Learn. Res. **10**(66-71), 13 (2009)
100. Várady, T., Martin, R.R., Cox, J.: Reverse engineering of geometric models - An introduction. CAD Comput. Aided Des. **29**(4), 255–268 (1997). https://doi.org/10.1016/s0010-4485(96)00054-1
101. Vergara, J.R., Estévez, P.A.: A review of feature selection methods based on mutual information. Neural Comput. Appl. **24**, 175–186 (2014)
102. Wang, A., Song, S., Huang, Q., Tsung, F.: In-plane shape-deviation modeling and compensation for fused deposition modeling processes. IEEE Trans. Autom. Sci. Eng. **17**(2), 968–976 (2017)
103. Wang, S., Pedrycz, W., Zhu, Q., Zhu, W.: Subspace learning for unsupervised feature selection via matrix factorization. Pattern Recogn. **48**(1), 10–19 (2015)
104. Weinberger, K.Q., Sha, F., Saul, L.K.: Learning a kernel matrix for nonlinear dimensionality reduction. In: Proceedings of the Twenty-First International Conference on Machine Learning, p. 106 (2004)
105. Wu, J., Zhang, C., Xue, T., Freeman, B., Tenenbaum, J.: Learning a probabilistic latent space of object shapes via 3d generative-adversarial modeling. In: Advances in Neural Information Processing Systems, pp. 82–90 (2016)
106. Xu, K., Kwok, T.H., Zhao, Z., Chen, Y.: A Reverse Compensation Framework for Shape Deformation Control in Additive Manufacturing. J. Comput. Inform. Sci. Eng. **17**(2), 021012 (2017). https://doi.org/10.1115/1.4034874
107. Yang, J., Li, H., Campbell, D., Jia, Y.: Go-ICP: A Globally Optimal Solution to 3D ICP Point-Set Registration. IEEE Trans. Pattern Anal. Mach. Intell. **38**(11), 2241–2254 (2016). https://doi.org/10.1109/TPAMI.2015.2513405
108. Yang, Y., Shen, H.T., Ma, Z., Huang, Z., Zhou, X.: 2, 1-norm regularized discriminative feature selection for unsupervised learning. In: IJCAI International Joint Conference on Artificial Intelligence (2011)
109. Zang, Y., Qiu, P.: Phase I monitoring of spatial surface data from 3D printing. Technometrics **60**(2), 169–180 (2018). https://doi.org/10.1080/00401706.2017.1321585
110. Zhang, X., Wang, J., Yamazaki, K., Mori, M.: A surface based approach to recognition of geometric features for quality freeform surface machining. Comput.-Aided Design **36**(8), 735–744 (2004). https://doi.org/10.1016/j.cad.2003.09.002
111. Zhao, G., Liu, S.: Estimation of discriminative feature subset using community modularity. Sci. Rep. **6**(1), 25040 (2016)
112. Zhao, Z., Liu, H.: Spectral feature selection for supervised and unsupervised learning. In: Proceedings of the 24th International Conference on Machine Learning, pp. 1151–1157 (2007)
113. Zhu, Z., Anwer, N., Huang, Q., Mathieu, L.: Machine learning in tolerancing for additive manufacturing. CIRP Ann. **67**(1) (2018). https://doi.org/10.1016/j.cirp.2018.04.119

Chapter 3
Small-Sample Learning and Prediction of 2D Geometric Shape Quality

Learning and prediction of geometric quality in AM aims to learn and predict geometric shape deviations of both built and untried product shapes based on a small set of training products. To effectively control shape deviations of new and untried product shapes in AM, we classify the modeling approaches as predicting modeling and prescriptive modeling. While traditional predictive modeling usual makes prediction within its experimental domains, e.g., a class or family of products, prescriptive modeling is able to make prediction of quality of new and untried categories of shapes beyond the experimental scope. This chapter will first introduce predictive modeling approach to predict geometric quality for products with simply 2D geometries. The modeling approach is further extended and generalized for prescriptive small-sample modeling and prediction for 2D freeform products.

3.1 Predictive Shape Quality Modeling for 2D Base Geometry

The "simplest" 2D object is a disk with a circular shape. Yet the circular shape is a base geometry connected with polygons. In 3D printing, a circular disk is the result of polygon approximation. In this section, experiments of printing disks with various radii were conducted. The shape deviation data of circular disks were measured for building the prediction model [11].

3.1.1 Experiment of Printing Circular Disks and Training Data Collection

The specific 3D printing technique used to build the products is called stereolithography (SLA) process. Note that the modeling and analysis approach is applicable to

Q. Huang, *Domain-informed Machine Learning for Smart Manufacturing*,
https://doi.org/10.1007/978-3-031-91631-1_3

other AM processes as well. In the experiment, a 3D model, e.g., a design model of circular disk, was first built by a CAD software. Next, a specialized program sliced the CAD model into several cross sections (STL format files) according to the predesignated thickness of each layer, so that the 3D printing machine could construct each layer sequentially. Each layer's construction is analogous to printing of an image with a particular thickness, explaining this technique's name. After all layers are printed, the final product will have the shape close to the original CAD model.

Patented in 1988, stereolithography (SLA) is the first commercialized technique for 3D printing [14], followed by the development of selective laser sintering (SLS), fused deposition modeling (FDM), and laminated object manufacturing (LOM). A variant of the SLA process, Mask-Image-Projection-based SLA (MIP-SLA) was used for methodology demonstration. As seen in Fig. 3.1, the MIP-SLA machine has liquid resin stored in a tank configured with a platform that can move vertically precisely. During the printing process, the surface of the resin is exposed to light, which triggers the resin solidification. Control of light exposure area and intensity is through a digital micro-mirror device that receives commands from STL files for each layer. The platform in the tank moves down with the predefined thickness for printing the next layer when the previous layer is solidified.

Specification of the SLA experiments done in a commercial MIP-SLA platform, the ULTRA® machine from EnvisionTEC, is given in Table 3.1.

Once the printing process was completed, the boundary of the final product was measured by the optical Micro-Vu vertex measuring center (a laser scanner can accomplish the same measurement), and the measurement data was converted to polar coordinates for deviation modeling. Four circular disks with radii $r_0 = 0.5''$, $1''$, $2''$, and $3''$ are manufactured and measured.

The plot of the deviation profiles of the four disks is illustrated in Fig. 3.2. For each deviation profile between 0 and 2π, one reading per 0.01 radian was sampled. It

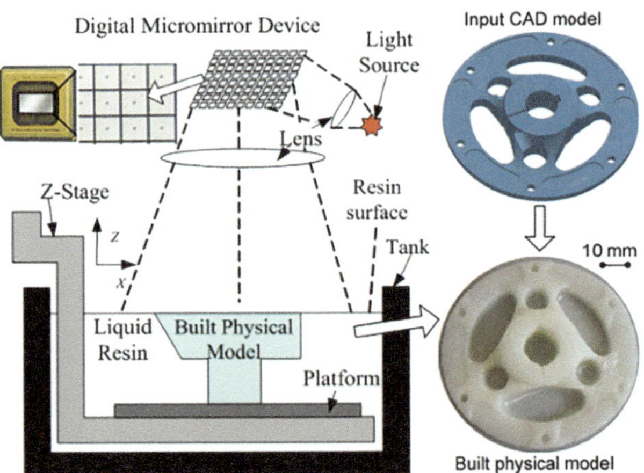

Fig. 3.1 Mask projection stereolithography (SLA) [39]

Table 3.1 The specific parameters of the MP-SLA process

Height of the product	0.5″ (inches)
Thickness of each layer	0.004″ (inches)
Resolution of the mask	1920 ∗ 1200
Dimension of each pixel	0.005″ (inches)
Illuminating time of each layer	9s
Waiting time between layers	15s
Type of the resin	SI500

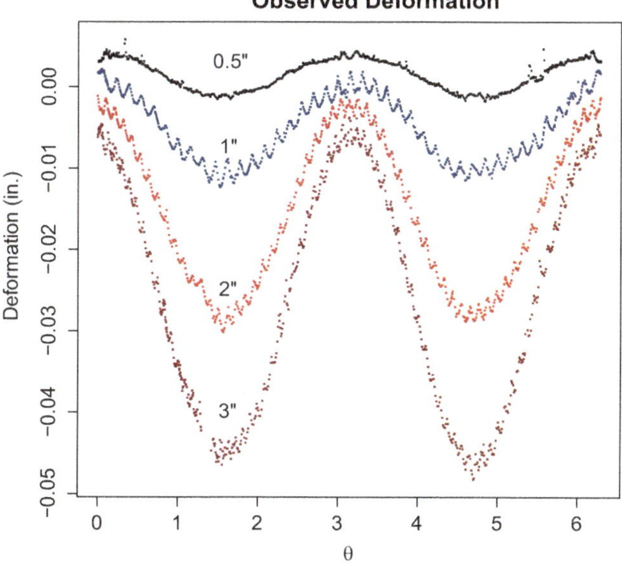

Fig. 3.2 Deviation profiles of four disks with $r_0 = 0.5''$, $1''$, $2''$, $3''$ [11]

can be observed that the smallest disk expanded with its boundary deviation profile largely above the zero line, while the other three disk shrank with their deviation profiles mainly below the zero line.

3.1.2   Statistical Modeling of Deviation Profiles of Circular Disks

Once the measurement of boundary deviations is transformed into deviation profiles, predictive modeling will establish the functional dependence of $y = \Delta r(\theta, r_0(\theta))$ at θ and the design input $r_0(\theta, z)$. To ensure stable causal relation, we assume that the AM process is more or less under control and the process is repeatable (i.e., the Stable Unit-Treatment Value Assumption under the Rubin causal framework [13]).

To capture local geometric deviations, the deviation profile $\Delta r(\theta, r_0(\theta))$ of a circular is decomposed into three components:

$$\Delta r(\theta, r_0(\theta)) = f_1(r_0(\theta)) + f_2(\theta, r_0(\theta)) + \epsilon_\theta. \tag{3.1}$$

The three components of the above equation are defined and interpreted as follows:

1. Function $f_1(r_0(\theta))$ represents average deviation or trend independent of location variable θ. Geometrically, the area or volume of a shape is obtained by integrating $r_0(\theta)$ over the space of θ, eliminating the location-dependent variable θ. Physically, it describes the uniform shrinkage/expansion of an AM-built product. The overall or average deformation can be related to the volumetric change of the product.
2. Function $f_2(\theta, r_0(\theta))$ is the location-dependent deviation in addition to the trend. Following the same reasoning as above, this location-dependent term is geometrically and physically related to $r_0(\theta)$.
3. Term ϵ_θ represents un-modeled terms such as surface waviness and roughness. It often manifests itself as high-frequency components adding on to the main harmonic trend.

We can also interpret $f_1(\cdot)$ as a lower-order term and $f_2(\cdot, \cdot)$ as a higher-order component of the deformation function.

For a circular disk shape, $r_0(\theta) = r_0$ for all θ. $f_1(\cdot)$ is a function of r_0 only, and $f_2(\cdot, \cdot)$ is a function of both r_0 and θ. Due to shape simplicity, the deviation profile of a circular shape can be expressed by a few Fourier basis terms, and Eq. (3.1) is reduced to the following form:

$$\Delta r(\theta, r_0(\theta)) = f_1(r_0) + \sum_k \{a_k \cos(k\theta) + b_k \sin(k\theta)\} + \epsilon_\theta \tag{3.2}$$

where $f_1(r_0(\theta)) = f_1(r_0)$ and $\{a_k\}$, $\{b_k\}$ are coefficients of a Fourier series expansion of $f_2(\cdot, \cdot)$. Note that $\{a_k\}$ and $\{b_k\}$ are functions of r_0. k is often small, e.g., $k = 2$.

3.1.3 Model Building and Bayesian Estimation of Model Parameters

Based on model (3.2), an observation of Fig. 3.2 (or a model selection process) leads to the choice of the following functional form to model disk boundary deviation as a special case of Eq. (3.2):

$$\Delta r(\theta, r_0(\theta)) = c_{r_0} + a_{r_0} \cos(2\theta) + \epsilon_\theta \tag{3.3}$$

for each disk of nominal radius r_0. Here, $\epsilon_\theta \sim N(0, \sigma^2)$ independently and represents high-frequency components adding on the main harmonic trend.

When we fit this model, we use the finite subset of angles $\theta_1, \ldots, \theta_n$ for each disk as described above, which may make our assumption regarding the independence of error terms more tenable if correlations among neighboring angles die out very quickly as a function of their separation (in radians).

In consideration of the location-independent model describing shape deviation caused by temperature and phase changes, the volumetric deviation should be proportional to the entire volume of the product based on the knowledge of heat transfer literature. Assuming that the height of the cylinder, h, in the z direction remains unchanged, the expected volumetric deviation is

$$h\left\{(r_0 + \Delta r)^2 - r_0^2\right\} \propto hr_0^2.$$

The radial deviation Δr is considerably less than the nominal radius r_0, leading to the approximation $\Delta r \propto r_0$. Thus, we model

$$c_{r_0} = \alpha r_0^a$$

where the parameter a should be approximately 1. Similarly, we model

$$a_{r_0} = \beta r_0^b$$

to describe the location-dependent geometric deviations, with b also approximately 1 as well. To summarize, the first parametric-shape deviation model we consider fitting for a disk product is

$$\Delta r(\theta, r_0) = \alpha r_0^a + \beta r_0^b \cos(2\theta) + \epsilon_\theta, \tag{3.4}$$

with α, β, a, b and σ, all independent of r_0.

As we possess prior engineering knowledge regarding parameters a and b, we implement a Bayesian procedure to draw inferences on all parameters α, β, a, b and σ. In particular, we assume

$$a \sim N(1, 2^2), \quad b \sim N(1, 1^2),$$

and we place flat priors on α, β, and $\log(\sigma)$, with all parameters independent a priori. We calculate the posterior distribution of the parameters by Markov Chain Monte Carlo (MCMC) and summarize the marginal posteriors by taking the mean, median, standard deviation, and 2.5% and 97.5% quantiles of the posterior draws.

The MCMC strategy used here is Hamiltonian Monte Carlo (HMC): the logarithm of the posterior is differentiable, and so an MCMC strategy such as HMC that uses the gradient of the log posterior can be expected to perform better than a generic Metropolis-Hastings or Gibbs algorithm, in terms of yielding high-quality draws with minimal tuning of the algorithm [19]. The mass matrix was chosen as the negative of the Hessian of the log posterior at the posterior mode, the leapfrog

Table 3.2 Summary of posterior estimation

	Mean	SD	2.5%	Median	97.5%	ESS
α	-0.0047	4.063×10^{-5}	-0.0048	-0.0047	-0.0047	7713.878
β	0.0059	6.847×10^{-5}	0.0058	0.0059	0.0060	8810.248
a	1.566	0.0084	1.5498	1.566	1.5819	7882.552
b	1.099	0.0120	1.0755	1.099	1.1232	8981.86
σ	0.0019	2.503×10^{-5}	0.00185	0.0019	0.00195	8513.814

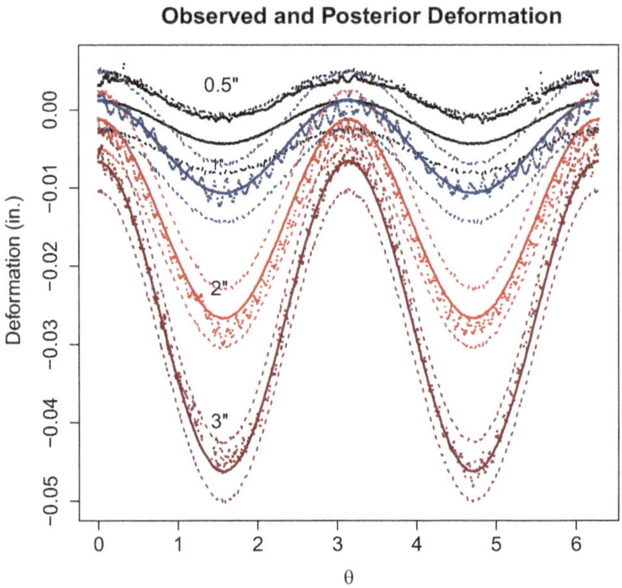

Fig. 3.3 Posterior predictive distribution for $r_0 = 0.5''$, $1''$, $2''$, $3''$ (inches) [11]

step size was 0.3, and the number of leapfrog steps was 50. We obtained 1000 draws from the posterior distribution of these parameters after a burn-in of 500 draws. Convergence was gauged by analysis of ACF and trace plots of the posterior draws, and the effective sample size (ESS) and Gelman-Rubin statistics [6] for these parameters were calculated by using ten independent chains of draws, each having 1000 draws after a burn-in of 500. Summary statistics in Table 3.2 below suggest that we have effectively sampled from the joint posterior of the parameters.

A simple comparison of the posterior predictive distribution of deviation profiles to the observed data (Fig. 3.3) demonstrates the fit for this model except for the smallest disk. In this figure, bold solid lines denote posterior means, and dashed lines denote the 2.5% and 97.5% posterior quantiles of the boundary deviation at each angle, with colors distinguishing disks with different radii. We see that this fit captures shape deviations for disks with $r_0 = 1''$, $2''$ and $3''$ fairly well, but does not provide a good fit for the disk with $r_0 = 0.5''$. The observed data for this particular

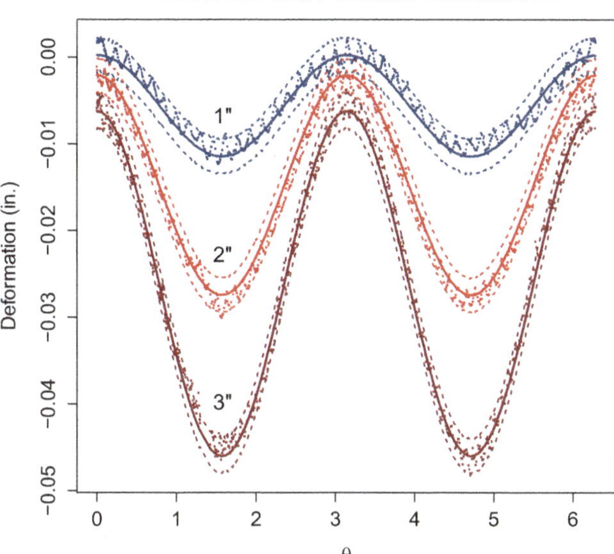

Fig. 3.4 Posterior predictive distribution for $r_0 = 1''$, $2''$, $3''$ [11]

Table 3.3 Summary of posterior estimation, excluding $0.5''$ radius cylinder

	Mean	SD	2.5%	Median	97.5%	ESS
α	-0.0056	2.792×10^{-5}	-0.0057	-0.0056	-0.00556	10002.67
β	0.0058	4.306×10^{-5}	0.00575	0.00584	0.00592	8908.08
a	1.400	0.00496	1.390	1.400	1.4098	8537.357
b	1.114	0.00767	1.100	1.1139	1.1301	8674.559
σ	0.00101	1.574×10^{-5}	0.000977	0.00101	0.00104	8997.69

disk is consistently located too far away from the posterior mean, in consideration of its posterior quantiles. It is largely attributed to the fact that the majority of the data demonstrates a shrinkage pattern, as opposed to the expansion pattern shown in the disk with $r_0 = 0.5''$.

In fact, if we fit the model in (3.4) for all disks except the $0.5''$ disk, we obtain a better fit for the remaining disks, as seen in Fig. 3.4. The posterior distribution of the parameters for this fit is summarized in Table 3.3. Note that the posterior mean and median of σ are now much smaller than before and that the posterior standard deviations of all parameters decrease. Furthermore, the observed data better correspond to the posterior quantiles in Fig. 3.4 as opposed to Fig. 3.3. These considerations suggest that the model does not capture the data for the $0.5''$ radius disk, so including this disk in the fitting procedure only served to increase the variance of model parameters. However, the posterior mean of a is 1.4, which doesn't correspond with our previous analytical considerations. A possible reason for this discrepancy, and an improvement of this model, is discussed in Sect. 3.1.4.

For now, we propose model (3.4) as the conjectured deviation model when no compensation is applied to a disk product, with the understanding that this model will be verified by more comprehensive physical experiments.

3.1.4 Modeling of Overexposure Phenomenon in AM Using Effect Equivalence Concept

A common phenomenon in AM processes involving light or laser is **over- or under-exposure**, where each region exposed to light or laser is bigger or smaller than its intended area due to imperfect light control over working space. Hereafter this phenomenon is referred to as overexposure because under-exposure is treated as the negative amount of overexposure. The overexposure can be viewed as a **lurking variable**, which will be discussed separately in a latter chapter.

One characteristic feature of overexposure is that expansion or shrinkage pattern of a small object is opposite to those of larger ones. As shown in Fig. 3.2, the $0.5''$ disk expands, while disks with $r_0 = 1.0''$, $2.0''$, and $3.0''$ shrink in sizes in general. These opposite deviation patterns of small and big objects, or more precisely, patterns with transition from expansion to shrinkage, create issues for modeling, prediction, and control, as partially illustrated in Fig. 3.3 when predicting accuracy for disks with radius smaller than $1.0''$.

Huang and co-authors [11] presented an efficient way to handle overexposure phenomenon, which applied the concept of effect equivalence. Wang and Huang [32–34] initially developed effect equivalence concept and methods intended to understand and explore the engineering phenomenon that different error sources generate identical error patterns. Following this concept, overexposure can be equivalently "reproduced" by design input error, that is, the CAD design model "instructs" light bulb or laser to expose a region larger or smaller than the target area. By transforming light exposure error into equivalent amount design error, we essentially create a "virtual" AM process with no light exposure errors. We could therefore only focus on the modeling of shape deviation without directly dealing with process errors such as overexposure, which greatly reduces the modeling complexity.

For the SLA experiments shown in Fig. 3.2, the amount of light overexposure for each pixel is hypothesized to be fixed along the product boundary (This assumption can be relaxed by imposing location dependence). This assumption implies that the same amount of overexposure will have larger impact on small objects, while the material shrinkage of larger objects will diminish the effect of expansion due to light overexposure. This explains the phenomena that small disks and large disks can have opposite shrinkage patterns.

A constant effect of overexposure for all disks is equivalent to a default design error x_0 applied to the original CAD model. The predictive deviation model given by Eq. (3.4) is modified as [11]:

$$\Delta r(\theta, r_0 + x_0) = r(\theta, r_0 + x_0) - r_0 \approx x_0 + \alpha(r_0 + x_0)^a$$

$$+ \beta(r_0 + x_0)^b \cos(2\theta) + \epsilon_\theta \tag{3.5}$$

This revised model (3.5) was fitted to data for disks with $r_0 = 0.5''$, $1''$, $2''$ and $3''$, where the earlier model becomes a special case with $x_0 = 0$. We maintain the same prior specification for α, β, a, b, and under the assumptions that $x_0 > 0$ and that x_0 is substantially smaller than any of the radii above. A weakly informative prior for x_0 of $\log(x_0) \sim N(0, 1)$ a priori is adopted. The same HMC strategy for parameter estimation is implemented to acquire draws from the posterior distribution of the parameters, which are summarized in Table 3.4, and the posterior predictive model for disks is presented in Fig. 3.5. We see that this new model provides a

Table 3.4 Summary of posterior estimation when modeling overexposure

	Mean	SD	2.5%	Median	97.5%
α	−0.0134	1.596×10^{-4}	−0.0137	−0.0134	−0.0131
β	0.0057	3.097×10^{-5}	0.00565	0.00571	0.0058
a	0.8606	0.00733	0.8465	0.8606	0.8752
b	1.1331	0.00546	1.123	1.1332	1.1442
x_0	0.00879	0.00015	0.008489	0.00879	0.00907
σ	0.000869	1.182×10^{-5}	0.000848	0.000869	0.000892

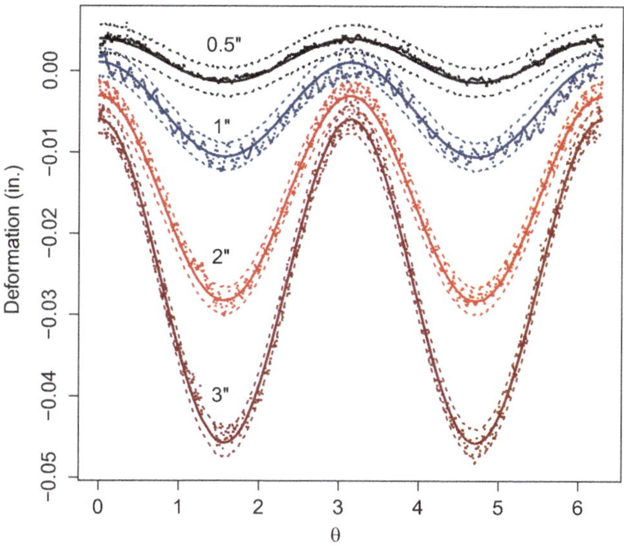

Fig. 3.5 Model considering overexposure: posterior predictive model for disks with $r_0 = 0.5''$, $1''$, $2''$, $3''$ (inches) [11]

substantially better fit for all observed data as compared to our original model (3.4), suggesting that we may have captured the conjectured overexposure phenomenon.

Table 3.4 also indicates that the posterior mean of parameter a decreased from 1.4 to 0.86, closer to the theoretical value of 1. The remaining difference is suspected to be associated with the deformation of thickness in the z direction, which is ignored in our analysis for simplicity. The estimated value of x_0 is slightly less than twice the pixel size (0.005″). This result corresponds with Zhou and Chen's previous work on pixel calibration [38] for the MIP-SLA process. All these considerations suggest that the new model that explicitly includes an overexposure effect provides a better physical interpretation and efficient means to model this engineering phenomenon using effect equivalence concept.

3.2 Predictive Shape Quality Modeling for Both Circular and Polygon Shapes

Although model (3.1) is quite generic and model (3.2) seems easily extendable beyond circular shape, polygon shapes bring complications to these models. First, straight edges and sharp corners of a polygon will require a large number of Fourier basis terms in Eq. (3.2) for better approximation. This will compromise the model robustness and flexibility for AM with frequent changes of product geometries. Second, when the number of polygon sides increase, the polygon model should asymptotically approach to the circular disk model defined in Eq. (3.5). Considering the limited number of test shapes, the connection among different shapes should be explored to facilitate prescriptive modeling and prediction. Therefore, model robustness, flexibility, and extendability for various shapes are of main modeling considerations for small-sample modeling in AM [9].

3.2.1 Experiment of Printing Polygon Plates and Training Data Collection

Polygon is another common type of 2D geometries with non-smooth boundaries. To build a model capable of predicting boundary deviations for both circular and polygon plates, experiments were conducted using the same SLA machine as for printing circular disk experiments. Specification of the SLA process is shown in Table 3.5, where the product height and layer thickness differ from circular disk experiments (Table 3.1). Another major experimental condition difference is that the SLA machine went through maintenance and repair after initial cylinder experimentation, which changed the product deviation patterns of the SLA machine.

The experimental factors considered for printing polygon plates include the number of sides and the size of a polygon in the $x - y$ plane. When the number of sides increases, the corner angles of the polygon shape vary. The polygon size

Table 3.5 The specific parameters in MP-SLA process

Height of the product	0.25″
Thickness of each layer	0.00197″
Resolution of the mask	1920 ∗ 1200
Dimension of each pixel	0.005″
Illuminating time of each layer	7s
Average waiting time between layers	15s
Type of the resin	SI500

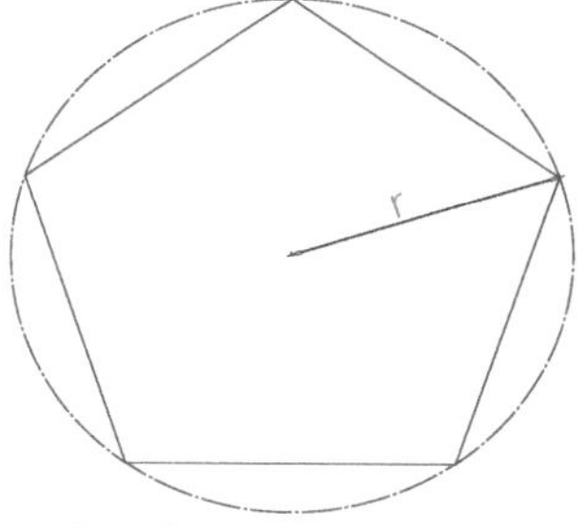

(a) Regular polygon with with size *r*

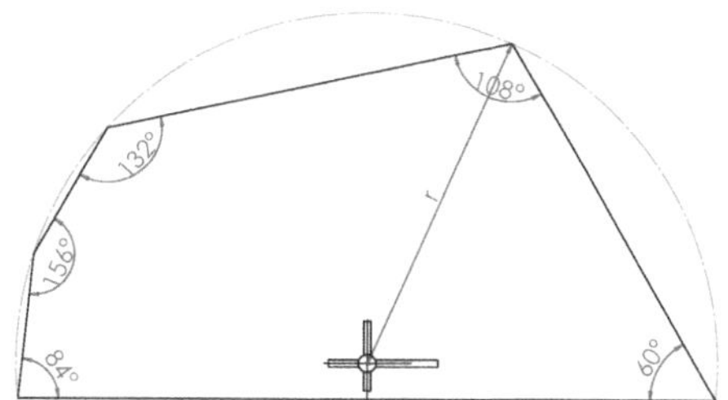

(b) Irregular polygon with size *r*

Fig. 3.6 Size of a polygon plate: circumcircle radius *r* [9]

Table 3.6 Experimental design of printing polygons

Cross-section geometry	Circumcircle radius	Process condition
Square	$r = 1''/\sqrt{2};\ 2''/\sqrt{2};\ 3''/\sqrt{2}$	Before machine repair
Square	$2''/\sqrt{2}$	After machine repair
Regular pentagon	$r = 1'';\ 3''$	After repair, settings changed
Regular dodecagon	$r = 3''$	After repair, settings changed
Circle	$r = 1'';\ 2''$	After repair, settings changed

is defined as the radius of its in-plane ($x - y$ plane) circumcircle (see examples in Fig. 3.6). The purpose of using a circumcircle is to connect the polygon experiments with circular disk experiments done before.

Three polygons with in-plane shapes of square, regular pentagon, and dodecagon were investigated. As shown in Table 3.6, four squares with side lengths of $1''$, $2''$ (before and after repair), and $3''$ (or circumcircle radii of $1''/\sqrt{2}$, $2''/\sqrt{2}$, and $3''/\sqrt{2}$), two regular pentagons with circumcircle radii $1''$ and $3''$, and one regular dodecagon (polygon with 12 sides) with circumcircle radius of $3''$ were designed and fabricated by using the MIP-SLA machine. The purpose of fabricating a dodecagon was to check how well a model can be generalized to different polygons and how well a polygon model can be approximated by a circular shape when the number of polygon sides is large. Since the MIP-SLA machine settings were changed after repair, new experiments of printing circular disks were also included for experimentation in this study with $r = 1''$ and $2''$.

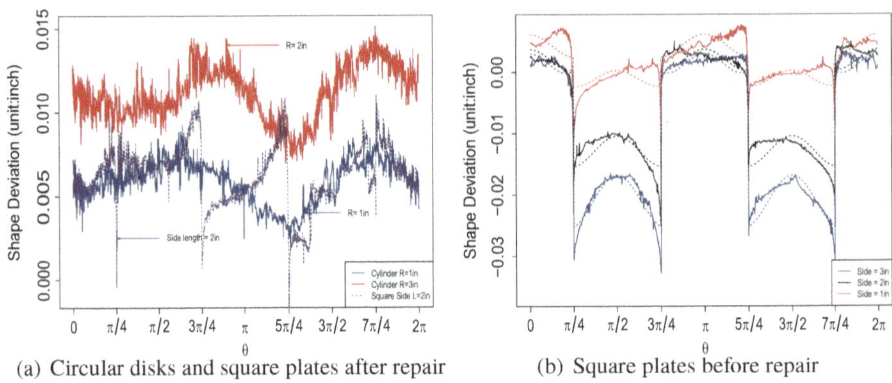

(a) Circular disks and square plates after repair   (b) Square plates before repair

Fig. 3.7 Boundary deviation profiles (solid lines) of disks with $r_0 = 1''$, $2''$ and square plates with side length = $1''$, $2''$, $3''$ (inches) [9]

To facilitate the identification of the orientation of test parts during or after the building process, a non-symmetric cross with line thickness of $0.02''$ was added on the top of them (see Fig. 3.6b). All test parts had a height of $0.25''$. The 3D CAD models were exported to STL format files, which were then sent to the SLA machine.

After the fabrication process, all test parts are measured using a Micro-Vu precision machine. In order to reduce human errors, the same measuring procedure is followed for each test part. For simplicity of measurement, we choose the center of the cross to be the origin of the measurement coordinate system in the Micro-Vu machine. The boundary profile is fitted using the splines in the metrology software associated with the Micro-Vu machine. The obtained measurement data are then converted to polar coordinates for deviation modeling and analysis. It must be noted that the part orientation is kept the same during the building and measurement processes. Figures 3.7, 3.8, and 3.9 show the measured boundary deviation profiles (solid lines) for three kinds of polygons and circular disks presented in the PCS.

By comparing the deviation profiles of circular disks and square plates before and after the machine repair (Fig. 3.2 vs. Fig. 3.7), it is clear that the repaired MIP-SLA machine tends to overcompensate the product shrinkage and lead to positive shape deviation; and the shape deviation patterns clearly differ before and after the machine repair. More importantly, the product deviation patterns vary with geometries. The fundamental question is whether a unified model can capture and predict the deviation patterns of circular disks and polygon plates. In the following section, we intend to identify basic deviation patterns consistent across different shapes and provide an integrated model for prediction and compensation.

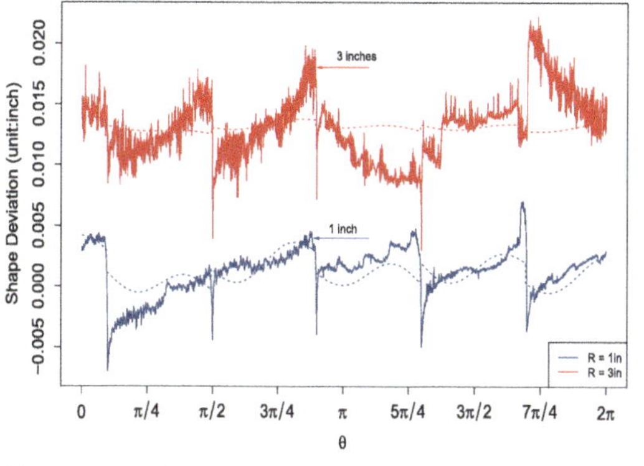

(a) Two regular pentagon deviation profiles with circumcircle radii = 1″, 3″ (solid lines)

(b) Printed regular pentagon with circumcircle radius = 3″

Fig. 3.8 Observed deviation profiles for pentagons [9]

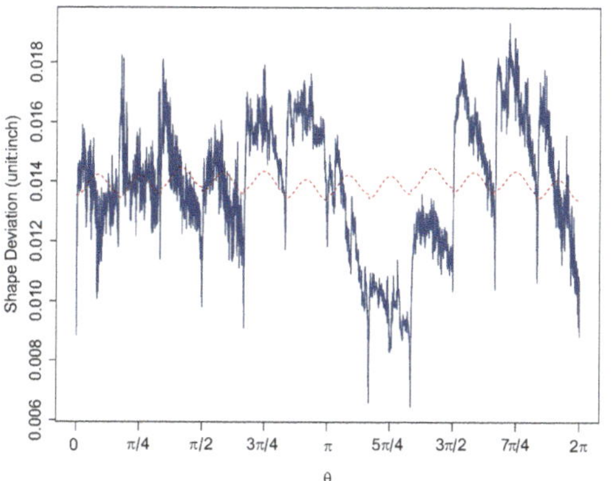

(a) Regular dodecagon deviation profile (solid line) with circumcircle radius = 3″

(b) Printed dodecagon with circumcircle radius = 3″

Fig. 3.9 Observed deviation profile for a dodecagon [9]

3.2.2 *A Unified Predictive Model for Both Circular and Polygon Shapes Under a Cookie-Cutter Modeling Framework*

Following the same notation in model (3.1), we denote $\Delta(\theta, r_0(\theta))$ as the in-plane shape deviation of 3D printed product presented in the polar coordinate system. Approximation of $\Delta(\theta, r_0(\theta))$ using Fourier series, though works well for circular disks, clearly encounters the difficulty of model fitting for polygon shapes, i.e., too many high-order Fourier basis terms have to be included in order to capture the sharp transition at the polygon vertices. It will compromise model robustness, flexibility, and extendability for various shapes.

In order to connect the circular shape to polygon shapes under a unified model, the so-called cookie-cutter modeling strategy is proposed. This modeling approach treats a polygon as being carved out from its circumcircle as shown in Fig. 3.6. This strategy implies that an integrated deviation model for both circular and polygon shapes contains at least two major basis functions: (1) basis model g_1 for circular shapes and (ii) g_2 for a cookie-cutter function. This concept extends our previous model in Eq. (3.1) into

$$\Delta(\theta, r_0(\theta)) = g_1(\theta, r_0(\theta)) + g_2(\theta, r_0(\theta)) + g_3(\theta, r_0(\theta)) + \epsilon_\theta. \tag{3.6}$$

Note that this extended model (3.6) serves for accuracy prediction of both circular and polygon shapes. If circular basis $g_1(\theta, r_0(\theta))$ and cookie-cutter basis $g_2(\theta, r_0(\theta))$ cannot fully capture the patterns in the deviation profiles, we could introduce additional $g_3(\theta, r_0(\theta))$ term if necessary. For example, individual sides of a polygon may have a higher-order deviation pattern after trimming by the cookie-cutter function. We could have another term $cos[n(\theta - \phi_0)]$ or $sin[n(\theta - \phi_0)]$, which also has the consistent limiting property when the number of polygon size n goes to infinity.

The circular basis model g_1 can take the form of model (3.2) or model (3.5) for SLA process with overexposure effect. The cookie-cutter model g_2 is first proposed in [9] with two alternatives functional forms: (i) square wave model and (ii) sawtooth wave model.

Square Wave Cookie-Cutter Function A square wave is a non-sinusoidal periodic waveform in which the amplitude alternates at a steady frequency between fixed minimum and maximum values. One of its functional form is $sign[cos(\theta)]$ with $sign[\cdot]$ being a sign function. Since the number of polygon sides n should change the period of the square wave function, the cookie-cutter function to trim a polygon from a circle is proposed as

$$g_2(\theta, r_0(\theta)) = \beta_2 \, r_0^\alpha \, sign[cos(n(\theta - \phi_0)/2)] \tag{3.7}$$

where ϕ_0 is a phase variable to shift the cutting position in the PCS. It is determined by the smallest angular distance from the vertex of a polygon to the axis of the PCS. For instance, ϕ_0 is $\pi/4$ for a square shape and $(1/2 - 2/5)\pi$ for a pentagon with a polar axis being horizontal in Fig. 3.6. Here coefficients β_2 and α are to be estimated.

We plot the cookie-cutter functions for polygons with $n = 4$, 5, 6, 8, 12, and 60 sides. Meanwhile, we scale the magnitudes of the deviation profiles observed for a square, pentagon, and dodecagon and superimpose the profiles on their corresponding cookie-cutter functions. As shown in Fig. 3.10, the cookie-cutter model in Eq.(3.7) perfectly capture the sharp transitions around the vertices of polygons in the deviation profiles. Furthermore, when n approaches to infinity and a polygon is supposed to become a circle, the cookie-cutter function approaches to white noise. The limiting property is preserved.

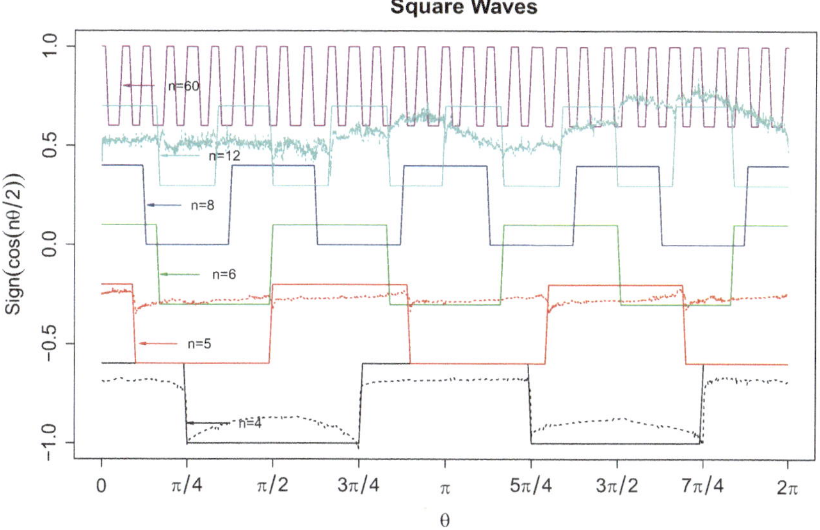

Fig. 3.10 Square wave functions with square, pentagon, and dodecagon deviation profiles

Sawtooth Wave Cookie-Cutter Function An alternative cookie-cutter model is sawtooth wave model. One observation from the pentagon and dodecagon deformation is the presence of some sort sawtooth-like wave form. We define the sawtooth function formally as:

$$saw.tooth(\theta - \phi_0) = (\theta - \phi_0) \text{ MOD } (2\pi/n)$$

where we use MOD function in the usual sense to obtain remainders (as in x MOD y = remainder of (x/y)).

Figure 3.11 illustrates examples of sawtooth functions. Comparing the definition of square wave function, it is clear that sawtooth wave function will also capture the sharp transitions around the vertices of polygons in the deviation profiles. It preserves the limiting property when n approaches to infinity.

To control the direction of each sawtooth (flipping upside down), we introduce an indicator function:

$$I(\theta - \phi_0) = \{sign[sin(n(\theta - \phi_0)/2)] + 1\}/2 \tag{3.8}$$

The alternative cookie-cutter basis using sawtooth wave function is therefore given as

$$g_2(\theta, r_0(\theta)) = \beta_2 \, r_0^{\alpha} \, I(\theta - \phi_0) \, saw.tooth(\theta - \phi_0)$$

$$= \beta_2 \, r_0^{\alpha} \, \{sign[sin(n(\theta - \phi_0)/2)] + 1\}/2 \, [(\theta - \phi_0) \text{ MOD } (2\pi/n)] \tag{3.9}$$

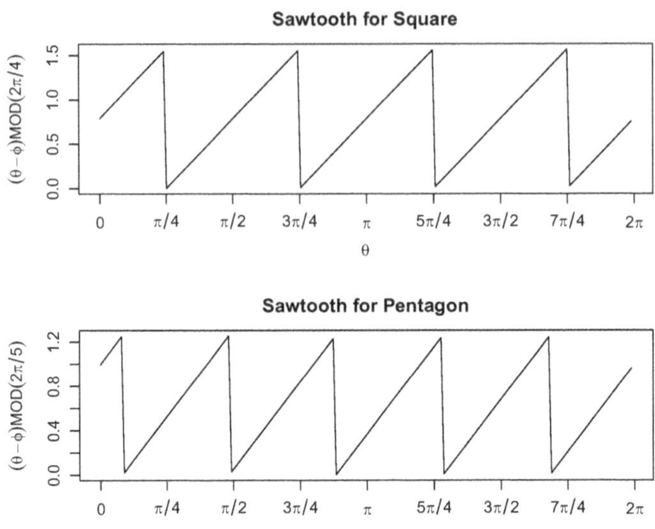

Fig. 3.11 Sawtooth wave functions

3.2.3 *Parameter Estimation for the Unified Predictive Deviation Model*

This section illustrates the estimation of model (3.6) for SLA experimentation on both circular and polygon shapes.

Due to the aforementioned SLA machine repair, cylindrical basis model $g_1(\theta, r_0(\theta))$ represented in Eq. (3.5) or $g_1(\theta, r_0(\theta)) = x_0 + \beta_0(r_0 + x_0)^a + \beta_1(r_0 + x_0)^b \cos(2\theta)$ has to be re-estimated using new data shown in Fig. 3.7. When extending this basis model to generic shape, there is a danger of *overparameterization*, that is, the total number of unknown parameters in g_1 and g_2 will be more than can be estimated from the limited data. For illustration purpose, we adopt the following model without considering overexposure term:

$$g_1(\theta, r_0(\theta)) = \beta_0 r_0^a + \beta_1 r_0^b \cos(2\theta). \tag{3.10}$$

Due to apparent pattern change presented in Fig. 3.7a, the circular basis model $g_1(\theta, r_0(\theta))$ after machine repair is expected to have different coefficients β_0 and β_1. On the other hand, we tend to keep a and b as the same values in Table 3.2 for two reasons. First, r_0^a and r_0^b in Eq. (3.10) represent the volumetric shrinkage varying with size r_0. For the same material, the change after MIP-SLA machine repair might be small. Second, concern of over-parametrization limits the number of unknown parameters in the model. Therefore, the circular basis function used is

$$g_1(\theta, r_0(\theta)) = \beta_0 r_0^{1.566} + \beta_1 r_0^{1.099} \cos(2\theta). \tag{3.11}$$

To test the proposed modeling strategy and cookie-cutter function, we first fit a statistical model for individual product shapes observed in the experiment. Given a cookie-cutter function, e.g., square wave function, model (3.6) can be simplified by merging variables as

$$\Delta(\theta, r_0(\theta)) = \beta_0 + \beta_1 cos(2\theta) + \beta_2 sign[cos(n(\theta - \phi_0)/2)] + \beta_3 cos(n\theta) + \epsilon$$

(3.12)

We simply apply the least square estimation (LSE) to estimate the individual models for three types of polygons. The estimated model parameters of Eq. (3.12) for fabricated product are shown in Table 3.7. The fitted models are shown as dashed lines in Figs. 3.7, 3.8, and 3.9, respectively. As can be seen in Fig. 3.7, with only three base functions, model (3.12) captures the deviation pattern of circular and square shapes very well (before machine repair), comparing with many model terms for Fourier series.

Table 3.7 Individual model estimation for polygons

Polygon	Parameter	Estimate	Std. error	P-value	Residual σ_ϵ
1″ square	β_0	1.5149e-3	1.313e-4	≈0	0.003207
	β_1	5.706e-4	3.571e-4	0.11	
	β_2	2.343e-3	2.173e-4	≈0	
	β_3	1.561e-3	1.733e-4	≈0	
2″ square	β_0	−5.470e-03	8.924e-05	≈0	0.002568
	β_1	−8.053e-04	2.618e-04	0.00217	
	β_2	7.856e-03	1.678e-04	≈0	
	β_3	2.071e-03	1.211e-04	≈0	
3″ square	β_0	−9.763e-03	8.771e-05	≈0	0.002597
	β_1	−2.776e-03	2.640e-04	≈0	
	β_2	1.266e-02	1.714e-04	≈0	
	β_3	2.797e-03	1.204e-04	≈0	
1″ pentagon	β_0	1.317e-03	2.160e-05	≈0	0.001812
	β_1	8.255e-04	3.173e-05	≈0	
	β_2	8.739e-04	2.314e-05	≈0	
	β_3	1.178e-03	3.245e-05	≈0	
3″ pentagon	β_0	1.310e-02	2.062e-05	≈0	0.002768
	β_1	3.584e-04	3.158e-05	≈0	
	β_2	2.546e-04	2.251e-05	≈0	
	β_3	2.012e-04	3.003e-05	≈0	
3″ dodecagon	β_0	1.391158e-02	1.496537e-05	≈0	0.002138664
	β_1	−1.417587e-04	2.089859e-05	≈0	
	β_2	7.680937e-05	1.494737e-05	2.7925e-07	
	β_3	3.338678e-04	2.033879e-05	≈0	

Table 3.8 Integrated model estimation for polygons

β_0	β_1	β_2	α	Residual σ_ϵ
0.00297	5.30065e-05	1.2315e-09	1.36459	0.00283

However, model (3.12) seems unable to explain pentagon and dodecagon well (Figs. 3.8 and 3.9). The potential reason can be attributed to process condition change, and pentagon and dodecagon products were fabricated after the repair of the MIP-SLA machine. The sawtooth pattern is more obvious after repair, which may cause the poor fitting for pentagon and dodecagon shapes based on model (3.12).

The initial model fitting for individual polygon shapes suggest the use of sawtooth wave function as cookie-cutter basis model after the machine repair. In addition, we attempt to fit one integrated model (3.6) for all polygons fabricated after machine repair. We will use square shape with side lengths $2''$, $1''$, and $3''$ pentagons to fit one model and validate the model using $3''$ dodecagon data. Specifically, the model after the machine repair is

$$\Delta(\theta, r_0(\theta)) = g_1(\theta, r_0(\theta)) + g_2(\theta, r_0(\theta)) + \epsilon$$

$$= \beta_0 r_0^{1.566} + \beta_1 r_0^{1.099} \cos(2\theta) + \beta_2 \, r_0^{\alpha} \, I(\theta - \phi_0) \, saw.tooth(\theta - \phi_0) + \epsilon \tag{3.13}$$

Here we start without high-order term $g_3(\theta, r_0(\theta))$.

To fit this nonlinear model, we use the Bayesian approach to find if parameter estimation will maximize the log-likelihood function. The optimization routine *optim*() in **R** was implemented. The estimated model parameters and residual standard deviation are given in Table 3.8.

The observed data and model prediction for three polygons are illustrated in Fig. 3.12a. Figure 3.12b shows the prediction for dodecagon shape. Note that dodecagon data was not used in the model training date set.

This section and the work in [9] demonstrated the feasibility of developing a shape quality model for heterogeneous 2D geometries. Though the model selection and estimation can be further improved for better prediction accuracy, the cooking-cutter modeling strategy enables small-sample learning of geometric shape deviations, which will be further generalized to 2D and 3D freeform geometries in latter sections.

3.3   Prescriptive Shape Quality Learning and Prediction for 2D Freeform Geometries

In contrast to predictive modeling, prescriptive modeling aims to predict quality of new and untried products beyond the experimental scope of test shapes. As shown in Fig. 3.13, a prescriptive model is expected to predict geometric shape quality of

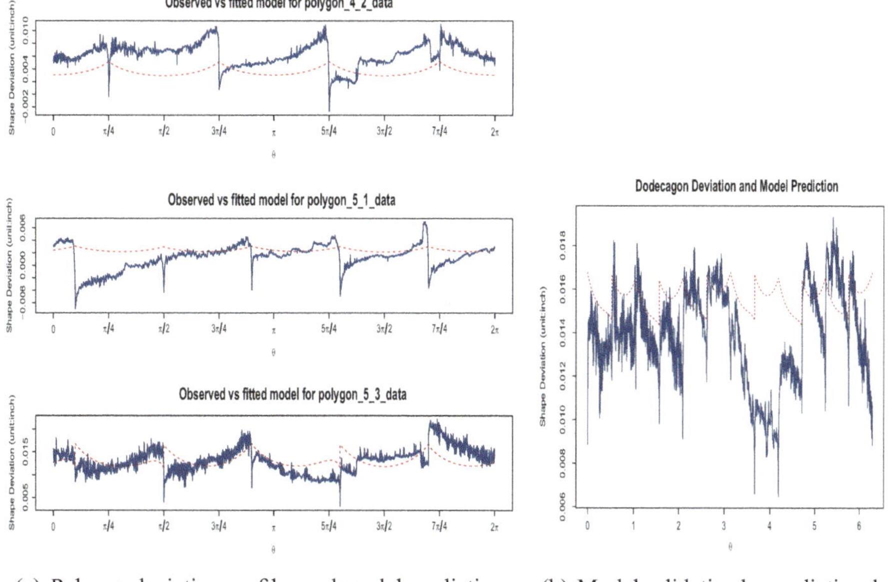

(a) Polygon deviation profiles and model predictions

(b) Model validation by predicting dodecagon deviation profile

Fig. 3.12 Integrated model prediction and validation

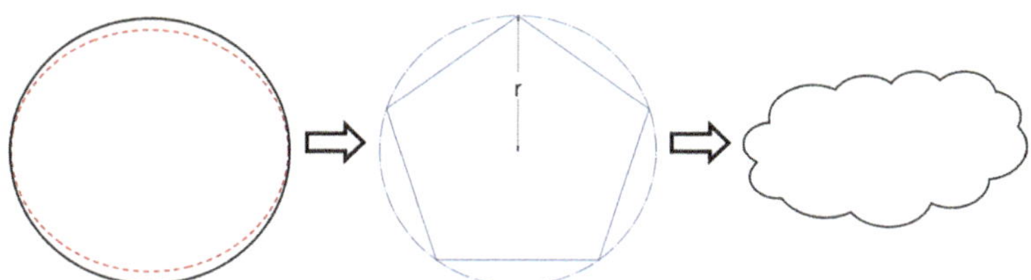

Fig. 3.13 From circles and polygons to 2D freeform geometries [7]

freeform shapes based on limited trials of simple shapes such as circular disks and polygon plates.

Research on solid freeform fabrication [2] have been focusing more on design, process planning, finite element modeling (FEM), materials, processes, and machines. For instance, the work in [3] proposes a feature-based design approach for modeling complex components with local composition control. To ensure geometric accuracy of freeform products, an algorithm is developed to calculate the correct amount of droplets along the boundary and the amount of proportional deflections. The work in [4] puts forward a closed-loop feedback control strategy to improve geometric quality by making real-time decision regarding droplet scale and locations of subsequent droplets. Yet these methods are more or less dependent on specific geometric shapes and processes.

This section generalizes the cookie-cutter modeling approach discussed in Sect. 3.2 [9] to predict shape accuracy of 2D freeform products with a prescriptive learning strategy independent of geometric and process complexities. The shape deviation of arbitrary 2D freeform shapes can be predicted based on limited test shapes of circular and polygon shapes as shown in Fig. 3.13.

3.3.1　Strategy of Prescriptive Modeling for Methodology Extension to 2D Freeform Shapes and Experimental Validation

The first challenging issue for extending the cookie-cutter modeling approach to 2D freeform shapes is the functional characterization of shape deviations for arbitrary 2D shapes. A close-form representation is essential for implementing the optimal compensation policy derived in [11]. Below we discuss the formulation of two possible strategies.

3.3.1.1　Strategy A: Polygon Approximation with Local Compensation (PALC)

One observation is that any in-plane 2D freeform shape can be approximated by a polygon (Fig. 3.14). Since we have predictive in-plane shape deviation models established for circular and polygon shapes [9, 11, 25, 31], one intuitive idea is therefore to approximate an arbitrary freeform shape by a polygon first and then improve the shape deviation model of that polygon through compensation and adjustment. This line of thinking results in the first strategy: polygon approximation with local compensation (PALC).

Denote the in-plane error model for the fitted polygon as $f(\theta, r(\theta))$ defined in the polar coordinate system (PCS). (Different polygonal approximation approaches will be discussed shortly.) At angle θ, the approximation error is denoted as $x(\theta)$, which is the amount of compensation to be applied to improve prediction (refer to Part III for topic on optimal compensation [7, 11]). The predictive in-plane shape

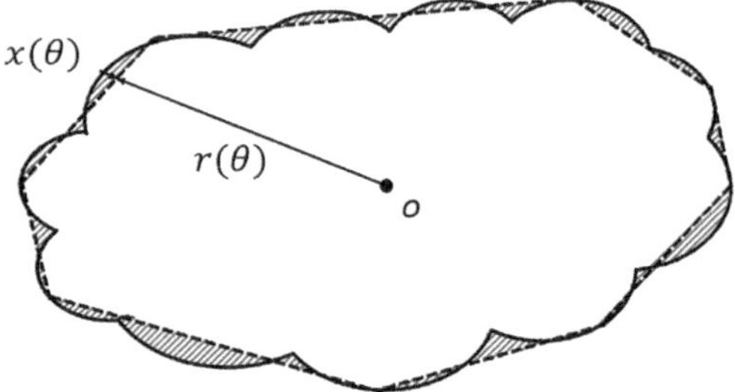

Fig. 3.14 Strategy of polygon approximation with local compensation (PALC) (solid line, freeform shape; dashed line, approximated polygon shape; shadow area, amount of compensation to approach the freeform from the polygon shape)

deviation model for the freeform can thus be derived as $f(\theta, r(\theta) + x(\theta))$. In this way, shape quality modeling for polygons in Sect. 3.2 [9] can be extended to arbitrary 2D freeform shapes.

3.3.1.2   Strategy B: Circular Approximation with Selective Cornering (CASC)

An alternative strategy, which is not that intuitive, is based on the observation that any in-plane 2D freeform shape can be approximated by the addition of a series of circular sectors with different radii (Fig. 3.15). To accommodate the potential sharp transitions or corners between adjacent sectors, we properly select corners and impose a cookie-cutter function [9] to the circular base functions. This second strategy is called circular approximation with selective cornering (CASC).

The first step of CASC strategy is to obtain a series of sectors with different $r_i(\theta_i)$ at θ_i in the PCS (Fig. 3.15). A natural approach to derive $r_i(\theta_i)$ is to fit a polygon with a large number of sides to the 2D freeform shape. The deviation of each circular sector can be predicted by the circular base model in Sect. 3.1 [11] using a section of a circle.

The second step of CASC is to properly select corners to catch the transition points along the boundary of the 2D freeform shape. The good approximation in step 1 of CASC results in a large number of sectors. Yet the sharp transition points along the 2D freeform can be less than the number of circular sectors. We will present a method to design the cookie-cutter function to catch the transition points from the CAD design. Therefore, the two-step CASC strategy presents another strategy to predict the shape quality of a 2D freeform product.

Polygon approximation to an in-plane freeform shape is clearly an important step in both strategies. The related research has been widely reported in image processing, computer vision, and pattern reorganization with methods classified into three groups: (1) split-based approaches [5, 37], (2) merge-based approaches [24, 35], and (3) dynamic-programming-based optimal approximation [22, 26]. Both the split- and merge-based approaches demonstrate good approximation of convex and concave freeform. Not a focus of this chapter, the merge-based approach is adopted for polygon approximation in this study.

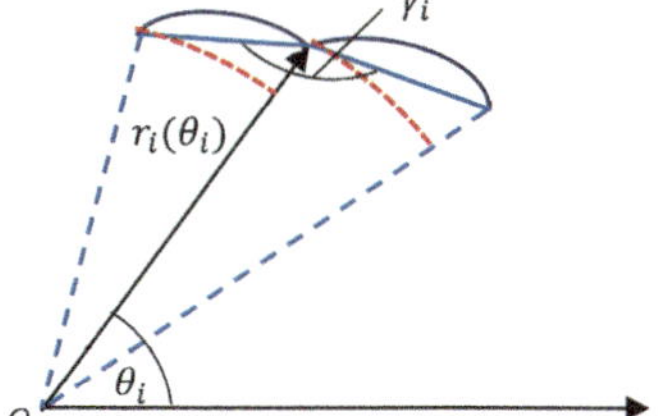

Fig. 3.15 Strategy of circular approximation with selective cornering (CASC) (solid arcs, actual radii of sectors; solid line segments, polygons sides)

3.3.1.3 Strategy Evaluation and Selection Through Experimental Investigation

To evaluate the two plausible strategies, experimental investigation was conducted for evaluation and selection.

Our null hypothesis is that the intuitive PALC strategy works better than CASC. If this hypothesis is true, the main shape deviation pattern of a 2D freeform product should be captured by the deviation pattern of a polygon that approximates the 2D freeform shape (cf. Fig. 3.14). Otherwise, CASC strategy has to be considered.

To test this hypothesis, one 2D freeform product and one polygon plate with its in-plane polygon shape approximating the in-plane freeform shape were built. As shown in Fig. 3.16, the freeform product takes arbitrary shape with "circumcircle" radius being $2''$. Here the radius of minimum circumcircle that contains a part shape is used to determine its size. Using the merge-based approach, a polygon was generated to approximate the freeform shape. Using the same polygon shape at a smaller size (circumcircle radius = $1''$), a polyhedron was built as well. All test parts have the same thickness of $0.25''$.

It should be noted that size of the polygon is not a concern to us because our previous studies [9] show that the deformation pattern of a polygon does not vary with its size, although the magnitude differs.

To facilitate the identification of the orientation of test parts during or after the building process, a non-symmetric sunk cross with line thickness of $0.02''$ was designed on the top surface. The 3D CAD models were exported to STL format files and then sent to the Mask-Image-Projection-based SLA (MIP-SLA) machine. The MIP-SLA platform in this experiment was the ULTRA® machine from EnvisionTEC. Specification of the manufacturing process is shown in Table 3.9.

All test parts were measured using a Micro-Vu precision machine. Same measurement procedure was followed for each test part to reduce errors. For convenience and consistency, the center of the cross was chosen as the origin of measurement coordinate system in the Micro-Vu machine. Boundary profile was fitted using splines in the metrology software associated with the machine. For the

(a) Freeform with circumcircle radius = 2 $''$

(b) Approximated similar polyhedron with circumcircle radius = 1 $''$

Fig. 3.16 Printed freeform shape and corresponding similar polyhedron

Table 3.9 MP-SLA process settings

Thickness of the product	0.25″
Thickness of each layer	0.00197″
Resolution of the mask	1920 × 1200
Dimension of each pixel	0.005″
Illuminating time of each layer	7s
Average waiting time between layers	15s
Type of the resin	SI500

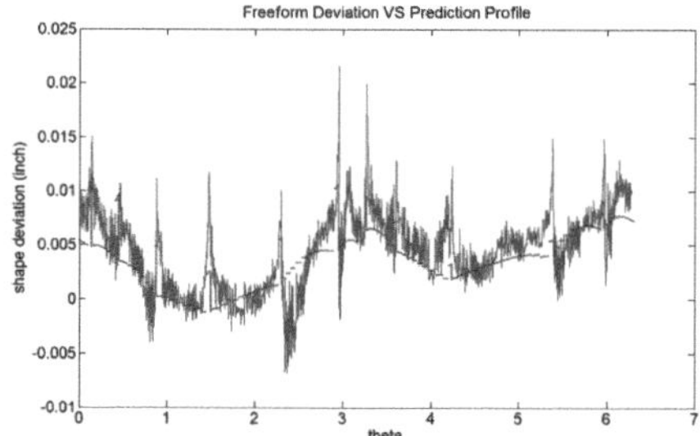

(a) Deviation profile (blue solid line) and model prediction (red dash line) of convex freeform shape with circumcircle radius = 2″

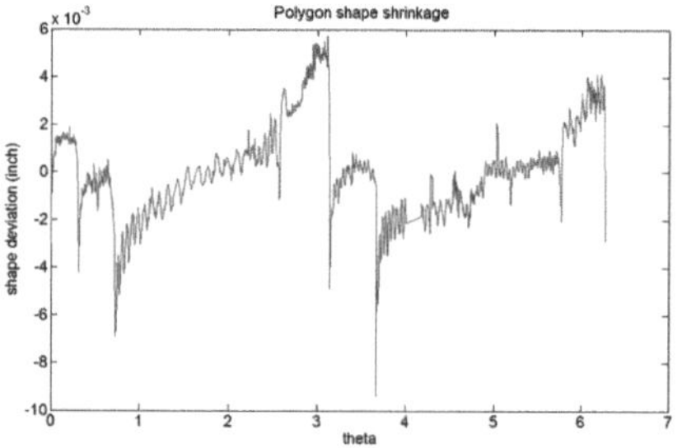

(b) Deviation profile of approximated convex polygon with circumcircle radius = 1 ″

Fig. 3.17 Deviation profiles of convex freeform and corresponding polygon approximated

two printed parts shown in Fig. 3.16, their deviation profiles defined in the PCS are shown as blue solid lines in Fig. 3.17.

Comparing these two deviation profiles, their basic deviation patterns show significant discrepancies in two aspects: (1) the trends of deviation profiles do not match; (2) the polygon deviation profile has more sharp transitions, and the positions of sharp changes are inconsistent with its freeform counterpart.

The immediate conclusion from the experimental investigation suggests that our null hypothesis is not true, i.e., the intuitive PALC strategy may not work well. We believe major reasons causing the discrepancies can be as follows:

- Comparing the two parts in Fig. 3.16, the amount of approximation errors vary with location θ. This different level of approximation accuracy can alter the trend of deviation profile for the approximated polygon.
- Although increasing the number of sides can improve the approximation, a more serious issue is that vertices in approximated polygon and its corresponding 2D freeform shape don't match. It is unavoidable that a vertex can be identified by the merged-based method at a smooth segment of the 2D freeform shape. Additionally, chances exist that a line segment passing through a vertex of the freeform can be fitted for the polygon.

Both these two reasons lead to vertex mismatch. Therefore, if we directly use the shape deviation model for polygons to represent the major shape deviation pattern of the freeform shape, large errors can be introduced. Compensation for sharp changes could be introduced at wrong locations, and real sharp transitions of the freeform shape may be missed.

Based on experimental result and analysis, CASC is deemed to be a more suitable strategy for methodology extension.

3.3.2 Prescriptive Modeling and Small-Sample Learning of 2D Freeform Shape Quality

The experimental disapproval of PALC strategy raises critical issues for CASC strategy:

i How to represent circular basic function appropriately to capture the correct trend pattern of 2D freeform shape deviation?
ii How to detect actual vertices or sharp corners in 2D freeform shapes in order to add cookie-cutter function at right locations?

3.3.2.1 Prescriptive Modeling of 2D Freeform Shape Deviation via CASC Strategy

It is worthwhile to first review the predictive model developed in Sect. 3.2 [9] for in-plane polygon shapes. Denote $f^P(\theta, r(\theta))$ as the deviation model of polygon deviation profiles defined in the PCS. According to Eq. (3.6), $f^P(\theta, r(\theta))$ is formulated as

$$f^P(\theta, r(\theta)) = g_1(\theta, r(\theta)) + g_2(\theta, r(\theta)) + g_3(\theta, r(\theta))$$

where g_1 depicts the deviation model of circular shapes, g_2 is the cookie-cutter function trimming the circular base shapes, and g_3 is the remaining deviation along each side not captured by g_1 and g_2.

Since the PALC strategy of directly applying $f^P(\theta, r(\theta) + x(\theta))$ to freeform prediction has been invalidated by experiments and analysis, we need to modify the model $f^P(\theta, r(\theta))$ so that it is consistent with strategy CASC.

According to CASC (Fig. 3.15), a large number of sectors with different radii $r_i(\theta_i)$ at θ_i will be adopted to capture the curvature and trend of 2D freeform deviation profiles. This suggests that, rather than using a single and whole circular base, we need to modify $g_1(\theta, r(\theta))$ as a combination of circular basis functions, with each of them only representing a portion or a circular sector, e.g., $g_1(\theta_i, r_i(\theta))$ for the ith sector falling between (θ_{i-1}, θ_i). The circular basis function $g_1(\theta, r(\theta))$ is generalized as:

$$g_1(\theta, r(\theta)) = g_1(\theta, r_i(\theta_i)) \tag{3.14}$$

where $\theta_{i-1} < \theta < \theta_i,\ \ 1 \leq i \leq n,\ \ \theta_0 = \theta_n.$

For example, the circular basis model $g_1(\theta, r(\theta))$ has been established for MIP-SLA process in [11]:

$$g_1(\theta, r(\theta)) = x_0 + \alpha(r_0 + x_0)^a + \beta(r_0 + x_0)^b \cos(2\theta) \qquad (3.15)$$

where $\cos(2\theta)$ term represents the dominant deformation pattern and x_0 is a constant effect of overexposure for the MIP-SLA process, which is equivalent to a default compensation x_0 applied to original CAD model.

When the MIP-SLA machine settings were changed after the experimentation in [11], experiments on circular disks were conducted again, and new deviation profiles clearly show the difference between the upper and lower half of the disks [9]. For the upper half ($\theta = 0 \sim \pi$) of the product, the circular basis model $g_1(\theta, r(\theta))$ is modified as:

$$g_1(\theta, r(\theta)) = x_0^u + \alpha^u(r_0 + x_0^u)^{a^u} + \beta^u(r_0 + x_0^u)^{b^u} \cos(2(\theta + \pi/8)) \qquad (3.16)$$

For the lower half ($\theta = \pi \sim 2\pi$) of the product, the circular basis model is modified as:

$$g_1(\theta, r(\theta)) = x_0^l + \alpha^l(r_0 + x_0^l)^{a^l} + \beta^l(r_0 + x_0^l)^{b^l}(-\sin(2\theta)) \qquad (3.17)$$

Note that superscripts u and l indicate parameters corresponding to upper and lower half of the product, respectively.

To demonstrate the CASC strategy, the upper circular basis function in Eq. (3.16) is generalized for 2D freeform shapes as:

$$g_1^F(\theta, r(\theta)) = x_0^u + \beta_0^u(r_i(\theta_i) + x_0^u)^{a^u}$$

$$+ \beta_1^u(r_i(\theta_i) + x_0^u)^{b^u} \cos(2(\theta + \pi/8))$$

$$for \ \theta_{i-1} < \theta < \theta_i, \ 1 \leq i \leq n^u, \ \theta_0 = \theta_n \qquad (3.18)$$

And the lower circular basis function in Eq. (3.17) is generalized for 2D freeform shapes as:

$$g_1^F(\theta, r(\theta)) = x_0^l + \beta_0^l(r_i(\theta_i) + x_0^l)^{a^l}$$

$$+ \beta_1^l(r_i(\theta_i) + x_0^l)^{b^l}(-\sin(2\theta))$$

$$for \ \theta_{i-1} < \theta < \theta_i, \ n^u + 1 \leq i \leq n, \qquad (3.19)$$

where n is the number of sides of the approximated polygon and n^u is the number of sides of the upper-half polygon (θ from 0 to π).

In order to obtain n, n^u and θ_i in Eqs. (3.18) and (3.19), the merge-based approach was adopted for polygon approximation. By linear scanning of a digital

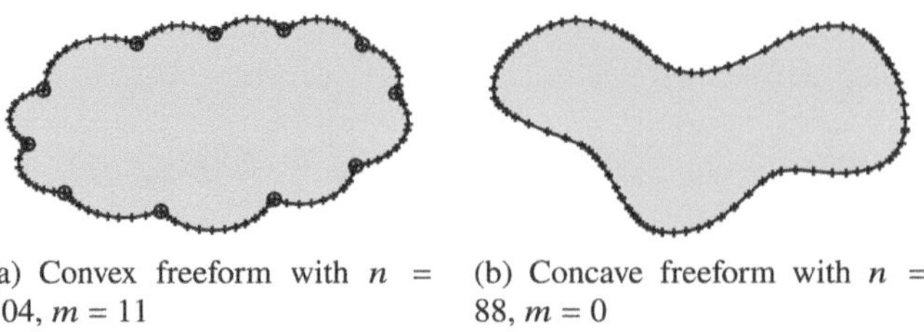

(a) Convex freeform with $n = 104$, $m = 11$

(b) Concave freeform with $n = 88$, $m = 0$

Fig. 3.18 Polygon approximation of 2D freeform shapes

curve, merged-based approach uses two thresholds regarding distance and area to determine whether each scanned point is a vertex of the approximated polygon. n is the number of all the selected vertices. When thresholds are small enough, the freeform boundary can be precisely approximated by a polygon with large number of sides (see in Fig. 3.18). θ_i can be obtained numerically as one outcome of the approximation procedure. In the approximation procedure, the boundary points are stored as digital coordinates. Therefore, the angle θ_i in the PCS could be easily derived once the ith vertex is determined. For large n, the radius of each sector can be approximated by the distance from ith vertex to the origin of circumcircle, which is denoted as $r_i(\theta_i)$ (as shown in Fig. 3.15). Figure 3.18 shows that both convex and concave freeform shapes can be well approximated.

With the generalized circular basis model, the second issue to be addressed for CASC strategy is the proper determination of vertices or corners with sharp transitions along the boundary of a 2D freeform shape. The cookie-cutter model or $g_2(\theta, r(\theta))$ in Eq. (3.5) was first developed in [9] to capture the sharp transitions of polygon shapes.

$$g_2(\theta, r(\theta)) = \beta_2(r_0 + x_0)^\alpha cookie.cutter(\theta - \phi_0) \tag{3.20}$$

One example of cookie-cutter functions is the square wave model:

$$sign[cos(n(\theta - \phi_0)/2)]$$

where n is the number of sides of a polygon and ϕ_0 is a phase variable to shift the cutting position in the PCS. The sawtooth cookie-cutter model is another alternative:

$$saw.tooth(\theta - \phi_0) = (\theta - \phi_0) \text{ MOD } (2\pi/n)$$

where x MOD y = remainder of (x/y).

These two cookie-cutter functions apply to regular polygons well where the transitions of the square wave or sawtooth wave can be easily identified. For a 2D freeform shape, however, we already point out that the vertices of the approximated polygons cannot correctly describe the transition points of the freeform shape boundary. In addition, the fitted polygons from merged-based approach can be irregular in general. We therefore have to develop a method to design the cookie-cutter function to catch the transition points of the 2D freeform from its CAD design.

Notice that when an interior angle γ_i of the fitted polygon (Fig. 3.15) is close to π or $|\gamma_i - \pi| \leq \delta_{critical}$, the corresponding ith vertex positioned at θ_i is not likely to produce a sharp transition in the deformation profile at location θ_i. Since each boundary point is stored as coordinates, the interior angle γ_i could be derived accordingly. Here, $\delta_{critical}$ is a threshold value, e.g., $\delta_{critical}$ is less than $(1/6)\pi$ from the experimental studies in [9].

Based on this observation and proposed criterion, only vertices with sharp transitions in the fitted polygon will be selected for the cookie-cutter function to alternate the function amplitude. Let m be the number of selected vertices for the cookie-cutter function with $m \ll n$ (the small circles in Fig. 3.18 show m vertices finally selected) and the angle of m vertices be ϑ_k, $k = 1, 2, \ldots, m$. Both m and ϑ_k are obtained by the polygon approximation procedure. Note that the cookie-cutter will only be applied to sectors whose vertices have sharp transitions. It is generalized as:

$$g_2(\theta, r(\theta)) = \begin{cases} g_2(\theta, r_i(\theta_i)), & if \ \theta_i = \vartheta_k \\ 0, & o/w \end{cases} \tag{3.21}$$

where $\theta_{i-1} < \theta < \theta_i$, $1 \leq i \leq n$, $\theta_0 = \theta_n$; ϑ_k, $k = 1, 2, \ldots, m, m \ll n$.

Then the square wave cookie-cutter function is extended for a 2D freeform shape as:

$$g_2^F(\theta, r(\theta)) = \begin{cases} \beta_2(r_i(\theta_i) + x_0)^\alpha sign[cos(\dfrac{(2 + (-1)^i)\pi\theta}{2\theta_i})], \ if \ \theta_i = \vartheta_k \\ 0, \ otherwise \end{cases} \tag{3.22}$$

where the term $(2 + (-1)^i)$ is adopted to guarantee that $sign(\cdot)$ changes between -1 and $+1$ alternatively to build the cookie-cutter function.

The sawtooth wave cookie-cutter model is also extended as:

$$g_2^F(\theta, r(\theta)) = \begin{cases} \beta_2(r_i(\theta_i) + x_0)^\alpha saw.tooth(\theta), \ if \ \theta_i = \vartheta_k \\ 0, \ otherwise \end{cases} \tag{3.23}$$

where $saw.tooth(\theta) = \dfrac{\pi(\theta - \theta_{i-1}) \text{ MOD } (\theta_i - \theta_{i-1})}{2(\theta_i - \theta_{i-1})}$.

We still adopt the sawtooth cookie-cutter function proposed in [9]. Note that not all freeform shapes have sharp transitions required for cookie-cutter functions. For example, the freeform in Fig. 3.18b doesn't contain sharp transition points, and the cookie-cutter function is not needed in prediction.

With the generalized circular basis model and extended cookie-cutter model, the prescriptive learning model $f^F(\theta, r(\theta))$ for 2D freeform shape deviation is extended from the polygon model $f^P(\theta, r(\theta))$ as

$$f^F(\theta, r(\theta)) = g_1^F(\theta, r(\theta)) + g_2^F(\theta, r(\theta)) \tag{3.24}$$

3.3.2.2 Prescriptive Model Estimation Based on Small-Sample Trial Geometries

Model (3.24) is deemed to be prescriptive if its model parameters can be estimated based on a small sample of trial shapes and it can be applied to predict the deformation of a freeform shape given the product CAD design. This section illustrates model parameter estimation based on six test parts: three circular disks and three polygon plates in various sizes.

Three circular disks are fabricated with circumcircle radii 0.5 $''$, 1.5 $''$, and 3 $''$, respectively. Figure 3.19a shows their deviation profiles. The maximum likelihood estimation (MLE) of circular basis model is given in Table. 3.10, while the predicted disk deviation profiles are shown as smooth lines in Fig. 3.19a. Note that we assume error term $\epsilon \sim N(0, \sigma^2)$, and the initial parameter values for the numerical MLE estimation are chosen as $\alpha^u = \alpha^l = 0.01$, $a^u = a^l = 0.5$, $\beta^u = \beta^l = 0.001$, $b^u = b^l = 0.5$, and $x_0^u = x_0^l = -0.01$, which are based on our previous analysis in [11].

Three regular polygons, that is, a 2$''$ by 2$''$ cube, a pentagon with circumcircle radius 1$''$, and a pentagon with circumcircle radius 3$''$, are fabricated to establish the in-plane polygon model. Figure 3.19b shows the in-plane shape deviation profiles of these polygons.

Since the polygons are regarded as being trimmed from circular disks, the circular model basis in polygon model should be consistent with the circular disk model. Therefore, we plug the six fitted parameters x_0^u, x_0^l, a^u, a^l, b^u, and b^l in the circular disk model to the polygon model directly. The MLE parameter estimation of the remaining parameters is shown in Table 3.11, and the smooth black lines in Fig. 3.19b show the predicted shape deformation profiles of three polygons. The initial parameter values for numerical iteration are set as: $\beta_0^u = \beta_0^l = 0.01$, $\beta_1^u = \beta_1^l = 0.01$, $\beta_2^u = \beta_2^l = 0.001$, $\alpha^u = \alpha^l = 1$ [9].

3.3.3 Prescriptive Model Validation Through Experimentation

The model (3.24) is obtained by learning from a limited number of tested shapes (circular disks, squares, and pentagons) in Sect. 3.3.2.2. The model is deemed to be prescriptive if it can predict new and untried products. To validate the prescriptive power of model (3.24) for arbitrary 2D shapes, we build one convex and one concave freeform shape with circumcircle radius 2$''$, which are shown in Figs. 3.16a, b,

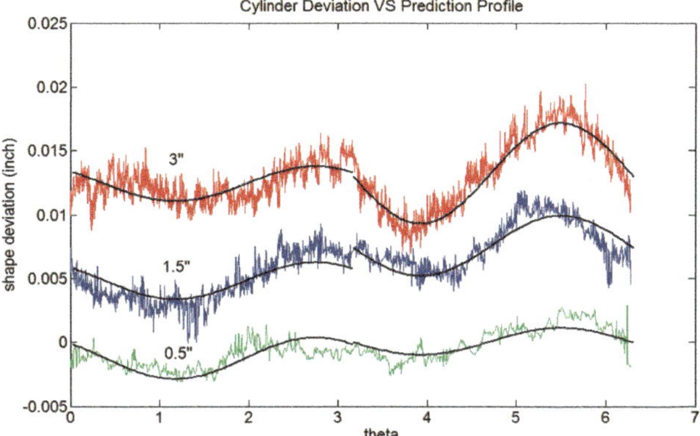

(a) Deviation profiles (red, blue and green lines) and prediction (smooth dark lines) of three disks

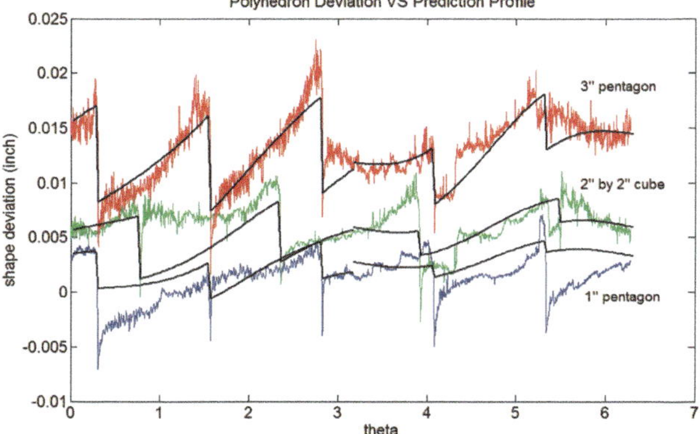

(b) Deviation profiles (red,blue and green lines) and prediction (dark lines) of three polygon plates

Fig. 3.19 Deviation and prediction profiles of training geometries

respectively. We will compare the prediction of model (3.24) to the measured shape deviation of two 2D freeform shapes.

To obtain model prediction of the 2D freeform shapes, we first plug the estimated parameters for the polygon model in Sect. 3.3.2.2 into the generalized circular basis model (3.18) and (3.19) and the extended cookie-cutter model (3.23). Then, following CASC strategy, the polygon approximation is applied to the 2D freeform CAD models to obtain n, m, θ_i, r_i, γ_i, and ϑ_k (Fig. 3.18), $i = 1, \ldots, n$, $k = 1, \ldots, m$. They are plugged into the extended 2D freeform model to predict the deviations for each sector. The combined prediction forms the predicted deviation profile of the 2D freeform shape. In the merge-based approach, the distance threshold is 3 pixel, and the area threshold is 10 pixel. The threshold for catching transition points is $\delta_{critical} = \pi/6$, which is not a strict constraint. For the convex 2D freeform shape (Fig. 3.16a), the approximated number of sides $n = 104$, and

Table 3.10 Estimated parameters for circular base model

Model of the upper half disk

Parameter	Estimated value	Standard deviation
α^u	0.0076	9.593×10^{-5}
a^u	0.7878	0.0069
β^u	0.0015	1.670×10^{-5}
b^u	−0.1043	0.0125
x_0^u	−0.0056	9.398×10^{-5}

Model of the lower half disk

Parameter	Estimated value	Standard deviation
α^l	0.0312	8.060×10^{-4}
a^l	0.2198	0.0054
β^l	0.0018	1.804×10^{-5}
b^l	0.7110	0.0096
x_0^l	−0.0264	7.931×10^{-4}

Table 3.11 Estimated parameters for polygon model

Model of the upper half polygon

Parameter	Estimated value	Standard deviation
β_0^u	0.0058	1.313×10^{-5}
β_1^u	0.0011	2.117×10^{-5}
β_2^u	0.0027	2.870×10^{-5}
α^u	0.8778	0.0094

Model of the lower half polygon

Parameter	Estimated value	Standard deviation
β_0^l	0.0291	2.141×10^{-5}
β_1^l	0.0012	1.004×10^{-5}
β_2^l	0.0009	2.712×10^{-5}
α^l	1.4067	0.0250

the number of vertices selected in cookie-cutter function $m = 11$ (Fig. 3.18a, which demonstrates that all the true vertices are founded without mistake. The concave 2D freeform shape (Fig. 3.20a) has parameters $n = 88$ and $m = 0$ (Fig. 3.18b). The smooth shape has no transition point selected.

The observed deviation profiles of the two freeform parts are shown as the solid blue lines in Figs. 3.17a and 3.20b, respectively. The predicted deviation profiles for two 2D freeform shapes using model (3.24) are represented as red dashed lines in each figure. Considering the limited data used for model fitting and errors involved in freeform shape measurement (e.g., curve registration and boundary approximation in measurement), the prescriptive prediction of model (3.24) is remarkably close to the measurement. The encouraging result suggests:

- The small-sample learning and prediction methodology of predicting 2D shape quality of AM-built product is generic, which can be directly extended from circular and polygon shape to 2D freeform shapes.

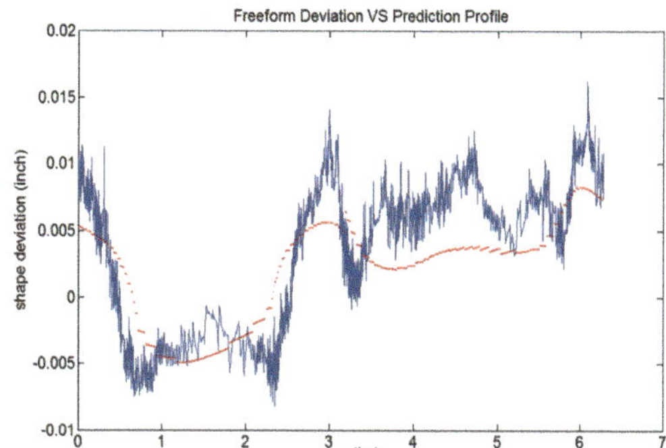

(a) Concave 2D freeform shape with circumcircle radius = 2 ″

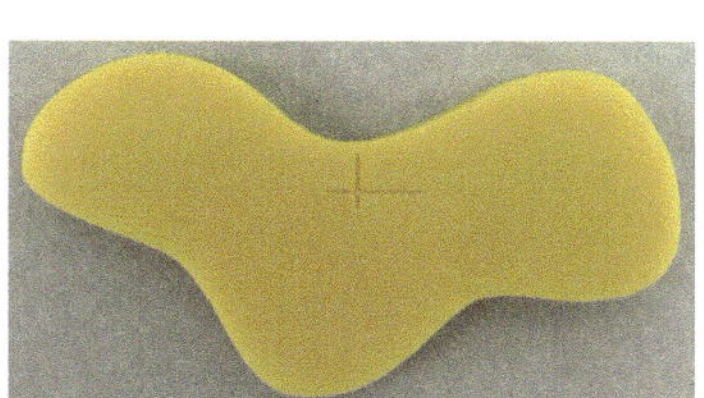

(b) Deviation profile (blue line) and model prediction (red dash line) of 2D concave freeform with circumcircle radius = 2″

Fig. 3.20 Deviation and prediction profiles for 2D concave freeform shape

- The prescriptive modeling methodology has the capability of predicting new and untried products by learning from a limited number of tested shapes.
- The foundation of this methodology is dimension reduction of 2D shape space, that is, representing a 2D freeform shape with shape primitives such as circular sectors and corners.

As a result, the deviation profile of any 2D arbitrary freeform can be derived directly from CAD design, though the prediction model can be further improved. This provides a great opportunity to implement the optimal compensation policy established in [11] to improve the geometric accuracy of AM-built products.

It should be mentioned that parameters in Table 3.11 are estimated from experimentation with regular polygons, while most of the approximated polygons are irregular. Our model prediction accuracy can be improved if irregular polygon data are available for model learning. This indirectly demonstrates the robustness of model (3.24).

Built upon predictive model for circular and polygon shapes, this section presents a circular approximation with selective cornering (CASC) strategy to extend the polygon quality prediction model to arbitrary 2D freeform shapes. This strategy essentially approximates a 2D freeform with a series of circular sectors with different radii first and then further improve the approximation by imposing a generalized cookie-cutter function. The series of circular sectors are modeled as generalized circular basis functions.

The prescriptive model is estimated/calibrated based on a limited number of trials shapes including disk, square plates, and pentagon plates. Experimental observations of both convex and concave 2D freeform shapes match the model prediction well. It demonstrates the prescriptive capability of predicting new and untried products by learning from a limited number of tested shapes. As a result, the 2D shape quality of any arbitrary 2D freeform can be derived directly from CAD design.

3.4   2D Finite Manufacturing Primitive Extraction and Learning for AM Quality

To reduce the complexity of AM operations, the finite manufacturing primitive (FMP) or printing primitive concept has been introduced in Sect. 2.5 to represent AM products in a low-dimensional space. This section introduces a 2D FMP extraction approach through a changepoint detection formulation. Using 2D shapes as an illustration, each layer geometry in AM will be represented by 2D FMPs including curve segments and corners, capturing smooth and non-smooth deviation patterns, respectively.

The developed changepoint detection algorithm automatically extracts 2D FMPs from any 2D freeform shape. Models of predicting 2D FMP quality are developed with consideration of covariates such as size, curvature, and location, which has not been addressed in [18].

3.4.1   *Automatic Extraction of 2D Finite Manufacturing Primitive (FMP) via Changepoint Detection Formulation*

Geometric features have been widely adopted to simplify shape representation across various fields, including engineering product design [23], reverse engineering [28], and object recognition [36]. They are typically defined as partial regions of the shape with primitive geometries or semantic meanings [23]. Complex shapes can be represented as combinations of a few types of pre-specified primitive geometric features, thereby enhancing the efficiency of design and manufacturing.

Extraction of geometric features from freeform shapes is typically based on geometric descriptors. For instance, geometric features of 2D freeform shapes, such as line segments and corner points [1, 30], are detected by identifying points with abnormal tangential angle changes. The 2D freeform shapes are then segmented at these points to form the features [20, 21, 27]. The decision variables, such as the threshold to determine the corner points, need to be specified for each design to avoid redundant small features.

However, in additive manufacturing (AM) of intricate products with complicated geometries, feature-based representation faces tremendous challenges because infinite features are needed to represent complex new designs [8, 29]. This significantly increases the difficulty of AM operations such as design representation, process planning, and quality control. For example, shape accuracy control for a new product design poses a unique challenge in AM. The feature-based representation should facilitate machine learning and inference of the new product quality based on the geometric features observed in sample prints [8]. However, a new topologically optimized design in AM more likely contains unknown features that make feature-based models ineffective. Furthermore, identical geometric features can exhibit different deviation patterns in AM due to covariates such as location and size

[12]. Luan and Huang (2017) [18] attempted to predict printing deviations of 2D freeform shapes using geometric primitives. The 2D layer geometry and its quality were represented and predicted with small circular or polygonal sectors extracted from disk and polygon shapes. Though the method in [18] was groundbreaking, the prediction model overlooked many covariates such as curvature and location. The prediction accuracy needs to be improved for complicated geometries too.

To address this limitation, a novel representation framework based on FMPs has been proposed for layer-wise 2D boundaries and 3D surfaces of AM products. [8]. For 2D products, three types of printing primitives were defined: circular sectors with constant curvature, line segments with zero curvature, and corners covering infinite curvature points. Unlike corner points, corner primitives were defined as regions on layer geometries to capture discontinuous deviation patterns. However, this work did not provide guidance on primitive extraction or modeling of their printing quality.

This section develops an automated printing primitive extraction and learning framework to reduce the complexity of AM operations, particularly to predict deviations of printed 2D freeform geometries in AM. To represent and predict the printing deviation of a freeform 2D geometry, we define $I = 2$ types of generalized 2D primitives: curve segments and corners, to capture smooth and non-smooth deviation patterns, respectively. In comparison to the primitives defined in [8], we allow smooth curvature variations for curve segments, as these variations typically do not affect the overall trend of the deviation patterns. Line segments are special cases of smooth curve segments with constant zero curvature. We define corners as small regions that compactly capture sharp changes in printed geometries.

Representing the boundary of a 2D layer geometry in the polar coordinate system (PCS) as (r, θ) [8, 12, 18], the contour $r(\theta)$ can be formulated as:

$$r(\theta) = \sum_{i=1}^{I} r_i(\theta)\mathbf{1}_{\theta \in \Theta_i},$$

where $r_i(\theta)$ is the ith-type primitive and Θ_i denotes its definition domain of θ.

In PCS, the deviation of a 2D shape is typically defined as the difference in radius between the actual print, r, and the design, r_0, along each angle θ. This is expressed as $\Delta r(\theta) = r(\theta) - r_0(\theta)$, where $\theta \in [0, 2\pi)$ [12]. Using 2D primitives, the deviation on shape $r(\theta)$ is formulated as:

$$\Delta r(\theta) = \sum_{i=1}^{I} \Delta(\theta)\mathbf{1}_{\theta \in \Theta_i}.$$

Since different types of primitives capture distinct deviation patterns, we propose constructing the deviation learning model for each type of primitive separately. A unique functional form is developed to represent the deviation profile of the ith-type primitive as $f_i(\theta, \boldsymbol{\beta}_i)$, where $\boldsymbol{\beta}_i$ contains the covariates of each primitive. The deviation of the printed 2D part can be described as:

$$\Delta r(\theta) = \sum_{i=1}^{I} f_i\left(\theta, \boldsymbol{\beta}_i\right) \mathbf{1}_{\theta \in \Theta_i} + \epsilon \tag{3.25}$$

Since the primitives are defined based on the smoothness of deviation patterns within a local region, we propose to extract them by detecting segment-wise deviation shifts. The boundary smoothness is intrinsically linked to the homogeneity of curvature values within a segment. Consequently, by identifying changes in curvature, we can effectively separate discontinuous and continuous segments in the deviation profile. The curvature distribution along the 2D shape can be represented as a curvature profile $\kappa(\theta)$, denoting the curvature values at each angle θ.

Representing $\kappa(\theta)$ in PCS may encounter multiple points corresponding to the same angle θ for freeform shapes, an issue that was not addressed in [18]. To overcome this issue, we first parameterize the contour by using unit arc length. Suppose s is the distance along the curve from some fixed starting point, i.e., the arc length. At "time" t, the length of the point traveled is $s(t) = \int_0^t \sqrt{x(u)^2 + y(u)^2} du$. The distance from a point at s to the origin is denoted as $r(s)$. Automatic extraction of 2D primitives is formulated as follows:

$$\min_{\tau_{1:J}} \sum_{j=1}^{J} \left[C(\kappa_{(\tau_{j-1}+1):\tau_j}) \right] + \lambda \mathcal{P}(J), \tag{3.26}$$

where $\tau_{1:J} = (\tau_1, \tau_2, \ldots, \tau_J)$ are the J changepoints along $s(t)$, C measures the homogeneity of curvatures within a segment, and $\lambda \mathcal{P}(J)$ denotes a penalty against overfitting.

The curvature κ at s in Eq. (3.26) is defined as $\kappa = \frac{d\boldsymbol{\phi}}{ds} = \left\| \boldsymbol{\phi}'(s) \right\|$, where $\boldsymbol{\phi}(t) = \frac{\mathbf{r}'(t)}{\|\mathbf{r}'(t)\|}$ is the unit tangent vector. It measures how much the curve deviates from a straight line at each point. For a 2D freeform shape approximated by n discrete points, the curvature of each point is estimated by approximating a second-order polynomial around it and is denoted as $\kappa_{1:n} = (\kappa_1, \kappa_2, \ldots, \kappa_n)$.

The cost function C measures the curvature homogeneity within a segment $\kappa_{(\tau_{j-1}+1):\tau_j} = (\kappa_{(\tau_{j-1}+1)}, \kappa_{(\tau_{j-1}+2)}, \ldots, \kappa_{\tau_j})$. A low value of $C(\kappa_{(\tau_{j-1}+1):\tau_j})$ indicates that the segment does not contain any changepoints [15]. Given that there are drastic changes in curvature values between adjacent primitives, we detect changes in the mean curvature on a segment-wise basis:

$$\bar{\kappa}_{(\tau_{j-1}+1):\tau_j} = \frac{1}{\tau_j - \tau_{j-1}} \sum_{r=\tau_{j-1}+1}^{\tau_j} \kappa_r. \tag{3.27}$$

Thus, the cost function C in Eq. (3.26) can be defined as:

$$C(\kappa_{(\tau_{j-1}+1):\tau_j}) = \sum_{r=\tau_{j-1}+1}^{\tau_j} \left[\kappa_r - \bar{\kappa}_{(\tau_{j-1}+1):\tau_j}\right]^2 \tag{3.28}$$

To automatically identify the optimal number of segments, we penalize the number of segments by setting $P(J) = J$ to avoid generating small segments. The penalty λ controls the sensitivity to the changes of κ. We employ a greedy search method to find the smallest λ value that ensures adjacent segments have curvature values of differing signs, satisfying the condition $\text{sign}(\bar{\kappa}_{(\tau_{j-1}+1):\tau_j})$ · $\text{sign}(\bar{\kappa}_{(\tau_j+1):\tau_{j+1}}) \leq 0, \forall j$. The objective function, Eq. (3.26), can be written as:

$$\min_{\tau_{1:J}} \sum_{j=1}^{J} \sum_{r=\tau_{j-1}+1}^{\tau_j} \left[\kappa_r - \bar{\kappa}_{(\tau_{j-1}+1):\tau_j}\right]^2 + \lambda J. \tag{3.29}$$

The detected changepoints include both inflection points between convex and concave curve segments and endpoints of corner regions. Different types of 2D primitives can be extracted simultaneously by segmenting at these points.

3.4.2 Prescriptive Modeling and Learning 2D FMP Quality

To predict deviations of 2D freeform shapes using heterogeneous printing samples, a cookie-cutter learning framework has been established in [10, 16] and was mathematically proved in [8]. However, the existing learning framework does not distinguish the global coordinate system of a part and the local coordinate systems of printing primitives. The consequence is that primitives of the same shape could end up with different "design radii" because of their printing locations, i.e., the confounding of location and size effects. Additionally, the framework has not yet incorporated the geometric curvature, an aspect that greatly impacts shape deviations and could enhance prediction accuracy. Furthermore, the existing non-smooth cookie-cutter functions may not adequately capture the heterogeneity among smooth curve segments.

To overcome these issues, this section provides a generalized approach to model shape primitives. Suppose that a smooth curve primitive (the segment with blue boundary in Fig. 3.21) is printed in the global polar coordinate system with center O. The local polar coordinate system (LPCS) of the primitive is centered at O_j, representing location of the printed primitive. To separate the location and size effects, the deviation profiles of primitives are represented and modeled in the LPCS, where a smooth curve primitive is viewed as being carved out from the covering circular sector through a cookie-cutter function. The model for smooth

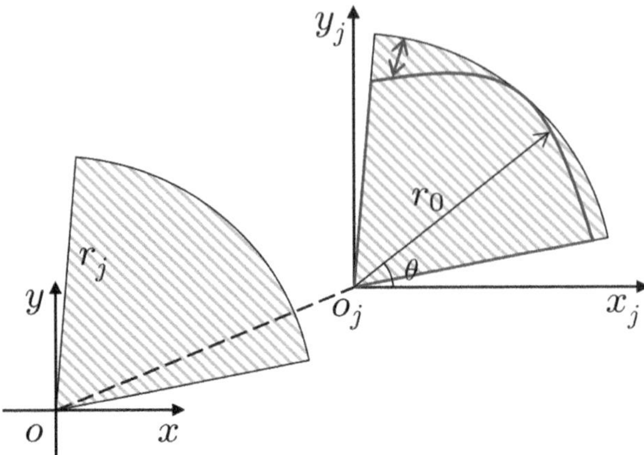

Fig. 3.21 A smooth curve primitive (in blue) under the global and local coordinate systems

curve primitives is therefore formulated as:

$$f\left(\theta, r_0(\theta), \kappa(\theta), r_j, x_j, y_j\right) =$$

$$f_1\left(\theta, r_j, x_j, y_j\right) \qquad \text{Circular basis function} \qquad (3.30)$$

$$+ f_2\left(\theta, r_0(\theta), \kappa(\theta), r_j\right) \quad \text{Cookie-cutter function}$$

where f_1 captures the deviation pattern of circular sectors and cookie-cutter function f_2 describes the deviations incurred by removing materials. Model for corner primitives can follow the same idea, but it is not discussed in this book.

Explanations of the covariate effects are as follows:

- Location (x_j, y_j) is the Cartesian coordinates of the LPCS origin. We model the location effect via the effect equivalence framework as in [17]. Specifically, the deviations induced at different locations are captured through a compensation term that modifies the design radius.
- Size $r_0(\theta)$ denotes the length from LPCS center O_j to the boundary point defined by its angle θ. With more material deposited along the θ direction, a larger scale of deviation is supposed to be observed, which coincides with the observations in [10, 12]. When $r_0 \to \infty$, the deviation should be bounded. Also, the deviation is expected to be around zero when $r_0 \to 0$. Note that based on the nominal design $(\theta, r_0(\theta))$ and segmentation results, each primitive can be covered by a circular sector with radius r_j for $j = 1, \ldots, J$.
- Curvature $\kappa(\theta)$ calculated in Sect. 3.4.1 represents how much the segment bends from a straight line around θ locally. Due to the uneven heat dissipation, more complicated deviation patterns are expected for larger κ. On the other hand, when $\kappa = 0$, a straight line segment is fabricated.

Since the deviation patterns of circular sectors are mostly continuous and smooth, a natural choice to model them is the Fourier basis expansion as in [12]. The formulation is written as:

$$f_1\left(\theta, r_j\right) = x_0 + \sum_{k=0}^{K} b_{k,0}\left(r_j + x_0\right)^{b_{k,1}} \varphi_k(\theta), \tag{3.31}$$

where $\varphi_k(\theta), k = 0, \ldots, K$ are Fourier bases, parameter x_0 is a compensation term capturing the equivalent effects such as location, and coefficients $b_{k,0}(r_j + x_0)^b_{k,1}$ capture the size effect [12].

For example, we could set $K = 2$ and rewrite Eq. (3.31) as

$$
\begin{aligned}
f_1\left(\theta, r_j, x_j, y_j\right) = {} & x_0(x_j, y_j) + b_{0,0}\left(r_j + x_0(x_j, y_j)\right)^{b_{0,1}} \\
& + b_{1,0}\left(r_j + x_0(x_j, y_j)\right)^{b_{1,1}} \cos\left(n_{1,1}\theta + \psi_{1,1}\right) \\
& + b_{2,0}\left(r_j + x_0(x_j, y_j)\right)^{b_{2,1}} \cos\left(n_{1,2}\theta + \psi_{1,2}\right).
\end{aligned}
\tag{3.32}
$$

A second-order polynomial is adopted to capture the equivalent effect of location [17]:

$$
\begin{aligned}
x_0\left(c_x, c_y\right) = {} & p_{0,0} + p_{1,0}c_x + p_{0,1}c_y \\
& + p_{2,0}c_x^2 + p_{1,1}c_x c_y + p_{0,2}c_y^2.
\end{aligned}
\tag{3.33}
$$

The cookie-cutter function in Eq. (3.30) captures the differences between the deviations of circular sectors and smooth curve segments. It is formulated as follows:

$$
\begin{aligned}
f_2\left(\theta, r_0(\theta), \kappa(\theta), r_j\right) = {} & f_{2,1}\left(\theta, r_0(\theta), r_j\right) \\
& + f_{2,2}\left(\theta, \kappa(\theta), r_j\right),
\end{aligned}
\tag{3.34}
$$

where $f_{2,1}$ captures the low-order deviation patterns caused by material extraction on a larger scale and $f_{2,2}$ captures the high-order deviation patterns within local, smaller regions.

Since the differences between circular sectors and smooth curve segments are smooth, we model $f_{2,1}$ and $f_{2,2}$ using smooth Fourier bases. The amount of material removed from a covering circular sector at angle θ is characterized as $(r_j - r_0(\theta))$. This is utilized as a covariate in $f_{2,1}$ to capture the low-order deviation patterns of f_2:

$$f_{2,1}\left(\theta, r_0(\theta), r_j\right) = a_0\left(r_j - r_0(\theta)\right)^{a_1} \cos\left(n_{2,1}\theta + \psi_{2,1}\right). \tag{3.35}$$

For the high-order deviations, we denote the curvature difference $(\kappa(\theta) - 1/r_j)$ between the smooth curve segment and the circular sector of radius r_j. The curvature difference is modeled as a covariate in $f_{2,2}$:

$$f_{2,2}\left(\theta, \kappa(\theta), r_j\right) = \left[e_0 - e_1 \exp\left(-e_2\left(\kappa(\theta) - 1/r_j\right)\right.\right.$$
$$\left.\left. +e_3\right)\right] \cos\left(n_{2,2}\theta + \psi_{2,2}\right). \tag{3.36}$$

We set $e_0 = e_1 \exp(e_3)$ so that the curvature effect is zero for circular segments. This form of $f_{2,2}$ ensures a larger curvature effect at points with larger curvatures.

To learning the prescriptive model, parameters of the circular basis and the rest covariate effects are estimated separately with the corresponding training data split. Circular sectors are chosen as the basis due to their constant curvature and that they require no material removal. We first assume a constant compensation term x_0 in Eq. (3.32) and estimate parameters in the following formulation via maximum likelihood estimation (MLE) with circular shape deviations:

$$\tilde{f}_1(\theta, r_j) = x_0 + b_{0,0}(r_j + x_0)^{b_{0,1}}$$
$$+ b_{1,0}(r_j + x_0)^{b_{1,1}} \cos\left(n_{1,1}\theta + \psi_{1,1}\right) \tag{3.37}$$
$$+ b_{2,0}(r_j + x_0)^{b_{2,1}} \cos\left(n_{1,2}\theta + \psi_{1,2}\right)$$

After obtaining $\hat{b}_{0,0}$, $\hat{b}_{0,1}, \hat{b}_{1,0}, \hat{b}_{1,1}, \hat{b}_{2,0}$, $\hat{b}_{2,1}$, $\hat{n}_{1,1}$, $\hat{n}_{1,2}$, $\hat{\psi}_{1,1}$, and $\hat{\psi}_{1,2}$ from Step 1, the remaining coefficients in Eq. (3.33), Eq. (3.35), and Eq. (3.36) are fitted simultaneously via MLE with the rest of the smooth curve deviations.

3.4.3　Validating 2D FMP Extraction and Prediction

We demonstrate FMP extraction from 2D freeform shapes and build FMP models for prediction under the global coordinate system for practical consideration.

Six shapes were employed, including disks with radii of 0.5 inches, 1.5 inches, and 3 inches, a 2-inch by 2-inch square, a dodecagon with a radius of 3 inches, as well as two freeform shapes resembling a cloud and a trilobe, as illustrated in Fig. 3.24a. These parts were printed using a SLA process.

In total, 19 primitives were extracted from these shapes and divided into two groups, where nine primitives form the training set and the remaining ten compose the validation set. Specifically, all the primitives of the dodecagon were assigned to the validation set to examine the ability to predict deviations of a new design shape. At least two circular disks are required in the training set to facilitate learning f_1 in Eq. (3.30), especially for estimating the size effect. We illustrated the FMP model using convex smooth curve segments, including line segments. The models for corner primitives and concave segments are left for future work.

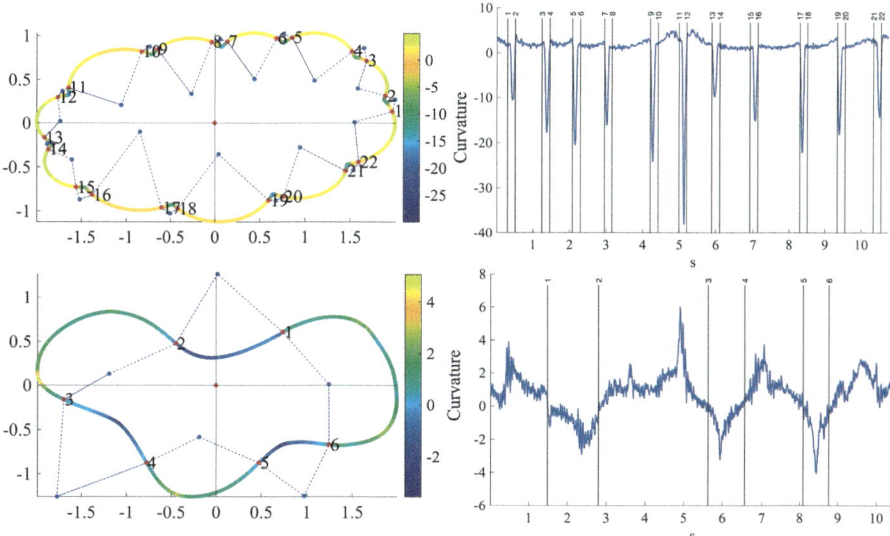

Fig. 3.22 Left column: extracted 2D primitives on a freeform cloud shape and a trilobe shape. Right column: detected changepoints on the curvature profiles

3.4.3.1 Extracting Primitives from 2D Freeform Shapes

The 2D primitives are extracted by segmenting the freeform shapes according to Eq. (3.26). The curvature profiles of the two freeform shapes and detected changepoints are shown in the right column of Fig. 3.22. The freeform cloud shape has been divided into 22 sections, which consist of 11 smooth curve segments and 11 corners. The freeform trilobe shape has been divided into six smooth curve segments.

The O_j of each smooth curve segment was determined by finding its circumcenter. The detected LPCS centers of the freeform cloud shape and the trilobe shape are shown in the left column of Fig. 3.22. The deviation profiles of the smooth curve segments in the LPCS are shown in the right column in Fig. 3.23. Compared to the deviation profiles defined in the global coordinate system, as shown in the left column of Fig. 3.23, deviation profiles in LPCS have wider angle ranges to sufficiently capture the primitive deviation patterns.

The polygonal shapes are segmented into line segments and corners. The deviation profiles of the line segments are represented in the global coordinate system, because an infinite number of circumcenters can be identified for line segments. The deviation profiles for all primitives, divided into training and validation sets, are shown in Fig. 3.24b–c.

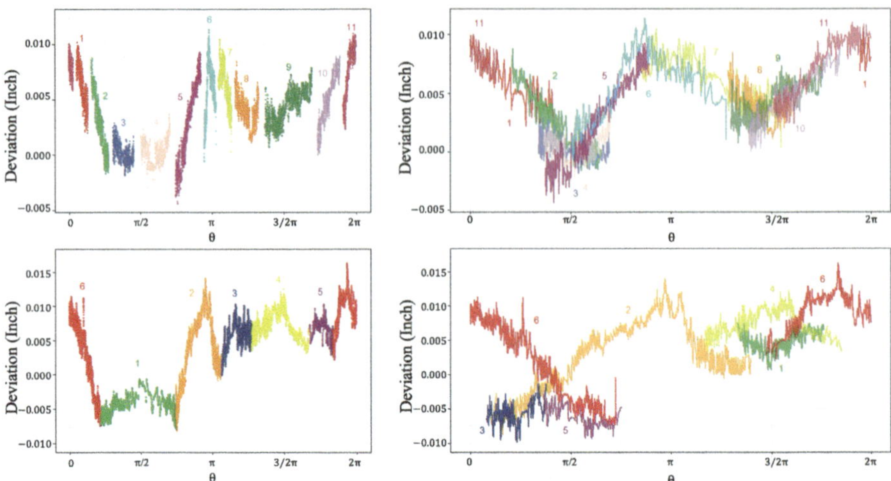

Fig. 3.23 Left column: deviation profiles of the freeform shapes in the global coordinate system. Right columns: deviation profiles of the primitives in the local coordinate system

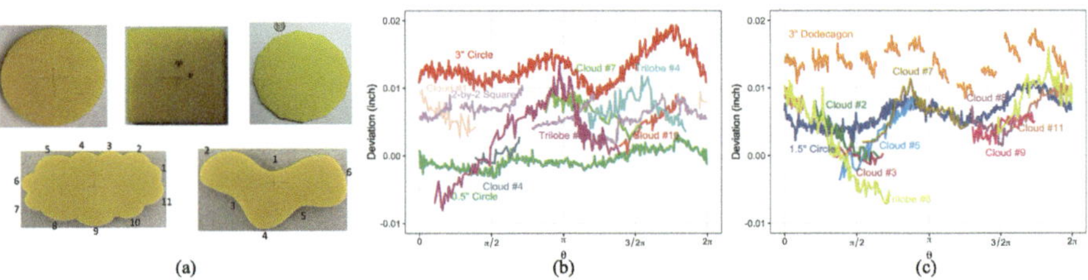

Fig. 3.24 (**a**) Printed parts. (**b**) Deviation profiles of primitives in the training set and (**c**) in the validation set

3.4.3.2 Modeling and Predicting the Deviations of 2D FMPs

To implement the learning framework in Eq. (3.25), the first step is to estimate Eq. (3.37). Its coefficients were estimated using the selected disks with 0.5-inch and 3-inch radii in the training dataset. Considering only the size effect, the estimated model is

$$
\begin{aligned}
f_1(\theta, r_j) =\ & 0.0079 + 0.0086(r_j + 0.0079)^{0.8307} \\
& - 0.0157(r_j + 0.0079)^{0.1154}\cos(0.1188\theta - 0.0546) \\
& + 0.0010(r_j + 0.0079)^{0.6941}\cos(2.4753\theta - 0.5081).
\end{aligned}
\tag{3.38}
$$

The fitted deviation profiles of the disks are shown in Fig. 3.25. The mean absolute error (MAE) is 0.731×10^{-3}, and the root mean square error (RMSE) is 0.955×10^{-3}.

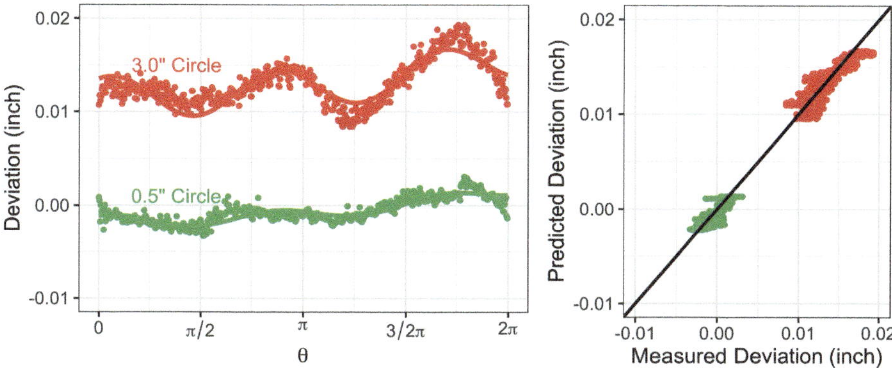

Fig. 3.25 Measured shape deviation against predicted deviation of the circular disks in the training set

Next, we estimated the coefficients of covariate effect $x_0(x, y)$ of f_1 in Eq. (3.32). The fitted location effect term $x_0(x, y)$ is

$$x_0(x, y) = 0.0099 + 0.0012x - 0.0022y$$
$$+ 0.0012x^2 + 0.0009xy + 0.0137y^2. \tag{3.39}$$

The coefficients of the cookie-cutter function f_2 were estimated using MLE. The fitted first term in f_2 in Eq. (3.34) is

$$f_{2,1}(\theta, r_0(\theta), r_j) = -0.0022(r_j - r_0(\theta))^{0.0108} \cos(1.2390\theta + 0.1432) \tag{3.40}$$

and the second term in f_2 is

$$f_{2,2}(\theta, r_j, \kappa(\theta)) = \left[0.0087 - 0.3835 \exp\left(-0.2491(\kappa(\theta) - 1/r_j\right)\right.$$
$$\left. -3.7878)\right] \cos(2.0305\theta + 0.2107), \tag{3.41}$$

Using the fitted model, the predicted deviations of the primitives in the training dataset are shown in Fig. 3.26. It is observed that the proposed model effectively captures the overall trends of the deviation patterns on the primitives. This demonstrates the capability of the proposed cookie-cutter function to accurately model the differences between circular sectors and smooth curve segments. The MAE is 1.708×10^{-3}, and the RMSE is 2.194×10^{-3} for the training dataset.

The trained model was applied to the validation dataset, and the prediction results are presented in Fig. 3.27. The MAE is 1.888×10^{-3}, and the RMSE is 2.932×10^{-3}, demonstrating a similar prediction accuracy as in the training dataset. Specifically, as shown in Fig. 3.27, the model can predict the deviations of line segments of the new shape, a 3-inch dodecagon. This demonstrates the effectiveness of the trained cookie-cutter function f_2 in capturing the differences between the circular sector

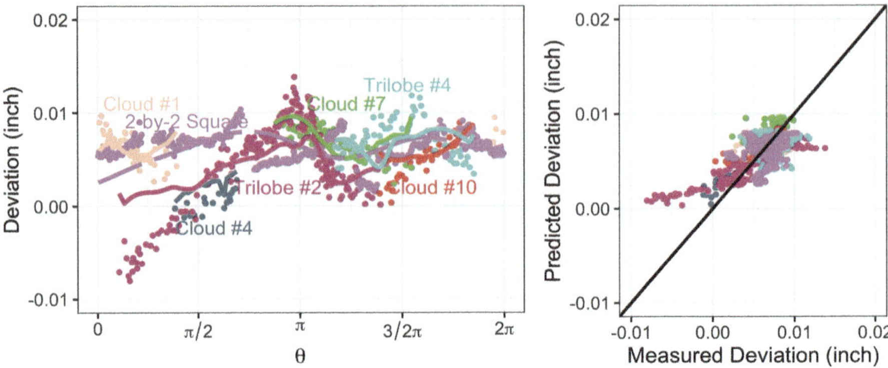

Fig. 3.26 Measured shape deviation against predicted deviation of the smooth curve segments in the training set

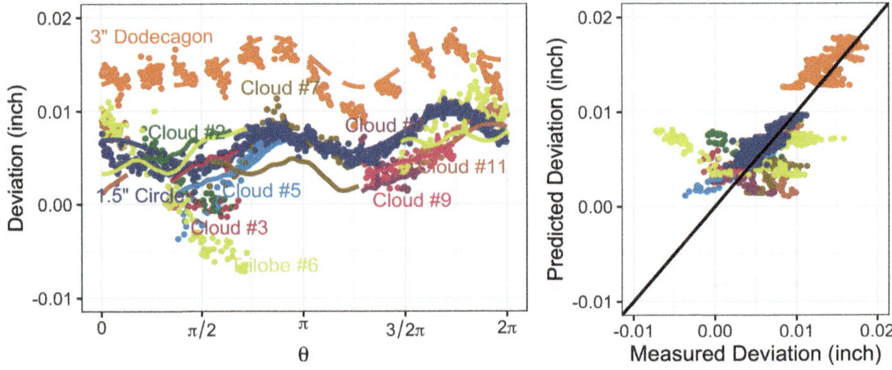

Fig. 3.27 Measured shape deviation against predicted deviation of the smooth curve segments in the validation set

and line segments. More importantly, since the entire 3-inch dodecagon was not seen in the training dataset, this result demonstrates the ability of the constructed framework to predict the deviations of a new design with small samples.

The prediction accuracy is relatively lower for the 6th primitive of the trilobe and the 7th primitive of the cloud shape. These primitives correspond to segments exhibiting curvature changes, as shown in Fig. 3.22. However, within the training set, noticeable curvature changes are present only in the 2nd and 4th primitives of the trilobe. Their curvature profiles exhibit more complex patterns than those observed in the cloud primitives. Consequently, the fitted $f_{2,2}$ might be overfitting to the complex curvature patterns. This accuracy could be improved by collecting additional data that includes a wider variety of curvature patterns.

Future work is needed to model the corner primitive and use the two types of 2D FMPs to predict the quality of 2D freeform shapes under the global coordinate system.

Acknowledgments The work in this chapter is partially supported by the US Office of Naval Research with grant # N000141110671 and by the US National Science Foundation with grant # CMMI-1333550. We thank Prof. Yong Chen and Mr. Kai Xu at the University of Southern California for providing insightful discussions and experimental support on SLA process.

References

1. Ansari, N., Delp, E.J.: On detecting dominant points. Pattern Recogn. **24**(5), 441–451 (1991). https://doi.org/10.1016/0031-3203(91)90057-C
2. Beaman, J.J., Barlow, J.W., Bourell, D.L., Crawford, R.H., Marcus, H.L., McAlea, K.P.: Solid Freeform Fabrication: A New Direction in Manufacturing, vol. 2061, pp. 25–49. Kluwer Academic Publishers, Norwell (1997)
3. Cho, W., Sachs, E.M., Patrikalakis, N.M., Troxel, D.E.: A dithering algorithm for local composition control with three-dimensional printing. Comput.-Aided Design **35**(9), 851–867 (2003)
4. Cohen, D.L., Lipson, H.: Geometric feedback control of discrete-deposition SFF systems. Rapid Prototyping J. **16**(5), 377–393 (2010)
5. Dinesh, R., Damle, S.S., Guru, D.: A split-based method for polygonal approximation of shape curves. In: Pattern Recognition and Machine Intelligence, pp. 382–387. Springer, Berlin (2005)
6. Gelman, A., Rubin, D.: Inference from iterative simulation using multiple sequences. Stat. Sci. **7**(4), 457–472 (1992)
7. Huang, Q.: An analytical foundation for optimal compensation of three-dimensional shape deformation in additive manufacturing. ASME Trans. J. Manuf. Sci. Eng. **138**(6), 061010 (2016)
8. Huang, Q.: An impulse response formulation for small-sample learning and control of additive manufacturing quality. IISE Trans. (2022). https://doi.org/10.1080/24725854.2022.2113186
9. Huang, Q., Nouri, H., Xu, K., Chen, Y., Sosina, S., Dasgupta, T.: Statistical predictive modeling and compensation of geometric deviations of three-dimensional printed products. ASME Trans. J. Manuf. Sci. Eng. **136**(6), 061008 (2014)
10. Huang, Q., Nouri, H., Xu, K., Chen, Y., Sosina, S., Dasgupta, T.: Statistical predictive modeling and compensation of geometric deviations of three-dimensional printed products. J. Manuf. Sci. Eng. **136**(6), 061008 (2014). https://doi.org/10.1115/1.4028510
11. Huang, Q., Zhang, J., Sabbaghi, A., Dasgupta, T.: Optimal offline compensation of shape shrinkage for 3d printing processes. IIE Trans. Quality Reliab. **47**(5), 431–441 (2015)
12. Huang, Q., Zhang, J., Sabbaghi, A., Dasgupta, T.: Optimal offline compensation of shape shrinkage for three-dimensional printing processes. IIE Trans. **47**(5), 431–441 (2015). https://doi.org/10.1080/0740817X.2014.955599
13. Imbens, G., Rubin, D.: Causal Inference for Statistics, Social, and Biomedical Sciences: An Introduction, 1st edn. Cambridge University Press, New York (2015)
14. Jacobs, P.: Rapid Prototyping & Manufacturing: Fundamentals of Stereolithography. SME (1992)
15. Killick, R., Fearnhead, P., Eckley, I.A.: Optimal detection of changepoints with a linear computational cost. J. Am. Stat. Assoc. **107**(500), 1590–1598 (2012). https://doi.org/10.1080/01621459.2012.737745
16. Luan, H., Grasso, M., Colosimo, B.M., Huang, Q.: Prescriptive data-analytical modeling of laser powder bed fusion processes for accuracy improvement. J. Manuf. Sci. Eng. **141**(1), 011008 (2018)
17. Luan, H., Grasso, M., Colosimo, B.M., Huang, Q.: Prescriptive data-analytical modeling of laser powder bed fusion processes for accuracy improvement. J. Manuf. Sci. Eng. **141**(1), 011008 (2019). https://doi.org/10.1115/1.4041709

18. Luan, H., Huang, Q.: Prescriptive modeling and compensation of in-plane shape deformation for 3-D printed freeform products. IEEE Trans. Autom. Sci. Eng. **14**(1), 73–82 (2017)

19. Neal, R.: Mcmc using hamiltonian dynamics. In: Brooks, S., Gelman, A., Jones, G.L., Meng, X.L. (eds.) Handbook of Markov Chain Monte Carlo, pp. 113–162. Chapman & Hall/CRC Press (2010)

20. Orrite, C., Lopez, J., Alcolea, A.: Curve segmentation by continuous smoothing at multiple scales. In: Proceedings of 3rd IEEE International Conference on Image Processing, vol. 3, pp. 579–582. IEEE, Lausanne (1996). https://doi.org/10.1109/ICIP.1996.560561

21. Pala, P., Del Bimbo, A., Berretti, S.: Retrieval by shape similarity with perceptual distance and effective indexing. IEEE Trans. Multimedia **2**(4), 225–239 (2000). https://doi.org/10.1109/6046.890058

22. Perez, J.C., Vidal, E.: Optimum polygonal approximation of digitized curves. Pattern Recogn. Lett. **15**(8), 743–750 (1994)

23. Pernot, J.P., Falcidieno, B., Giannini, F., Léon, J.C.: Incorporating free-form features in aesthetic and engineering product design: state-of-the-art report. Comput. Ind. **59**(6), 626–637 (2008)

24. Pikaz, A., et al.: An algorithm for polygonal approximation based on iterative point elimination. Pattern Recogn. Lett. **16**(6), 557–563 (1995)

25. Sabbaghi, A., Dasgupta, T., Huang, Q., Zhang, J.: Inference for deformation and interference in 3d printing. Ann. Appl. Stat. **8**(3), 1395–1415 (2014)

26. Salotti, M.: An efficient algorithm for the optimal polygonal approximation of digitized curves. Pattern Recogn. Lett. **22**(2), 215–221 (2001)

27. Silkan, H., Tmiri, A., Ouatik, S.E.A., Lachkar, A.: A new shape descriptor 2D for content based image retrieval. In: 2014 Second World Conference on Complex Systems (WCCS), pp. 670–674. IEEE, Agadir (2014). https://doi.org/10.1109/ICoCS.2014.7060903

28. Sunil, V., Pande, S.: Automatic recognition of features from freeform surface CAD models. Comput.-Aided Design **40**(4), 502–517 (2008). https://doi.org/10.1016/j.cad.2008.01.006

29. Thompson, M.K., Moroni, G., Vaneker, T., Fadel, G., Campbell, R.I., Gibson, I., Bernard, A., Schulz, J., Graf, P., Ahuja, B., et al.: Design for additive manufacturing: trends, opportunities, considerations, and constraints. CIRP Ann. **65**(2), 737–760 (2016)

30. Ventura, J.A., Chen, J.M.: Segmentation of two-dimensional curve contours. Pattern Recogn. **25**(10), 1129–1140 (1992). https://doi.org/10.1016/0031-3203(92)90016-C

31. Wang, A., Song, S., Huang, Q., Tsung, F.: In-plane shape-deviation modeling and compensation for fused deposition modeling processes. IEEE Trans. Autom. Sci. Eng. **17**(2), 968–976 (2017)

32. Wang, H., Huang, Q.: Error cancellation modeling and its application in machining process control. IIE Trans. Quality Reliab. **38**, 379–388 (2006)

33. Wang, H., Huang, Q.: Using error equivalence concept to automatically adjust discrete manufacturing processes for dimensional variation control. ASME Trans. J. Manuf. Sci. Eng. **129**, 644–652 (2007)

34. Wang, H., Huang, Q., Katz, R.: Multi-operational machining processes modeling for sequential root cause identification and measurement reduction. ASME Trans. J. Manuf. Sci. Eng. **127**, 512–521 (2005)

35. Wu, J.S., Leou, J.J.: New polygonal approximation schemes for object shape representation. Pattern Recogn. **26**(4), 471–484 (1993)

36. Xia, S., Chen, D., Wang, R., Li, J., Zhang, X.: Geometric primitives in LiDAR point clouds: a review. IEEE J. Sel. Top. Appl. Earth Observ. Remote Sensing **13**, 685–707 (2020). https://doi.org/10.1109/JSTARS.2020.2969119

37. Yin, P.Y.: A discrete particle swarm algorithm for optimal polygonal approximation of digital curves. J. Vis. Commun. Image Represent. **15**(2), 241–260 (2004)

38. Zhou, C., Chen, Y.: Calibrating large-area mask projection stereolithography for its accuracy and resolution improvements. In: International Solid Freeform Fabrication Symposium. The University of Texas at Austin (2009)
39. Zhou, C., Chen, Y.: Additive manufacturing based on optimized mask video projection for improved accuracy and resolution. J. Manuf. Process. **14**, 107–118 (2012)

Part II

Fabrication-Aware Machine Learning for Additive Manufacturing: Three-Dimensional Shape Quality Learning and Prediction

Chapter 4
Representation of Three-Dimensional (3D) Geometric Quality

The final product geometries in AM are often deformed or distorted. The deviations of three-dimensional (3D) shapes from their intended designs can be represented as 2D surfaces in a $\mathbb{R}^3$ space, which constitutes a complicated set of data for representing, learning, and predicting geometric quality. Patterns of deviation surfaces vary with shape geometries, sizes/volumes, materials, and AM processes.

It is pointed out in Chap. 2 that the representation of product geometric quality is determined by the selection of regions of interest on a product and characterization of the deviation patterns in selected regions. The representation scheme needs to consider the challenges of lack of knowledge and right data in AM. This chapter introduces the representation of 3D geometric quality in AM, which is a consistent extension of the representation scheme for 2D geometric quality introduced in Chap. 2.

4.1 Representation of Geometric Quality for Individual 3D Objects

We adopt the spherical coordinate system (SCS) (r, θ, φ) to depict 3D shape deviations, including both the in-plane and out-of-plane (z direction) shape deviations [3]. Under a unified formulation, it facilitates the representation of the out-of-plane error in the same way as the in-plane error.

Denote $r(\theta, \varphi, r_0(\theta, \varphi))$ as the shape boundary of an AM-built product, with $r_0(\theta, \varphi)$ being the nominal shape. The geometric shape quality or output y_u is the difference between the actual printed product $r(\theta, \varphi, r_0(\theta, \varphi))$ and its intended design $r_0(\theta, \phi)$:

$$y_u = d(R(\mathbb{P}_{u_D}), R(\mathbb{P}_{u_A})) = \Delta r(\theta, \phi) = r(\theta, \varphi, r_0(\theta, \varphi)) - r_0(\theta, \phi) \tag{4.1}$$

Q. Huang, *Domain-informed Machine Learning for Smart Manufacturing*,
https://doi.org/10.1007/978-3-031-91631-1_4

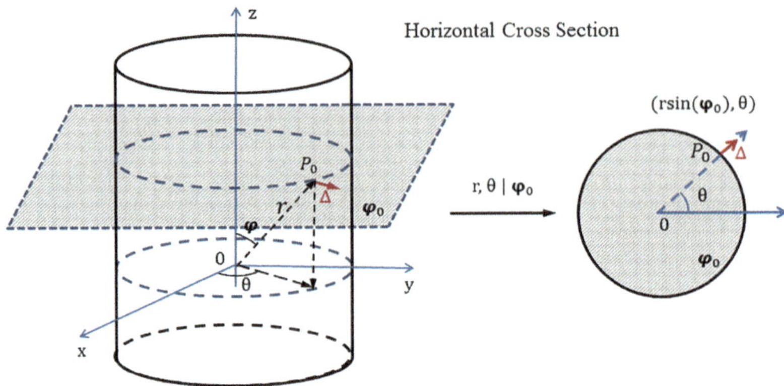

Fig. 4.1 In-plane shape deviation representation

To illustrate the representation scheme, we define the in-plane error and out-of-plane error in the SCS using the simple cylindrical shape, though it is applicable to 3D freeform shapes. As shown in Fig. 4.1, for an arbitrary point $P_0(r_0, \theta_0, \varphi_0)$ at a given height $\varphi = \varphi_0$ or $z = r_0(\theta, \varphi)\cos(\varphi_0)$, the horizontal cross-sectional view of the product passing P_0 is given as $(r_0(\theta, \varphi)\sin(\varphi_0), \theta \mid \varphi_0)$, whose shape deviation $\Delta r(\theta, r_0(\theta) \mid \varphi_0)$ represents the in-plane geometric error. Its model formulation has been developed in Sect. 2.4 [5, 7].

Denote the in-plane deviation model $\Delta r(\theta, r_0(\theta, \varphi) \mid \varphi)$ as $h(r, \theta \mid \varphi)$. Let us define the expectation of the in-plane shape deviation $h(r, \theta \mid \varphi)$ over all φ:

$$\int_{\varphi} h(r, \theta \mid \varphi) d\varphi \tag{4.2}$$

Model (4.2) represents the average in-plane deviation over all layers. Models developed for circular disks, polygon plates, and 2D freeform plates in Chap. 3 can be viewed as the average in-plane deviation when out-of-plane deviation is not considered.

The out-of-plane error, which is the error in the vertical direction, can be represented in the vertical cross section containing P_0 (Fig. 4.2). Any point P_0 on the boundary of the vertical cross section is given as $(r, \varphi \mid \theta_0)$.

Denote the out-of-plane deviation model $\Delta r(\varphi, r_0(\theta, \varphi) \mid \theta)$ as $v(r, \varphi \mid \theta)$. Let us define the expectation of the out-of-plane shape deviation $v(r, \varphi \mid \theta)$ over all θ in the vertical cross section as:

$$\int_{\theta} v(r, \varphi \mid \theta) d\theta \tag{4.3}$$

which can be interpreted as the average out-of-plane shape deviation in a similar fashion.

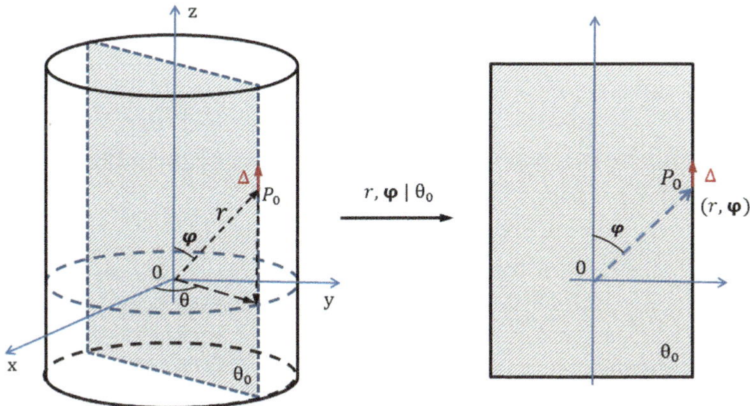

Fig. 4.2 Out-of-plane shape deviation representation

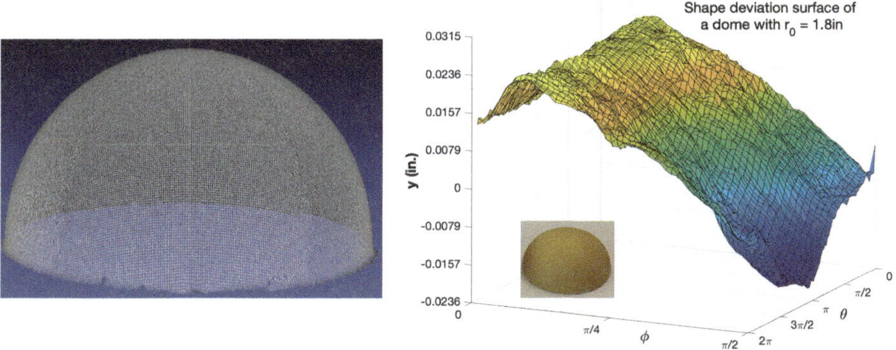

Fig. 4.3 The scan point cloud data of a dome and representation of the dome deviation under the SCS [6]

Mathematically, modeling of $v(r, \varphi|\theta)$ is essentially equivalent to modeling of $h(r, \varphi|\theta)$. This suggests that the mathematical formulation developed in Chap. 3 for the in-plane errors can be borrowed under this new formulation in the SCS. In this way, 3D geometric errors can be described in a unified framework.

However, it should be noted that $v(r, \varphi|\theta)$ and $h(r, \theta|\varphi)$ may differ even if the horizontal and vertical cross-sectional views share the same shape. Different from the in-plane shape deviation whose representation is along the horizontal direction, the out-of-plane shape deviation is defined along the vertical direction, which is influenced by extra factors such as layer interactions and gravity, resulting in different deviation patterns [8].

Figure 4.3 shows a dome shape built by a SLA machine, its point cloud data generated by a laser scanner, and its shape deviation $y_u = d(R(\mathbb{P}_{u_D}), R(\mathbb{P}_{u_A}))$

(Eq. 2.1) under the SCS, which is a 2D deviation surface. The shape deviation pattern is clearly visible after the transformation, which facilitates the modeling.

4.2 Fabrication-Aware Representation of Geometric Quality for 3D Freeform Objects

It has been pointed out in Sect. 2.5 that domain-informed dimension reduction is needed to address the challenge of representing heterogeneous (manifold) data for the high-dimensional product variety space. A fabrication-aware input and output representation scheme was introduced for 2D freeform objects. Finite manufacturing primitive (FMP) method was introduced to represent the input and output for 2D geometries. 2.5 [4]. This section introduces the representation scheme for 3D geometries based on FMPs.

Fabrication-Aware Input Representation for 3D Freeform Objects
The definition of primitive manufacturing input (2.1) states that a 3D object is a stack-up of individual layers with boundaries composed by 2D shape primitives. The 2D design input is represented by Eq. (2.3) as $u(\theta) = \sum_{i=0}^{I} \mathbf{1}_{\theta \in [\theta_i, \theta_{i+1})}\, r_i(\theta)$. The 3D design input u is represented as a stack-up of 2D design input as

$$u(\theta, \phi) = \sum_{j=0}^{J} \sum_{i=0}^{I} \mathbf{1}_{\theta \in [\theta_i, \theta_{i+1})} \mathbf{1}_{\{\phi:\, r(\theta,\phi)\cos\phi = z_j\}}\, r_i(\theta, \phi) \sin\phi \qquad (4.4)$$

where 3D shapes are represented in the SCS [3] and a layer indexed by j with height z_j.

Purely from the geometric point of view, it is more convenient to adopt plane surface patches, spherical patches, and surface edges as 3D shape primitives to represent and construct 3D shapes. However, it is still an open issue to segment a 3D shape into 3D shape primitives for AM [13]. In particular, prediction of the shape deviation in AM relies on description of how products are fabricated. Since 3D shape primitives are constituents of a 3D shape, they are the results of primitive manufacturing inputs. We hence define 3D shape primitives for AM with the schematic plot shown in Fig. 4.4.

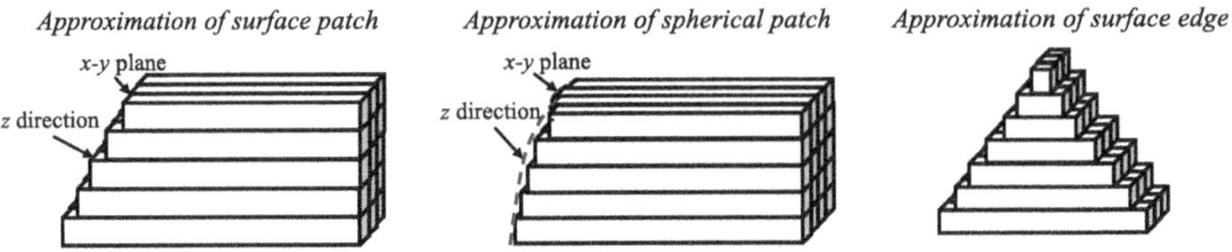

Fig. 4.4 3D geometric shape primitives for layer boundary approximation

In essence the definitions of primitive manufacturing input and 3D shape primitive intend to establish a fabrication-aware formulation that enables understanding of the deviations of 3D shape primitives. As such, deviation prediction of a 3D freeform shape product can be achieved by predicting the deviation of the shape constructed by 3D shape primitives. A freeform 3D shape can be constructed as:

$$u(\theta, \phi) = \sum_{j=0}^{J} \sum_{i=0}^{I} \mathbf{1}_{\{\theta \in [\theta_i, \theta_{i+1}), \ \phi \in [\phi_j, \phi_{j+1})\}} \, r_{ij}(\theta, \phi) \tag{4.5}$$

where $r_{ij}(\theta, \phi)$ is either a plane surface patch, spherical patch, or a surface edges.

Definition 4.1 (3D Shape Primitive)   The 3D shape primitives such as plane surface patches, spherical patches, and surface edge, are the result of layer stack-up with layer boundaries composed by 2D shape primitives.

Output Representation Based on Transformation of Point-Cloud Data

The fabrication-aware output representation adopts the expression in Eq. (4.1), that is, $y_u = \Delta r(\theta, \phi) = r(\theta, \varphi, r_0(\theta, \varphi)) - r_0(\theta, \phi)$, for 3D shapes. The output y_u is the difference between the actual printed product $r(\cdot)$ and its intended design $r_0(\cdot)$, not the input u to the fabrication process. The fabrication input u is only a close approximation of design $r_0(\cdot)$ in the digital manufacturing. Figure 4.5 shows an example of representing the shape deviation (2D deviation surface) of a 3D thin wall in half disk shape. The products were printed in a stereolithography process (SLA) [13].

The proposed input-output representation approaches reveal the following information:

- Comparing to the scanned data in point cloud format, the output representation in the SCS and PCS uncovers shape deviation patterns for both 2D and 3D shapes.

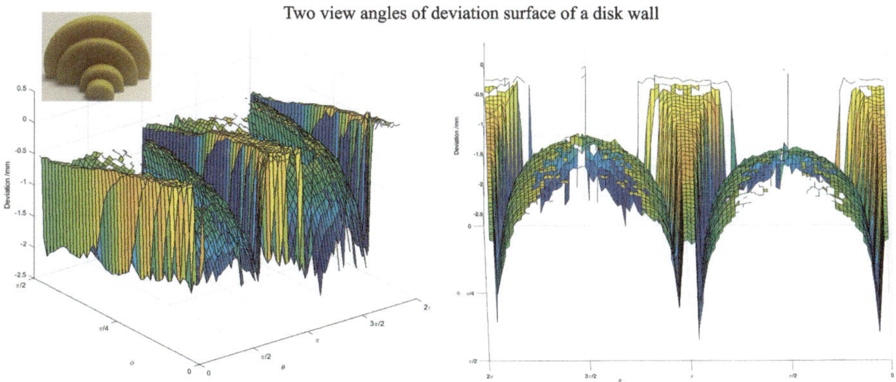

Fig. 4.5   Shape deviation surface of a thin wall in half-disk shape [13]

- Shape deviation patterns vary with shapes and sizes (pls refer to additional example in [9, 10]). Learning deviation patterns based on shapes alone is not sufficient and effective for small-sample learning.
- The shape primitives can be utilized to learn and construct the shape deviation patterns of many different shapes with small training data.
- The shape deviation pattern of a 3D shape is not a simple linear summation or extension of the deviation patterns of its constructive layers. For example, the pattern of the dome in Fig. 4.3 is not composed by simply stacking up the patterns of disks along the z direction. But the deviation profile of a 3D shape at a given height ϕ (i.e., a horizontal section) partially resembles the one of its 2D counterpart. It is particularly remarkable to see in Fig. 4.5 that the horizontal section of the thin walls in half-disk shapes has a rectangular shape. At a given ϕ (Fig. 4.5), the shape deviation resembles the ones for square plates to a large degree (Fig. 2.7).
- The connection between 2D and 3D shape deviation patterns implies that shape deviation models for 2D shapes should be special cases of those for 3D shapes. The 2D shape deviation models can be viewed as a projection of its 3D counterpart onto a subspace. The physical underpinning of projection relation is the nonlinear stack-up of 2D layers into 3D shapes.

For complicated geometries, shape segmentation may be necessary to accurately represent shape deviations with deviation primitives. Shape segmentation has long been applied in 3D object recognition [1, 2], reverse engineering [12], and freeform surface metrology [11]. It involves the process of dividing the original point set into subsets corresponding to natural surfaces. Segmentation for the purpose of deformation identification and modeling is worthy of further investigation. Little research has been reported.

References

1. Besl, P.J., Jain, R.C.: Three-dimensional object recognition. ACM Comput. Surv. **17**(1), 75–145 (1985). https://doi.org/10.1145/4078.4081
2. Besl, P.J., Jain, R.C.: Segmentation through variable-order surface fitting. IEEE Trans. Pattern Anal. Mach. Intell. **10**(2), 167–192 (1988). https://doi.org/10.1109/34.3881
3. Huang, Q.: An analytical foundation for optimal compensation of three-dimensional shape deformation in additive manufacturing. ASME Trans. J. Manuf. Sci. Eng. **138**(6), 061010 (2016)
4. Huang, Q.: An impulse response formulation for small-sample learning and control of additive manufacturing quality. IISE Trans. (2022). https://doi.org/10.1080/24725854.2022.2113186
5. Huang, Q., Nouri, H., Xu, K., Chen, Y., Sosina, S., Dasgupta, T.: Statistical predictive modeling and compensation of geometric deviations of three-dimensional printed products. ASME Trans. J. Manuf. Sci. Eng. **136**(6), 061008 (2014)
6. Huang, Q., Wang, Y., Lyu, M., Lin, W.: Shape deviation generator (SDG)—a convolution framework for learning and predicting 3d printing shape accuracy. IEEE Trans. Autom. Sci. Eng. **17**(3), 1486–1500 (2020)

7. Huang, Q., Zhang, J., Sabbaghi, A., Dasgupta, T.: Optimal offline compensation of shape shrinkage for 3d printing processes. IIE Trans. Quality Reliab. **47**(5), 431–441 (2015)
8. Jin, Y., Qin, S., Huang, Q.: Out-of-plane geometric error prediction for additive manufacturing. In: 2015 IEEE International Conference on Automation Science and Engineering (CASE 2015), Gothenburg (2015)
9. Jin, Y., Qin, S., Huang, Q.: Offline predictive control of out-of-plane geometric errors for additive manufacturing. ASME Trans. Manuf. Sci. Eng. **138**(12), 121005 (2016)
10. Jin, Y., Qin, S., Huang, Q.: Modeling inter-layer interactions for out-of-plane shape deviation reduction in additive manufacturing. IISE Trans. Design Manuf. **52**(7), 721–731 (2020)
11. Savio, E., De Chiffre, L., Schmitt, R.: Metrology of freeform shaped parts. CIRP Ann. **56**(2), 810–835 (2007)
12. Várady, T., Martin, R.R., Cox, J.: Reverse engineering of geometric models - an introduction. CAD Comput. Aided Design **29**(4), 255–268 (1997). https://doi.org/10.1016/s0010-4485(96)00054-1
13. Wang, Y., Ruiz, C., Huang, Q.: Learning and predicting shape deviations of smooth and non-smooth 3d geometries through mathematical decomposition of additive manufacturing. IEEE Trans. Autom. Sci. Eng. (2022). https://doi.org/10.1109/TASE.2022.3174228

Chapter 5
Small-Sample Learning and Prediction of 3D Geometric Quality

Three-dimensional (3D) shape accuracy is a critical performance measure for products built via AM. With advances in computing and increased accessibility of AM product data, machine learning for AM (ML4AM) has become a viable strategy for enhancing AM. Proper description of the 3D shape formation through the layer-by-layer fabrication process is critical to incorporate process domain knowledge into ML4AM. The physics-based modeling and simulation approaches present voxel-level description of an object formation from points to lines, lines to surfaces, and surfaces to 3D shapes. However, this computationally intensive modeling framework does not provide a clear structure for machine learning of AM data. Chapter 5 introduces domain-informed small-sample learning and prediction of shape accuracy of 3D objects.

5.1 Shape Quality Modeling for 3D Basis Geometry

With increased accessibility of AM data, predicting AM built accuracy, or product shape accuracy in particular, has become a focal issue in ML4AM [9, 15, 34, 36, 42, 50, 52, 67, 68, 70, 83, 98]. Empirical and statistical methods have been applied to the investigation and modeling of AM processes [5, 31, 45, 97]. Factors such as layer thickness and flow rate are varied to discover optimal settings for quality control. Data-driven surrogate models are established to reduce computational costs. More recently, machine learning approaches are applied to 3D scanned data for geometric accuracy quantification and classification [42, 70] and distortion control based on deep learning [23].

A series of work [10, 32, 34, 36, 39, 50, 66, 68, 85] establish a prescriptive approach to model, predict, and compensate 2D shape deviations based on geometric measurement of AM-built products. In these studies, models are learned from a limited number of tested shapes, and optimal compensation plans for new and untried products are derived and validated. However, there has been a lack of progress on learning 3D shape data for improving 3D printing accuracy [35]. Describing the 3D shape formation through the layer-by-layer fabrication process has been a daunting task. Physics-based modeling and simulation approaches present voxel-level description of the 3D object formation from points to lines, lines to surfaces, and surfaces to 3D shapes [4, 43]. However, this computationally intensive modeling framework does not provide a clear structure or framework for machine learning of AM simulation or measurement data. The predominant strategy is to obtain physical understandings of critical process variables, such as temperature field and melt pool geometry, as proxies of product quality through their correlation with product geometries [21, 44, 56, 57].

Since the understanding of 3D shape formation in AM is the foundation for the subsequent accuracy control activities such as compensation, it is imperative to establish a data-analytical framework that not only provides an insightful description of 3D shape formation in AM but also enables machine learning of simulation or measurement data.

It is worthy to note the recent ML advances in deep representation learning that aim to synthesize and reconstruct realistic and novel 3D shapes [1, 8, 90] (and reference therein). Various shape representation methods such as meshes, skeletons, pre-training shape templates, voxel grids, multi-view images, point clouds, or surface patches have been developed to generate digital 3D objects with less issues of low-resolution outputs, overly smoothed or discontinuous surfaces, and topological irregularities [8]. By contrast, digital 3D models in this study are given from engineering design. Our objective is to learn and predict how the *physical* 3D shapes are generated and built in AM processes and, more importantly, how the physical 3D objects are different from their digital counterparts. As can be found out in this study, complication and variations in physical AM processes will generate different shape deviation patterns even for the same shapes. This has been outside the scope of consideration for the general ML community.

For the purpose of predicting, learning, and compensating 3D shape deviations based on data, this section introduces the shape deviation generator (SDG), a convolution formulation of the 3D shape deviation generation process.

5.1.1 A Convolution Formulation of Layer-by-Layer Shape Deviation Generation

This section will first introduce the rationale of deriving a convolution formulation to describe the 3D shape deviation generation process. This formulation leads to the definition of a set of subproblems suitable for statistical or machine learning methods. Applicable to a wide variety of AM processes, the 3D SDG formulation is illustrated with examples from SLA processes.

The Layer-by-Layer Fabrication Process and Mathematical Integration Since a typical AM process builds 3D objects layer by layer (Fig. 5.1), a mathematical

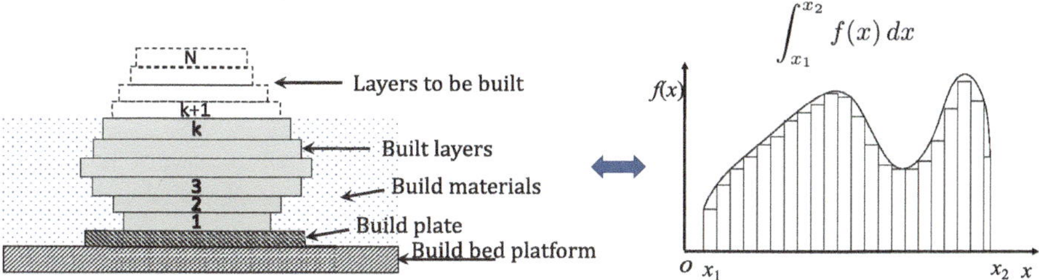

Fig. 5.1 The layer-by-layer fabrication process and math integral

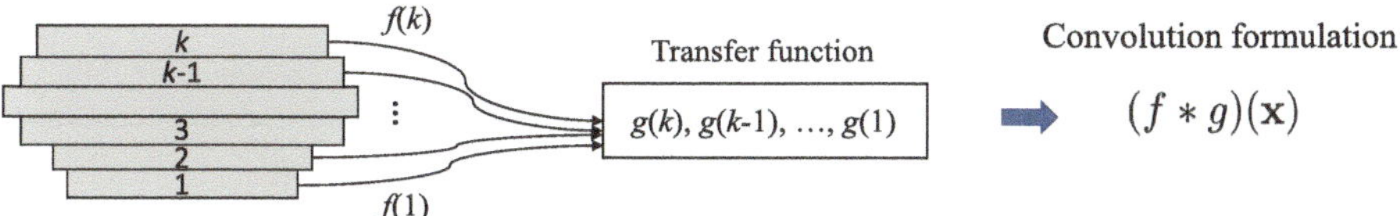

Fig. 5.2 System inputs, stack-up function, and convolution formulation

formulation of the shape generation in AM can naturally consider the *integration* along the direction of layer stack-up. Equivalently, the layer-by-layer fabrication process can be viewed as a *physical realization of a mathematical integration over a closed interval*. Then the key question is what is the proper form of the integrand?

Individual Layers and Layer Stack-up Function In the layer-by-layer shape formation process, individual layers can be viewed as inputs $f(k)'s$ to a system with stack-up or interaction function $g(k)'s$, $k = 1, 2, \ldots, N$ layers. The layer stack-up or interaction function $g(k)$ not only processes the input layer at the current layer k but also modifies previous layers due to the heat exchange and thermal stress between layers. Therefore, there is a "time shift" involved in the stack-up function and an "accumulation" of interactions leading to the final shape formation, as illustrated in Fig. 5.2. A natural integral formulation that reflects the concepts of stack-up function, time shift, and accumulation effects is the convolution integral.

Convolution Integral for SDG in AM We therefore propose the following convolution formulation for the AM SDG:

$$y(\mathbf{x}) = (f * g)(\mathbf{x}) + \epsilon \tag{5.1}$$

where $y(\mathbf{x})$ represents the shape deviation (1D curves or 2D surfaces for 2D or 3D shapes, respectively), $\mathbf{x}$ are parameters that describe the shape deviation, and ϵ is the model error term that may contain spatial correlations.

Boundary Conditions of SDG Two boundary conditions are required to satisfy engineering constraints and limit the scope of parameter optimization and selection.

B1. If $\mathbf{x} = r(\theta, \varphi) \to 0$, $y(\mathbf{x}) \to 0$.

   If the size of a built part tends to zero, the shape deviation approaches zero as well. The system defined in Eq. (5.1) is therefore a linear causal system.

B2. If $\mathbf{x} = r(\theta, \varphi) \to \infty$, $y(\mathbf{x})$ is bounded.

   Even if the size of the built part tends to infinity or becomes very large, the shape deviation of the part has to be bounded because of physical constraints on the materials and process.

Relation with State Space Representation of AM Another intuitive formulation of the AM process is the state space equation with the time index k being the layer index:

$$\mathbf{x}(k) = \mathbf{A}\mathbf{x}(k-1) + \mathbf{B}\mathbf{u}(k)$$

$$\mathbf{y}(k) = \mathbf{C}\mathbf{x}(k) + \mathbf{D}\mathbf{u}(k)$$

where $\mathbf{x}(k)$ represents the state vector of the kth layer and $\mathbf{y}(k)$ is the system response.

The solution to the state space equation through the Laplace transformation is $\mathbf{Y}(s) = [\mathbf{D} + \mathbf{C}(s\mathbf{I} - \mathbf{A})^{-1}\mathbf{B}]\mathbf{U}(s)$. By defining the stack-up function $\mathbf{G}(s) = \mathbf{D} + \mathbf{C}(s\mathbf{I} - \mathbf{A})^{-1}\mathbf{B}$ and conducting the inverse Laplace transformation of $\mathbf{Y}(s) = \mathbf{G}(s)\mathbf{U}(s)$, we also end up with a convolution formulation consistent with Eq. (5.1):

$$\mathbf{y}(k) = (\mathbf{g} * \mathbf{u})(k)$$

Compared to the state space model, the convolution model (5.1) has less unknown terms to be identified and learned. The compromise, however, is that the convolution model (5.1) does not enable the dynamic system representation and real-time feedback control.

Note that under this convolution formulation, $f(\mathbf{x})$ and $g(\mathbf{x})$ represent the features of 3D shape deviation and the characteristics of the layer-by-layer fabrication process. This provides opportunities for developing change detection and correction procedures.

To realize the SDG formulation in model (5.1), we will discuss (1) description of shape deviation $y(\mathbf{x})$, (2) identification and learning of $f(\mathbf{x})$, and (3) identification and learning of stack-up function $g(\mathbf{x})$.

5.1.1.1   Description of Shape Deviation $y(\mathbf{x})$

The system response y represents the 3D shape (boundary) deviation of an AM-built object. The measurement data of a 3D object is usually in the form of point cloud data defined in the Cartesian coordinate system (CCS). Shape description or representation has been extensively investigated in the field of computer vision. It is the first stage of shape analysis such as shape matching, shape deformation, shape correspondence, and shape registration for 3D objects [37, 54, 55, 84]. *Depending on applications*, shape representation generally looks for effective and perceptually important shape features such as point sets, curves, surfaces, level sets, or deformable templates [75, 95]. For accuracy control of AM-built products, the contour-based feature representations such as continuous and discrete approaches are more relevant. A feature vector derived from the integral boundary, e.g., Fourier descriptors [93], is typically applied among continuous approaches. Discrete approaches tend to break the shape boundary into segments for approximation [60]. Representing a shape as diffeomorphism of the unit circle to itself through conformal mapping has been a popular method for shape classification [72]. Invariant to translation and scaling, this representation approach allows one to move back and forth computationally between shapes and their diffeomorphisms. Popular 3D shape representations include medial and skeletal representation, deformable templates, and geometric descriptors [71].

In AM literature, the shape deviation representation methods can be generally classified in two categories: global summary and feature extraction of shape deviation and detailed characterization of whole shape deviations. The first category includes volumetric shrinkage to capture overall size distortion [62, 88] and dimensional and form errors [5, 28, 91] to characterize feature distortion of specific shapes. To enable detailed description of shape deviations for arbitrary shapes in AM, the second category provides three main methods: point-wise deviation representation [42, 80–82, 92], transformation of point cloud [32, 39], and mesh-based deviation representation [17, 94].

Since the shape representation in AM should facilitate the development of a generic model for quality prediction of arbitrary shapes, one key consideration of describing the system response or the shape deviation $y(\mathbf{x})$ is to decouple the geometric shape complexity from the modeling of $y(\mathbf{x})$. We transform the 2D shape deviations under the CCS into deviation profiles in the polar coordinate system (PCS) [36] and 3D shape deviations under the CCS into deviation surfaces in the spherical coordinate system (SCS) [32, 39]. This representation approach is closely related to those in computer vision, e.g., spherical harmonic description [41, 71]. The main difference here is that we focus on shape deviations, as opposed to shape itself because specific shapes are known by design.

For 2D shape deviations, the system response $y(\mathbf{x})$ is defined by Eq. (2.2):

$$y(\mathbf{x}) = r(\theta, r_0(\theta)) - r_0(\theta)$$

where $r(\theta, r_0(\theta))$ and $r_0(\theta)$ are the measured shape and design shape represented in the PCS, respectively. Here $\mathbf{x} = (r_0(\theta), \theta)$. Examples of deviation profiles for various 2D shapes can be found in Fig. 2.7.

For 3D shapes, point cloud data under the CCS is transformed into the observed radial distance $r(\theta, \varphi, r_0(\theta, \varphi))$ for a point with polar angle θ and azimuth angle φ. The system response y is defined by Eq. (4.1):

$$y(\mathbf{x}) = r(\theta, \varphi, r_0(\theta, \varphi)) - r_0(\theta, \phi)$$

to measure the difference between measurement $r(\theta, \varphi, r_0(\theta, \varphi))$ and the nominal design radial distance $r_0(\theta, \varphi)$, Here $\mathbf{x} = \big(r_0(\theta, \varphi), \theta, \varphi\big)$.

Figure 4.3 shows a dome shape built by a SLA machine, its point cloud data generated by a laser scanner, and its shape deviation y under the SCS. The shape deviation pattern is clearly visible after the transformation, which facilitates the modeling.

5.1.1.2 Identification of $f(\mathbf{x})$

To identify the proper form of individual layer inputs $f(\mathbf{x})$, we first consider a simplest case of building a single-layer horizontal disk as shown in Fig. 5.3. If we further assume that the stack-up function $g(\mathbf{x})$ only varies along the z direction, i.e., $g(\mathbf{x}) = g(z)$ or $g(\varphi)$, then $g(\varphi)$ only has definition at $z \approx 0$ or $\varphi \approx 0$ because of the layer thickness along the z direction.

One general form of this type of functions is Dirac's delta function or $\delta(\mathbf{x})$. Under the SDG formulation in Eq. (5.1) and the property of convolving with a delta function, we have

$$(f * \delta)(\mathbf{x}) = f(\mathbf{x}) \tag{5.2}$$

Together with Eq. (5.1), Eq. (5.2) implies that individual layer input $f(\mathbf{x})$ essentially represents the deviation of the 2D horizontal plate (in $x - y$ plane) with a shape defined by the input design. Note that $f(\mathbf{x})$ is not affected by inter-layer interactions. This also suggests an experimental design strategy for establishing the SDG defined in Eq. (5.1), i.e., horizontal plates with selected 2D shapes should be built first to understand $f(\mathbf{x})$.

One example of $f(r_0(\theta), \theta)$ for cylindrical disks built by a SLA process is

$$f\big(r_0(\theta), \theta\big) = c_1(r_0) + c_2(r_0) \cos(2\theta)$$

where coefficients $c_1(r_0)$ and $c_2(r_0)$ can be obtained through estimation based on data shown in Fig. 2.7 (left panel for circular disks with various sizes) [36].

Learning $f(r_0(\theta), \theta)$ for 2D freeform shapes from a small set of training shapes has been introduced in Chap. 3 [34, 36, 50, 66, 68, 85]. Note that different statistical and machine learning methods can be introduced to obtain $f(\mathbf{x})$ in general.

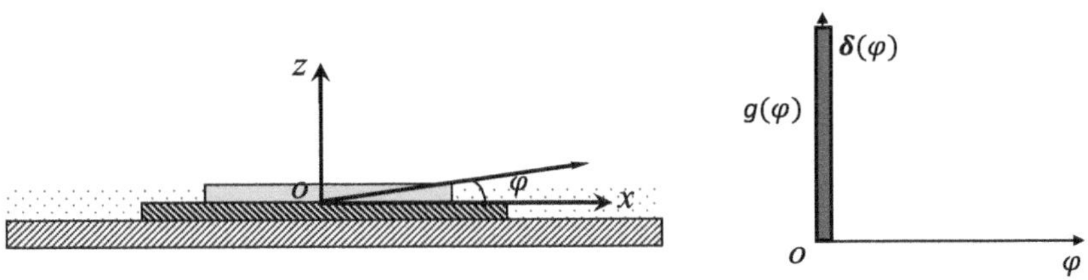

Fig. 5.3 Identification of $f(\mathbf{x})$: horizontal 2D disk

5.1.1.3   Identification of Stack-Up Function $g(\mathbf{x})$

The layer buildup along the z direction incurs the inter-layer interaction problem in AM. Complete inter-layer bonding implies full density and continuous interface between successively fabricated layers, while poor bonding may cause delamination and defects like balling or beading across a layer [13]. Li and Gu [46] find that the heat accumulation effect and the remelting phenomenon due to laser energy penetration will lead to the increase of the temperature and size of melt pools when the laser beam moves from bottom to top layers. Denlinger [20] decomposes the overall distortion along the z direction to individual layers through in situ distortion measurement and experimentally investigates the relationship between the inter-layer dwell time and accumulation of distortions for different materials.

The stack-up function $g(\mathbf{x})$ in the convolution formulation (5.1) intends to provide a data-analytical description of the effect of inter-layer interactions and error accumulation on build accuracy.

For simplicity of methodology illustration and development, we first consider a simple case of building half disks along the z direction (Fig. 5.4), where the $g(\mathbf{x})$ ends up as a univariate function assigning weights for individual layer inputs. Then we will extend to a 3D case of building domes with various sizes to model the multivariate stack-up function $g(\mathbf{x})$.

5.1.2   *Convolution Modeling and Learning for Vertically Printed Half Disks*

Since the half disk can be approximated as a 2D shape, Eq. (5.1) can be simply written as $y(\varphi) = (f * g)(\varphi) + \epsilon$. Here we introduce the size of the disk, i.e., nominal radius r_0, as an additional covariate because the size is directly proportional to the

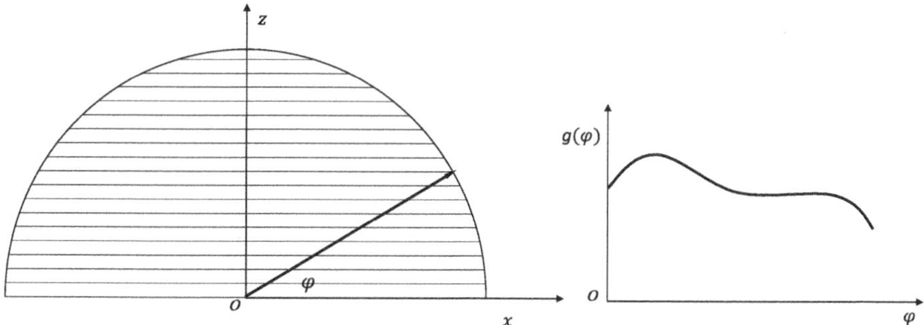

Fig. 5.4   Half disks built vertically and identification of $g(\mathbf{x})$

number of layers. The SDG model is rewritten as:

$$y(r_0, \varphi) = \alpha(r_0)(f * g)(r_0, \varphi) + \beta(r_0) + \epsilon$$
$$= \alpha(r_0) \int_0^\varphi f(r_0, \tau)g(r_0, \varphi - \tau)\, d\tau + \beta(r_0) + \epsilon \qquad (5.3)$$

where $\alpha(r_0)$ and $\beta(r_0)$ are scaling and location parameters depending on r_0.

Function $f(r_0, \varphi)$ has been identified to be the deviation function of the same shape built horizontally (Eq. 5.2). Study has to be done first to obtain $f(r_0, \varphi)$. Given $f(r_0, \varphi)$ and measurement data $y(r_0, \varphi)$, identifying $g(\mathbf{x})$ is a classical deconvolution problem [7]. Signal processing, statistical model estimation, and machine learning (including neural networks) can be applied to address the deconvolution problem in this context. Here we present a model-informed estimation approach.

5.1.2.1 Normalizing $f(\mathbf{x})$ and $g(\mathbf{x})$ for Convolution Integral in SDG

First, by following the results in in Sect. 3.1 [36] for SLA processes, we can take $f(r_0, \varphi) = c_1(r_0) + c_2(r_0)\cos(2\varphi)$. Since coefficients $c_1(r_0)$ and $c_2(r_0)$ relating to size r_0 can be absorbed by $\alpha(r_0)$ and $\beta(r_0)$ in Eq. (5.3), it is equivalent for us to take $f(r_0, \varphi) = \cos(2\varphi)$, i.e., the "normalized" functional basis. Since the normalization idea applies to $g(\mathbf{x})$ as well, a natural choice of basis function for $g(\mathbf{x})$, given $f(r_0, \varphi) = \cos(2\varphi)$, is a Fourier base:

$$g(r_0, \varphi) = \sin\left[n(r_0)\varphi + \psi(r_0)\right] \qquad (5.4)$$

where $\psi(r_0)$ is a phase variable and $n(r_0)$ is a real number. Note that both ψ and n are potentially related to the size covariate r_0.

The SDG model (5.3) can thus be rewritten as

$$y(r_0, \varphi) = \alpha(r_0) \int_0^\varphi \cos(2\tau) \sin\left[n(r_0)(\varphi - \tau) + \psi(r_0)\right] d\tau + \beta(r_0) + \epsilon \qquad (5.5)$$

Let us define $h(r_0, \varphi) = \int_0^\varphi \cos(2\tau) \sin\left[n(r_0)(\varphi - \tau) + \psi(r_0)\right] d\tau$. The SDG model (5.3) can be conveniently expressed as

$$y(r_0, \varphi) = \alpha(r_0)h(r_0, \varphi) + \beta(r_0) + \epsilon \qquad (5.6)$$

It is interesting to compare this SDG model (5.6) for disks built vertically with the one for horizontal disks: $c_1(r_0) + c_2(r_0)\cos(2\theta)$ established in [36]. By Eq. (5.2), the model for horizontal disk is a special case of model (5.6) with $g(\mathbf{x}) = \delta(\mathbf{x})$ in the convolution integral of $h(r_0, \varphi)$.

After performing integration, $h(r_0, \varphi)$ becomes

$$
\begin{aligned}
h(r_0, \varphi) &= \int_0^\varphi \cos(2\tau) \sin\left[n(r_0)(\varphi - \tau) + \psi(r_0)\right] d\tau \\
&= \frac{-n(r_0)}{n(r_0)^2 - 4} \cos\left[n(r_0)\varphi + \psi(r_0)\right] \\
&\quad + \frac{1}{2n(r_0) + 4} \cos\left[2\varphi - \psi(r_0)\right] \\
&\quad + \frac{1}{2n(r_0) - 4} \cos\left[2\varphi + \psi(r_0)\right]
\end{aligned}
\tag{5.7}
$$

5.1.2.2   Experimentation and Data Collection

A commercial mask-image-projection-based stereolithography apparatus is used to build four half cylindrical disks vertically. This SLA process uses a digital micromirror device to project a set of mask images onto the resin surface to cure layers. After solidification of each layer, the building platform moves down at a predefined amount for the next layer. Process parameters and the design of four parts are shown in Table 5.1.

Figure 5.5 shows the four half disks and their deviation profiles under the PCS. Comparing with the repeatable deviation patterns of horizontally built disks illustrated in Fig. 2.7 (left panel for circular disks with various sizes), the effect of the layer buildup and inter-layer interactions along the z direction is clearly visible in the sense that deviation patterns are not consistent when the disk size increases.

5.1.2.3   SDG Model Estimation and Domain-Informed Sequential Model Refinement

Initial estimation of model (5.5) is accomplished through maximum likelihood estimation (MLE). Four separate models of (5.5) are obtained for half disks built vertically with $r_0 = 0.5', 0.8'', 1.5''$, and $2.0''$ shown in Fig. 5.5. The results of estimated parameters are presented in Table 5.2, and predicted shape deviations are

Table 5.1 Specifications of SLA process and design parameters

Resolution of the mask	1920×1200
Dimension of each pixel	$0.005''$
Thickness of each layer	$0.00197''$
Illuminating time of each layer	10–15s
Average waiting time between layers	15s
Type of the resin	Perfactory SI 500
Radii of half cylinders	$0.5'', 0.8'', 1.5'', 2.0''$

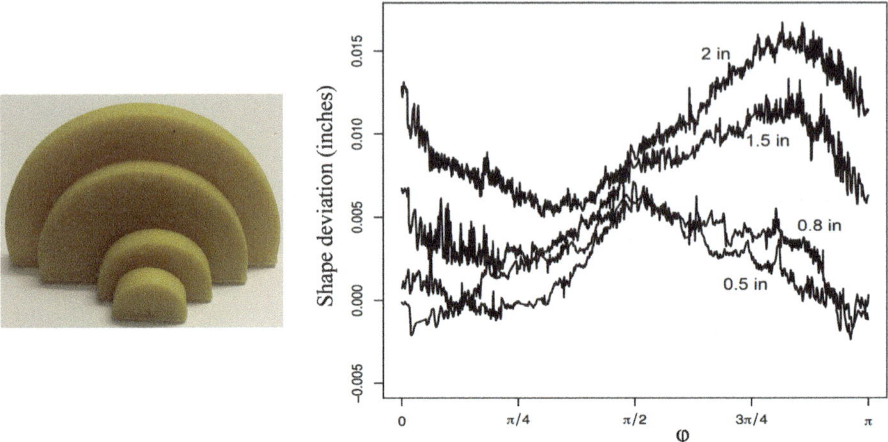

Fig. 5.5 Four half disks and corresponding shape deviation profiles in the PCS

Table 5.2 Initial model estimation through MLE

r_0	$\alpha(r_0)$	$n(r_0)$	$\psi(r_0)$	$\beta(r_0)$	σ
0.5″	0.0220	0.4493	−0.0928	0.0002	0.0007
0.8″	0.0069	0.9158	−0.7619	0.0013	0.0008
1.5″	0.0056	1.1916	−2.0898	0.0054	0.0008
2.0″	0.0055	1.9238	−3.2883	0.0087	0.0008

shown in Fig. 5.6 as dashed lines. We can find that the predicted deviation profiles fit the data (solid lines) well, and modeling error σ is small and consistent.

This initial model estimation in Table 5.2 also suggests that $\alpha(r_0)$, $\beta(r_0)$, $n(r_0)$ and $\psi(r_0)$ potentially vary with covariate r_0. To build one consistent model for all disks and discover process insights, we will first focus on $n(r_0)$ and $\psi(r_0)$, which are inside the convolution integral $h(r_0, \varphi)$, and conduct model estimation through a sequential model refinement procedure informed by physical knowledge.

The following two conjectures are proposed for $n(r_0)$ and $\psi(r_0)$.

- Periodicity of stack-up function $g(\mathbf{x})$ and length of interaction window: Parameter $n(r_0)$ determines the period of stack-up function $g(\mathbf{x})$ defined in Eq. (5.4). On one hand, as the number of layers increases, the interaction of adjacent layers will be more complicated due to, for example, energy penetration and remelting. $n(r_0)$ therefore is expected to increase with the number of layers or r_0 because a long period or small $n(r_0)$ will assign similar weights to adjacent layers. On the other hand, inter-layer interactions only play within a window with certain length defined by a certain number of consecutive layers. For instance, the depth of energy penetration or remelting is limited by the thermal properties of materials and energy input. With this process understanding and the initial model fitting

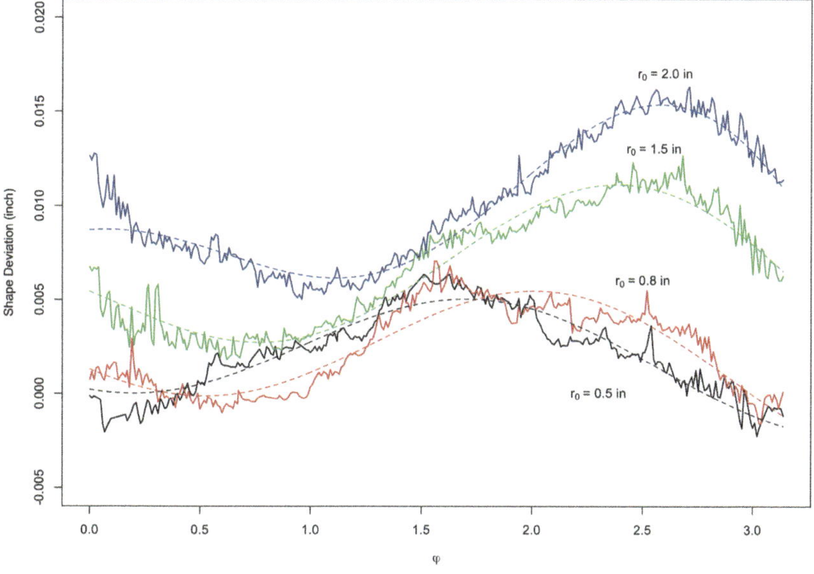

Fig. 5.6 Deviation profiles and model prediction for half disks built vertically: initial model fitting

results in Table 5.2, $n(r_0)$ is hypothesized to be

$$n(r_0) = \frac{r_0}{r^*},\tag{5.8}$$

with r_0 being the nominal radius of the disk and r^* being the theoretical length of the interaction window. Notice that introducing a proportional coefficient such as $a\, r_0/r^*$ could cause over-parameterization problem in model estimation because a can be absorbed by r^*. The same consideration applies to the second conjecture.

- Phase shift of stack-up function $g(\mathbf{x})$ and change of layer areas: Other than the effect of the number of layers, the change of layer area between adjacent layers will also affect stack-up function $g(\mathbf{x})$. If all adjacent layers within an interaction window are identical, $g(\mathbf{x})$ defined in Eq. (5.4) is expected to have $\psi(r_0) = 0$ because only the number of layers varies. We hypothesize that $\psi(r_0)$ is related to the ratio of the area of the half disk to the area of the theoretical "window" shown in Fig. 5.7. Since the estimated values of $\psi(r_0)$ in Table 5.2 are all negative, we finalize the conjecture as

$$\psi(r_0) = -\frac{r_0^2}{r^{*2}}\tag{5.9}$$

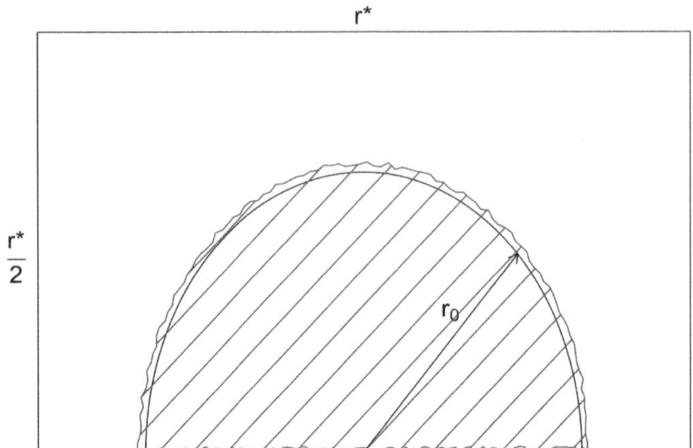

Fig. 5.7 Area ratio between a half disk and the theoretical window

With the above two conjectures, $h(r_0, \varphi)$ in Eq. (5.7) is updated as:

$$
\begin{aligned}
h(r_0, \varphi) &= \int_0^\varphi \cos(2\tau) \sin\left[\frac{r_0}{r^*}(\varphi - \tau) - \frac{r_0^2}{r^{*2}}\right] d\tau \\
&= -\frac{r_0 r^*}{r_0^2 - 4r^{*2}} \cos\left[\frac{r_0}{r^*}\varphi - \frac{r_0^2}{r^{*2}}\right] \\
&\quad + \frac{r^*}{2r_0 + 4r^*} \cos\left[2\varphi + \frac{r_0^2}{r^{*2}}\right] \\
&\quad + \frac{r^*}{2r_0 - 4r^*} \cos\left[2\varphi - \frac{r_0^2}{r^{*2}}\right]
\end{aligned}
\tag{5.10}
$$

With updated $h(r_0, \varphi)$ in Eq. (5.10), we pool data of four half disks together to estimate one single SDG model. The model estimation through MLE is shown in Table 5.3 and Fig. 5.8. All the coefficients are significant at the 0.0001 level. Note that $\hat{r}^* = 1.116$ and the upper bound of r_0 for experimentation is $2.0''$, which avoid the issue of $2r_0 - 4r^* = 0$ in Eq. (5.10) because model extrapolation itself is problematic.

Next we aim to determine the functional forms of $\alpha(r_0)$ and $\beta(r_0)$, which are dictated by both the model fitting results in Table 5.3 and the two boundary conditions listed at the beginning of this section for SDG: $y(r_0, \varphi) = \alpha(r_0)h(r_0, \varphi) + \beta(r_0) + \epsilon$.

When $r_0 \to 0$, the first term of $h(r_0, \varphi)$ in Eq. (5.10) tends to 0, while the second and the third term cancel each other. One constraint for $\alpha(r_0)$ is that $\alpha(r_0)r_0 \to 0$ when $r_0 \to 0$. The remaining term of the expected value of $y(r_0, \varphi)$ is $\beta(r_0)$. Therefore, $\beta(r_0) \to 0$ when $r_0 \to 0$, because $\beta(r_0)$ determines the mean shape deviation. It must be close to 0 if a product with negligible size is built. So the

Table 5.3 Sequential model refinement with conjectures on $n(r_0)$ and $\psi(r_0)$

Parameters	Estimate	Standard error
r^*	1.11589962	0.00401990
$\alpha_{0.5''}$	0.01771531	0.00055778
$\alpha_{0.8''}$	0.00957536	0.00027614
$\alpha_{1.5''}$	0.00534351	0.00010980
$\alpha_{2.0''}$	0.00555896	0.00010031
$\beta_{0.5''}$	0.00050797	0.00007831
$\beta_{0.8''}$	0.00103725	0.00006886
$\beta_{1.5''}$	0.00551954	0.00006387
$\beta_{2.0''}$	0.00883869	0.00006383
σ_ϵ	0.00103552	0.00002059

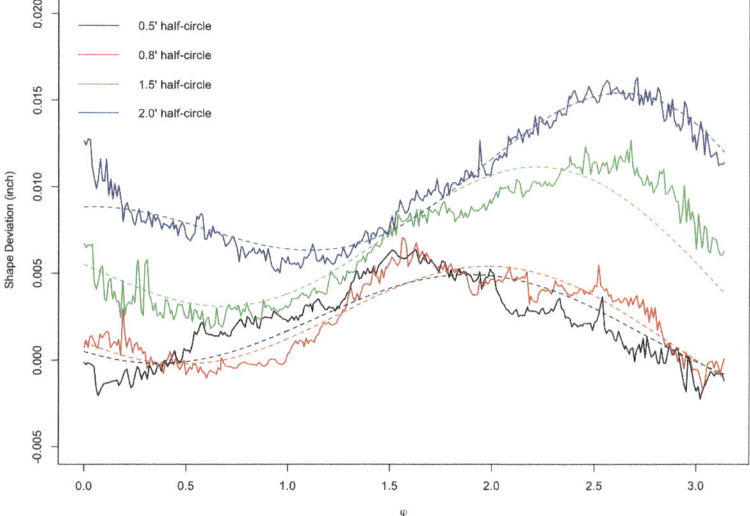

Fig. 5.8 Sequential model refinement with conjectures on $n(r_0)$ and $\psi(r_0)$

boundary condition one will be satisfied for $r_0 \to 0$. When $r_0 \to \infty$, the first three terms of $h(r_0, \varphi)$ in Eq. (5.10) all tend to 0. $\alpha(r_0)/r_0 \to constant$ when $r_0 \to \infty$. The expected value of $y(r_0, \varphi)$ will tend to $\beta(r_0)$, which has to be bounded. To summarize, the constraints for $\alpha(r_0)$ and $\beta(r_0)$ are

$$r_0\alpha(r_0) \to 0, \ \beta(r_0) \to 0; \quad \text{if} \quad r_0 \to 0$$

$$\frac{\alpha(r_0)}{r_0} \to constant, \ \beta(r_0) \to constant; \quad \text{if} \quad r_0 \to \infty \tag{5.11}$$

To obtain proper forms of $\alpha(r_0)$ and $\beta(r_0)$ satisfying constraints listed in Eq. (5.11), we still adopt the physics-informed sequential model refinement strategy. Since the model estimates of $\alpha(r_0)$ in Table 5.3 show a strong and clear pattern

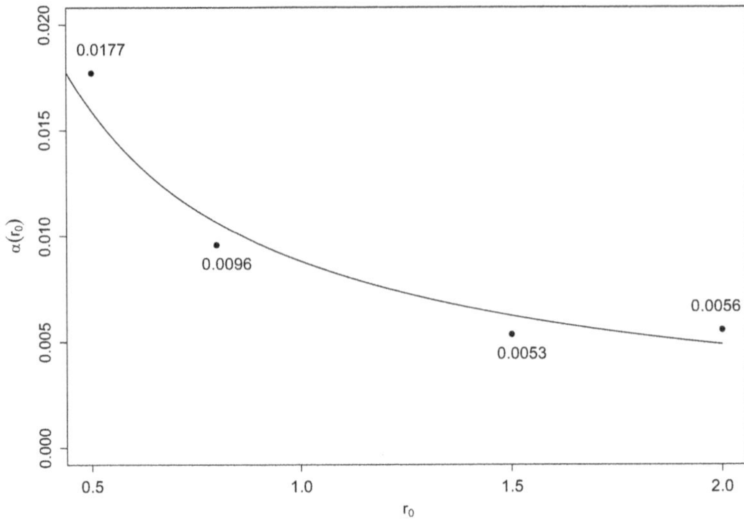

Fig. 5.9 Estimates of $\alpha(r_0)$

Table 5.4 Sequential model refinement with conjectures on $n(r_0)$, $\psi(r_0)$, and $\alpha(r_0)$

Parameters	Estimate	Standard error
r^*	1.11636406	0.00460214
a_1	−0.85252348	0.02892385
a_2	0.00880909	0.00012932
$\beta_0.5''$	0.00042821	0.00007754
$\beta_0.8''$	0.00053241	0.00006941
$\beta_1.5''$	0.00515338	0.00006512
$\beta_2.0''$	0.00870620	0.00006698
σ_ϵ	0.00110081	0.00002261

illustrated in Fig. 5.9, our conjecture starts with $\alpha(r_0)$ as

$$\alpha(r_0) = a_2 r_0^{a_1} \tag{5.12}$$

Then the refined model (5.6) takes the form of $y(r_0, \varphi) = a_2 r_0^{a_1} h(r_0, \varphi) + \beta(r_0) + \epsilon$ with $h(r_0, \varphi)$ defined in Eq. (5.10). The MLE estimation of the updated model is shown in Table 5.4 and Fig. 5.10. All the coefficients are significant at the 0.001 level. Notice that $a_1 = -0.853$. The boundary constraints for $\alpha(r_0)$ listed in Eq. (5.11) are satisfied as well.

The model estimates of $\beta(r_0)$ in Table 5.4 are illustrated in Fig. 5.11. By observing the subtle pattern and considering the boundary conditions of $\beta(r_0)$ in Eq. (5.11), we propose a sigmoid function with a logistic function base $S(x) = 1/(1 + e^{-x})$. In addition, we introduce a shift, an intercept, and a scaling term, i.e., $b_3 S(x + b_1) + b_2$. By satisfying $\beta(0) = 0$ and grouping unknown coefficients, the final form of $\beta(r_0)$ is

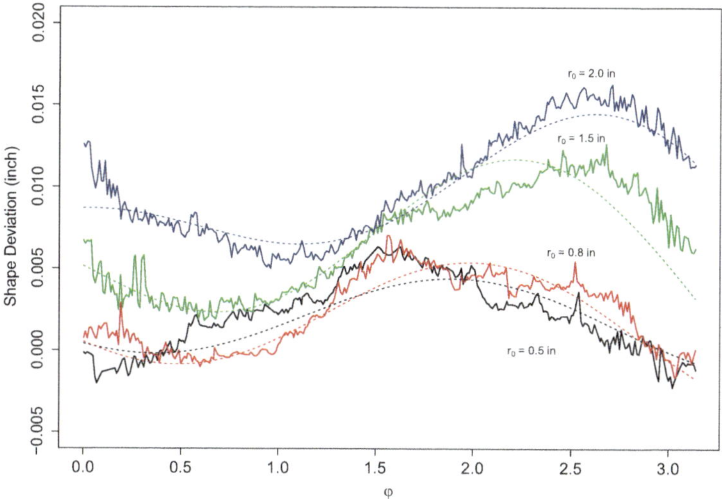

Fig. 5.10 Sequential model refinement with conjectures on $n(r_0)$, $\psi(r_0)$, and $\alpha(r_0)$

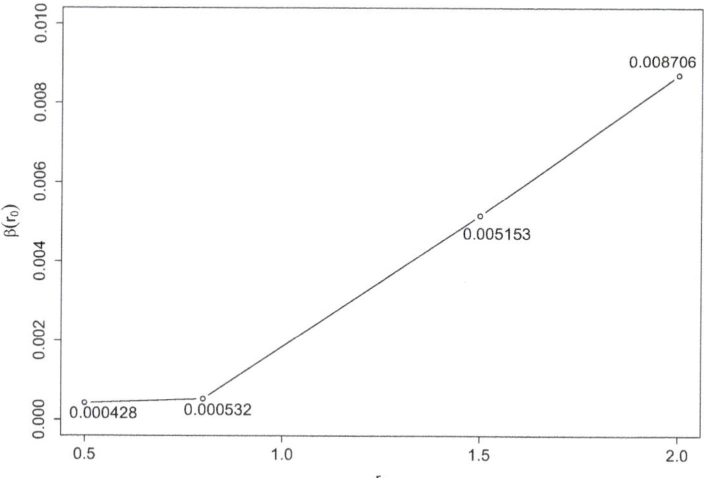

Fig. 5.11 Estimates of $\beta(r_0)$

$$\beta(r_0) = \frac{c_2}{c_1 + e^{-r_0}} - \frac{c_2}{c_1 + 1} \tag{5.13}$$

By substituting Eq. (5.13) into Eq. (5.6), the finalized SDG model for half disks built vertically is

$$y(r_0, \varphi) = a_2 r_0^{a_1} h(r_0, \varphi) + \frac{c_2}{c_1 + e^{-r_0}} - \frac{c_2}{c_1 + 1} + \epsilon \tag{5.14}$$

Table 5.5 Sequential model refinement with conjectures on $n(r_0)$, $\psi(r_0)$, $\alpha(r_0)$, and $\beta(r_0)$

Parameters	Estimate	Standard error
r^*	1.12310491	0.00528744
a_1	−0.83753308	0.02648314
a_2	0.00852358	0.00012209
c_1	−0.01077476	0.00385099
c_2	0.00130300	0.00004232
σ_ϵ	0.00118232	0.00002412

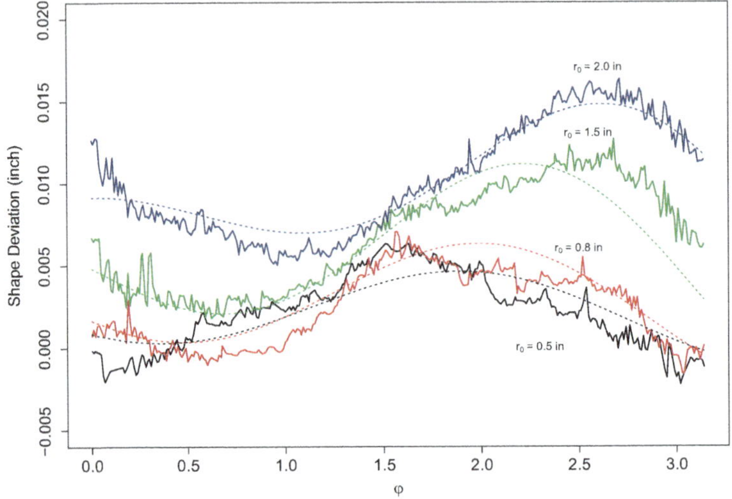

Fig. 5.12 Sequential model refinement with conjectures on $n(r_0)$, $\psi(r_0)$, $\alpha(r_0)$, and $\beta(r_0)$

The final model estimation through MLE is shown in Table 5.5 and Fig. 5.12. All the coefficients are significant at the 0.001 level.

5.1.2.4 Process Insights Derived from SDG Model for Vertically Built Disks

The proposed convolution formulation (5.1) not only provides a framework to learn models of predicting shape deviations, but it also facilitates the understanding of process insights. Using the model in Eq. (5.6) for vertically built half disks as an example, the essential deviation patterns are defined by the convolution integral $h(r_0, \varphi)$ in Eq. (5.7). We can further define and analyze the three terms in $h(r_0, \varphi)$:

$$h_1(r_0, \varphi) = -\frac{r_0 r^*}{r_0^2 - 4r^{*2}} \cos\left[\frac{r_0}{r^*}\varphi - \frac{r_0^2}{r^{*2}}\right] \tag{5.15}$$

$$h_2(r_0, \varphi) = \frac{r^*}{2r_0 + 4r^*} \cos\left[2\varphi + \frac{r_0^2}{r^{*2}}\right] \tag{5.16}$$

$$h_3(r_0, \varphi) = \frac{r^*}{2r_0 - 4r^*} \cos\left[2\varphi - \frac{r_0^2}{r^{*2}}\right] \tag{5.17}$$

Given $r_0 = 2.0$, $h_1(r_0, \varphi)$, $h_2(r_0, \varphi)$, and $h_3(r_0, \varphi)$ are superimposed in Fig. 5.13, clearly $h_2(r_0, \varphi)$ has least influence among the three terms because of its smaller weights assigned to base function $f(r_0, \varphi)$ in the convolution.

Figure 5.14 further shows $h_1(r_0, \varphi)$, $h_2(r_0, \varphi)$, and $h_3(r_0, \varphi)$ by (i) varying r_0 and φ (left panel) and (ii) varying r_0 and φ with $\psi(r_0) = 0$ (right panel). By observing Figs. 5.14 and 5.13, we postulate the following interpretation which can guide further investigations:

- $h_1(r_0, \varphi)$ is deemed to be the main descriptor of inter-layer interaction effects. First, $h_1(r_0, \varphi)$ change sharply with radius r_0, e.g., much smaller weights for disks with r_0=0.5″ and 0.8″ and larger weights for disk with $r_0 = 2.0″$ (in blue). Second, the period of $h_1(r_0, \varphi)$ varies with r_0 as well, slow for disks with less number of layers (similar weights for adjacent layers) and faster for disks with more layers.
- $h_2(r_0, \varphi)$ is likely due to the gravity effect. First, it has least influence as indicated in Fig. 5.13. Second, its period does not change with r_0 or the number of layers. Further experimentation can be conducted to verify this hypothesis.
- $h_3(r_0, \varphi)$ is deemed to be the effect of shape deviation for the same shape built horizontally. In another word, $h_3(r_0, \varphi)$ is mainly determined by $f(\mathbf{x})$. First, it has the same period with $f(\mathbf{x})$ and base function $\cos(2\varphi)$. Second, as expected, the larger the shape deviations in the horizontal plane, the bigger the influence on the same shape built vertically.

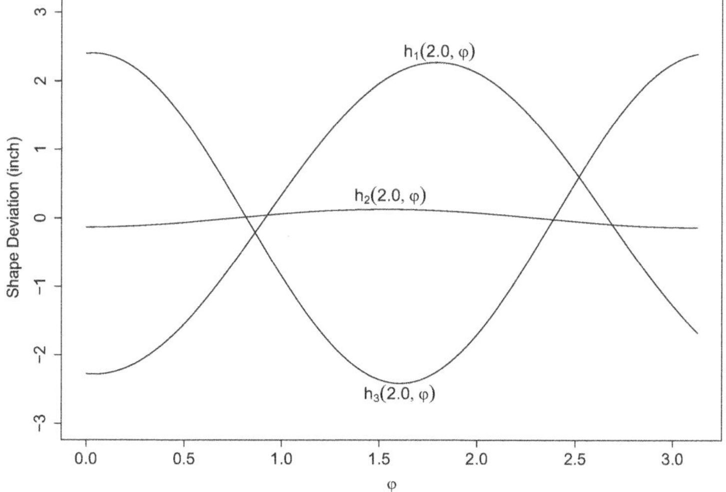

Fig. 5.13 $h_1(2.0, \varphi)$, $h_2(2.0, \varphi)$, and $h_3(2.0, \varphi)$

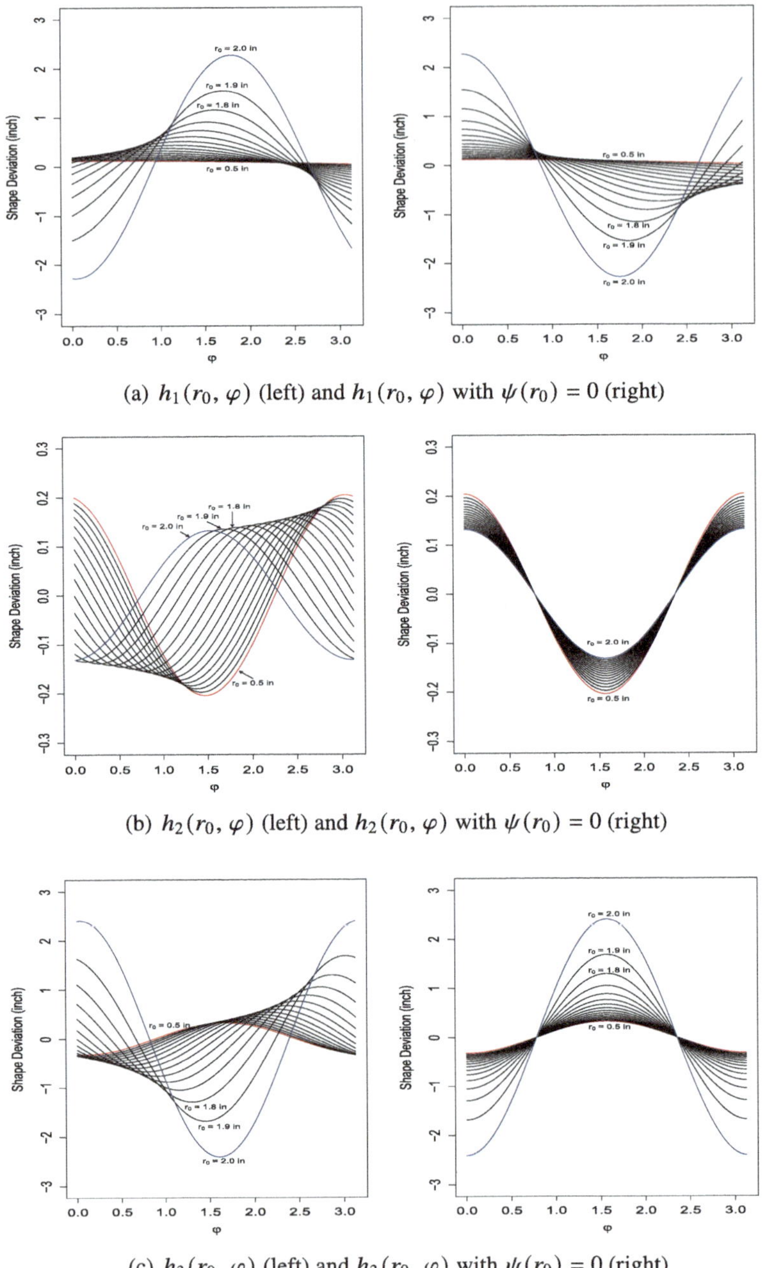

(a) $h_1(r_0, \varphi)$ (left) and $h_1(r_0, \varphi)$ with $\psi(r_0) = 0$ (right)

(b) $h_2(r_0, \varphi)$ (left) and $h_2(r_0, \varphi)$ with $\psi(r_0) = 0$ (right)

(c) $h_3(r_0, \varphi)$ (left) and $h_3(r_0, \varphi)$ with $\psi(r_0) = 0$ (right)

Fig. 5.14 $h_1(r_0, \varphi)$, $h_2(r_0, \varphi)$, and $h_3(r_0, \varphi)$

5.1.2.5   Model Improvement Through Gaussian Process Regression

Though the error term in model (5.6) is assumed to be noise $\epsilon \sim N(0, \sigma^2)$, the shape deviations tend to be spatially correlated. Figure 5.15a confirms the spatial correlations by showing the model residuals of four vertically printed half disks. One way to improve the model fitting is to adopt Gaussian process regression (GPR) [64] to model the residual shape deviations:

$$\epsilon = \mathcal{GP}(0, k(\cdot, \cdot)) + \epsilon' \tag{5.18}$$

where $\mathcal{GP}(0, k(\cdot, \cdot))$ is a Gaussian process with kernel function $k(\cdot, \cdot)$ and $\epsilon' \sim N(0, \sigma^2)$. The kernel function adopted in this study is the squared exponential kernel:

$$k(x, x') = \tau_f^2 \exp\left(-\frac{1}{2\tau_l^2}||x - x'||^2\right) \tag{5.19}$$

with x being angle φ in the modeling of vertically printed half disks.

The GPR of model residuals using MATLAB function `fitrgp` produces the result in Fig. 5.15a (the thick line). The length scale $\tau_l = 0.2335$ suggests that the predicted values of the Gaussian process do not change slowly (i.e., not smooth) in the interval $[0, \pi]$. The scaling factor or standard deviation $\tau_f = 0.0006$ is small, which indicates that the predicted values of the Gaussian process will be close to its mean (zero). The residual standard deviation $\sigma_{\epsilon'} = 0.001$, as expected, is smaller than the one in model (5.14). The prediction from the updated SDG model, i.e., $\alpha(r_0)h(r_0, \varphi) + \beta(r_0) + \mathcal{GP}(0, k(\cdot, \cdot))$, is given in Fig. 5.15b. Comparing to Fig. 5.12, the prediction of local shape deviation by the updated SDG model is improved.

5.1.3   Convolution Modeling and Learning for Domes

5.1.3.1   SDG Model for 3D Shapes

The SDG model for general 3D shapes will take the following form:

$$\begin{aligned}
y\big(r_0(\theta, \varphi), \theta, \varphi,\big) &= (f * g)\big(r_0(\theta, \varphi), \theta, \varphi\big) + \epsilon \\
&= \int_{\tau_2}\int_{\tau_1} f(\tau_1, \tau_2)g(\theta - \tau_1, \varphi - \tau_2)\, d\tau_1\, d\tau_2 \\
&\quad + \mathcal{GP}(0, k(\cdot, \cdot)) + \epsilon'
\end{aligned} \tag{5.20}$$

where $\mathcal{GP}(0, k(\cdot, \cdot))$ is a 2D Gaussian process and $\epsilon' \sim N(0, \sigma_{\epsilon'}^2)$.

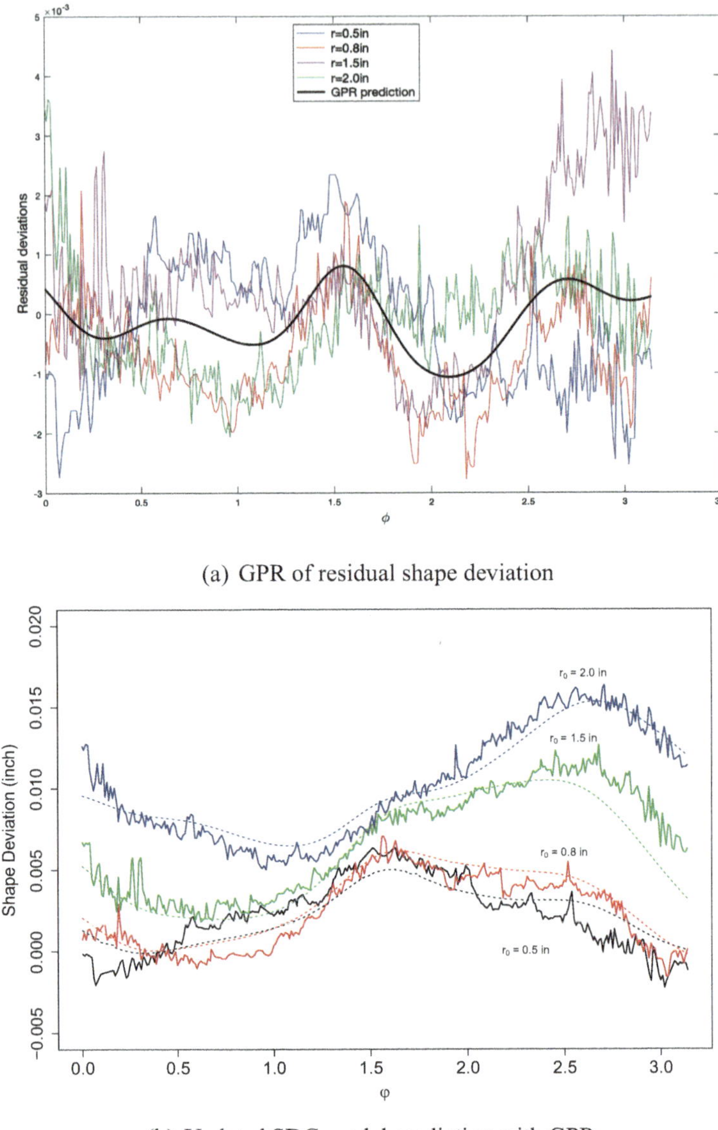

(a) GPR of residual shape deviation

(b) Updated SDG model prediction with GPR

Fig. 5.15 SDG model prediction with Gaussian process regression of residuals

Since establishing the convolution model for freeform 3D shapes deserves dedicated efforts, this study illustrates the proposed framework using dome shapes with varying sizes (Fig. 5.16) and focuses on the extraction of stack-up function to understand inter-layer interactions. Enlightened by the SDG model (5.6) for half disks and Eq. (5.6), we propose the SDG model for dome shape as

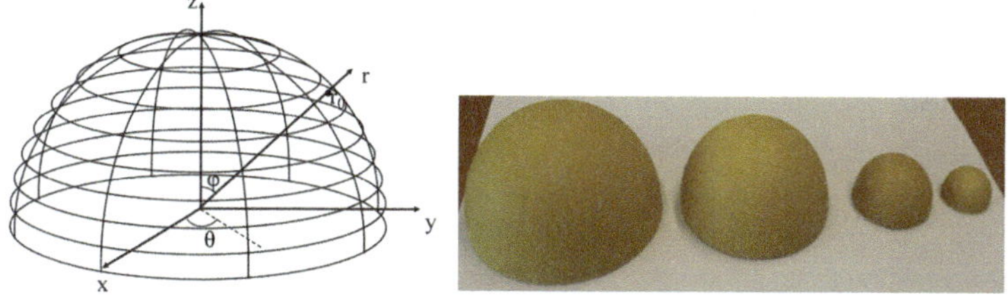

Fig. 5.16 Dome shape and four domes printed in a SLA process

$$
\begin{aligned}
y(r_0, \theta, \varphi) &= \alpha(r_0)h(r_0, \theta, \varphi) + \beta(r_0) + \epsilon \\
&= \alpha(r_0) \int_{\tau_2} \int_{\tau_1} f(\tau_1, \tau_2) g(\theta - \tau_1, \varphi - \tau_2)\, d\tau_1\, d\tau_2 \\
&\quad + \beta(r_0) + \mathcal{GP}(0, k(\cdot, \cdot)) + \epsilon'
\end{aligned}
\tag{5.21}
$$

where $f(\theta, \varphi)$ and $g(\theta, \varphi)$ are both normalized basis functions.

Note that the SDG model (5.21) for dome shapes contains the SDG model (5.6) for 2D half disks. Model (5.6) can be viewed as a special case of model (5.21) with $\theta = \pi/2$ and $\theta = 3\pi/2$. The stack-up function $g(\theta, \varphi)$ can be viewed as $g(\theta, \varphi) = \delta_{\theta=\pi/2} g(\varphi)$, and when it convolves with $f(\theta, \varphi)$ in the convolution integral $h(r_0, \theta, \varphi)$, the 3D model (5.21) degenerates into a 2D model (5.6).

With the result in Eq. (5.2), $f(\theta, \varphi)$ in Eq. (5.21) represents the shape deviation of horizontal disks at φ, without inter-layer interaction effects. The effect of φ shows on the layer radius, i.e., $r_0 \sin(\varphi)$. Combining with the in-plane deviation basis function $\cos(2\theta)$ used previously, $f(\theta, \varphi)$ in Eq. (5.21) is proposed to be

$$
f(\theta, \varphi) = \cos(2\theta) \sin(\varphi)
\tag{5.22}
$$

where $\sin(\varphi)$ reflects the impact of radius change along the build direction.

5.1.3.2   Stack-Up Function Identification Through Deconvolution and Model Selection via LASSO

The deconvolution problem is to identify $g(\theta, \varphi)$ given $f(\theta, \varphi)$ and measurement data $y(r_0, \theta, \varphi)$. Following the rationale for 2D stack-up function defined in Eq. (5.4), $g(r_0, \theta, \varphi)$ for dome shape can be expressed as a combination of 2D Fourier bases, i.e.,

$$
g(\theta, \varphi) = \sum_{n=0}^{\infty} \sum_{m=0}^{\infty} c_{n,m} \cos(n\theta + \psi_n) \cos(m\varphi + \omega_m)
$$

where $n(r_0)$ and $m(r_0)$ determine the periods along θ and φ, respectively. ψ_n) and ω_m) are phase variables.

Furthermore, $g(\theta, \varphi)$ likely has a sparse representation with features selected from a large 2D Fourier bases. Among different regularization methods, LASSO [79] is adopted in this study because it not only reduces model variances but also makes the model more interpretable with sparse solutions [96]. Different regularization terms can be added depending on applications. For example, to track anomalies of network traffic volumes, Mardani et al. [53] decompose the traffic flow data into low-rank normal flow, sporadic abnormal flow, and noise by adopting a nuclear norm, L1-norm, and Frobenius norm, respectively.

Let g_j denote the jth Fourier base with coefficient c_j for $f * g_j$. For each dome, we conduct feature screening through the LASSO formulation:

$$\min_{C} \frac{1}{N} \sum_{i=1}^{N} (y_i - \sum_{j} c_j f * g_j(\theta_i, \varphi_i))^2 + \gamma \|C\|_1 \tag{5.23}$$

where N represents the total number of sampled points on each dome.

Four domes with radii of $0.5''$, $0.8''$, $1.5''$, and $1.8''$ are built in a SLA process. The measurement data of dome shape deviation is presented in the SCS (Fig. 5.17). The $0.5''$, $0.8''$ and $1.8''$ domes are regarded as the training set, and the $1.5''$ dome is left for model validation. Similar to the phenomenon shown in Fig. 5.5, the deviation patterns of four domes also vary with their sizes due to layer buildup and interactions. This poses a challenging issue for model learning because models can make poor prediction of shape accuracy even with a simple change of shape sizes.

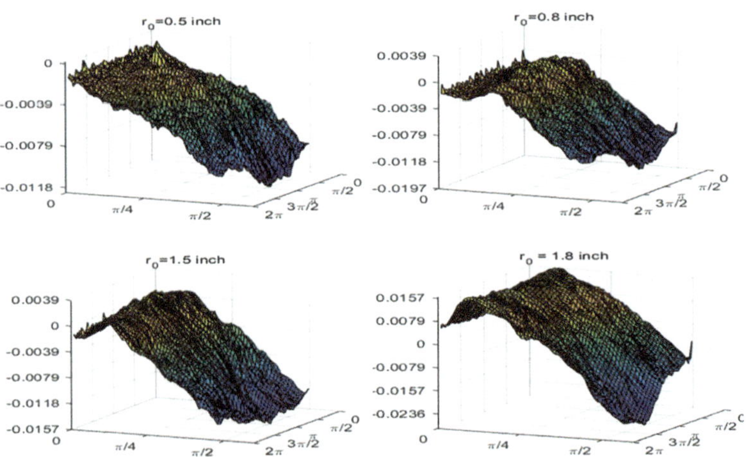

Fig. 5.17 Shape deviation measurement of four domes presented in the SCS

Significant terms shared across different domes are selected, and the resulting set of features (Fourier bases) includes $\cos(n_1\theta + \psi_1)$, $\cos(n_2\varphi + \psi_2)$, and $\cos(n_1\theta + \psi_1)\cos(n_2\varphi + \psi_2)$. A further MLE estimation of stack-up function g for individual domes chooses the following form for g:

$$g(\theta, \varphi) = \cos(n_1\varphi)[1 + \cos(n_2\theta + \psi)] \tag{5.24}$$

where n_1 and n_2 determine the periods along φ and θ, respectively, and ψ is a phase variable.

It is interesting to notice that the stack-up function is determined not only by the height of layers (defined by $\cos(n_1\varphi)$) but also by its interaction with shape deviation within that layer (defined by $\cos(n_2\theta + \psi)$).

With the stack-up function g identified in Eq. (5.24), the convolution integral $h(r_0, \theta, \varphi)$ in Eq. (5.21) becomes

$$\begin{aligned}
h(\theta, \varphi) = \frac{\cos(\varphi) - \cos(n_1\varphi)}{2(n_1^2 - 1)} \cdot \Bigg[&\frac{1}{n_2 + 2}\sin(2\theta - \psi) \\
&- \frac{1}{n_2 - 2}\sin(2\theta + \psi) \\
&+ \frac{2n_2}{n_2^2 - 4}\sin(n_2\theta + \psi) + \sin(2\theta) \Bigg]
\end{aligned} \tag{5.25}$$

5.1.3.3 SDG Model Estimation

To obtain one unified SDG model for training shapes, we follow the same physics-informed sequential model refinement process as we did for the vertically printed half disks (omitted). The data suggests that

- $\alpha(r_0) = a_1 + a_2 r_0$
- $\beta(r_0) = b_1 + b_2 r_0$
- $n_1(r_0) = c_1 + c_2 r_0$
- n_2 and ψ: unknown constants

With the training set (three domes), Table 5.6 gives the initial MLE estimation of the SDG model without considering the Gaussian process, i.e., $(a_1 + a_2 r_0)\, h(r_0, \theta, \varphi) + (b_1 + b_2 r_0) + \epsilon$. The measured shape deviations (response, black point cloud) and the SDG model predictions (blue and red point cloud) are superimposed together in Fig. 5.18. Due to the difficulty of pattern visualization in the 3D space, we plot the shape deviation by point index, which is sorted by φ and θ. So the first quarter from the left corresponds to the 0.5″ dome, the second quarter represents the shape deviation surface of the 0.8″ dome, and so on. Only three domes marked in blue are used to train the SDG model, and the red point cloud (the third quarter) is the validation set. Apart from the predicted triangular shape on the upper right corner of each dome, the trends are well captured by the model.

Table 5.6 Initial MLE estimation without Gaussian process model term

Parameters	Estimate	Standard error
n_2	0.3319	0.00155
ψ	3.3897	0.00162
a_1	−0.1137	0.00237
a_2	−0.1683	0.00210
b_1	0.0082	0.00008
b_2	0.0062	0.00007
c_1	0.0551	0.05747
c_2	−0.0324	0.04382
σ	0.0046	0.00002

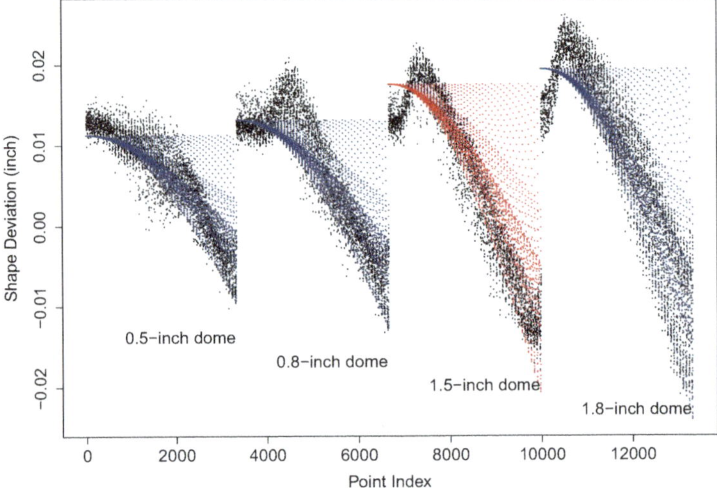

Fig. 5.18 Measured shape deviation (black dots) and the SDG model prediction (blue and red dots)

To improve the model prediction, the GPR is conducted to capture the spatial correlation in the model residuals, i.e., $\epsilon = \mathcal{GP}(0, k(\cdot, \cdot)) + \epsilon'$, which is more critical for 3D shapes. With the square exponential kernel function, the GPR of dome model residuals is conducted with optimized parameter estimates $\tau_f = 0.0097$, $\tau_l = 0.9886$, and $\sigma_{\epsilon'} = 0.0029$. The length scale $\tau_l = 0.9886$ suggests that the predicted values of the Gaussian process vary moderately in the area $[0, 2\pi] \times [0, \pi/2]$. The scaling factor $\tau_f = 0.0097$ is significantly larger than the residual standard deviation $\sigma_{\epsilon'} = 0.0029$, which indicates spatial correlation and variation is significant for 3D shapes.

Figure 5.19 shows the measured shape deviations and the updated SDG model predictions with GPR. The prediction covers most of the measurement point cloud. Figure 5.20 illustrates the predicted shape deviations versus the measurement data. In both figures, the validation outcomes of 1.5″ dome are as good as the training cases.

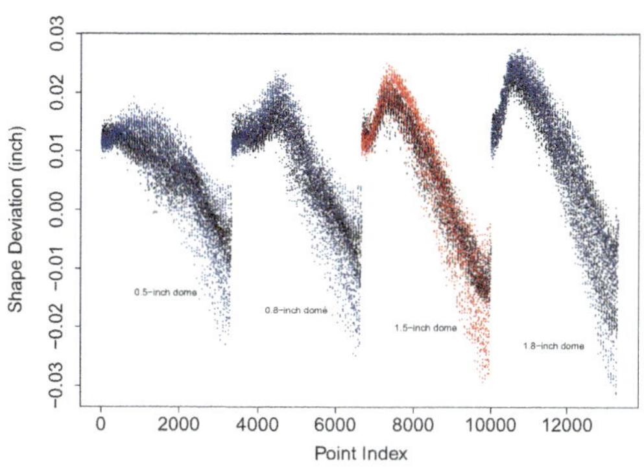

Fig. 5.19 Measured shape deviation (black dots) and the updated SDG model prediction with GPR (blue and red dots)

Fig. 5.20 Final model prediction of shape deviations against the measured shape deviations

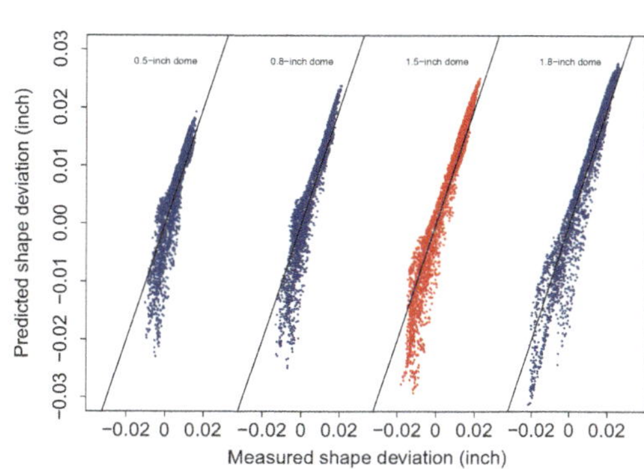

The root mean square errors (RMSE) of the training set and validation set are 0.0036361″ and 0.0036379″, respectively. And the mean absolute errors (MAE) for the two sets are 0.002636″ and 0.002637″ for the training and validation sets, respectively.

5.1.3.4   Process Insights Derived from SDG Model for Domes

We define and analyze the four terms in $h(r_0, \theta, \varphi)$ as:

$$h_0(r_0, \varphi) = \frac{\cos(\varphi) - \cos[n_1(r_0)\varphi]}{2[n_1(r_0)^2 - 1]} \tag{5.26}$$

$$h_1(\theta) = \frac{1}{n_2 + 2} \sin(2\theta - \psi) - \frac{1}{n_2 - 2} \sin(2\theta + \psi) \tag{5.27}$$

$$h_2(\theta) = \frac{2n_2}{n_2^2 - 4} \sin(n_2\theta + \psi) \tag{5.28}$$

$$h_3(\theta) = \sin(2\theta) \tag{5.29}$$

Fig. 5.21 $h_1(\theta)$, $h_2(\theta)$, and $h_3(\theta)$

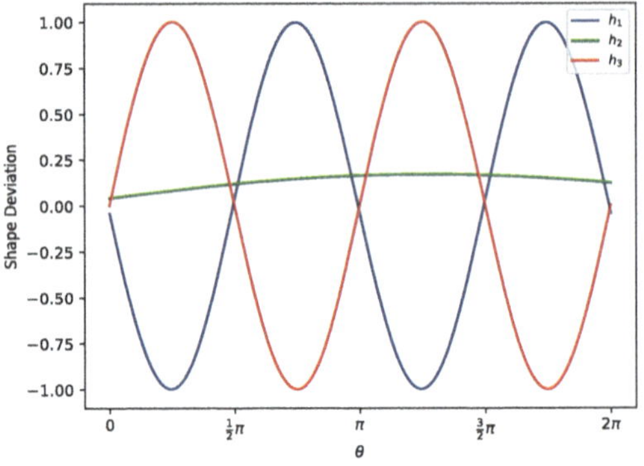

Fig. 5.22
$h_0(r_0, \varphi)$, $r_0 \in [0.5, 1.8]$

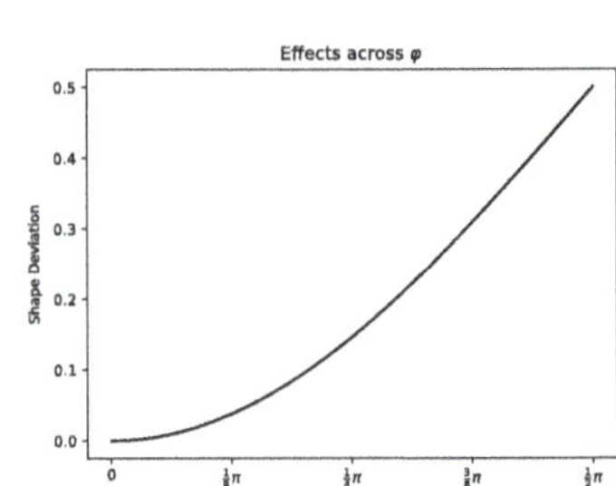

Then

$$h(r_0, \theta, \varphi) = h_0(r_0, \varphi)[h_1(\theta) + h_2(\theta) + h_3(\theta)]$$

Notice that the term $h_0(r_0, \varphi)$ is decided by r_0 and φ, while the other three terms only depend on θ. To compare the magnitude of each term, we plot $h_{1,2,3}(\theta)$ in Fig. 5.21.

Three functions $h_{1,2,3}(\theta)$ have the same interpretation as those in the half disk case, that is, they represent inter-layer interaction effect, the gravity effect, and the effect of input in-plane shape deviations, respectively. $h_3(\theta)$ is not due to interaction effect because it is the result of convolving in-plane deviation pattern $\cos(2\theta)$ in $f(\theta, \varphi)$ with the constant one in $g(r_0, \theta, \varphi) = \cos(n_1(r_0)\varphi)[1 + \cos(n_2\theta + \psi)]$. Note that we separate θ and φ. One difference from the half disk case is that all three effects will be influenced by the height of the layer defined by φ. We plot $h_0(r_0, \varphi)$ in Fig. 5.22. One interesting fact is that the function values are almost the same when we variate r_0 from $0.5''$ to $1.8''$ in $h_0(r_0, \varphi)$.

Combine the term h_0 with $h_{1,2,3}$ respectively, we visualize the inter-layer interaction effect, gravity effect, and the input in-plane shape effect in Fig. 5.23. The interpretation is consistent with the case for half disks.

As can be seen, the shape deviation generator (SDG) provides a data-analytical framework to learn geometric measurement data of AM-built products. Under a convolution framework, SDG enables a consistent description of 3D shape formation

Fig. 5.23 $h_0h_1(r_0, \theta, \varphi)$, $h_0h_2(r_0, \theta_0, \varphi)$, and $h_0h_3(r_0, \theta_0, \varphi)$

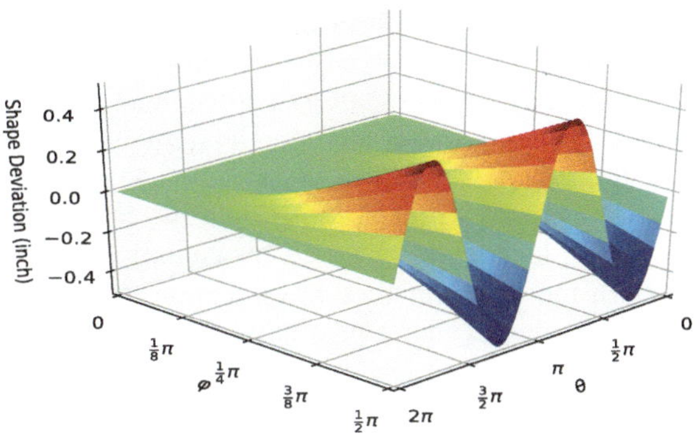

(a) $h_0h_2(r_0, \theta, \varphi)$

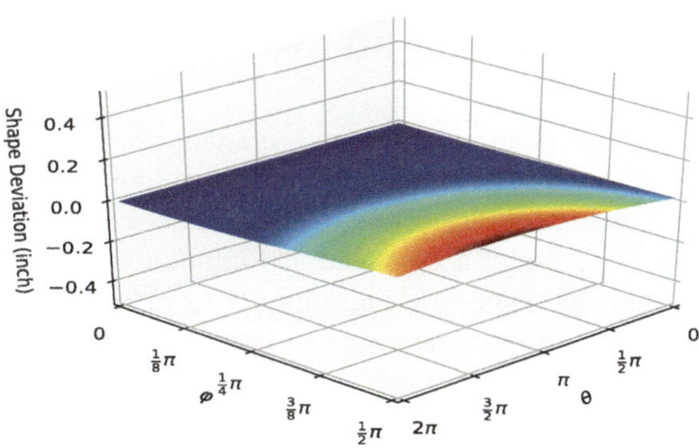

(b) $h_0h_2(r_0, \theta, \varphi)$

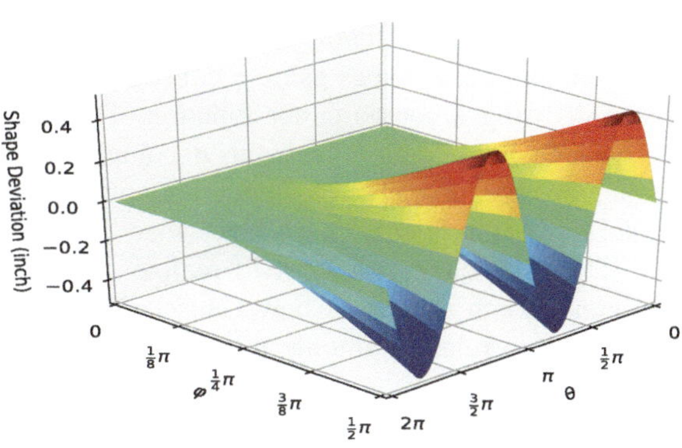

(c) $h_0h_3(r_0, \theta, \varphi)$

in layer-by-layer fabrication processes, from horizontally built disks, vertically built disks, to fully 3D domes. This is achieved through proper representation of shape deviation data, modeling of individual layer input, and stack-up function derived to capture inter-layer interactions. The domain-informed sequential model estimation and refinement strategy leads to efficient learning and better understanding of process insights. Effects due to inter-layer interactions, gravity, and deviation of individual layers are separated for guiding further experimentation and validation. Though the initial methodology demonstration applies to simple shapes such as disks and domes, the convolution framework allows input functions (f) to take complicated geometries for each layer and to convolute with the stack-up function (g) to form complicated 3D shapes.

5.2 Shape Quality Modeling for Both Smooth and Non-smooth 3D Geometries

Section 5.1 has established a domain-informed convolution framework to learn shape deviation from a small set of training products built with the same materials and process. It incorporates the characteristics of the layer-wise shape forming process through a convolution formulation and the size factor for a category of smooth 3D shapes such as domes or cylinders. This section extends this fabrication-aware learning framework to a larger class of products including both smooth and non-smooth surfaces (polyhedral shapes).

The key idea of the methodology extension is to learn heterogeneous deviation surface data by establishing the association between the deviation profiles of smooth base shapes and those of non-smooth polyhedral shapes. The association, which is characterized by a novel 3D cookie-cutter function, views polyhedral shapes as being carved out from smooth base shapes. In essence, the AM process of building non-smooth shapes is mathematically decomposed into two steps: additively fabricate smooth base shapes using a convolution framework, and then subtract extra materials using a cookie-cutter function. The proposed joint learning framework of shape deviation data reflects this decomposition by adopting a sequential model estimation procedure. The model learning procedure first establishes the convolution model to capture the effects of layer-wise fabrication and sizes and then estimates the 3D cookie-cutter function to realize geometric differences between smooth and non-smooth shapes. A new Gaussian process model is proposed to consider the spatial correlation among neighboring regions within a 3D shape and across different shapes. The case study demonstrates the feasibility and prospects of prescriptive learning of complex 3D shape deviations in AM and extension to broader engineering surface data.

As a motivating example, four domes and three thin walls with half-cylindrical shapes were vertically printed through an AM process using the same material (Fig. 5.24). After collecting the point cloud data on the product surfaces and compar-

Fig. 5.24 (a) Domes with 0.5, 0.8, 1.5, and 1.8 inches radii. (b) Thin walls with 0.8, 1.5, and 2.0 inches radii

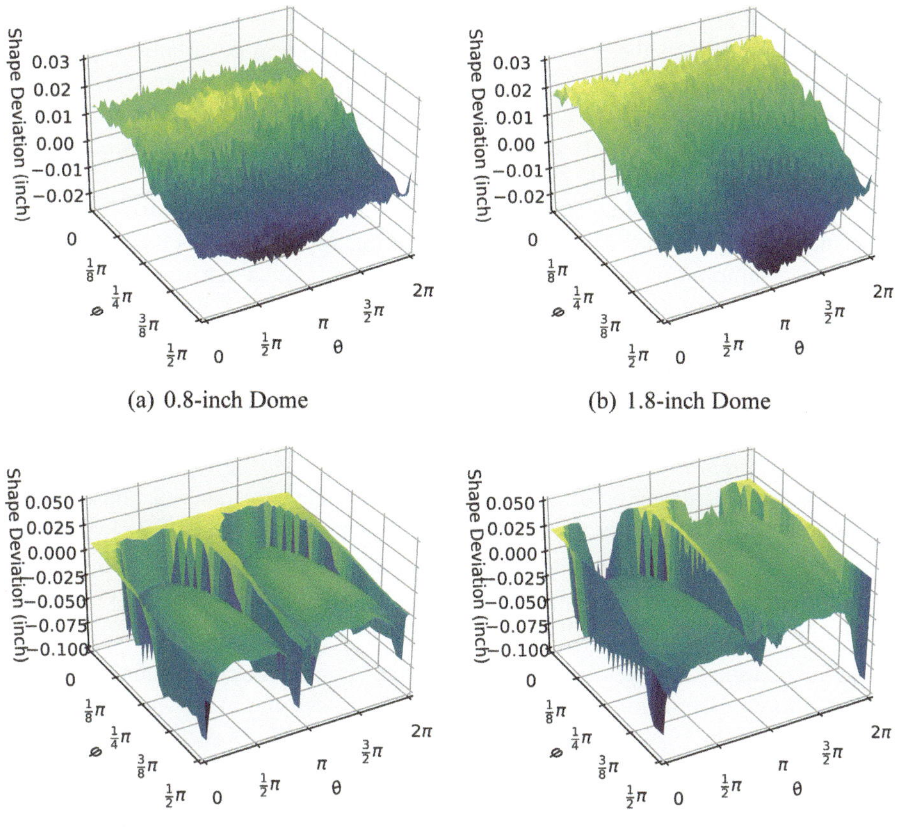

(a) 0.8-inch Dome

(b) 1.8-inch Dome

(c) 0.8-inch Thin Wall

(d) 2.0-inch Thin Wall

Fig. 5.25 Shape deviation measurements of two dome and two thin walls presented in the SCS (darker-purple and lighter-yellow color for negative and positive deviation, respectively)

ing with the designs, we present their shape deviations as functional surfaces [35] in the spherical coordinate system (SCS) as shown in Fig. 5.25. The deviation surfaces of the thin walls are quite different from those of the domes. Thin walls present sharp increments in the deviation around the corners/edges, while the deviation surfaces of domes are smooth everywhere. Furthermore, the shape deviation profiles and their patterns can vary with the size of the same target shape.

To learn and predict surface deviations of 3D shapes, we face several fundamental challenges:

- *Heterogeneity.* Deviation patterns are affected by multiple factors including geometries, sizes, materials, and process parameters. This often produces heterogeneous data even under the same process settings of an AM machine. As observed in Fig. 5.25, thin walls show distinct deviation patterns from domes. Different sizes/volumes would affect not only the magnitude of the deviation surface but also the locations of sharp transitions for the thin walls. Taking the deviation profile as the response, there are three types of covariates: (1) location covariates defined with respect to the printing and registration center [18, 34]; (2) process parameters such as the type of AM process, printing materials, printing temperature, etc. [24, 41, 67]; and (3) size and shape information specified by the designs of AM products [22, 34, 69].
- *Limited samples.* A key advantage of AM is the easy customization of the printed parts. This one-of-a-kind manufacturing approach limits the applicability of large-sample machine learning methods and classical statistical methods for modeling shape deviations. With a limited number of training samples and various covariates, it is challenging to establish a prescriptive model, estimate its parameters, and validate assumptions on model specifications and parameters, especially for non-parametric models such as random forest [18] and Dirichlet process [63].
- *Spatial correlation within and among shapes.* In AM, the layer-by-layer fabrication process exacerbates the quality issues related to material phase change and heat penetration observed in traditional manufacturing [12, 27, 48], leading to stronger spatial correlation among neighboring regions of a product. Joint learning of different 3D shapes presents an additional challenge of defining spatial correlation among different geometries.

In the literature, physics-based and data-driven approaches have been extensively reported to predict printing quality. A common strategy in physics-based modeling is the use of finite element analysis to simulate the printing process and reveal the relationship among process, geometric structure, and material properties [47, 76]. These models can achieve voxel-level accuracy for describing 3D products during the printing process but at a high computational cost. Extensive expert knowledge is required to develop such models, and they are typically restricted to a specific material, geometry, and AM process. In the data-driven AM research, regression models [81, 87], design of experiments techniques [73, 97], and machine learning strategies [58, 86] have been applied. However, most data-driven AM models suffer from lack of interpretability and overfitting issues.

To learn heterogeneous data with limited sample sizes, efforts have been devoted to transfer learning across materials, manufacturing processes, and part designs. Sabbaghi and Huang [67] proposed the effect equivalence framework to calibrate the effects of lurking variables in different AM processes through a base factor, which was adopted by Francis et al. [24] to accomplish the model transfer from Ti-6Al-4V to 316L stainless steel. Chen et al. [11] achieved knowledge transfer across different shapes by first decomposing the geometric error into shape-independent and shape-specific components and then fixing the global shape-independent parameters and shape features. Ferreira et al. [22] employed a Bayesian extreme learning machine methodology to automatically predict the shape deviation profiles of 2D freeform shapes under different printing processes. However, the applicability of these transfer learning methods has not been demonstrated for 3D geometries.

Engineering-informed machine learning approaches have been proposed to model and predict 2D shape deviation using a limited number of training samples. Huang et al. [36] developed a prescriptive statistical modeling approach to predict and compensate the 2D shape deviation of circular shapes considering the size effect and lurking variables such as overexposure. It serves as a base model to predict the deviation of circular shapes, regular polygons, and ultimately freeform shapes. A key challenge to model polygonal shapes is that the deviation pattern changes dramatically on the sharp corners due to high residual stresses and thermal gradients [59]. To link the circular shape deviations to those of regular polygons, Huang et al. [34] viewed polygons as being mathematically carved out from their circumcircles (base shapes) through cookie-cutter functions. To predict the shape deviation profiles of freeform shapes, Luan and Huang [51] approximated the freeform design with a series of piecewise circular sectors or polygonal segments.

The modeling and control challenges faced in printing 2D shapes are exacerbated for 3D geometries, since deviation patterns can change from layer to layer due to complex inter-layer interactions, residual stresses, and heat dissipation profiles [38, 40]. To incorporate the layer-by-layer fabrication mechanism into modeling and learning, Huang et al. [35] proposed a convolution learning framework to describe the 3D deviation patterns as the result of the 2D shape deviation of each layer convolved with a layer interaction function that captures inter-layer interactions. Although this framework was validated with spherical shapes, it still requires fundamental work to predict the complex deviation patterns of both smooth and non-smooth shapes in a consistent unified modeling framework. Due to the smoothing effect of convolution operations, this approach alone is inadequate to capture the sharp transitions in the shape deviation profiles of non-smooth polyhedral geometries such as in thin-wall shapes. The main reason to study these particular shapes is that 3D freeform shapes can be approximated as a combination of smooth and non-smooth patches [78]. Understanding these two basic categories of 3D shapes enables 3D freeform shape deviation prediction similar to the significant advances achieved for the 2D case [34, 36, 51].

To tackle these challenges, we mathematically decompose the fabrication of non-smooth 3D shapes into two steps: (1) additively building the smooth base shapes and (2) subtractively craving out the non-smooth polyhedral shapes. An additive model is proposed to connect smooth and non-smooth 3D geometries through a new 3D cookie-cutter function with an engineering-informed sequentially model estimating strategy. Spatial correlation among regions within a product and across different product geometries is modeled by a Gaussian process (GP) with a novel distance metric integrating both geodesic and geometric information.

5.2.1 Mathematical Decomposition of AM Through an Additive Model for 3D Shape Deviation Modeling

This section proposes a mathematical decomposition of AM as a general additive model to learn and predict shape deviation surfaces of 3D geometries. Adopting smooth shapes as the model baseline, we propose a new class of cookie-cutter functions to link the shape deviation of 3D smooth and non-smooth convex shapes. Lastly, a GP is employed to capture the spatial correlation with a novel distance metric.

To learn insights from the heterogeneous data (Fig. 5.25), domain knowledge must be incorporated. As shown in Fig. 5.26, the process of building a 3D shape can be mathematically decomposed into two steps: (1) additive step, which fabricates a smooth base shape that bounds the target geometry, and (2) subtractive step, which removes extra materials to form non-smooth edges and corners. Thus, the smooth and non-smooth 3D shapes are connected through an association or cookie-cutter function that captures the shape deviation caused by the subtractive step.

Here, the shape smoothness is defined with respect to the smoothness of curves on the surface of the product design. If any curve along the shape surface is smooth, i.e., there exists a common tangent direction at any point on the surface, then the shape is smooth; otherwise, it is non-smooth. For example, the curved surfaces of spherical and cylindrical shapes are smooth, while the thin wall and cuboid

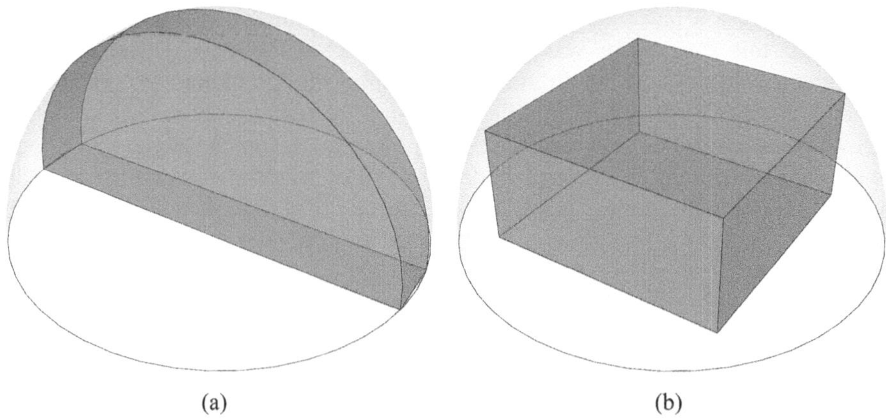

(a)   (b)

Fig. 5.26 Mathematical decomposition of AM to additively build the smooth base shape (outer dome shape) and subtractively carve out non-smooth shapes with sharp corners such as (**a**) thin wall shape and (**b**) cuboid shape

shapes in Fig. 5.26 are non-smooth since any curve crossing the edges would be non-smooth. In general, non-smooth shapes contain edges and corners, which are directly associated with sharp changes in their deviation surfaces.

The proposed mathematical decomposition procedure leads to an additive model. Due to the flexibility and interpretability of additive models, they are extensively used in diverse applications such as ecology [89], healthcare [6, 29], machine learning [14], and geomorphology [26]. For shape deviation at the ith location of part j, a general additive model can be defined as

$$y(\boldsymbol{x}_i, \boldsymbol{s}_j, \boldsymbol{p}_j) = \mu(\boldsymbol{x}_i, \boldsymbol{s}_j, \boldsymbol{p}_j) + \eta(\boldsymbol{x}_i, \boldsymbol{s}_j, \boldsymbol{p}_j) + \epsilon, \tag{5.30}$$

where y is the shape deviation, $\boldsymbol{x}$ are location covariates, $\boldsymbol{p}_j$ are the process parameters or part-independent covariates such as materials and process characteristics, $\boldsymbol{s}_j$ represents the geometric information or part-dependent covariates, μ models the mean pattern of the shape deviation, η is a zero-mean random field that captures the spatial correlation, and ϵ is the measurement error. In the rest of this work, we drop $\boldsymbol{p}_j$ for notation simplicity since only one specific printing process is investigated in this work.

For 2D shape deviations, Huang et al. [34] considered the regular polygons as being carved out from their circumscribed circles through the additive model

$$\mu(\boldsymbol{x}_i, \boldsymbol{s}_j) = f_1(\boldsymbol{x}_i) + f_2(\boldsymbol{x}_i, \boldsymbol{s}_j) + f_3(\boldsymbol{x}_i, \boldsymbol{s}_j), \tag{5.31}$$

where f_1 describes the shape deviation of disks, f_2 is a 2D cookie-cutter function that links the deviation profiles of circular and polygonal shapes, and f_3 is a high-order term for the remaining pattern.

To extend the model (5.31) to 3D cases, the mean pattern μ for 3D shape deviation follows the same decomposition formulation in Eq. (5.31). The main difference is that f_1 now describes the deviation profile of the smooth 3D base shape, f_2 is a 3D cookie-cutter function that connects smooth and non-smooth geometries, and f_3 is a high-order term. For example, we can use f_1 to model the shape deviation of the smooth base shape (e.g., domes), f_2 to capture the differences between the smooth base and non-smooth geometries such as thin walls (Fig. 5.26), and f_3 to illustrate the high-order pattern.

5.2.2 Convolution Framework as a Baseline for Smooth Shape Deviation Modeling

A convolution learning framework proposed in Sect. 5.1 [35] can be used to identify the baseline function for shape deviation modeling of convex smooth shapes as

$$y(\boldsymbol{x}) = (f * g)(\boldsymbol{x}) + \eta(\boldsymbol{x}) + \epsilon, \tag{5.32}$$

where $\mu = f * g$, f is the input function describing the 2D shape deviation in a horizontal layer, g is the interaction function that models complex layer-to-layer interactions, η is a zero-mean GP capturing spatial correlations in the deviation profile, and ϵ is the measurement error following a zero-mean normal distribution. The variable x is the spatial location of the points in SCS, i.e., $x = (r, \theta, \varphi)$, where θ is the polar angle and φ is the azimuth angle. The printed shape of a product is treated as the functional response $r(\theta, \varphi)$, and the nominal shape is denoted as $r_0(\theta, \varphi)$; then shape deviation is defined as $y(x) = r(\theta, \varphi) - r_0(\theta, \varphi)$, where $\theta \in [0, 2\pi)$ and $\varphi \in [0, \pi/2]$.

To specify μ, the first step is to identify the input function f. As the most common smooth geometries, spherical shapes (as shown in Fig. 5.24a) are chosen as the base geometries, where each horizontal layer is a circular disk. Shape deviation for a disk of radius r_0 can often be modeled with a few Fourier basis functions due to its geometric simplicity, for example, $f(r_0, \theta) = c_1(r_0) + c_2(r_0) \cos(2\theta)$ in an SLA process studied by [36]. Using this formulation, Huang et al. [35] modeled the deviation of a dome shape as

$$\mu(r_0, \theta, \varphi) = \alpha_0(r_0) + \alpha_1(r_0)(f * g)(\theta, \varphi). \tag{5.33}$$

The size factors $c_1(r_0)$ and $c_2(r_0)$ can be absorbed in $\alpha_0(r_0)$ and $\alpha_1(r_0)$ in Eq. (5.33). Since each layer of a dome shape has radius $r_0 \sin \varphi$, the input function for the domes can be normalized, for example, as

$$f(\theta, \varphi) = \cos(2\theta) \sin \varphi. \tag{5.34}$$

For the layer interaction or stack-up function $g(x)$, lasso regression was adopted for model selection [30]:

$$\min_{c} \quad \frac{1}{N} \sum_{i=1}^{N} \left(y_i - \sum_{j} c_j (f * g_j)(\theta_i, \varphi_i) \right)^2 + \gamma ||c||_1, \tag{5.35}$$

where N is the number of sampled points, $g_j(\theta, \varphi)$ is a 2D Fourier basis, and c_j is the coefficient of the basis function g_j. Significant terms shared among all domes were selected resulting in the layer interaction function

$$g(\theta, \varphi) = \cos(n_1 \varphi)[1 + \cos(n_2 \theta + \psi)]. \tag{5.36}$$

5.2.3 Association Between Smooth and Non-smooth Geometries: 3D Cookie-Cutter Function

The convolution operator alone is inadequate to capture the sharp transitions observed in the deviation profiles around the corners of non-smooth polyhedral shapes (Fig. 5.25). We propose to use a 3D cookie-cutter function to subtractively

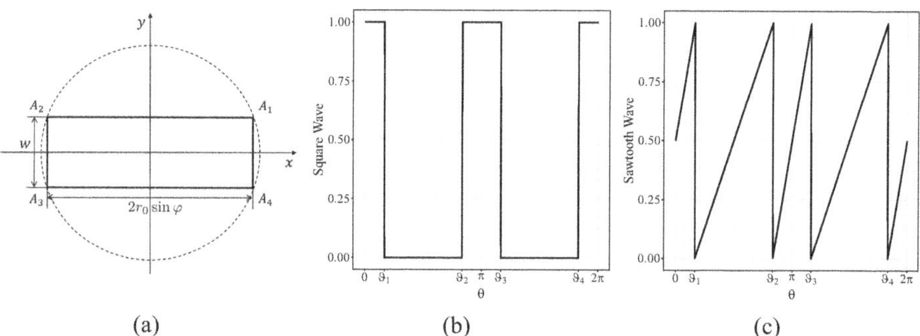

Fig. 5.27 (a) A rectangle cut from its circumcircle and corresponding (b) 2D square wave cookie-cutter function and (c) 2D sawtooth wave cookie-cutter function

carve out the deformation profile of a non-smooth shape from that of a smooth baseline shape. For example, a polygon is cut out from its minimum bounding circle for each horizontal layer as shown in Fig. 5.27a. Similar to [34] and [69], the minimum bounding circle is employed as the smooth base shape since the circumcircle, which passes through all vertices of the polygon, may not exist for an arbitrary polygon.

To represent the proposed learning framework as an additive model, we define the basis functions: $h_1(x) = (f * g)(x)$ describing smooth base shape deviation, $h_2(x)$ being the 3D cookie-cutter function linking the smooth and non-smooth geometries, and a high-order term $h_3(x)$ for the remaining pattern. Then, the learning framework can be written as

$$\mu(x) = \alpha_0(x) + \sum_{j=1}^{3} \alpha_j(x) h_j(x) \tag{5.37}$$

where $\alpha_j, j = 0, \ldots, 3$ represent size effects, h_1 can be learned from smooth products as in [35], and both h_2 and h_3 are fully determined by the geometries of AM-fabricated products. Note that because we use convex smooth shapes as the bases, we can infer the shape deviation of convex polyhedra, while the shape distortion of concave geometries (e.g., pentagram in 2D) needs to be studied using a concave smooth geometry as the baseline.

For the 2D case, [34] applied two candidate 2D cookie-cutter functions h_2 as the association function to carve out regular polygons from their circumcircles: the square wave function

$$sq(\theta) = \text{sign}\{\cos[n(\theta - \phi_0)/2]\}, \tag{5.38}$$

and the sawtooth wave function

$$sw(\theta) = (\theta - \phi_0)\text{MOD}(2\pi/n), \tag{5.39}$$

where n is the number of sides and ϕ_0 is a phase term to shift the cutting position. These functions allow sharp transitions on the deformation patterns near the corners.

A generalization of the sawtooth wave function was established in [51] as

$$sw(\theta) = \frac{\pi(\theta - \vartheta_{j-1})\mathrm{MOD}(\vartheta_j - \vartheta_{j-1})}{2(\vartheta_j - \vartheta_{j-1})} \tag{5.40}$$

for selected angles ϑ_j, $j = 1, \ldots, n$. Note that such function is only needed when the interior angle of a corner is less than $\pi/6$ according to their experimental studies.

For the 3D case with more complex geometries, we apply the 2D cookie-cutter function in each horizontal layer defined by φ in SCS by modifying the frequency of the square wave or sawtooth wave function such that the amplitudes alternate at the sharp corners defined by $\vartheta_j(\varphi)$, $j = 1, \ldots, n$. Thus, one candidate for h_2 is the 3D square wave function

$$sq(\theta, \varphi) = \frac{1}{2}\left\{\mathrm{sign}\left[\sin\left(\frac{(-1)^{j+1}\pi\theta}{\vartheta_j(\varphi)}\right)\right] + 1\right\}, \tag{5.41}$$

and the other alternative is 3D sawtooth wave function

$$sw(\theta, \varphi) = \frac{\theta - \vartheta_{j-1}(\varphi)}{\vartheta_j(\varphi) - \vartheta_{j-1}(\varphi)}, \tag{5.42}$$

for $\vartheta_{j-1}(\varphi) \leq \theta < \vartheta_j(\varphi)$, $j = 1, \ldots, n + 1$, where $\vartheta_j(\varphi)$, $j = 1, \ldots, n$ are the polar angles of sharp transitions with $\vartheta_0(\varphi) = 0$ and $\vartheta_{n+1}(\varphi) = 2\pi$.

The proposed 3D cookie-cutter functions can be regarded as the stack of 2D cookie-cutter functions over the φ direction, where each layer could have sharp transitions at different angles according to the designed geometry. As the number of corners increases and the polyhedron approaches a sphere, the sharp corners effectively banish, and h_2 is approximately constant in both definitions.

In the motivating example, a thin wall with a half-cylindrical shape has radius r_0 and thickness w (Fig. 5.28), where each horizontal layer is a rectangle with length $2r_0 \sin\varphi$ and width w. To cut out a rectangle defined by the corner points (A_1, A_2, A_3, A_4) from its circumcircle as shown in Fig. 5.27a, we first find the angles of each corner as $\vartheta_1, \vartheta_2, \vartheta_3$ and ϑ_4; then the corresponding 2D square wave function is shown in Fig. 5.27b, and 2D sawtooth wave function is shown in Fig. 5.27c. Since the rectangle sizes in each horizontal layer defined by φ are different, the sharp transitions for each layer happen at different polar angles, which are purely defined by the geometry of the product. The scatter plots of proposed 3D square wave and sawtooth wave cookie-cutter functions are shown in Fig. 5.29.

While the same thin-wall parts are treated as 2D shapes in [35], they are regarded as 3D shapes in this work. Unlike the deviation profiles $y(\varphi)$ for 2D cases, shape deviations of 3D non-smooth thin walls show deviation surfaces $y(\theta, \varphi)$ in SCS (Fig. 5.25). Furthermore, to predict the shape deviation of these thin walls on the front, back, and curved top surfaces, each thin wall is regarded as the stack of

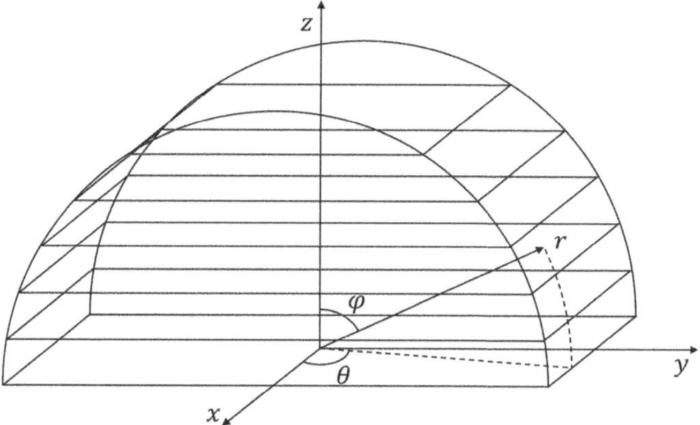

Fig. 5.28 A thin wall fabricated by stacking rectangles

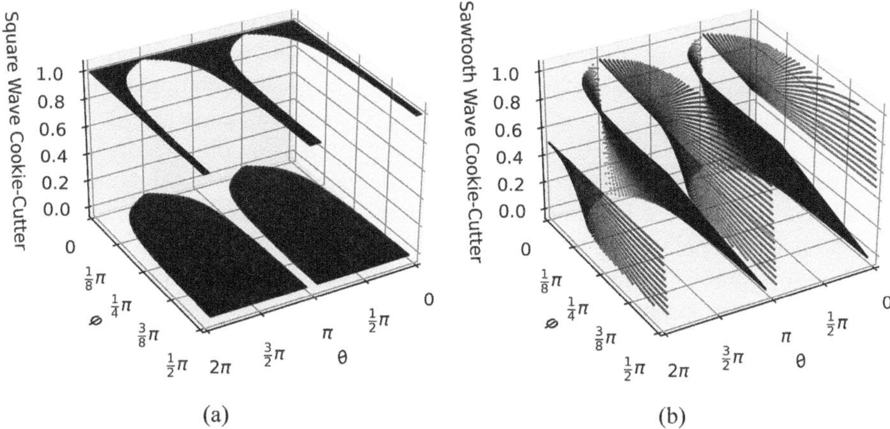

(a)   (b)

Fig. 5.29 (**a**) 3D square wave function and (**b**) 3D sawtooth wave function for the 0.8-inch thin wall

rectangles of different sizes (Fig. 5.28), which requires the 3D cookie-cutter function in Eqs. (5.41) and (5.42).

5.2.4   *Spatial Correlation Modeling with a Novel Distance Metric for Heterogeneous Shape Data*

The random field η in model (5.30) is intended to capture the spatial correlations among deformed regions. It is frequently assumed to be a zero-mean GP with a squared-exponential kernel [65]

$$k(\boldsymbol{x}_i, \boldsymbol{x}_j) = \exp\left(-\frac{d(\boldsymbol{x}_i, \boldsymbol{x}_j)^2}{\delta}\right), \tag{5.43}$$

where $d(\boldsymbol{x}_i, \boldsymbol{x}_j)$ is the distance between $\boldsymbol{x}_i$ and $\boldsymbol{x}_j$.

The challenge, however, is to learn the spatial correlation not only among regions of the same product but also across heterogeneous shapes.

For convenience, the spatial correlation between two sample points in the $\mathbb{R}^3$ space is described in the Cartesian coordinate system, i.e., $\boldsymbol{x}_i = (x_i, y_i, z_i)$ rather than the SCS. Note that it is feasible to establish the GP in the SCS as well. The coordinates are scaled to the current coordinate over the maximum in each direction. Note that $(0, 0, 0)$ corresponds to the center of the printing bed, and the maximum height was chosen as to eliminate the effect of different scales in the z direction.

Due to the physics involved in generating and measuring the shape deviation, Sun et al. [77] and Castillo et al. [16] pointed out that geodesic distance, which is defined as the shortest distance between points on a 3D surface, is a better measure of the spatial correlation in the same additively manufactured part rather than the Euclidean distance. A thorough review of geodesic paths and distances on the surface of triangle meshes can be found in [3]. Due to the high computational cost of the geodesic distance, we employ the as-rigid-as-possible parameterization to first explore the surface into a 2D plane, and then the point-to-point geodesic distance can be approximated by the corresponding Euclidean distance [74]. However, one challenge is that there is no clear definition of geodesic distance among different parts since such path along the surface does not exist.

For two points $\boldsymbol{x}_i$ and $\boldsymbol{x}_j$ lying on the surfaces of two shapes s_i and s_j, respectively, we propose a new distance metric

$$d(\boldsymbol{x}_i, \boldsymbol{x}_j) = \frac{1}{2}\Big[d_e(\boldsymbol{x}_i, \boldsymbol{x}'_i) + d_g(\boldsymbol{x}'_i, \boldsymbol{x}_j) \\ + d_e(\boldsymbol{x}_j, \boldsymbol{x}'_j) + d_g(\boldsymbol{x}'_j, \boldsymbol{x}_i)\Big] \tag{5.44}$$

where $\boldsymbol{x}'_i$ is the projection of point $\boldsymbol{x}_i$ onto shape s_j, i.e., $\boldsymbol{x}'_i$ and $\boldsymbol{x}_i$ have the same angles (θ_i, φ_i); then $\boldsymbol{x}_j$ and $\boldsymbol{x}'_i$ are points on the same part with a properly defined geodesic distance $d_g(\boldsymbol{x}'_i, \boldsymbol{x}_j)$. Similarly, $\boldsymbol{x}'_j$ and $\boldsymbol{x}_j$ are on the line defined by the angles (θ_j, φ_j), and d_e is the standard Euclidean distance.

For example, considering the side view of a thin wall and a dome shape as shown in Fig. 5.30, the proposed distance is the combination of four red paths, where the solid curves denote the geodesic distances d_g and the dashed ones are the Euclidean distances d_e. By adding the projection distance $d_e(\boldsymbol{x}'_j, \boldsymbol{x}_j)$ and $d_e(\boldsymbol{x}'_i, \boldsymbol{x}_i)$, we complete a circuit from $\boldsymbol{x}_i$ to $\boldsymbol{x}_j$ and back, which ensures that d is a valid distance measure. Another advantage of Eq. (5.44) is that if two shapes s_i and s_j are the same, the projection distances are zero, and then the proposed distance is the standard geodesic distance.

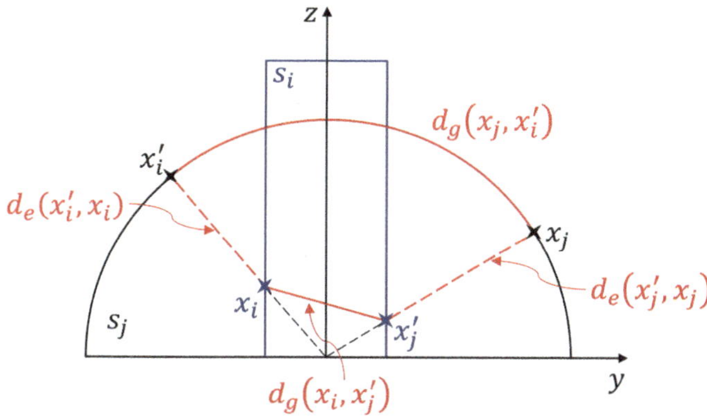

Fig. 5.30 Proposed distance metric between x_i on shape s_i (in blue) and x_j on shape s_j (in black)

5.2.5   Sequential Model Estimation Procedure for the United Modeling Framework

After specifying each component of the additive model, a sequential model fitting strategy is proposed to efficiently estimate the model parameters, mitigate overfitting, and transfer the knowledge from smooth base shape to non-smooth polyhedral shapes.

In general, the parameter estimation procedure follows a similar strategy to the boosted models [25] of fitting models sequentially based on the residuals of earlier models. As specified in [35], the convolution h_1 describes the shrinkage of smooth geometries during printing caused by material phase changes, inter-layer interactions, and gravity effects. On the other hand, the deviation profiles of non-smooth shapes exhibit sharp transitions introduced by uneven thermal stresses around the corners [59]. By regarding the non-smooth shape as being cut from a set of smooth patches, we apply the 3D cookie-cutter function h_2 to capture the sharp transitions and h_3 to model the difference in deviation profiles between smooth and non-smooth geometries. Thus, the smooth shape deviation is modeled first to build the baseline model, while non-smooth parts are included later to assess the association and remaining pattern.

As for η, we use GP regression (GPR) on the parametric model residuals in the final step for the following reasons: (1) estimating GP parameters is computationally expensive; (2) the full training set is involved since spatial correlation affects all shapes; (3) due to the flexibility of GP, the main effect μ could be confounded, which compromises the insights from the parametric model.

To be more specific, after dividing the data $\mathcal{D} = (y, x)$ into $\mathcal{D}^0$ for smooth shapes and $\mathcal{D}^1$ for non-smooth shapes, there are three steps to fit the model sequentially. First, we estimate the parameters in α_0, α_1, and h_1 of Eq. (5.37) using the data in $\mathcal{D}^0$. Since only smooth shapes are involved, the model is reduced to $y = \alpha_0 + \alpha_1 h_1 + \epsilon$.

Algorithm 1: Process-informed sequential model fitting strategy

---

Split data $\mathcal{D}$ into $\mathcal{D}^0$ (smooth) and $\mathcal{D}^1$ (non-smooth) according to design geometry;

Initialize the parameters γ for h_1;

Fit the parameters for the model $y(\boldsymbol{x}) = \alpha_0(\boldsymbol{x}) + \alpha_1(\boldsymbol{x})h_1(\boldsymbol{x}) + \epsilon, \forall (y, \boldsymbol{x}) \in \mathcal{D}^0$;

Calculate residuals

$\tilde{y} = y(\boldsymbol{x}) - \left(\hat{\alpha}_0(\boldsymbol{x}) + \hat{\alpha}_1(\boldsymbol{x})\hat{h}_1(\boldsymbol{x}) \right), \forall (y, \boldsymbol{x}) \in \mathcal{D}$;

Fit the parameters for the model $\tilde{y} = \alpha_2(\boldsymbol{x})h_2(\boldsymbol{x}) + \alpha_3(\boldsymbol{x})h_3(\boldsymbol{x}) + \epsilon, \forall (y, \boldsymbol{x}) \in \mathcal{D}^1$;

Calculate residuals

$\tilde{\tilde{y}} = \tilde{y} - \left(\hat{\alpha}_2(\boldsymbol{x})h_2(\boldsymbol{x}) + \hat{\alpha}_3(\boldsymbol{x})h_3(\boldsymbol{x}) \right), \forall (y, \boldsymbol{x}) \in \mathcal{D}$;

Calculate pairwise distance $d(\boldsymbol{x}_i, \boldsymbol{x}_j), \forall \boldsymbol{x}_i, \boldsymbol{x}_j \in \mathcal{D}$

Fit a GP for model $\tilde{\tilde{y}} \sim \mathcal{N}\left(0, k(d(\boldsymbol{x}_i, \boldsymbol{x}_j))\right)$

Second, non-smooth shape deviations in $\mathcal{D}^1$ are used to estimate the parameters in α_2 and α_3 with respect to the residuals $\tilde{y} = y - (\hat{\alpha}_0 + \hat{\alpha}_1 \hat{h}_1)$. Recall that h_2 and h_3 are determined by the geometry, and the model is $\tilde{y} = \alpha_2 h_2 + \alpha_3(\boldsymbol{x})h_3 + \epsilon$ for the non-smooth shape deviations. Lastly, to model the spatial correlation, GPR techniques are applied to model the residuals $\tilde{\tilde{y}} = \tilde{y} - (\hat{\alpha}_2 h_2 + \hat{\alpha}_3 h_3)$ from all data with $h_2 = 0$ and $h_3 = 0$ for smooth parts, i.e., $\tilde{\tilde{y}} \sim \mathcal{N}(0, k(\cdot, \cdot))$. By learning these three components sequentially, we can predict the shape deviation of untried products. The proposed parameter estimation strategy is summarized in Algorithm 1.

Furthermore, if $\alpha_k, k = 0, \ldots, 3$, have a parametric form, the model parameters can be estimated using the profile likelihood approach. Without loss of generality, assume that $\alpha_k(\boldsymbol{x})$ is a basis expansion of the form $\alpha_k(\boldsymbol{x}) = \sum_{j=1}^{J_k} q_k^j(\boldsymbol{x})\beta_k^j$. Then, the model can be expressed as

$$y = \mathbf{H}(\gamma)\boldsymbol{\beta} + \boldsymbol{\eta} + \boldsymbol{\epsilon}, \tag{5.45}$$

where

$$\mathbf{H}(\gamma) = \left(q_0^1(\boldsymbol{x}), \ldots, q_0^{J_0}(\boldsymbol{x}), q_1^1(\boldsymbol{x})h_1(\gamma), \ldots, q_3^{J_3}(\boldsymbol{x})h_3 \right)$$

and

$$\boldsymbol{\beta} = \left(\beta_0^1, \ldots, \beta_0^{J_0}, \beta_1^1, \ldots, \beta_3^{J_3} \right)^{\mathsf{T}}.$$

The maximum likelihood estimates (MLE) are obtained by solving

$$\min_{\gamma} \quad \frac{1}{2\sigma^2} ||y - \mathbf{H}_1(\gamma)\hat{\boldsymbol{\beta}}'||^2 + \frac{n}{2} \ln\left(\sigma^2\right)$$

$$\text{s.t.} \quad \hat{\boldsymbol{\beta}}' = \left(\mathbf{H}_1^{\mathsf{T}}(\gamma)\mathbf{H}_1(\gamma)\right)^{-1} \mathbf{H}_1^{\mathsf{T}}(\gamma)y \tag{5.46}$$

where $\boldsymbol{y}$ is the shape deviation of observations included in $\mathcal{D}^0$, n is the total number of points in $\mathcal{D}^0$, $\mathbf{H}_1$ is the matrix of the first $J_0 + J_1$ columns of $\mathbf{H}$, and $\boldsymbol{\beta}' = (\boldsymbol{\beta}_0, \boldsymbol{\beta}_1)$. For the parameters of α_2 and α_3, the MLE is

$$\hat{\boldsymbol{\beta}}'' = \left(\mathbf{H}_2^\mathsf{T} \mathbf{H}_2\right)^{-1} \mathbf{H}_2^\mathsf{T} (\boldsymbol{\gamma}) \tilde{\boldsymbol{y}}$$

where $\tilde{\boldsymbol{y}}$ is as defined in Algorithm 1 for observations in $\mathcal{D}^1$, $\mathbf{H}_2$ is the matrix of the last $J_2 + J_3$ columns of $\mathbf{H}$, and $\boldsymbol{\beta}'' = (\boldsymbol{\beta}_2, \boldsymbol{\beta}_3)$.

5.2.6   Validation Through Shape Deviation Prediction for Domes and Thin Walls

We revisit the motivating example at the beginning of this section to demonstrate the capability of the proposed learning framework for modeling and predicting the shape deviation patterns of a wide variety of geometries containing both smooth and non-smooth features in AM. Seven parts (Fig. 5.24) were printed through the mask image projection stereolithography (MIP-SLA) process. After the printing process, a ROMER absolute arm with RS4 laser scanner is used to collect the measurements as point clouds, which are then registered using the constrained iterative closest point algorithm [19]. The printing quality is evaluated by the shape deviation surface of each part [35] as illustrated in Fig. 5.25. We employ all domes and 0.8-inch- and 2.0-inch thin walls as the training set and leave the 1.5-inch thin wall as the validation set.

To implement the learning framework in Eq. (5.37), the first step is to identify the input function f. Note that, due to machine repair, the pattern of shape deviation of 2D circular disks changed from what was presented in [36], i.e., the input function $f(\theta, \varphi)$ should have a different pattern, and we need to fit the spherical shape model for the new data. Luan and Huang [51] defined the new pattern as

$$f(\theta) = \cos\left(2\theta + \frac{\pi}{3}\right) \mathbf{1}_{\theta \in [0,\pi)} - \sin(2\theta) \mathbf{1}_{\theta \in [\pi, 2\pi)}. \tag{5.47}$$

Recall that the form of input function was changed to Eq. (5.34) by multiplying it by $\sin\varphi$ because the radius of each layer is $r_0 \sin\varphi$ for the dome shape. Similarly, we have the input function $f(\theta, \varphi)$ for spherical shapes as

$$f(\theta, \varphi) = f(\theta) \sin\varphi. \tag{5.48}$$

If the radius is r_0 and the thickness is w for the thin-wall shape, the circumcircle radius for each horizontal layer φ is

$$\max_{\theta}\{r_0(\theta, \varphi)\} = \sqrt{r_0^2 \sin^2\varphi + \left(\frac{w}{2}\right)^2} \approx r_0 \sin\varphi, \tag{5.49}$$

since w is much smaller than r_0 for the thin products. Then, we can compute the convolution explicitly and regard the thin walls as cut from the domes with the same radii, and the same input function $f(\theta, \varphi)$ as in Eq. (5.48) can be applied. For the layer interaction function $g(\theta, \varphi)$, Eq. (5.36) is used since the printing mechanism is the same and only the shapes and sizes change. Thus, h_1 is fully specified.

Next, we need to specify the sharp transition angles ϑ_j and n used in the 3D cookie-cutter function h_2 in Eq. (5.41). As shown in Fig. 5.27, $n = 4$, and the angles are $\vartheta_1 = \arctan(w/(2r_0 \sin \varphi))$, $\vartheta_2 = \pi - \vartheta_1$, $\vartheta_3 = \pi + \vartheta_1$, and $\vartheta_4 = 2\pi - \vartheta_1$ according to the geometry of the thin walls.

To capture the arch pattern of thin walls presented in the deviation profiles in Fig. 5.25, we choose h_3 as

$$
h_3(\theta, \varphi) = \left\{ \sin\left(\frac{n}{4}\theta\right) \mathbf{1}_{\theta \in [\vartheta_1, \vartheta_2)} \right.
$$
$$
\left. + \sin\left[\frac{n}{4}(\theta - \pi)\right] \mathbf{1}_{\theta \in [\vartheta_3, \vartheta_4)} \right\} \sin \varphi, \tag{5.50}
$$

where n is the number of sides. For the thin walls, we have $n = 4$, and when $n \to \infty$, this term becomes white noise.

Due to the limited number of samples, we follow similar linear assumptions as in [35] to incorporate the size effect. Denoting $x = (x, y, z, r_0, \theta, \varphi)$, which contains the location information of Cartesian coordinates and spherical coordinates, the conjectures are:

1. $n_1(x) = c_1 + c_2 r_0$.
2. n_2 and ψ are unknown constants.
3. $\alpha_0(x) = \beta_{0,1} + \beta_{0,2} r_0$.
4. $\alpha_1(x) = \beta_{1,1} + \beta_{1,2} r_0$.
5. $\alpha_2(x) = \beta_{2,1} + \beta_{2,2}\left[\sqrt{r_0^2 \sin^2 \varphi + w^2/4} - r_0(\theta, \varphi)\right]$.
6. $\alpha_3(x) = \beta_{3,1} + \beta_{3,2} r_0$.

Note that n_1, n_2, and ψ are the parameters in the layer interaction function g as in Eq. (5.36), while $\alpha_i, i = 0, \ldots, 3$, are the coefficients describing the size effect in Eq. (5.37). The first four conjectures are for the baseline model, and the last two would affect the 3D cookie-cutter function. For simplicity, we assume the layer-to-layer interactions change for different sizes mainly along the φ direction and are related to the number of layers printed; thus n_1 deciding the period over φ is assumed to be proportional to the size, while n_2 and ψ control the period, and phase in θ-direction is assumed to be constant.

To achieve better model interpretability, a linear relationship to the size of the product is imposed on the coefficients of α_0, α_1 and α_3. As the coefficient of cookie-cutter term, α_2 is proportional to the cutting width, i.e., the difference between the circumcircle radius and polygonal shape at each angle, which is $\sqrt{r_0^2 \sin^2 \varphi + w^2/4} - r_0(\theta, \varphi)$. Under these assumptions, at least two samples of each shape are required to estimate the model parameters. Because a limited number

Table 5.7 Parameter estimates and standard error (SE) for the deviation of dome shapes

Parameters	Estimate	SE
n_2	0.6462	0.006591
ψ	4.0793	0.019610
c_1	0.0034	0.230790
c_2	−0.0016	0.169782
$\beta_{0,1}$	0.0068	0.000210
$\beta_{0,2}$	0.0047	0.000166
$\beta_{1,1}$	0.0063	0.000601
$\beta_{1,2}$	0.0158	0.000476
σ	0.0065	0.000046

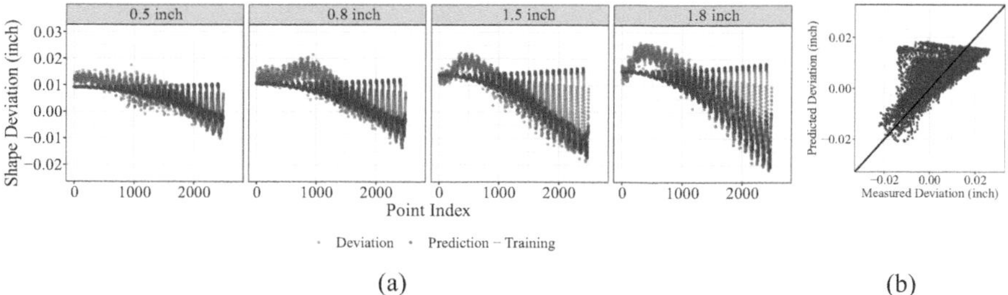

Fig. 5.31 Measured shape deviation (in gray) and model prediction (in blue) for domes

of training samples are usually provided in AM, more complex relationships for α require knowledge of the material and process interactions.

The MLE procedure described in Sect. 5.2.5 is employed to fit the spherical shape deviation model through the *mle2* function in R package *bbmle* with randomized initialization in the parameter space, and the results are given in Table 5.7 and Fig. 5.31. The mean absolute error (MAE) is 0.0048, and the root mean square error (RMSE) is 0.0065. We plot the shape deviation by point index due to the difficulty of comparing model fitting performance in 3D space. The four blocks from the left correspond to the 0.5-inch, 0.8-inch, 1.5-inch, and 1.8-inch domes, respectively. The predictions are close to the actual deviation measurements, except at the upper-right corner. Note that the results are different from [35] since the input function $f(\theta, \varphi)$ is changed to Eq. (5.48). The estimates of c_1 and c_2 are not statistically different from zero, and the layer interaction function can be simplified as $g(\theta, \varphi) = 1 + \cos(n_2\theta + \psi)$.

Next, the thin walls with radii of 0.8 inch and 2.0 inches are used as the training set, and the 1.5-inch thin wall is left as the validation set. The estimated model parameters for α_2 and α_3 are shown in Table 5.8, and the measured deviations versus model predictions are presented in Fig. 5.32 for both square wave and sawtooth wave cookie-cutter functions. The model performance metrics are summarized in Table 5.9. From both figures and performance metrics, the square wave function is better than the sawtooth wave function for modeling the shape deviation of thin walls. Thus, we use the square wave function for the remainder of the chapter. Since

Table 5.8 Parameter estimates and standard error (SE) for cookie-cutter and high-order terms

Parameters	Square wave		Sawtooth wave	
	Estimate	SE	Estimate	SE
$\beta_{2,1}$	−0.0387	0.00106	−0.0041	0.00123
$\beta_{2,2}$	0.0002	0.00004	−0.0003	0.00004
$\beta_{3,1}$	0.0124	0.00175	−0.0040	0.00184
$\beta_{3,2}$	0.0139	0.00074	0.0137	0.00093

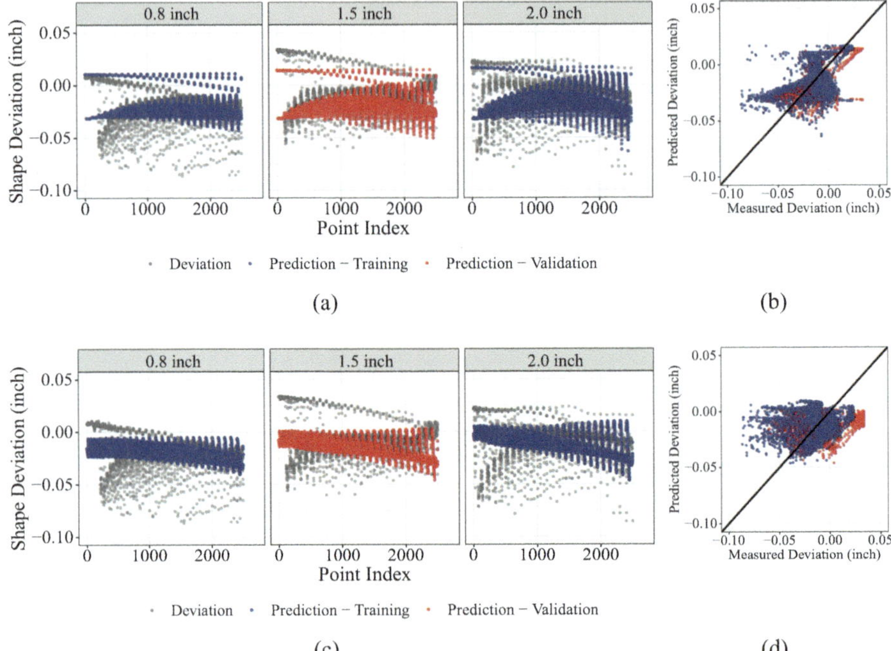

Fig. 5.32 Measured shape deviation (in gray), training set prediction (in blue), and validation set prediction (in red) for thin walls applying (**a** and **b**) square wave and (**c** and **d**) sawtooth wave functions

Table 5.9 Thin wall model performance applying different cookie-cutter functions

Cookie-cutter	Training		Validation	
	MAE	RMSE	MAE	RMSE
Square wave	0.0122	0.0160	0.0120	0.0158
Sawtooth wave	0.0156	0.0200	0.0164	0.0205

both $\beta_{2,2}$ and $\beta_{3,2}$ are positive, we can infer that the effects of the sharp transition and arch pattern on the deviation profile increase with the part's size. However, there are some remaining spatial patterns to be captured, so the residuals are fitted through GPR with squared-exponential kernel in Eq. (5.43) using Euclidean distance between points through the *gam* function in R package *mgcv*.

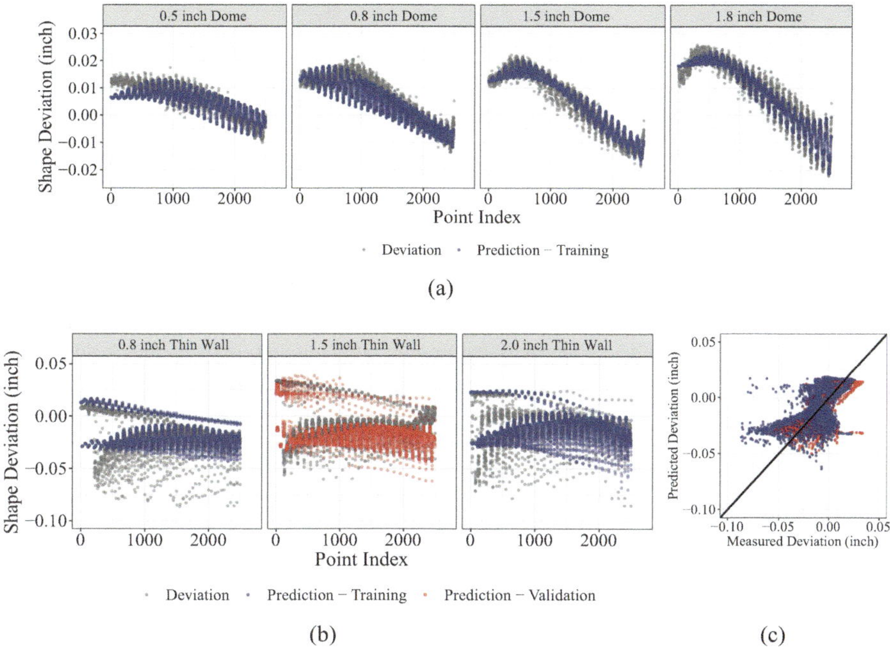

Fig. 5.33 Measured shape deviation (in gray), training set prediction (in blue), and validation set prediction (in red) after GPR with Euclidean distance

Table 5.10 Model performance comparison

Model	Training		Validation	
	MAE	RMSE	MAE	RMSE
Parametric	0.0122	0.0160	0.0120	0.0158
Parametric + Euclidean Distance	0.0050	0.0093	0.0107	0.0148
Parametric + Proposed Distance	0.0060	0.0095	0.0087	0.0124

The measurements and model predictions are presented in Fig. 5.33 and Table 5.10. For the dome parts, not only the upper-right corners in Fig. 5.31 have been offset by GP, but the expansion of the larger domes are correctly predicted. Predictions for thin walls are improved from the results in Fig. 5.32, and the performance metrics (MAE and RMSE) are reduced for both training and validation sets.

Since the GP combines the location information of all shapes together, and points on the smaller domes are closer to those on the thin walls, their deviation predictions are greatly affected by the deviation pattern of thin walls. Thus, the 0.5-inch and 0.8-inch domes show a wider variation of the deviation prediction compared with the 1.5-inch and 1.8-inch domes. This suggests that the spatial correlation among different shapes is overestimated, and we can improve the spatial correlation modeling by modifying the distance metric in the squared-exponential kernel as specified in Sec. 5.2.4.

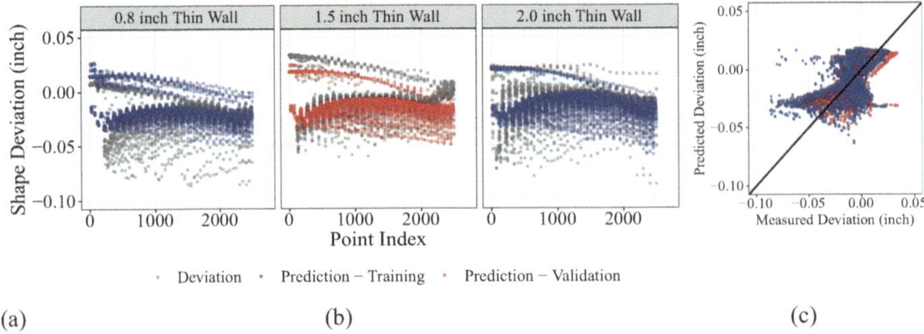

Fig. 5.34 Measured shape deviation (in gray), training set prediction (in blue), and validation set prediction (in red) after GPR with the proposed distance

After computing the proposed geodesic distance among all shapes and estimating the spatial correlation with the *mKrig* function in R package *fields*, the updated model prediction and performance are shown in Fig. 5.34 and Table 5.10. With the proposed distance metric considering both geographic and geometric information, the GP accurately captures the spatial correlation. Though the deviations of the first few points in domes are overestimated, predictions cover most of the actual deviation measurements in the training set. With slightly worse training set performance, the performance on the validation set improved around 20% using the proposed distance metric. The measured shape deviation, predicted deviation surface, and residuals of the final model with GPR using the proposed distance metric are presented in Fig. 5.35.

This section establishes a unified learning framework for shape deviation modeling of smooth and non-smooth 3D geometries in AM. The AM process is mathematically decomposed into two stages to enable the learning of heterogeneous shape deviation data. In stage one, a smooth base shape is additively built, while in stage two, non-smooth shapes are subtractively carved out. A unified predictive shape accuracy model is developed to combine the baseline deviation of smooth shapes, sharp transitions caused by the subtractive step, and the remaining spatial correlation. The previously proposed convolution framework serves as the baseline model for smooth shapes. The proposed 3D cookie-cutter function effectively captures the unique shape deviation pattern of sharp corners of non-smooth convex shapes and enables joint learning of deviation patterns on smooth and non-smooth geometries. A novel distance measure is proposed to model the spatial correlation among heterogeneous shapes by combining the local geodesic distance between points in the same 3D object and the Euclidean distance between projected points across 3D objects.

A case study shows that the unified model can successfully predict the shape deviation of convex smooth (domes) and non-smooth (thin walls) shapes. Since 3D freeform shapes can be approximated as a combination of smooth patches and sharp corners, the proposed learning framework builds a foundation to further extend

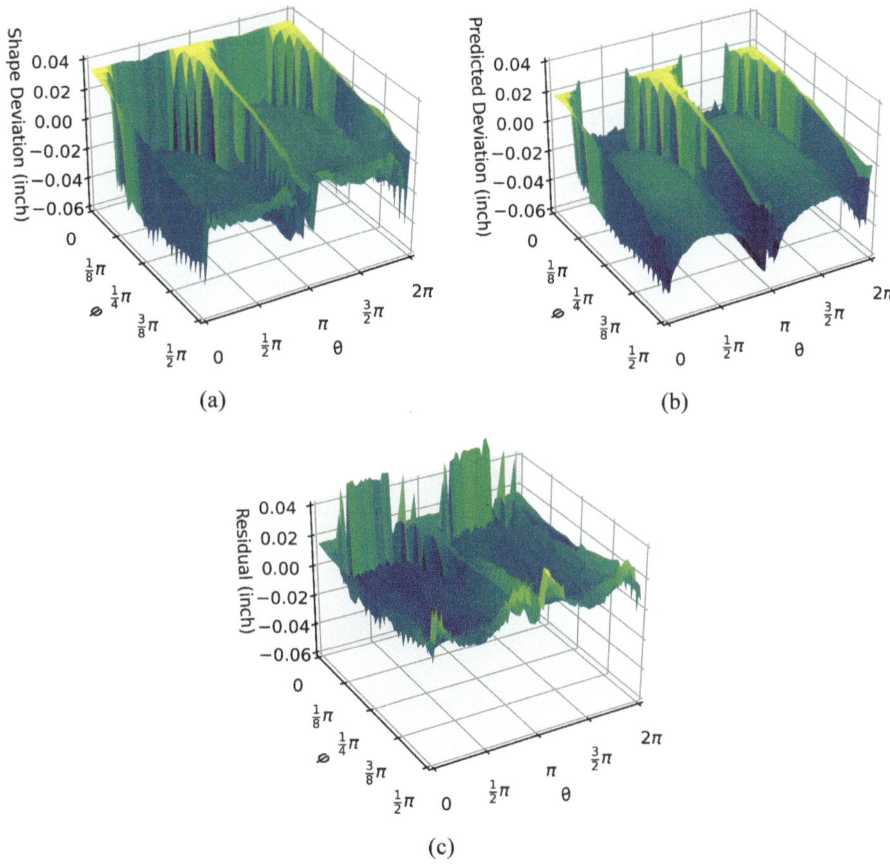

Fig. 5.35 (**a**) Measured shape deviation, (**b**) predicted shape deviation, and (**c**) residuals for the 1.5-inch thin wall (darker/lighter color for negative/positive deviation)

this modeling approach to predict the quality of 3D freeform shapes. Furthermore, the strategy of connecting engineering surface data through decomposition of data generation processes can be adopted in other domains for engineering-informed learning of heterogeneous data.

5.3   Shape Quality Modeling for 3D Freeform Geometries Through Domain-Informed Dimension Reduction

It is worthwhile to restate the offline learning problem and few-shot learning subproblem defined in Sect. 1.5:

- **Offline learning problem (OLP)**: The offline learning objective is to establish functional mapping $f : U \rightarrow Y_U$, that is, to generate a model $f(u)$ to predict

y_u by learning from a small set of training data $D_S = \{(u, y_u)|(u_i, y_{u_i}), i = 1, 2, ..., n\}$.

- **Model $f(u)$ generation and few-shot learning (OLP.2)**: Model $f(u)$ is expected to predict product quality of infinite variety by learning from small data D_S. This "1-to-∞" or few-shot learning problem prefers $f(u)$ being constructed with a limited number of building blocks, primitives, or basis functions for robustness and flexibility. Model interpretability is another key consideration to ensure better understanding of physical systems.

The fabrication-aware dimension reduction and finite manufacturing primitive (FMP) representation are introduced in Sects. 2.5 and 4.2 [33] to address the large product variety challenge in AM. The premise of dimension reduction is that a freeform product built in AM can be viewed as an assembly of a finite set of basic manufacturing objects called FMPs. The product quality can be represented by a finite set of functions, where each function represents the quality model of one type of FMPs. By operating in a low-dimensional space spanned by FMPs, the FMP method will enable learning from a small sample of 3D product varieties to infer and control the quality of new 3D product varieties. The small-sample machine learning of product-variety problem will be transformed into large-sample learning of manufacturing-primitive problem.

This section addresses the OLP.2 problem and formalizes the FMP-based modeling approach.

5.3.1 An Impulse Response Formulation for Model Generation

The proposed input-output representation enables the modeling of *model primitive*, that is, to construct $f(u)$ with a small set of basis functions that characterize deformation patterns of shape primitives. Small-sample few-shot learning therefore becomes feasible.

Definition 5.1 (2D Model Primitive) A 2D model primitive is defined as a functional model that predicts the deviation of a 2D shape primitive, which is either a circular sector, a line segment, or a corner.

Since a line segment can be viewed as a special case of a corner with one edge being zero, we only need two types of 2D model primitives for circular sectors and corners, denoted as function $\eta_1(\cdot)$ and $\eta_2(\cdot)$, respectively.

Definition 5.2 (3D Model Primitive) A 3D model primitive is defined as a functional model that predicts the deviation of a 3D shape primitive, which is either spherical patch or a surface edge with the plane surface patches being a special case of the surface edge.

2D model primitives are connected with 3D model primitives through definition of 3D shape primitives. the strategy to establish 3D model primitives is to model primitive manufacturing inputs, that is, 2D shape primitives that form the boundaries of layers and layer stack-up. Therefore, 2D model primitives have to be established first.

5.3.1.1 Impulse Response Modeling for 2D Model Primitives

Our work in [34, 50] innovated the shape deviation primitive concept for data-driven modeling of 2D shape deviation. The input representation, particularly primitive inputs to the fabrication process shown in Fig. 2.10, inspires the impulse response modeling developed in this work.

Theorem 5.1 (Impulse Response Characterization of 2D Model Primitives) *The 2D model primitive $\eta_{\{1,2\}}(\theta)$ for a 2D primitive shape $u(\theta)$ can be characterized by an impulse response function $h_{\{1,2\}}(\theta)$ through a convolution formulation:*

$$\eta_{\{1,2\}}(\theta) = (h_{\{1,2\}} * u)(\theta) = \int_0^\theta h_{\{1,2\}}(\theta - \tau)u(\tau)d\tau \qquad (5.51)$$

Proof As shown in Fig. 2.11, a 2D shape primitive is a piecewise constant input, which can be represented as a sum of 1D step signals represented in the PCS with θ being equivalent to the time variable t. Once this critical connection is established, the proof follows an excellent control reference along Fig. 5.36 [in chapter 6] [2].

Let the input $u(t)$ in Fig. 5.36 represent a primitive angular corner input with $t = \theta$. The model primitive $\eta_{\{1,2\}}(\theta)$ represents the output or the deviation of the primitive shape. Let $H_{\{1,2\}}(\theta)$ be the response to a unit step applied at $\theta = 0$, and assume $H_{\{1,2\}}(0) = 0$. The responses to a series of step inputs are $H_{\{1,2\}}(\theta - \theta_0)u(\theta_0)$, $H_{\{1,2\}}(\theta - \theta_1)(u(\theta_1) - u(\theta_0))$, and so on. The output $\eta_{\{1,2\}}(\theta)$ is the sum of individual responses [2]:

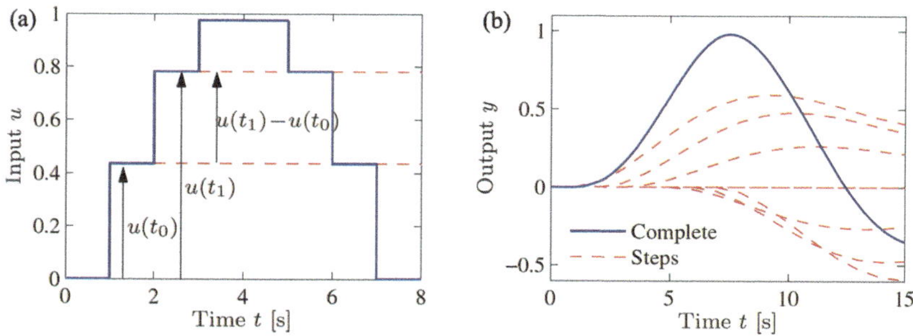

Fig. 5.36 Response to (**a**) piecewise constant input as a summation of step inputs and (**b**) output as the sum of individual output [2], ©Princeton University Press

$$\eta_{\{1,2\}}(\theta) = H_{\{1,2\}}(\theta - \theta_0)u(\theta_0) + H_{\{1,2\}}(\theta - \theta_1)(u(\theta_1) - u(\theta_0)) + \cdots$$

$$= (H_{\{1,2\}}(\theta - \theta_0) - H_{\{1,2\}}(\theta - \theta_1))u(\theta_0) +$$

$$(H_{\{1,2\}}(\theta - \theta_1) - H_{\{1,2\}}(\theta - \theta_2))u(\theta_1) + \cdots$$

$$= \lim_{n \to \infty} \sum_{k=1}^{n} (H_{\{1,2\}}(\theta - \theta_{k-1}) - H_{\{1,2\}}(\theta - \theta_k))u(\theta_{k-1}) +$$

$$H_{\{1,2\}}(\theta - \theta_n)u(\theta_n)$$

$$= \lim_{n \to \infty} \sum_{k=1}^{n} \frac{H_{\{1,2\}}(\theta - \theta_{k-1}) - H_{\{1,2\}}(\theta - \theta_k)}{\theta_k - \theta_{k-1}} u(\theta_{k-1}) * (\theta_k - \theta_{k-1}) +$$

$$H_{\{1,2\}}(\theta - \theta_n)u(\theta_n)$$

$$= \int_0^\theta H'_{\{1,2\}}(\theta - \tau)u(\tau)d\tau$$

$$= \int_0^\theta h_{\{1,2\}}(\theta - \tau)u(\tau)d\tau$$

$$= (h_{\{1,2\}} * u)(\theta)$$

where $h_{\{1,2\}}(\theta) = \frac{\mathrm{d}H_{\{1,2\}}(\theta)}{\mathrm{d}\theta}$, the derivative of the step response, is commonly known as the impulse response function (IRF) in control theory. $\square$

Based on the definition of 2D model primitives, we have IRFs $h_1(\theta)$ and $h_2(\theta)$ corresponding to $\eta_1(\theta)$ and $\eta_2(\theta)$ for circular sectors and corners, respectively.

5.3.1.2 Impulse Response Modeling for 3D Model Primitives

We first postulated a convolution formulation of 3D shape generation in AM without proof [35]. Based on the definition of 3D shape primitive and Theorem 5.1, this work provides a theoretical justification of the shape deviation generator model established in [35].

Theorem 5.2 (Impulse Response Characterization of 3D Model Primitives)
The 3D model primitive for a 3D primitive shape $u(\theta, \phi)$ can be characterized by a 2D impulse response function $g(\theta, \phi)$ through a 2D convolution formulation:

$$(\eta_{\{1,2\}} * *g_{\{1,2\}})(\theta, \phi) = \int_0^\phi \int_0^\theta g_{\{1,2\}}(\theta - \tau_1, \phi - \tau_2)\eta_{\{1,2\}}(\tau_1, \tau_2)d\tau_1 d\tau_2$$

$$(5.52)$$

Proof As shown in Figs. 4.4 and 2.10a, a 3D shape primitive can be viewed as an extension of 1D piecewise constant input to 2D case. In addition to the time variable

$\tau_1 = \theta$ in $x - y$ plan for 2D shape primitives, there is additional time variable z or equivalently $\tau_2 = \phi$ along the z direction to stack up 2D shape primitives at time τ_2. Since a 3D shape primitive is the result or output of stacking up 2D shape primitives, the true step input to a 3D primitive at time τ_2 is therefore $\eta_{\{1,2\}}(\theta, \tau_2)$, that is, deviation of a 2D shape primitive. (Here we remove the nominal input of 2D shape primitives for simplification of notation.)

Let $G_{\{1,2\}}(\theta, \tau_2)$ be the response to the 2D unit step applied at $\tau_2 = 0$, and assume $G_{\{1,2\}}(\theta, 0) = 0$. The responses to a series of 2D step inputs are $G_{\{1,2\}}(\theta - \theta_0, \phi - \phi_0)\eta_{\{1,2\}}(\theta_0, \phi_0)$, $G_{\{1,2\}}(\theta - \theta_1, \phi - \phi_1)(\eta_{\{1,2\}}(\theta_1, \phi_1) - \eta_{\{1,2\}}(\theta_0, \phi_0))$, and so on. Following the same procedure, but working in the 2D time space τ_1, τ_2, we will have the output or the 3D model primitive as

$$\int_0^\phi \int_0^\theta \frac{\partial G_{\{1,2\}}}{\partial \theta, \phi} \eta_{\{1,2\}}(\tau_1, \tau_2) d\tau_1 d\tau_2$$

And let $g_{\{1,2\}}(\theta, \phi) = \frac{\partial G_{\{1,2\}}}{\partial \theta, \phi}$, the partial derivative of the 2D step response, or the 2D IRF. Denoting this 2D convolution integral as $(\eta_{\{1,2\}} * g_{\{1,2\}})(\theta, \phi)$, we have the proof for Theorem 5.2. $\qquad\square$

Remark 5.1 As we pointed out in [35], if we only print a single layer of the 3D primitive shape, IRF $g_{\{1,2\}}(\theta, \phi)$ is degenerated to a Dirac's delta function $\delta(\theta, z \approx 0)$. Then

$$(\eta_{\{1,2\}} * *\delta)(\theta, z \approx 0) = \eta_{\{1,2\}}(\theta, z \approx 0) = \eta_{\{1,2\}}(\theta)$$

which is the 2D model primitive in Eq. (5.51). Clearly, Theorem 5.2 is an extension of Theorem 5.1.

5.3.2 Model Construction and Few-Shot Learning Based on Model Primitives

Theorems 5.1 and 5.2 establish fabrication-aware, control-theoretic formulation of model building blocks that can be learned from a small training sample D_S. Furthermore, two theorems enable model construction to predict the quality of freeform shapes in theory. In this work, we use simple examples to show the feasibility and potentials of applying the two theorems. Certainly, significant efforts are needed to achieve ultimate "1-to-∞" learning for complex 3D freeform shapes.

(a) **Few-shot learning to predict shape deviation of 2D freeform products**: We have achieved data-driven few-shot learning to predict shape deviation of 2D freeform products in [15, 34, 36, 50, 85]. By taking a fresh look under the control-theoretic formulation, we intend to discover not only the process dynamics but also principles for generalization.

Corollary 5.1 (Shape Deviation Model for 2D Freeform Products) *Based on Theorem 5.1, Eq. (5.51), and Eq. (2.3), the shape deviation of a freeform 2D product or layer can be derived as $y(u(\theta)) = f(u(\theta)) + \epsilon$ with $f(u(\theta))$:*

$$f(u(\theta)) = \sum_{i=0}^{I} \mathbf{1}_{\theta \in [\theta_i, \theta_{i+1})} \left[\mathbf{1}_c(r_i(\theta))(h_1 * r_i)(\theta) + (1 - \mathbf{1}_c(r_i(\theta)))(h_2 * r_i)(\theta) \right]$$

$$= \sum_{i=0}^{I} [\mathbf{1}_c(r_i(\theta))\eta_1(\theta, r_i(\theta)) + (1 - \mathbf{1}_c(r_i(\theta)))\eta_2(\theta, r_i(\theta))]$$

$$\tag{5.53}$$

where $\mathbf{1}_c(r_i(\theta)) = 1$ if $r_i(\theta)$ is a circular sector and zero otherwise. Notation $\eta_{\{1,2\}}(\theta, r_i(\theta))$ specifies the model primitive $\eta_{\{1,2\}}(\theta)$ has shape primitive $r_i(\theta)$ as input. The error term ϵ can be noise or can impose a correlation structure to capture the interaction among neighboring shape primitives, depending on AM processes.

However, modeling and computation can be complicated and inefficient when the number of shape primitives I is large. One improvement of model (5.53) is provided.

Corollary 5.2 (Cookie-Cutter Modeling for 2D Freeform Shapes) *The shape deviation model $f(u(\theta))$ for 2D freeform products in Eq. (5.53) can be well approximated as*

$$f(u(\theta)) = \sum_{i=0}^{I} \eta_1(\theta, \tilde{r}_i(\theta)) + \sum_{i=0}^{I} \eta_2(\theta, r_i(\theta) - \tilde{r}_i(\theta)) \tag{5.54}$$

where $\tilde{r}_i(\theta)$ is the smallest circular sector that covers a line segment or a corner (Fig. 5.37a).

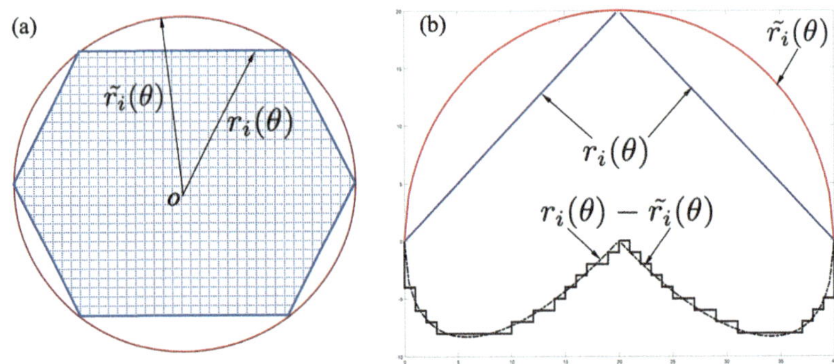

Fig. 5.37 Simplification of deformation model (5.53) for 2D freeform shapes

Proof We can find a smallest circular sector $\tilde{r}_i(\theta)$ to approximate each shape primitive $r_i(\theta)$. Apparently, $r_i(\theta) - \tilde{r}_i(\theta) = 0$ if $r_i(\theta))$ is a circular sector. Then Eq. (5.53) can be rewritten as:

$$f(u(\theta)) = \sum_{i=0}^{I} \eta_1(\theta, \tilde{r}_i(\theta)) + (1 - \mathbf{1}_c(r_i(\theta))) \sum_{i=0}^{I} [\eta_2(\theta, r_i(\theta)) - \eta_1(\theta, \tilde{r}_i(\theta))]$$

When $r_i(\theta)) - \tilde{r}_i(\theta) \neq 0$, as shown in Fig. 5.37b, the shape of $r_i(\theta)) - \tilde{r}_i(\theta)$ can still be approximated by a corner or a line segment with a higher-order approximation error. Therefore, $\eta_2(\theta, r_i(\theta)) - \eta_1(\theta, \tilde{r}_i(\theta)) \approx \eta_2(\theta, r_i(\theta) - \tilde{r}_i(\theta))$. When $r_i(\theta) - \tilde{r}_i(\theta) = 0$, clearly $\eta_2(\theta, r_i(\theta) - \tilde{r}_i(\theta)) = 0$, so we can drop the indicator $(1 - \mathbf{1}_c(r_i(\theta)))$ and obtain Eq. (5.54). $\square$

Remark 5.2 This revised model (5.54) essentially treats a 2D shape as being carved out from a circular disk or segments of circular disks using cookie-cutters. In [34] we empirically established the so-called "cookie-cutter" modeling framework. Corollary 5.2 provides a theoretical justification and opportunity to further enhance the model identification.

Learning 2D model primitives can now be achieved by a small training sample consisting of disks and polygon shapes with different sizes. From our study of different AM processes [10, 34, 36, 49, 85]), $\eta_1(\theta, \tilde{r}_i(\theta))$ for circular sector can be well approximated by low-order Fourier bases:

$$\eta_1(\theta, \tilde{r}_i(\theta)) = \sum_{k=0}^{K} \alpha_k(\eta_1(\theta, \tilde{r}_i(\theta))) cos(k\theta + \varphi_k) \tag{5.55}$$

For example, four disks (Fig. 3.2) printed in a SLA process can be well modeled by $\eta_1(\theta, r) = -0.0134(r + 0.0088)^{0.86} + 0.0057(r + 0.0088)^{1.13} \cos(2\theta)$ with r being the disk radius [36]. This data-driven model provides complete description of circular sectors at different location θ.

Candidate choice of $\eta_2(\theta, r_i(\theta) - \tilde{r}_i(\theta))$ is proposed in [34], for example, for three square plates (Fig. 3.7b), which can be a square wave function,

$$\eta_2(\theta, \tilde{r}(\theta)) = \beta(\tilde{r}(\theta)) \, sign[cos(n(\theta - \phi_0)/2)]$$

or a sawtooth wave function

$$\eta_2(\theta, \tilde{r}(\theta)) = \beta(\tilde{r}(\theta)) \, \{sign[sin(n(\theta - \phi_0)/2)] + 1\} * [(\theta - \phi_0) \text{ MOD } (2\pi/n)]$$

where n represents the number of polygon sides, MOD function obtains remainders, and ϕ_0 is phase variable.

Apparently if a segment of 2D shape boundary can be approximated by a circular sector or a corner of a n-side polygon (including a line segment), the corresponding model primitive can be utilized for prediction. This deviation primitive concept was first postulated in [50] through a unified data-driven model to predict deformation of 2D freeform shapes. Now model in Eq. (5.54) essentially provides a theoretical justification and model-driven formulation.

(b) **Few-shot learning to predict shape deviation of 3D freeform products**: Few-shot learning for 3D freeform shapes has been an open issue due to the lack of fabrication-aware modeling and learning theories [35]. Theorem 5.2 has shown that the deviation of a spherical patch on a 3D shape can be modeled as $(\eta_1 * g_1)(\theta, \phi)$ and the deviation of a surface edge (including the plane surface patch) can be modeled as $(\eta_2 * g_2)(\theta, \phi)$. With a 3D shape defined in Eq. (4.5), the deformation of a freeform 3D shape is provided by the following corollary.

Corollary 5.3 (Cookie-Cutter Modeling for 3D Freeform Shapes)

$$y(u(\theta, \phi)) =$$

$$\sum_{j=0}^{J}\sum_{i=0}^{I} \mathbf{1}_{\{\theta \in [\theta_i, \theta_{i+1}), \ \phi \in [\phi_j, \phi_{j+1})\}}[\mathbf{1}_s(r_{ij}(\theta, \phi))(\eta_1 * *g_1)(\theta, \phi, r_{ij}(\theta, \phi))+$$

$$(1 - \mathbf{1}_s(r_{ij}(\theta, \phi)))(\eta_2 * *g_2)(\theta, \phi, r_{ij}(\theta, \phi))] + \psi(\theta, \phi, r_{ij}(\theta, \phi)) + \epsilon \qquad (5.56)$$

where $\mathbf{1}_s(r_{ij}(\theta, \phi)) = 1$ *if* $r_{ij}(\theta, \phi)$ *is a spherical patch and zero otherwise.* $\psi(\theta, \phi, r_{ij}(\theta, \phi))$ *models the spatial correlations among shape primitives, and* ϵ *is the noise term.*

If we denote $y(u(\theta, \phi)) = f(u(\theta, \phi)) + \psi(\theta, \phi, r_{ij}(\theta, \phi)) + \epsilon$. *Model* $f(u(\theta, \phi))$ *in (5.56) can be further improved through approximation as*

$$f(u(\theta, \phi)) = \sum_{j=0}^{J}\sum_{i=0}^{I}(\eta_1 * *g_1)(\theta, \phi, \tilde{r}_{ij}(\theta, \phi))+$$

$$\sum_{j=0}^{J}\sum_{i=0}^{I}(\eta_2 * *g_2)(\theta, \phi, r_{ij}(\theta, \phi) - \tilde{r}_{ij}(\theta, \phi)) \qquad (5.57)$$

where $\tilde{r}_i(\theta, \phi)$ *is the smallest spherical patch that covers a plane surface patch or a surface edge.*

The proof of this corollary is omitted because it is an extension from the 2D case presented in Corollary 5.1 and 5.2.

In Sect. 5.1 [35] we studied the dome shapes (Fig. 5.17) and first proposed a convolution formulation of AM processes. The model takes the form of:

$$y(\theta, \phi, r) = \alpha(r) \int_0^{\pi/2} \int_0^{2\pi} g_1(\theta - \tau_1, \phi - \tau_2) * \eta_1(\tau_1, \tau_2) d\tau_1 d\tau_2 +$$

$$\beta(r) + \psi(\theta, \phi, r) + \epsilon \qquad (5.58)$$

where $\eta_1(\theta, \phi) = \cos(2\theta) \sin(\phi)$ and $g_1(\theta, \phi) = \cos(n_1 \phi)[1 + \cos(n_2 \theta + \phi_0)]$ are found suitable for the studied SLA process (pls refer to [35] for a detailed modeling and model estimation).

Apparently the model for dome shapes in Eq. (5.58) is a special case of the model in Eq. (5.57). It provides meaningful understanding of the deviation of spherical patches. Further modeling and few-shot learning based on the generalized model (5.57) will make the quality prediction of 3D freeform shapes feasible with small training data and cost-effective experimentation.

Our initial data-driven few-shot learning [15, 34–36, 50, 85] is consistent with the theoretic framework established in this study. The control-theoretic framework enables the exciting and more powerful model-based machine learning, for example, impulse response estimation through kernel methods in system identification and machine learning [61].

5.3.3   Transfer Function and Block Diagram Representation of AM Systems

Transfer functions and block diagrams have been powerful tools in control theory to design, learn, identify, and analyze system dynamics. Compared to data-driven models [15, 34–36, 50, 85], the control-theoretic framework established in the study allows the utilization of transfer functions to find new interpretation of AM processes.

5.3.3.1   Transfer Function Characterization of AM Process Dynamics

Laplace transform maps function $f : \mathbb{R}^+ \to \mathbb{R}$ to function $F = \mathcal{L}_1[f] : \mathbb{C} \to \mathbb{C}$ of a complex variable s. Double Laplace transform of a 2D f is denoted as $\mathcal{L}_2[f]$. By the property of Laplace transform of convolution, i.e., $\mathcal{L}[f * g] = F(s)G(s)$, the prediction model in Eq. (5.54) for 2D shapes and prediction model in Eq. (5.57) for 3D shapes can be transformed as

$$F(s) = H_1(s)\tilde{R}(s) + H_2(s)[U(s) - \tilde{R}(s)]$$

$$F(s_1, s_2) = H_1(s_1, s_2)G_1(s_1, s_2)\tilde{R}(s_1, s_2) +$$

$$H_2(s_1, s_2)G_2(s_1, s_2)[U(s_1, s_2) - \tilde{R}(s_1, s_2)] \qquad (5.59)$$

where $\hat{Y}$ represents the prediction without considering the error terms.

Remark 5.3 Clearly, the input/output dynamics of the AM processes can be captured by transfer functions $H_1(s)$ and $H_2(s)$ for printing 2D shapes (plates) and by transfer functions $H_1(s_1, s_2)$, $H_2(s_1, s_2)$, $G_1(s_1, s_2)$, and $G_2(s_1, s_2)$ for printing 3D shapes. Identification of transfer functions for AM processes and systems is therefore a critical area that can enable principled design, control, and generalization of process knowledge.

There are generally two strategies to identify transfer functions: data-driven system identification and model-based machine learning [61]. There is no reported research for AM quality control in this regard.

Here we present an example of data-driven method to identify transfer functions when printing 2D shapes. For the SLA process that printed the four disks (Fig. 3.2) we obtained $\eta_1(\theta, r) = -0.0134(r + 0.0088)^{0.86} + 0.0057(r + 0.0088)^{1.13} \cos(2\theta)$ [36]. By Laplace transform of Eq. (5.51), we have $\mathcal{L}_1[\eta_{\{1,2\}}] = H_{\{1,2\}}(s)U(s)$. As to the true input $u(\theta)$ to the SLA process of study, as we pointed out in [36], is not the nominal design r, but $u(\theta) = (r + 0.0088)$ due to overexposure effect.

Since the input to printing disks is r, then $H_1(s)$ can be obtained as

$$
\begin{aligned}
H_1(s) &= \frac{\mathcal{L}_1[\eta_1(\theta, r)]}{\mathcal{L}_1[r + 0.0088]} \\
&= -\frac{0.0134(r + 0.0088)^{0.86}}{r + 0.0088} + \frac{0.0057(r + 0.0088)^{1.13}}{r + 0.0088} \frac{s^2}{s^2 + 4} \\
&= -0.0134(r + 0.0088)^{-0.14} + 0.0057(r + 0.0088)^{0.13} \frac{s^2}{s^2 + 4}
\end{aligned}
$$

Apparently, the radius/size of the disk has an impact on the process dynamics that causes the deformation of disk. The transfer function also indicates that the SLA process needs improvement on two aspects: tuning the process to be independent of size (dimensionless) and to be close to zero gain.

Following the same procedure, we can also obtain $H_2(s)$ by Laplace transform of the square wave function or sawtooth function. For example, by expressing the square wave function in term of Heaviside's function, the $H_2(s)$ for the SLA process of study is

$$
H_2(s) = \frac{\beta(\tilde{r})}{s} \tanh \frac{\pi}{n} s
$$

and $H_2(s)$ is

$$
H_2(s) = \beta(\tilde{r})\Big[\frac{n}{2\pi s^2} - \frac{1}{se^{\frac{2\pi}{n}s} - s}\Big]
$$

for the sawtooth wave function.

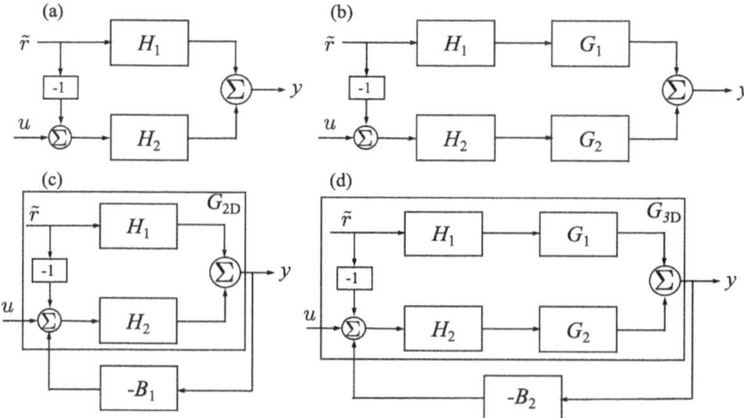

Fig. 5.38 Block diagrams of AM processes. (**a**) Printing 2D shapes, (**b**) printing 3D shapes, (**c**) feedback control for printing 2D shapes, and (**d**) feedback control for printing 3D shapes

5.3.3.2 Block Diagram Representation of AM Processes and Systems

The transfer function expression of AM processes in (5.59) can be conveniently represented in block diagrams. Figure 5.38a, b shows the block diagrams for AM processes to print 2D and 3D shapes, respectively. With obtained transfer functions, system dynamics can be analyzed and simulated with, e.g., MATLAB Simulink.

Another important application of the block diagram algebra for AM is represent, design, and analyze feedback control systems (Fig. 5.38c, d). Since most AM machines adopt standard settings for different materials and designs, a common feedback control strategy is to change the design if errors occur in the manufacturing [32, 36], as opposed to changing process settings. Following the block diagram algebra for feedback control, we first define the equivalent transfer functions for 2D and 3D printing in Fig. 5.38a, b as:

$$G_{2D} = [H_1 - H_2)]\frac{\tilde{R}}{U} + H_2$$

$$G_{3D} = [H_1 G_1 - H_2 G_2]\frac{\tilde{R}}{U} + H_2 G_2 \tag{5.60}$$

Let $\delta(s)$ represent the feedback adjustment. From the block diagram algebra, we know $\delta(s) = -Y(s)B_{\{1,2\}}(s)$. Our work in [32] has shown that

$$\delta^*(u) = -\frac{f(\theta, \phi, r(\theta, \phi))}{1 + \frac{\mathrm{d}f(\theta,\phi,r(\theta,\phi))}{\mathrm{d}r(\theta,\phi)}} = -\frac{f(u)}{1 + f'(u)}$$

where $f(u)$ is given in either Eq. (5.54) or Eq. (5.57). (Examples of $\delta^*(u)$ can be found in [32].)

Through Laplace transform, we will have $\delta^*(s) = \mathcal{L}[\delta^*(u)]$. For system robustness, we replace $Y(s)$ with $F(s)$. Then the transfer function for feedback adjustment is

$$B_{\{1,2\}}(s) = \mathcal{L}\left[\frac{f(u)}{1 + f'(u)}\right]/F(s) \tag{5.61}$$

With this design, the transfer function of the feedback control system will be

$$G_{fb} = \frac{G_{\{2D,3D\}}}{1 + G_{\{2D,3D\}}B_{\{1,2\}}} \tag{5.62}$$

The predicted process response after feedback control will be $\mathcal{L}^{-1}[G_{fb}u]$. System robustness and stability can be analyzed through powerful tools in control theory (e.g., Nyquist's criterion).

By establishing a fabrication-aware, impulse response formulation and modeling framework, this work attempts to bridge exciting research at the intersection of machine learning, control theory, and 3D printing. Though data-driven machine learning models have achieved notable progress in AM, this established theoretical foundation enables model-driven machine learning, which provides not only guided model formulation but also new perspectives and insights into process dynamics. The well-established control theory can thus be readily applied to AM and advance the intelligent quality control in AM.

This control-theoretic framework is built upon the primitives AM inputs, that is, sliced layers to be stacked up and individual layers with boundaries composed by 2D shape primitives. Two theorems are developed to characterize AM processes using impulse response functions. AM process models in convolution formulation are then derived based on the theorems. Transfer function theory and block diagram algebra are then applied to AM process representation, design, and analysis. Examples from SLA processes are given to demonstrate the proposed the control-theoretic framework, including design and analysis of feedback control of AM processes.

There are many other critical categories of ML4AM problems. These open problems include, but are not limited to, verification problem (VP) that aims to define, measure, characterize, evaluate, and verify the quality of 3D printed products; detection and diagnosis problem (DDP) with goals to detect process changes and identify root causes in AM; model adaptation and transfer learning (DTL) problem that aims to quickly generate new models for new AM processes or process conditions; and optimal design for system identification (ODSI) that aims to design an optimal set of training dataset that maximizes system identification, etc. Further research is imperative to advance the control-theoretic foundation and computational research for intelligent AM.

Acknowledgments The work in this chapter was partially supported by the US Office of Naval Research with grant no. N000141110671 and the US National Science Foundation Cyber-Physical Systems Program under grant no. CMMI-1544917 and by the University of Southern California (USC) Pratt & Whitney Institute for Collaborative Engineering Board (PWICE) Project.

References

1. Achlioptas, P., Diamanti, O., Mitliagkas, I., Guibas, L.: Learning representations and generative models for 3D point clouds. In: Dy, J., Krause, A. (eds.) Proceedings of the 35th International Conference on Machine Learning. Proceedings of Machine Learning Research, vol. 80, pp. 40–49. PMLR, Stockholmsmässan, Stockholm (2018)
2. Åström, K.J., Murray, R.M.: Feedback Systems: An Introduction for Scientists and Engineers, 2nd edn. Princeton University Press, Princeton (2021)
3. Bose, P., Maheshwari, A., Shu, C., Wuhrer, S.: A survey of geodesic paths on 3d surfaces. Comput. Geom. **44**(9), 486–498 (2011)
4. Bugeda Miguel Cervera, G., Lombera, G.: Numerical prediction of temperature and density distributions in selective laser sintering processes. Rapid Prototyping J. **5**(1), 21–26
5. Campanelli, S., Cardano, G., Giannoccaro, R., Ludovico, A., Bohez, E.L.: Statistical analysis of the stereolithographic process to improve the accuracy. Comput.-Aided Design **39**(1), 80–86 (2007)
6. Cederbaum, J., Pouplier, M., Hoole, P., Greven, S.: Functional linear mixed models for irregularly or sparsely sampled data. Stat. Model. **16**(1), 67–88 (2016)
7. Chaudhuri, S., Velmurugan, R., Rameshan, R.: Blind Deconvolution Methods: A Review, pp. 37–60. Springer, Berlin (2014). https://doi.org/10.1007/978-3-319-10485-0_3
8. Chen, Z., Zhang, H.: Learning implicit fields for generative shape modeling. In: Proceedings of the IEEE Conference on Computer Vision and Pattern Recognition, pp. 5939–5948 (2019)
9. Cheng, L., Tsung, F., Wang, A.: A statistical transfer learning perspective for modeling shape deviations in additive manufacturing. IEEE Robot. Autom. Lett. **2**(4), 1988–1993 (2017)
10. Cheng, L., Wang, A., Tsung, F.: A prediction and compensation scheme for in-plane shape deviation of additive manufacturing with information on process parameters. IISE Trans. **50**(5), 394–406 (2018)
11. Cheng, L., Wang, K., Tsung, F.: A hybrid transfer learning framework for in-plane freeform shape accuracy control in additive manufacturing. IISE Trans. **53**(3), 298–312 (2020)
12. Colosimo, B.M., Grasso, M.: Spatially weighted PCA for monitoring video image data with application to additive manufacturing. J. Quality Technol. **50**(4), 391–417 (2018)
13. Das, S.: Physical aspects of process control in selective laser sintering of metals. Adv. Eng. Mater. **5**(10), 701–711 (2003)
14. De Bock, K.W., Coussement, K., Van den Poel, D.: Ensemble classification based on generalized additive models. Comput. Stat. Data Anal. **54**(6), 1535–1546 (2010)
15. de Souza Borges Ferreira, R., Sabbaghi, A., Huang, Q.: Automated geometric shape deviation modeling for additive manufacturing systems via bayesian neural networks. IEEE Trans. Autom. Sci. Eng. **17**(2), 584–598 (2020). https://doi.org/10.1109/TASE.2019.2936821
16. del Castillo, E., Colosimo, B.M., Tajbakhsh, S.D.: Geodesic gaussian processes for the parametric reconstruction of a free-form surface. Technometrics **57**(1), 87–99 (2015)
17. Decker, N., Huang, Q.: Geometric accuracy prediction for additive manufacturing through machine learning of triangular mesh data. In: ASME 14th International Manufacturing Science & Engineering (MSEC) Conference. Erie, PA, USA (2019)
18. Decker, N., Lyu, M., Wang, Y., Huang, Q.: Geometric Accuracy Prediction and Improvement for Additive Manufacturing Using Triangular Mesh Shape Data. J. Manuf. Sci. Eng. **143**(6), 1–12 (2021). https://doi.org/10.1115/1.4049089
19. Decker, N., Wang, Y., Huang, Q.: Efficiently registering scan point clouds of 3d printed parts for shape accuracy assessment and modeling. J. Manuf. Syst. **56**, 587–597 (2020)
20. Denlinger, E.R., Heigel, J.C., Michaleris, P., Palmer, T.: Effect of inter-layer dwell time on distortion and residual stress in additive manufacturing of titanium and nickel alloys. J. Mater. Processing Technol. **215**, 123–131 (2015)
21. Denlinger, E.R., Irwin, J., Michaleris, P.: Thermomechanical modeling of additive manufacturing large parts. ASME Trans. J. Manuf. Sci. Eng. **136**(6), 061007 (2014)

22. Ferreira, R.D.S.B., Sabbaghi, A., Huang, Q.: Automated Geometric Shape Deviation Modeling for Additive Manufacturing Systems via Bayesian Neural Networks. IEEE Trans. Autom. Sci. Eng. **17**(2), 584–598 (2020). https://doi.org/10.1109/TASE.2019.2936821
23. Francis, J., Bian, L.: Deep learning for distortion prediction in laser-based additive manufacturing using big data. Manuf. Lett. **20**, 10–14 (2019)
24. Francis, J., Sabbaghi, A., Ravi Shankar, M., Ghasri-Khouzani, M., Bian, L.: Efficient distortion prediction of additively manufactured parts using Bayesian model transfer between material systems. J. Manuf. Sci. Eng. **142**(5), 1–16 (2020)
25. Friedman, J.H.: Stochastic gradient boosting. Comput. Stat. Data Anal. **38**(4), 367–378 (2002)
26. Goetz, J.N., Guthrie, R.H., Brenning, A.: Integrating physical and empirical landslide susceptibility models using generalized additive models. Geomorphology **129**(3–4), 376–386 (2011)
27. Guo, S., Guo, W., Bain, L.: Hierarchical spatial-temporal modeling and monitoring of melt pool evolution in laser-based additive manufacturing. IISE Trans. **52**(9), 977–997 (2020)
28. Ha, S., Ransikarbum, K., Han, H., Kwon, D., Kim, H., Kim, N.: A dimensional compensation algorithm for vertical bending deformation of 3d printed parts in selective laser sintering. Rapid Prototyping J. **24**(6), 955–963 (2018)
29. Hastie, T., Tibshirani, R.: Generalized additive models for medical research. Stat. Methods Med. Res. **4**(3), 187–196 (1995)
30. Hastie, T., Tibshirani, R., Friedman, J.: The Elements of Statistical Learning: Data Mining, Inference, and Prediction. Springer, Berlin (2009)
31. Hossain, M.S., Espalin, D., Ramos, J., Perez, M., Wicker, R.: Improved mechanical properties of fused deposition modeling-manufactured parts through build parameter modifications. ASME Trans. J. Manuf. Sci. Eng. **136**(6), 061002 (2014)
32. Huang, Q.: An analytical foundation for optimal compensation of three-dimensional shape deformation in additive manufacturing. ASME Trans. J. Manuf. Sci. Eng. **138**(6), 061010 (2016)
33. Huang, Q.: An impulse response formulation for small-sample learning and control of additive manufacturing quality. IISE Trans. **0**(ja), 1–27 (2022). https://doi.org/10.1080/24725854.2022.2113186
34. Huang, Q., Nouri, H., Xu, K., Chen, Y., Sosina, S., Dasgupta, T.: Statistical predictive modeling and compensation of geometric deviations of three-dimensional printed products. ASME Trans. J. Manuf. Sci. Eng. **136**(6), 061008 (2014)
35. Huang, Q., Wang, Y., Lyu, M., Lin, W.: Shape deviation generator (sdg) – a convolution framework for learning and predicting 3d printing shape accuracy. IEEE Trans. Autom. Sci. Eng. **17**(3), 1486–1500 (2020)
36. Huang, Q., Zhang, J., Sabbaghi, A., Dasgupta, T.: Optimal offline compensation of shape shrinkage for 3d printing processes. IIE Trans. Quality Reliab. **47**(5), 431–441 (2015)
37. Huber, D., Kapuria, A., Donamukkala, R., Hebert, M.: Parts-based 3d object classification. In: Computer Vision and Pattern Recognition, 2004. CVPR 2004. Proceedings of the 2004 IEEE Computer Society Conference on, vol. 2, pp. II–II. IEEE, Piscataway (2004)
38. Jin, Y., Joe Qin, S., Huang, Q.: Offline predictive control of out-of-plane shape deformation for additive manufacturing. J. Manuf. Sci. Eng. **138**(12), 121005 (2016)
39. Jin, Y., Qin, S., Huang, Q.: Offline predictive control of out-of-plane geometric errors for additive manufacturing. ASME Trans. Manuf. Sci. Eng. **138**(12), 121005 (2016)
40. Jin, Y., Qin, S.J., Huang, Q.: Modeling inter-layer interactions for out-of-plane shape deviation reduction in additive manufacturing. IISE Trans. **52**(7), 721–731 (2020)
41. Kazhdan, M., Funkhouser, T., Rusinkiewicz, S.: Rotation invariant spherical harmonic representation of 3 d shape descriptors. In: Symposium on Geometry Processing, vol. 6, pp. 156–164 (2003)
42. Khanzadeh, M., Rao, P., Jafari-Marandi, R., Smith, B.K., Tschopp, M.A., Bian, L.: Quantifying geometric accuracy with unsupervised machine learning: using self-organizing map on fused filament fabrication additive manufacturing parts. J. Manuf. Sci. Eng. **140**(3), 031011 (2017)

43. King, W., Anderson, A.T., Ferencz, R.M., Hodge, N.E., Kamath, C., Khairallah, S.A.: Overview of modelling and simulation of metal powder bed fusion process at lawrence livermore national laboratory. Mater. Sci. Technol. **31**(8), 957–968 (2015)
44. King, W.E., Anderson, A.T., Ferencz, R.M., Hodge, N.E., Kamath, C., Khairallah, S.A., Rubenchik, A.M.: Laser powder bed fusion additive manufacturing of metals; physics, computational, and materials challenges. Appl. Phys. Rev. **2**(4) (2015)
45. Lanzotti, A., Martorelli, M., Staiano, G.: Understanding process parameter effects of reprap open-source three-dimensional printers through a design of experiments approach. ASME Trans. J. Manuf. Sci. Eng. **137**(1), 011017 (2015)
46. Li, Y., Gu, D.: Parametric analysis of thermal behavior during selective laser melting additive manufacturing of aluminum alloy powder. Mater. Design **63**, 856–867 (2014)
47. Lindgren, L.E., Lundbäck, A., Fisk, M., Pederson, R., Andersson, J.: Simulation of additive manufacturing using coupled constitutive and microstructure models. Addit. Manuf. **12**, 144–158 (2016)
48. Liu, J., Liu, C., Bai, Y., Rao, P., Williams, C.B., Kong, Z.: Layer-wise spatial modeling of porosity in additive manufacturing. IISE Trans. **51**(2), 109–123 (2019)
49. Luan, H., Grasso, M., Colosimo, B., Huang, Q.: Prescriptive data-analytical modeling of laser powder bed fusion processes for accuracy improvement. J. Manuf. Sci. Eng. Trans. ASME **141**(1) (2019). https://doi.org/10.1115/1.4041709
50. Luan, H., Huang, Q.: Prescriptive modeling and compensation of in-plane shape deformation for 3-d printed freeform products. IEEE Trans. Autom. Sci. Eng. **14**(1), 73–82 (2017)
51. Luan, H., Huang, Q.: Prescriptive modeling and compensation of in-plane shape deformation for 3-D printed freeform products. IEEE Trans. Autom. Sci. Eng. **14**(1), 73–82 (2017)
52. Luan, H., Post, B., Huang, Q.: Statistical process control of in-plane shape deformation for additive manufacturing. pp. 1274–1279. 2017 IEEE International Conference on Automation Science and Engineering (CASE 2017), Xi'an, China (2017)
53. Mardani, M., Mateos, G., Giannakis, G.B.: Dynamic anomalography: Tracking network anomalies via sparsity and low rank. IEEE J. Sel. Top. Signal Process. **7**(1), 50–66 (2012)
54. Marr, D., Nishihara, H.K.: Representation and recognition of the spatial organization of three-dimensional shapes. Proc. R. Soc. Lond. B **200**(1140), 269–294 (1978)
55. Montagnat, J., Delingette, H., Ayache, N.: A review of deformable surfaces: topology, geometry and deformation. Image Vision Comput. **19**(14), 1023–1040 (2001)
56. Mori, K., Osakada, K., Takaoka, S.: Simplified three-dimensional simulation of non-isothermal filling in metal injection moulding by the finite element method. Eng. Comput. **13**(2), 111–121 (1996)
57. Pal, D., Patil, N., Zeng, K., Stucker, B.: An integrated approach to additive manufacturing simulations using physics based, coupled multiscale process modeling. ASME Trans. J. Manuf. Sci. Eng. **136**(6), 061022 (2014)
58. Pan, J., Zi, Y., Chen, J., Zhou, Z., Wang, B.: Liftingnet: A novel deep learning network with layerwise feature learning from noisy mechanical data for fault classification. IEEE Trans. Ind. Electron. **65**(6), 4973–4982 (2017)
59. Paul, R., Anand, S., Gerner, F.: Effect of thermal deformation on part errors in metal powder based additive manufacturing processes. J. Manuf. Sci. Eng. **136**(3), 031009 (2014)
60. Pavlidis, T.: Algorithms for Graphics and Image Processing. Springer, Berlin (2012)
61. Pillonetto, G., Dinuzzo, F., Chen, T., De Nicolao, G., Ljung, L.: Kernel methods in system identification, machine learning and function estimation: a survey. Automatica **50**(3), 657–682 (2014)
62. Raghunath, N., Pandey, P.: Improving accuracy through shrinkage modelling by using taguchi method in selective laser sintering. Int. J. Mach. Tools Manuf. **47**(6), 985–995 (2007)
63. Rao, P.K., Liu, J.P., Roberson, D., Kong, Z.J., Williams, C.: Online real-time quality monitoring in additive manufacturing processes using heterogeneous sensors. J. Manuf. Sci. Eng. **137**(6), 061007 (2015)
64. Rasmussen, C.E., Williams, C.K.I.: Gaussian Processes for Machine Learning. The MIT Press, Cambridge (2006)

65. Rasmussen, C.E., Williams, C.K.I.: Gaussian Processes for Machine Learning. MIT Press, Cambridge (2006)
66. Sabbaghi, A., Dasgupta, T., Huang, Q., Zhang, J.: Inference for deformation and interference in 3d printing. Ann. Appl. Stat. **8**(3), 1395–1415 (2014)
67. Sabbaghi, A., Huang, Q.: Model transfer across additive manufacturing processes via mean effect equivalence of lurking variables. Ann. Appl. Stat. **12**(4), 2409–2429 (2018)
68. Sabbaghi, A., Huang, Q., Dasgupta, T.: Bayesian model building from small samples of disparate data for capturing in-plane deviation in additive manufacturing. Technometrics **60**(4), 532–544 (2018). https://doi.org/10.1080/00401706.2017.1391715
69. Sabbaghi, A., Huang, Q., Dasgupta, T.: Bayesian model building from small samples of disparate data for capturing in-plane deviation in additive manufacturing. Technometrics **60**(4), 532–544 (2018)
70. Samie Tootooni, M., Dsouza, A., Donovan, R., Rao, P.K., Kong, Z.J., Borgesen, P.: Classifying the dimensional variation in additive manufactured parts from laser-scanned three-dimensional point cloud data using machine learning approaches. J. Manuf. Sci. Eng. **139**(9), 091005 (2017)
71. Saupe, D., Vranić, D.V.: 3d model retrieval with spherical harmonics and moments. In: Joint Pattern Recognition Symposium, pp. 392–397. Springer, Berlin (2001)
72. Sharon, E., Mumford, D.: 2d-shape analysis using conformal mapping. Int. J. Comput. Vis. **70**(1), 55–75 (2006)
73. Sood, A., Ohdar, R., Mahapatra, S.: Improving dimensional accuracy of fused deposition modelling processed part using grey Taguchi method. Mater. Design **30**(10), 4243–4252 (2009)
74. Sorkine, O., Alexa, M.: As-rigid-as-possible surface modeling. In: Symposium on Geometry processing, vol. 4, pp. 109–116 (2007)
75. Srivastava, A., Turaga, P., Kurtek, S.: On advances in differential-geometric approaches for 2d and 3d shape analyses and activity recognition. Image Vis. Comput. **30**(6-7), 398–416 (2012)
76. Steuben, J.C., Birnbaum, A.J., Michopoulos, J.G., Iliopoulos, A.P.: Enriched analytical solutions for additive manufacturing modeling and simulation. Addit. Manuf. **25**, 437–447 (2019)
77. Sun, X., Rosin, P.L., Martin, R.R., Langbein, F.C.: Noise analysis and synthesis for 3d laser depth scanners. Graph. Models **71**(2), 34–48 (2009)
78. Thiery, J.M., Guy, É., Boubekeur, T.: Sphere-meshes: Shape approximation using spherical quadric error metrics. ACM Trans. Graph. **32**(6), 1–12 (2013)
79. Tibshirani, R.: Regression shrinkage and selection via the lasso. J. Roy. Stat. Soc.: B (Methodological) **58**(1), 267–288 (1996)
80. Tong, K., Joshi, S., Lehtihet, E.: Error compensation for fused deposition modeling (fdm) machine by correcting slice files. Rapid Prototyping J. **14**(1), 4–14 (2008)
81. Tong, K., Lehtihet, E., Joshi, S.: Parametric error modeling and software error compensation for rapid prototyping. Rapid Prototyping J. **9**(5), 301–313 (2003)
82. Tootooni, M.S., Dsouza, A., Donovan, R., Rao, P.K., Kong, Z.J., Borgesen, P.: Classifying the dimensional variation in additive manufactured parts from laser-scanned three-dimensional point cloud data using machine learning approaches. J. Manuf. Sci. Eng. **139**(9), 091005 (2017)
83. Tsung, F., Zhang, K., Cheng, L., Song, Z.: Statistical transfer learning: A review and some extensions to statistical process control. Quality Eng. **30**(1), 115–128 (2018)
84. Van Kaick, O., Zhang, H., Hamarneh, G., Cohen-Or, D.: A survey on shape correspondence. In: Computer Graphics Forum, vol. 30, pp. 1681–1707. Wiley Online Library (2011)
85. Wang, A., Song, S., Huang, Q., Tsung, F.: In-plane shape-deviation modeling and compensation for fused deposition modeling processes. IEEE Trans. Autom. Sci. Eng. **17**(2), 968–976 (2017)
86. Wang, C., Tan, X., Tor, S., Lim, C.: Machine learning in additive manufacturing: State-of-the-art and perspectives. Addit. Manuf. **36**, 101538 (2020)
87. Wang, W., Cheah, C., Fuh, J., Lu, L.: Influence of process parameters on stereolithography part shrinkage. Mater. Design **17**(4), 205–213 (1996)
88. Wang, X.: Calibration of shrinkage and beam offset in SLS process. Rapid Prototyping J. **5**(3), 129–133 (1999)

89. Wood, S.N.: Generalized Additive Models: An Introduction with R. CRC Press, Boca Raton (2017)
90. Wu, J., Zhang, C., Xue, T., Freeman, B., Tenenbaum, J.: Learning a probabilistic latent space of object shapes via 3d generative-adversarial modeling. In: Advances in Neural Information Processing Systems, pp. 82–90 (2016)
91. Xu, K., Chen, Y.: Mask image planning for deformation control in projection-based stereolithography process. J. Manuf. Sci. Eng. 137(3), 031014 (2015)
92. Xu, K., Kwok, T.H., Zhao, Z., Chen, Y.: A reverse compensation framework for shape deformation control in additive manufacturing. J. Comput. Inform. Sci. Eng. 17(2), 021012 (2017)
93. Zahn, C.T., Roskies, R.Z.: Fourier descriptors for plane closed curves. IEEE Trans. Comput. C-21(3), 269–281 (1972)
94. Zha, W., Anand, S.: Geometric approaches to input file modification for part quality improvement in additive manufacturing. J. Manuf. Process. 20, 465–477 (2015)
95. Zhang, D., Lu, G.: Review of shape representation and description techniques. Pattern Recogn. 37(1), 1–19 (2004)
96. Zhao, P., Yu, B.: On model selection consistency of lasso. J. Mach. Learn. Res. 7, 2541–2563 (2006)
97. Zhou, J., Herscovici, D., Chen, C.: Parametric process optimization to improve the accuracy of rapid prototyped stereolithography parts. Int. J. Mach. Tools Manuf. 40(3), 363–379 (2000)
98. Zhu, Z., Anwer, N., Huang, Q., Mathieu, L.: Machine learning in tolerancing for additive manufacturing. CIRP Ann. 67(1) (2018). https://doi.org/10.1016/j.cirp.2018.04.119

Process-Informed Optimal Compensation for Additive Manufacturing Quality Control

Chapter 6
Foundation of Offline Optimal Compensation of 3D Shape Deviation

6.1 Inverse Problem: Minimize Shape Deviation Through Compensation

While predicting shape deviation belongs to the forward problem, minimizing shape deviation of AM-built products is a challenging inverse problem due to geometric complexity, product varieties, material phase changing and shrinkage, interlayer bonding, and limited sample data. Various methods and strategies have been developed to improve geometric quality of AM processes. We summarize the related work in Table 6.1 with a sample literature given for each category.

Given measurement data regarding the shape deviation of an AM-built product, one viable and efficient approach for accuracy control is through compensation of the product design to offset the geometric shape deviation. As shown in Fig. 6.1, once the deviation between the actual and nominal boundaries or profiles of a product is known through either measurement or model prediction, the adjustment of the CAD model can be applied in such a way that the new actual profile will match the original nominal profile as much as possible. The key issue is therefore to solve the inverse problem and determine the optimal amount of compensation based on measured or predicted shape deviation or deviation for both 2D and 3D geometries.

The prevalent method of determining compensation in practice is shrinkage compensation factor approach, which takes its root in the material shrinkage study in casting and injection molding processes. This approach applies a shrinkage compensation factor uniformly to the entire product or different factors to the CAD model for each section of a product [6]. This method implicitly assumes the shape deviation is uniform in the section where the compensation factor is applied. Since products built via AM often have complex shapes, this assumption does not hold for general cases. The strategy of applying section-wise compensation may have detrimental effects on the overall shape due to "carryover effects" or interference

© The Author(s), under exclusive license to Springer Nature Switzerland AG 2025

Q. Huang, *Domain-informed Machine Learning for Smart Manufacturing*,

https://doi.org/10.1007/978-3-031-91631-1_6

Table 6.1 Geometric quality control methods in additive manufacturing

	Method and strategy	Sample literature
1.	Simulation study based on the first principles	[15, 18, 21]
2.	Offline optimization of process settings through experimentation	[20, 28, 32]
3.	Machine calibration through building test parts	[14, 23, 24, 28, 29, 32]
4.	Part geometry calibration through extensive trial build	[6]
5.	Adjustment of product design and process planning	[3, 10, 11, 14, 16, 17, 23, 24, 30, 31]
6.	Feedback control and online monitoring	[4, 5, 7, 8, 12, 19, 22]

Fig. 6.1 Adjust CAD model

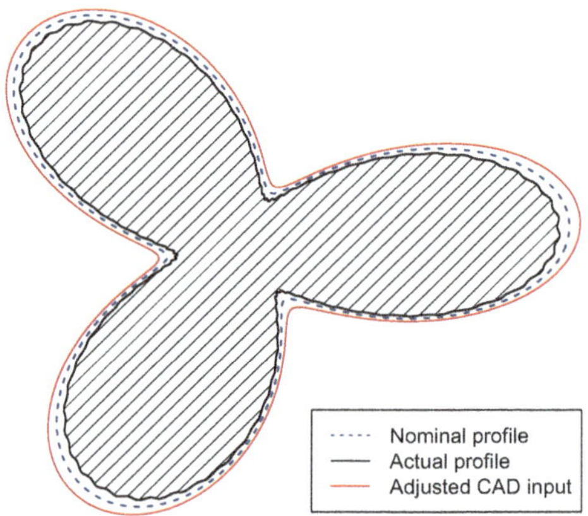

between adjacent sections [17]. Compensation factor approach is thus far from being optimal for AM.

Huang and co-authors [11] establish an optimal compensation policy for in-plane ($x - y$ plane) shape deviation. The optimal amount of compensation for reducing deviations at each point on the product boundary is derived for high-precision AM. This strategy apparently differs from the previous approaches where shrinkage compensation factors are applied, at best, to a limited number of sections of a product. Experimental validation shows the compensation method is able to improve accuracy by an order of magnitude for circular disks [11], by at least 75% for polygons [10] and by at least 50% for 2D freeform shapes [13]. However, the study in [11] does not analyze the property of the proposed optimal compensation policy. More importantly, no optimal policy or the optimal amount of compensation is provided to compensate 3D shape deviation.

This chapter provides an analytical foundation to achieve optimal compensation for high-precision AM [9]. We first present the optimal compensation policy or the optimal amount of compensation for 2D shape deviation. By analyzing its optimality property, we propose the minimum area deviation (MAD) criterion to offset 2D shape deviation. This result is then generalized by establishing the minimum volume deviation (MVD) criterion and by deriving the optimal amount of compensation for 3D shape deviation. Furthermore, MAD and MVD criteria provide convenient quality measure or quality index for AM-built products that facilitate online monitoring and feedback control of shape geometric accuracy.

6.2 Optimal Compensation of 2D Geometric Shape Deviation

In real applications, a thin product or a section of a product with small thickness can be approximated with a 2D shape where only in-plane geometric error is of major concern. Therefore, it is meaningful for us to start with the discussion of optimal compensation for 2D shape deviation.

Let $\Delta r(\theta, r_0(\theta)) = f(\theta, r_0(\theta)) + \varepsilon$, with $f(\theta, r_0(\theta))$ being the model predicting shape deviation with error ε. Modeling of $f(\theta, r_0(\theta))$ can be achieved either through finite element modeling simulation or through data-driven approaches. In this chapter, $f(\theta, r_0(\theta))$ is assumed to be known.

6.2.1 Optimal Compensation to Minimize 2D Shape Deviation

We aim to reduce deviation of manufactured products by direct compensation to the CAD model. Specifically, we revise the CAD model according to prediction of deviation, obtained through an understanding of the effect of compensation to the boundary of the CAD model. Under the polar coordinates system, a compensation of $x(\theta)$ units at location θ can be represented as an extension of the product's radius by $x(\theta)$ units in that specific direction θ. Clearly, we want an optimal compensation function that results in elimination of systematic shrinkage at all angles. To obtain such a function, we need to extend model (3.1) to accommodate the effect of compensation.

We first generalize the notation in (2.2). Let $r(\theta, r_0(\theta), x(\theta))$ denote the actual radius at angle θ when compensation $x(\theta)$ is applied at that location. Since the amount of compensation is relatively small in comparison to the nominal radius $r_0(\theta)$ (e.g., 1% to 2%), we can reasonably assume dynamics of the manufacturing and deviation process remains the same under compensation as compared to the entire process without compensation. Therefore, the predictive model $f(\cdot, \cdot)$ is still valid, given the new design input $r_0(\theta) + x(\theta)$. Hence, we have

$$r(\theta, r_0(\theta), x(\theta)) - (r_0(\theta) + x(\theta)) = f(\theta, r_0(\theta) + x(\theta))$$

Note that $f(\theta, r_0(\theta))$ without compensation is a special case of $f(\theta, r_0(\theta) + x(\theta))$ with $f(\theta, r_0(\theta) + 0) + 0$, and $\Delta r(\theta, r_0(\theta))$ is $\Delta r(\theta, r_0(\theta), 0)$.

Since the reference or nominal shape remains to be $r_0(\theta)$, the shape deviation at an angle θ is

$$\Delta r(\theta, r_0(\theta), x(\theta)) = r(\theta, r_0(\theta), x(\theta)) - r_0(\theta)$$

$$= f(\theta, r_0(\theta) + x(\theta)) + x(\theta). \tag{6.1}$$

The optimal amount of compensation $x^*(\theta)$ will minimize $\Delta r(\theta, r_0(\theta), x(\theta))$ or $\Delta r(\theta, r_0(\theta), x(\theta)) = 0$ for any θ. To find the solution of $x^*(\theta)$, the Taylor series expansion of $f(\theta, r_0(\theta) + x(\theta))$ at $r_0(\theta)$ is applied in [11]. We have from (6.1) that

$$\Delta r(\theta, r_0(\theta), x(\theta)) \approx f(\theta, r_0(\theta)) + f'(\theta, r_0(\theta))x(\theta) + x(\theta). \tag{6.2}$$

where $f'(\theta, r_0(\theta))$ is the derivative with respect to $r_0(\theta)$.

By equating $\Delta r(\theta, r_0(\theta), x(\theta))$ to zero, the optimal compensation function $x^*(\theta)$ can be obtained as

$$x^*(\theta) = -\frac{f(\theta, r_0(\theta))}{1 + f'(\theta, r_0(\theta))} \tag{6.3}$$

Remark 6.1 The result in (6.3) shows that the optimal compensation is not simply to apply the negative value of the observed or predicted shape deviation, i.e., $x^*(\theta) = -f(\theta, r_0(\theta)) = -\Delta r(\theta, r_0(\theta))$, which is essentially the shrinkage compensation factor approach widely used in practice. The compensation factor approach can only be optimal when $f'(\theta, r_0(\theta)) = 0$, i.e., deviation is uniform everywhere.

In consideration of accuracy control for AM, we view AM as one-of-a-kind manufacturing with frequent change of designs. Under this problem setting, accuracy control for products never being built before will be frequently encountered. Therefore, we only take the first-order Taylor series expansion for robustness consideration.

6.2.2 Minimum Area Deviation (MAD) Criterion and Optimality

This subsection aims to understand the optimality of the optimal compensation policy given in (6.3). First, we introduce the following optimality criterion.

Definition 6.1 (Minimum Area Deviation (MAD) Criterion) For a 2D shape deviating from its intended design model, the minimum area deviation (MAD) criterion is satisfied if the total absolute area change of the deformed shape is the smallest.

The optimal amount of compensation given in (6.3) intends to minimize the deviation of every point along the boundary of a product. Intuitively, we can expect the optimal compensation will result in the minimum area change of the nominal shape. Below we give a theorem and its proof.

Theorem 6.1 (Minimum-Area-Deviation Compensation) *The optimal compensation policy in (6.3) satisfies MAD criterion.*

Proof As shown in Fig. 6.2a, the area deviation at location θ due to $\Delta r(\theta, r_0(\theta))$ is

$$\Delta S(r(\theta, r_0(\theta)), \theta) \approx [r_0(\theta)\Delta\theta]\Delta r(\theta, r_0(\theta)) \approx r_0(\theta)f(\theta, r_0(\theta))d\theta$$

The total area deviation before compensation is therefore

$$\Delta S = \int_0^{2\pi} r_0(\theta)|f(\theta, r_0(\theta))|d\theta$$

Absolute value $|f(\theta, r_0(\theta))|$ is applied because shape deviation can be positive or negative at different locations.

After applying compensation $x(\theta)$ as shown in Fig. 6.2b, the total area deviation becomes

$$\Delta S(x(\theta)) = \int_0^{2\pi} r_0(\theta)\Big|f(\theta, r_0(\theta) + x(\theta)) + x(\theta)\Big|d\theta \tag{6.4}$$

Note that $x(\theta)$ in Fig. 6.2b is an unknown quantity to be determined. Its optimal value can be positive or negative, depending on the shape deviation at location θ. ΔS before compensation is just a special case with $x(\theta) = 0$ or $\Delta S(x(\theta) = 0)$.

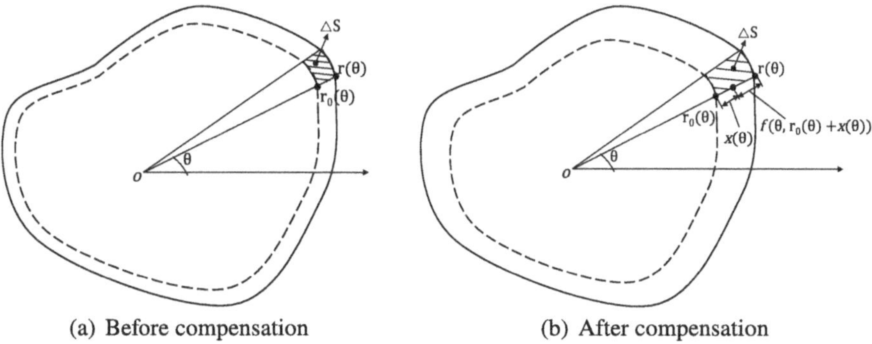

(a) Before compensation   (b) After compensation

Fig. 6.2 Shape area deviation before and after compensation

The optimal compensation $x^*(\theta)$ will minimize the area deviation from the nominal shape, i.e.,

$$\min_{x(\theta)} \quad \Delta S(x(\theta)) \tag{6.5}$$

$$\text{for all} \quad 0 \le \theta \le 2\pi.$$

And the solution clearly has to satisfy

$$f(\theta, r_0(\theta) + x(\theta)) + x(\theta) = 0$$

The Taylor series expansion of $f(\theta, r_0(\theta) + x(\theta))$ at $r_0(\theta)$ will yield the same result in (6.3) and complete the proof.

Theorem 6.1 is not just a reinterpretation of the previously developed compensation policy (6.3). It has two important implications by providing (1) a convenient measure of in-plane error and (2) a stepping stone to extend the 2D compensation policy to 3D case.

We first discuss the quality measure and defer the 3D extension to the next section.

Definition 6.2 (Equivalent Amount of Compensation) Suppose an error source, either from design or manufacturing, causes a product deviating from its intended shape $r_0(\theta)$ to $r(\theta, r_0(\theta))$ by $\Delta r(\theta, r_0(\theta))$. The same amount of deviation $\Delta r(\theta, r_0(\theta))$ can be reproduced if we apply the correct amount of compensation $x(\theta)$ under normal condition (i.e., condition without assignable causes). Here, $x(\theta)$ is defined as the equivalent amount of compensation to the error source.

This definition is based on the concept of error equivalence we developed for traditional manufacturing [25–27]. Two error sources are equivalent if they can generate the same error pattern. Once we establish the equivalence between two error sources, we can use one error source to compensate the other.

For AM processes, we represent the overexposure error in the stereolithography process [11] by the equivalent amount of compensation. This concept will greatly simplify the predictive modeling and take the best use of available model $f(\theta, r_0(\theta))$. Given an error source and its equivalent amount of compensation $x(\theta)$, the product shape deviation can be predicted by $f(\theta, r_0(\theta) + x(\theta)) + x(\theta)$. Additionally, this concept can be utilized for error source management through deviation compensation.

Lemma 6.1 (In-Plane Quality Measure) $\Delta S(x(\theta))$ *is a quality measure or quality index of an in-plane shape deviation.*

Proof Based on the definition of equivalent amount of compensation, the product shape deviation or the quality of two units can be represented as $f(\theta, r_0(\theta) + x_1(\theta)) + x_1(\theta)$ and $f(\theta, r_0(\theta) + x_2(\theta)) + x_2(\theta)$. From Theorem 1 and its proof, the shape deviation of two units will end up with area deviations $\Delta S(x_1(\theta))$

and $\Delta S(x_2(\theta))$. $x_1(\theta)$ is more optimal than $x_2(\theta)$ if $\Delta S(x_1(\theta)) \leq \Delta S(x_2(\theta))$. Therefore, $\Delta S(x(\theta))$ is a quantitative measure of shape quality; the smaller, the better.

Note that the quality measure based on (6.3) requires to compare the deviation of every point along the boundary of a product, which ends up with a deviation profile. $\Delta S(x(\theta))$ is a much more convenient quality measure with theoretical support from Theorem 6.1.

Remark 6.2 (MinMax Quality Criterion) Another candidate quality measure is the maximum shape deviation of AM-built products, which is equivalent to the Hausdorff distance between the profile of AM-built product and the profile of nominal design shape. As an example of the deviation profiles of circular disk shapes shown in Fig. 3.2, the maximum deviation for the small disk with $r_0 = 0.5''$ is the peak value (expansion), while the maximum deviation for the rest of disks is at the valleys (shrinkage).

The optimal compensation policy aiming to minimize the maximum shape deviation has to follow the MinMax criterion. Note that MinMax compensation policy is not simply to compensate the maximum shape deviation at one location because neighboring locations could become new peaks or valleys after compensation.

The optimal amount of compensation given in (6.3) is clearly not a MinMax policy because equal weights are assigned to minimize all deviations on the deviation profile. New optimal compensation policy in addition to (6.3) has to be developed if MiniMax quality criterion is adopted. MAD criterion will minimize the overall or total shape deviation, not necessarily the maximum shape deviation.

6.2.3 *Experimental Validation of Optimal Compensation Algorithm for 2D Cases*

6.2.3.1 Optimal Compensation of Printed Circular Disks

Section 3.1 introduced the deviation modeling of circular disk shape and experimentation with SLA process [11]. Four circular disk parts with different sizes were built, and their in-plane shape deviation profiles are shown in Fig. 3.2. This section demonstrates the optimal compensation policy with and without consideration of the overexposure effect.

Given the deviation model (3.4) without considering the overexposure effect, $f(\theta, r_0(\theta))$ for disks with different sizes has the form:

$$f(\theta, r_0(\theta)) = \hat{\alpha} r_0^{\hat{a}} + \hat{\beta} r_0^{\hat{b}} \cos(2\theta)$$

where $\hat{\alpha}$, $\hat{\beta}$, $\hat{a}$ and $\hat{b}$ denote the Bayes estimators of model parameters α, β, a and b, respectively (Table 3.3). The hat sign $\hat{\ }$ will be omitted hereafter for simplicity of presentation without losing clarity.

Substituting this $f(\theta, r_0)$ into the general compensation model (6.1), we have the predicted deviation as

$$f(\theta, r_0, x(\theta)) = x(\theta) + \alpha(r_0 + x(\theta))^a + \beta(r_0 + x(\theta))^b \cos(2\theta), \qquad (6.6)$$

Further approximation by the first and the second terms of the Taylor expansion at point r_0 yields

$$f(\theta, r_0, x(\theta)) \approx \alpha r_0^a + \beta r_0^b \cos(2\theta) + (1 + a\alpha r_0^{a-1} + b\beta r_0^{b-1} \cos(2\theta))x(\theta). \qquad (6.7)$$

Equation (6.7) technically serves as a description of the predicted deviation of disks at θ when compensation $x(\theta)$ applied to the boundary of the CAD model.

By setting $f(\theta, r_0, x(\theta))$ to zero, we have the closed-form expression for the optimal compensation of disks with different sizes r_0:

$$x^*(\theta) = -\frac{\alpha r_0^a + \beta r_0^b \cos(2\theta)}{1 + a\alpha r_0^{a-1} + b\beta r_0^{b-1}\cos(2\theta)} \qquad (6.8)$$

To validate the effectiveness of the optimal compensation strategy in Eq. 6.3, a disk with nominal radius $1.0''$ is built with adjustment of CAD design based on Eq. (6.8), that is, $1.0'' + x^*(\theta)$. Command *law curve* in the CAD software *UG* is employed to construct the compensated CAD model according to (6.8). All manufacturing and measuring specifications remain the same as in the case of uncompensated disks. Parameters α, β, a, and b are set as the mean values in Table 3.3.

The result for this disk is shown in Fig. 6.3. The shape deviation of the uncompensated disk is represented by the black line, and deviation of the compensated disk is the red line. The nominal value 0 is plotted as the dashed line. Obviously, absolute deviation has significantly decreased under compensation. The sinusoidal pattern of the original deviation has also been eliminated. We compute the average and standard deviation of deviation for both products. As can be seen in Table 6.2, the average and standard deviation of deviation have decreased to 10% of the original. This demonstrates that the compensation method has effectively increased the accuracy of the product. However, we notice that the deviation under compensation is still above the desired value of 0, which indicates an overall bias of the compensation method. The source and solution to this bias is attributed to overexposure, which is discussed in Sect. 3.1.4.

The error source of light overexposure causes expansion of the illuminated shape due to the spread of light beams on the boundary of the product. Based on the definition of equivalent amount of compensation, the constant effect of overexposure for all disks is equivalent to a default compensation x_0 applied to every angle in the original CAD model. According to Eq. (6.1), the predicted shrinkage model $f(\theta, r_0(\theta) + x(\theta)) + x(\theta)$ becomes

$$f(\theta, r_0(\theta) + x(\theta)) + x(\theta) = \alpha(r_0 + x_0)^a + \beta(r_0 + x_0)^b \cos(2\theta) + x_0 \qquad (6.9)$$

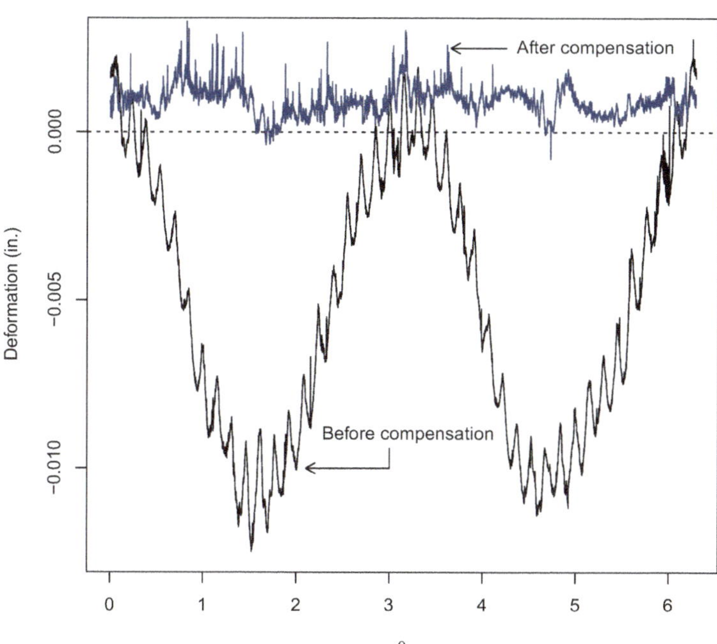

Fig. 6.3 Compensated $1.0''$ disk: without exposure term

Table 6.2 Deviation statistics for $1.0''$ disk, before and after compensation (in inches)

	Mean	SD
Before compensation	-5×10^{-3}	4×10^{-3}
After compensation	9×10^{-4}	4×10^{-4}

The optimal compensation level under this new model is then derived as

$$x^*(\theta) = -\frac{\alpha r_0^a + \beta r_0^b \cos(2\theta)}{1 + a\alpha r_0^{a-1} + b\beta r_0^{b-1}\cos(2\theta)} - x_0 \tag{6.10}$$

This derivation acknowledges the fact that the amount of compensation x_0 will always be automatically added afterward.

Alternatively, or more rigorously, we could view the nominal process input as $r_0 + x_0$ and perform the Taylor expansion at $r_0 + x_0$ instead of r_0. In this case, the compensation strategy will be

$$x^*(\theta) = -\frac{x_0 + \alpha(r_0 + x_0)^a + \beta(r_0 + x_0)^b \cos(2\theta)}{1 + a\alpha(r_0 + x_0)^{a-1} + b\beta(r_0 + x_0)^{b-1}\cos(2\theta)} \tag{6.11}$$

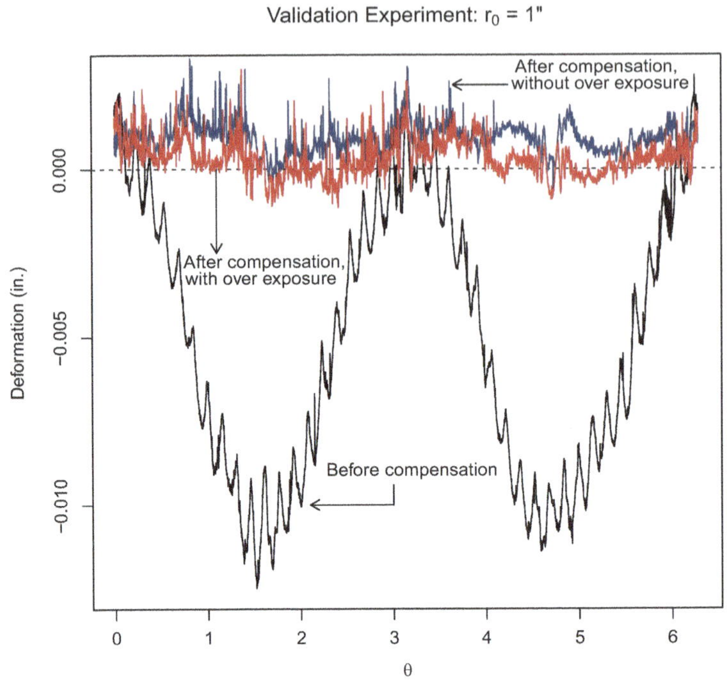

Fig. 6.4 Compensated $1.0''$ disk: with exposure term

A comparison of the compensations in (6.10) and (6.11) shows effectively no difference (details omitted). Consequently, we adopt the compensation strategy given by (6.11).

To validate the improved model with the overexposure effect, $1''$ and $2.5''$ disks are built with CAD design adjusted based on the optimal compensation plan in Eq. (6.11). The measured deviation results are shown in Figs. 6.4 and 6.5, respectively. In Fig. 6.4, a comparison of the uncompensated disk, compensated disk ignoring overexposure, and compensated disk considering overexposure is demonstrated. Although both compensation methods decrease deviation substantially, the product compensated according to overexposure apparently has uniformly smaller deviation: its deviation curve effectively shifted down closer to the nominal value 0, resolving the compensation bias problem discussed earlier.

Figure 6.5 shows the compensation effect for a disk with size $r_0 = 2.5''$ and its comparison with uncompensated disks with size $r_0 = 2''$, $3''$. Obviously, deviation has been dramatically decreased, and the significant sinusoidal pattern has been eliminated. Note that the $2.5''$ disk has not been constructed before, and so this experiment demonstrates great predictability of our compensation model.

The three shaded areas in Fig. 6.5 also illustrate the quality measure $\Delta S(x(\theta))$ proposed in this chapter. Since $r_0(\theta)$ is a constant for a disk, $\Delta S(x(\theta))$ in (6.4) becomes

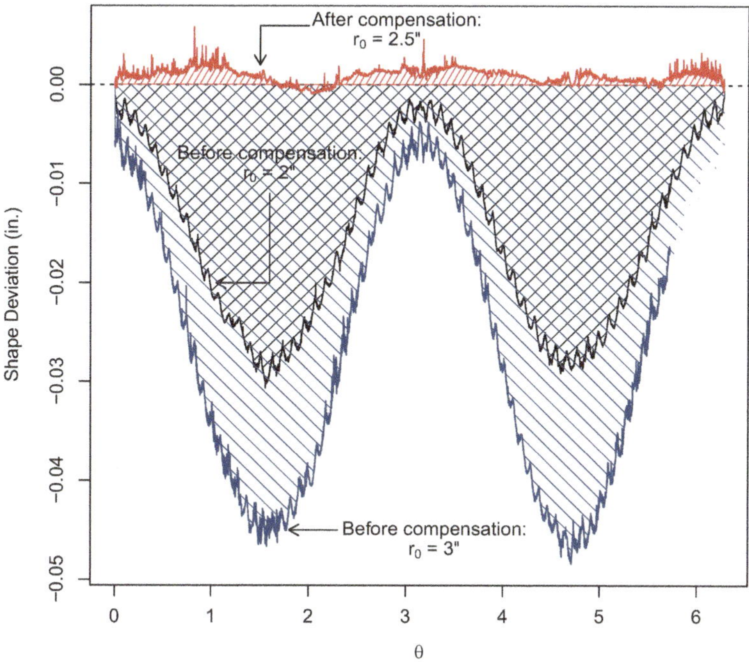

Fig. 6.5 Predictive validation for 2.5″ disk

$$\Delta S(x(\theta)) = r_0 \int_0^{2\pi} \left| f(\theta, r_0(\theta) + x(\theta)) + x(\theta) \right| d\theta$$

with $\int_0^{2\pi} \left| f(\theta, r_0(\theta) + x(\theta)) + x(\theta) \right| d\theta$ corresponding to three shaded areas, respectively.

The quality measure for the disk of size $r_0 = 2''$, for instance, $\Delta S(r_0 = 2'', x(\theta) = 0)$ is simply the mesh area multiplied by size $r_0 = 2''$, a convenient index to assess product quality.

6.2.3.2 Prescriptive Compensation of Printed 2D Freeform Shapes

One direct way to take advantage of the prescriptive model (3.24) is to improve printing accuracy by compensating the CAD design model of an arbitrary freeform shape even before fabricating the product. Work in [23, 24] first put forward the machine parametric error model to evaluate the part dimensional accuracy and model the parametric error functions. The negative values of predicted errors will be added to CAD design directly to compensate the product deviation. References [1, 2] further extend this error compensation method through employing conical sockets for probe measurement. The compensation strategy, however, is not optimal.

We first establish the optimal compensation policy for 2D shapes in [11] and 3D shapes in [9]. Based on the result in [11], the optimal amount of compensation $x^*(\theta)$ for 2D freeform shape is:

$$x^*(\theta) = -\frac{g_1^F(\theta, r(\theta)) + g_2^F(\theta, r(\theta))}{1 + g_1^{F'}(\theta, r(\theta)) + g_2^{F'}(\theta, r(\theta))} \tag{6.12}$$

Therefore, with the freeform model and its input information, the optimal compensation for each approximated sector is calculated and then combined together to achieve the whole compensation plan.

Two new experiments are conducted to fabricate two freeform products by modifying original freeform CAD designs based on model (6.12). Figure 6.6 compares the deviation profiles of two freeform products before (blue lines) and after (red lines) compensation is implemented. We also generate Fig. 6.7 by removing the sharp spikes in Fig. 6.6.

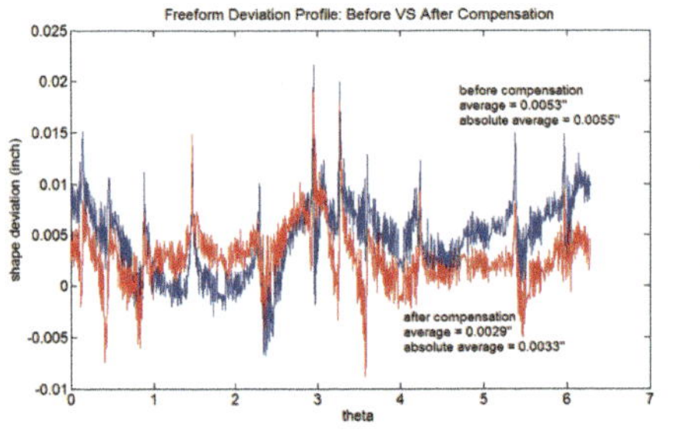

(a) Deviation profile of convex freeform shape: before and after compensation

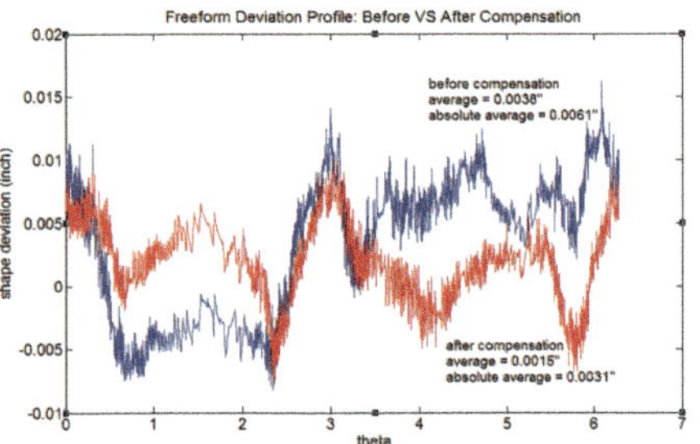

(b) Deviation profile of concave freeform shape: before and after compensation

Fig. 6.6 Freeform shape deviation profiles: before (blue lines) and after (red lines) compensation

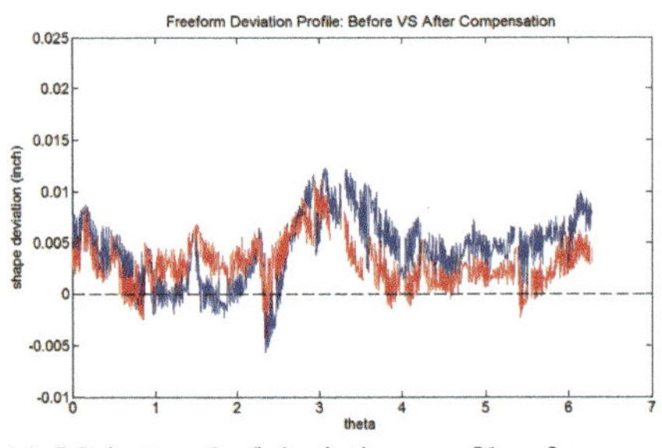

(a) Main trend of deviation profile of convex freeform shape

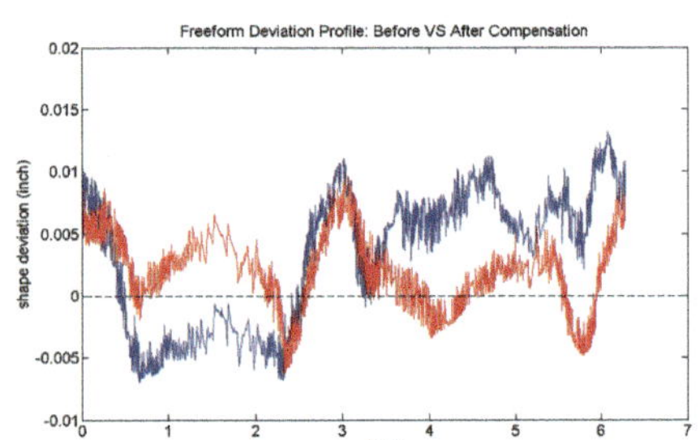

(b) Main trend of deviation profile of concave freeform shape

Fig. 6.7 Trend of freeform shape deviation profiles: before and after compensation

Table 6.3 Compensation performance

Convex freeform shape			
Deviation	Before	After	Reduction
Average	$0.0053''$	$0.0029''$	45%
Absolute average	$0.0055''$	$0.0033''$	40%
Concave freeform shape			
Deviation	Before	After	Reduction
Average	$0.0038''$	$0.0015''$	61%
Absolute average	$0.0061''$	$0.0031''$	49%

It is clear from these two figures that the whole deviation profiles after compensation are more centering around zero line. Table 6.3 summarizes the compensation performance. The average deviation is reduced nearly 60%, and the absolute average deviation is reduced nearly 50%. Therefore, on average, the deviation reduction is remarkable, considering the limited data used to predict the complicated freeform shapes.

6.3 Optimal Compensation of 3D Geometric Shape Deviation

To derive an optimal compensation policy for 3D shape deviation, we first briefly summarize the formulation of spatial shape deviation presented in Sect. 4.1. We adopt the spherical coordinate system (SCS) (r, θ, φ) to depict both the in-plane and out-of-plane (z direction) deviation. Denote $r(\theta, \varphi, r_0(\theta, \varphi))$, the boundary shape of an AM-built product with $r_0(\theta, \varphi)$ being the nominal shape. At a given height $\varphi = \varphi_0$ or $z = r_0(\theta, \varphi) \cos(\varphi_0)$, $\Delta r(\theta, r_0(\theta)|\varphi_0) = h(r, \theta|\varphi)$ represents the in-plane geometric deviation of the horizontal cross-sectional view of the product. For the vertical cross-section of the product, denote the out-of-plane deviation $\Delta r(\varphi, r_0(\theta, \varphi)|\theta)$ as $v(r, \varphi|\theta)$.

6.3.1 Minimum Volume Deviation (MVD) Criterion and Optimality

Let us define the spatial deviation $\Delta r(\theta, \varphi, r_0(\theta, \varphi))$ as

$$\Delta r(\theta, \varphi, r_0(\theta, \varphi)) = r(\theta, \varphi, r_0(\theta, \varphi)) - r_0(\theta, \varphi) \tag{6.13}$$

Let $f(\theta, \varphi, r_0(\theta, \varphi))$ be the given model predicting $\Delta r(\cdot)$, i.e., $\Delta r(\theta, \varphi, r_0(\theta, \varphi)) = f(\theta, \varphi, r_0(\theta, \varphi)) + \varepsilon$. Denote $r(\theta, \varphi, r_0(\theta, \varphi), x(\theta, \varphi))$ the actual radius at (θ, φ) when compensation $x(\theta, \varphi)$ is applied at that location. By extending Theorem 6.1's MAD criterion to minimum volume deviation (MVD) criterion, we have the following result.

Theorem 6.2 (Minimum Volume Deviation)   *The optimal compensation policy or the optimal amount of compensation $x^*(\theta, \varphi)$ for spatial shape deviation reduction is*

$$x^*(\theta, \varphi) = -\frac{f(\theta, \varphi, r_0(\theta, \varphi))}{1 + f'(\theta, \varphi, r_0(\theta, \varphi))}, \tag{6.14}$$

which minimizes the volume deviation from its nominal shape, that is, it follows the minimum volume deviation (MVD) criterion.

Proof   As shown in Fig. 6.8, the volume deviation at location (θ, φ) due to $\Delta r(\theta, \varphi, r_0(\theta, \varphi))$ is

$$\Delta V(r, \theta, \varphi) \approx \Big(r_0(\theta, \varphi)sin(\varphi)d\varphi\Big)\Big(r_0(\theta, \varphi)d\theta\Big)\Delta r(\theta, \varphi, r_0(\theta, \varphi))$$

$$\approx r_0^2(\theta, \varphi)\, sin(\varphi)\, f(\theta, \varphi, r_0(\theta, \varphi))\, d\theta d\varphi$$

The total volume deviation before compensation is therefore

$$\Delta V = \iint r_0^2(\theta, \varphi)\Big|sin(\varphi)\, f(\theta, \varphi, r_0(\theta, \varphi))\Big|\, d\theta d\varphi$$

Again, we consider absolute volume change by taking $|sin(\varphi) f(\theta, \varphi, r_0(\theta, \varphi))|$ in the integral.

After applying compensation $x(\theta, \varphi)$, the shape deviation at (θ, φ) is

$$\Delta r(\theta, \varphi, r_0(\theta, \varphi), x(\theta, \varphi)) = r(\theta, \varphi, r_0(\theta, \varphi), x(\theta, \varphi)) - r_0(\theta, \varphi)$$

$$= f\Big(\theta, \varphi, r_0(\theta, \varphi) + x(\theta, \varphi)\Big) + x(\theta, \varphi)$$

Then the total volume deviation becomes

$$\Delta V(x(\theta, \varphi)) =$$

$$\iint r_0^2(\theta, \varphi)\Big|sin(\varphi)\Big[f\Big(\theta, \varphi, r_0(\theta, \varphi) + x(\theta, \varphi)\Big) + x(\theta, \varphi)\Big]\Big|\, d\theta d\varphi \tag{6.15}$$

Fig. 6.8 Volume deviation at location (θ, φ)

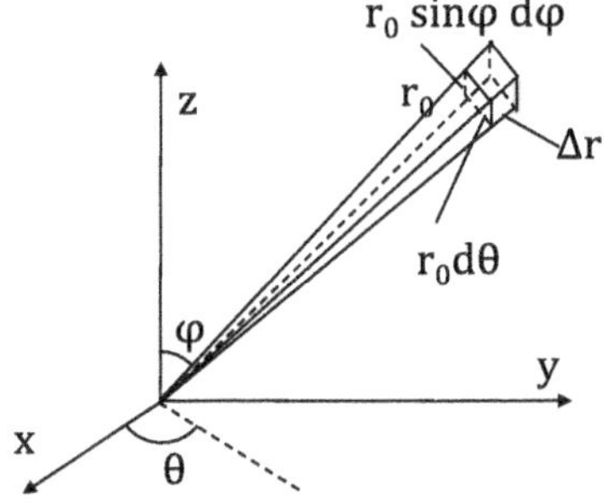

Note that ΔV before compensation is a special case with $x(\theta, \varphi) = 0$ or $\Delta V(x(\theta, \varphi)) = 0$.

The optimal compensation $x^*(\theta, \varphi)$ will minimize the volume deviation from the nominal shape, i.e.,

$$\min_{x(\theta,\varphi)} \quad \Delta V(x(\theta, \varphi))$$
$$\text{s.t.} \quad \theta_{min} \leq \theta \leq \theta_{max}; \text{ and } \varphi_{min} \leq \varphi \leq \varphi_{max} \tag{6.16}$$

A solution to minimize the volume deviation is

$$f\left(\theta, \varphi, r_0(\theta, \varphi) + x(\theta, \varphi)\right) + x(\theta, \varphi) = 0$$

The Taylor series expansion of $f\left(\theta, \varphi, r_0(\theta, \varphi) + x(\theta, \varphi)\right)$ at $r_0(\theta, \varphi)$ will yield the result in (6.14) and complete the proof.

Remark 6.3 The result (6.14) based on the MVD criterion essentially requires to minimize the deviation of every point on the boundary of the product, which is a strong condition. This is a direct extension from (6.3) for the compensation of the in-plane error.

Lemma 6.2 (3D Shape Quality Measure) $\Delta V(x(\theta, \varphi))$ *is a quality measure of 3D shape deviation.*

By extending the equivalent amount of compensation to the 3D case, the proof can be obtained by following Theorem 6.2.

6.3.2 Optimal Policy for Online Compensation of 3D Deviation

Although Theorem 6.2 nicely extends the optimal compensation to the 3D case, it relies on the model $f(\theta, \varphi, r_0(\theta, \varphi))$ to predict spatial deviation. Clearly, it is generally more challenging to establish $f(\theta, \varphi, r_0(\theta, \varphi))$ than the in-plane error model. Additionally, it is also demanding for the AM machine to apply compensation three-dimensionally because a product is built layer by layer. To overcome this issue, we derive the following Lemma.

Lemma 6.3 (A Weaker Condition) *The optimal compensation policy $x^*(\theta, \varphi)$*

$$x^*(\theta, \varphi)sin(\varphi) = -\frac{f(\theta, \varphi, r_0(\theta, \varphi))sin(\varphi)}{1 + f'(\theta, \varphi, r_0(\theta, \varphi))} \tag{6.17}$$

is a weaker condition to minimize the volume deviation.

Proof Other than the stronger condition of $f\left(\theta, \varphi, r_0(\theta, \varphi) + x(\theta, \varphi)\right) + x(\theta, \varphi) = 0$ to minimize $\Delta V(x(\theta, \varphi)) = 0$ in Eq. (6.15), a weaker condition below will also minimize the volume deviation $\Delta V(x(\theta, \varphi))$:

$$sin(\varphi) \left[f\Big(\theta, \varphi, r_0(\theta, \varphi) + x(\theta, \varphi)\Big) + x(\theta, \varphi) \right] = 0$$

The Taylor series expansion of $f\Big(\theta, \varphi, r_0(\theta, \varphi) + x(\theta, \varphi)\Big)$ at $r_0(\theta, \varphi)$, and keeping the term $sin(\varphi)$ will give (6.17).

Remark 6.4 The result (6.17) actually has powerful implications. Note that $x^*(\theta, \varphi)sin(\varphi)$ means the in-plane compensation by projecting $x^*(\theta, \varphi)$ to the $x - y$ plane at height φ. Following the notation in Fig. 4.1, let us denote $x^*(\theta, \varphi)sin(\varphi)$ as $x^*(\theta|\varphi)$, i.e., the optimal compensation for layer at height φ.

Similarly, $f(\theta, \varphi, r_0(\theta, \varphi))sin(\varphi)$ means the in-plane deviation by projecting spatial deviation to the $x - y$ plane at height φ, which is $h(r, \theta|\varphi)$ or $\Delta r(\theta, r_0(\theta, \varphi)|\varphi)$ in Fig. 4.1. $f(\theta, \varphi, r_0(\theta, \varphi))$ can be expressed as $h(r, \theta|\varphi)/sin(\varphi)$. Then (6.17) can be rewritten as:

$$x^*(\theta|\varphi) = -\frac{h(r, \theta|\varphi)}{1 + h'(r, \theta|\varphi)/sin(\varphi)} \tag{6.18}$$

This suggests that the optimal spatial compensation can be transformed to the in-plane compensation and can be implemented layer by layer. This is more consistent with the mechanism of AM itself and still satisfies the MVD criterion. It thus provides a theoretical base for online condition monitoring and online feedback control of 3D geometric shape deviation.

Remark 6.5 It is interesting to compare the transformed optimal in-plane compensation (6.18) for spatial shape deviation with the optimal 2D shape deviation compensation in (6.3). Note that the result in Eq. (6.18) is expected to reduce both in-plane and out-of-plane errors, while Eq. (6.3) only considers in-plane errors. The two results will agree only when $\varphi = \pi/2$. This can be illustrated by a vertical cross-section of a pyramid example in Fig. 6.9. When $\varphi = \pi/2$, the AM process just builds

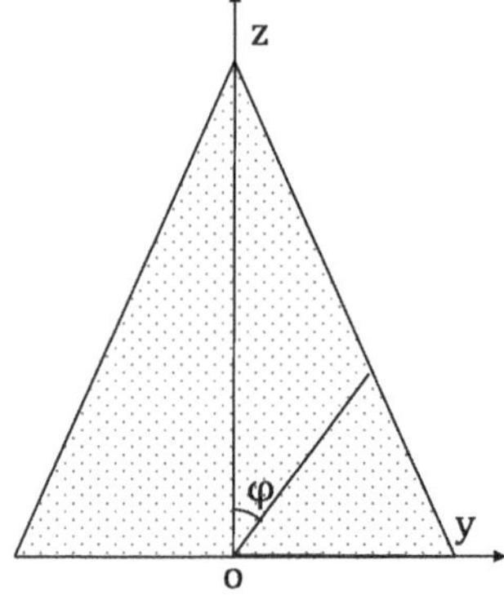

Fig. 6.9 Compensation at $\varphi = 0, \pi/2$

the first layer of the product, and this thin layer of product can be approximated as a 2D slab, with out-of-plane errors being ignored. Therefore, the spatial optimal compensation literally degenerates to 2D case. This also proves that (6.3) is a special case of (6.18).

It is also worthwhile to investigate the compensation at location $\varphi = 0$. Since $sin(\varphi = 0) = 0$, then $x^*(\theta|\varphi = 0) = 0$ at the tip, i.e., there is no need for compensation if the built pyramid has a tip. If there is a missing tip, then $\varphi > 0$, and the optimal compensation $x^*(\theta|\varphi) \neq 0$.

This chapter introduces the theoretical foundation of 2D and 3D shape deviation compensation for AM. We theorize the optimal compensation result for 2D shape deviation by proposing the minimum area deviation (MAD) criterion. By extending the MAD criterion to the minimum volume deviation (MVD) criterion, we derive the optimal amount of compensation for 3D shape deviation in a consistent framework. Furthermore, MAD and MVD criteria provide convenient quality measures for AM-built products that facilitate online monitoring and feedback control of shape geometric accuracy. The established analytical foundation fills the gap for the quality improvement in AM.

Acknowledgments The work in this chapter was partially supported by the US Office of Naval Research with grant no. N000141110671 and the US National Science Foundation Cyber-Physical Systems Program under grant no. CMMI-1544917.

References

1. Cajal, C., Santolaria, J., Samper, D., Velazquez, J.: Efficient volumetric error compensation technique for additive manufacturing machines. Rapid Prototyp. J. **22**(1), 2–19 (2016)
2. Cajal, C., Santolaria, J., Velazquez, J., Aguado, S., Albajez, J.: Volumetric error compensation technique for 3d printers. Procedia Eng. **63**, 642 – 649 (2013)
3. Cho, W., Sachs, E.M., Patrikalakis, N.M., Troxel, D.E.: A dithering algorithm for local composition control with three-dimensional printing. Comput. Aided Des. **35**(9), 851–867 (2003)
4. Cohen, D.L., Lipson, H.: Geometric feedback control of discrete-deposition sff systems. Rapid Prototyp. J. **16**(5), 377–393 (2010)
5. Heralic, A., Christiansson, A.K., Lennartson, B.: Height control of laser metal-wire deposition based on iterative learning control and 3d scanning. Opt. Lasers Eng. **50**(9), 1230–1241 (2012)
6. Hilton, P., Jacobs, P.: Rapid Tooling: Technologies and Industrial Applications. CRC Press, New York (2000)
7. Hu, D., Kovacevic, R.: Sensing, modeling and control for laser-based additive manufacturing. Int. J. Mach. Tools Manuf. **43**(1), 51–60 (2003)
8. Hu, D., Mei, H., Kovacevic, R.: Improving solid freeform fabrication by laser-based additive manufacturing. Proc. Inst. Mech. Eng. B J. Eng. Manuf. **216**(9), 1253–1264 (2002)
9. Huang, Q.: An analytical foundation for optimal compensation of three-dimensional shape deformation in additive manufacturing. J. Manuf. Sci. Eng. Trans. ASME **138**(6), 1–8 (2016)
10. Huang, Q., Nouri, H., Xu, K., Chen, Y., Sosina, S., Dasgupta, T.: Statistical predictive modeling and compensation of geometric deviations of three-dimensional printed products. ASME Trans. J. Manuf. Sci. Eng. **136**(6), 061008–061018 (2014)

11. Huang, Q., Zhang, J., Sabbaghi, A., Dasgupta, T.: Optimal offline compensation of shape shrinkage for 3d printing processes. IIE Trans. Qual. Reliab. **47**(5), 431–441 (2015)
12. Lu, L., Zheng, J., Mishra, S.: A layer-to-layer model and feedback control of ink-jet 3-d printing. IEEE/ASME Trans. Mechatron. **20**(3), 1056–1068 (2015)
13. Luan, H., Huang, Q.: Prescriptive modeling and compensation of in-plane shape deformation for 3-D printed freeform products. IEEE Trans. Autom. Sci. Eng. **14**(1), 73–82 (2017)
14. Lynn-Charney, C., Rosen, D.W.: Usage of accuracy models in stereolithography process planning. Rapid Prototyp. J. **6**(2), 77–87 (2000)
15. Mori, K., Osakada, K., Takaoka, S.: Simplified three-dimensional simulation of non-isothermal filling in metal injection moulding by the finite element method. Eng. Comput. **13**(2), 111–121 (1996)
16. Moroni, G., Syam, W.P., Petrò, S.: Towards early estimation of part accuracy in additive manufacturing. Procedia CIRP **21**, 300–305 (2014)
17. Sabbaghi, A., Dasgupta, T., Huang, Q., Zhang, J.: Inference for deformation and interference in 3d printing. Ann. Appl. Stat. **8**(3), 1395–1415 (2014)
18. Secondi, J.: Modeling powder compaction from a pressure-density law to continuum mechanics. Powder Metallurgy **45**(3), 213–217 (2002)
19. Song, L., Mazumder, J.: Feedback control of melt pool temperature during laser cladding process. IEEE Trans. Control Syst. Technol. **19**(6), 1349–1356 (2011)
20. Sood, A., Ohdar, R., Mahapatra, S.: Improving dimensional accuracy of fused deposition modelling processed part using grey taguchi method. Mater. Des. **30**(10), 4243–4252 (2009)
21. Storåkers, B., Fleck, N., McMeeking, R.: The viscoplastic compaction of composite powders. J. Mech. Phys. Solids **47**, 785–815 (1999)
22. Tapia, G., Elwany, A.: A review on process monitoring and control in metal-based additive manufacturing. ASME Trans. J. Manuf. Sci. Eng. **136**(6), 060801–060810 (2014)
23. Tong, K., Joshi, S., Lehtihet, E.: Error compensation for fused deposition modeling (fdm) machine by correcting slice files. Rapid Prototyp. J. **14**(1), 4–14 (2008)
24. Tong, K., Lehtihet, E., Joshi, S.: Parametric error modeling and software error compensation for rapid prototyping. Rapid Prototyp. J. **9**(5), 301–313 (2003)
25. Wang, H., Huang, Q.: Error cancellation modeling and its application in machining process control. IIE Trans. Qual. Reliab. **38**, 379–388 (2006)
26. Wang, H., Huang, Q.: Using error equivalence concept to automatically adjust discrete manufacturing processes for dimensional variation control. ASME Trans. J. Manuf. Sci. Eng. **129**, 644–652 (2007)
27. Wang, H., Huang, Q., Katz, R.: Multi-operational machining processes modeling for sequential root cause identification and measurement reduction. ASME Trans. J. Manuf. Sci. Eng. **127**, 512–521 (2005)
28. Wang, W., Cheah, C., Fuh, J., Lu, L.: Influence of process parameters on stereolithography part shrinkage. Mater. Des. **17**(4), 205–213 (1996)
29. Wang, X.: Calibration of shrinkage and beam offset in sls process. Rapid Prototyp. J. **5**(3), 129–133 (1999)
30. Xu, K., Chen, Y.: Mask image planning for deformation control in projection-based stereolithography process. J. Manuf. Sci. Eng. **137**(3), 031014 (2015)
31. Zhou, C., Chen, Y., Waltz, R.A.: Optimized mask image projection for solid freeform fabrication. ASME J. Manuf. Sci. Eng. **131**(6), 061004 (2009)
32. Zhou, J., Herscovici, D., Chen, C.: Parametric process optimization to improve the accuracy of rapid prototyped stereolithography parts. Int. J. Mach. Tools Manuf. **40**(3), 363–379 (2000)

Chapter 7
Applications of Process-Informed Optimal Compensation

This chapter introduces applications of the optimal compensation method and algorithm to reduce the shape deviation in metal AM process and to statistically monitor the part-to-part changes in AM.

7.1 Optimal Compensation of Product Shape Deviation in Metal Wire and Arc AM

Wire and arc additive manufacturing (WAAM) has become an economically viable way for fast fabrication of large near-net metal products using high-value materials in the aerospace and petroleum industries. However, a wide adoption of WAAM technologies has been limited by low shape accuracy, high surface roughness, and poor reproducibility. Since WAAM part quality is affected by a multitude of factors related to part geometries, materials, and process parameters, experimental characterization or physics-based simulation for WAAM process optimization can be cost prohibitive, particularly for new part designs. This section introduces optimal compensation approach to reduce shape deviation in WAAM through domain-informed statistical learning.

7.1.1 Wire and Arc AM and Product Geometric Shape Quality

In recent years, metal additive manufacturing (AM) technologies have attracted significant interest to meet the growing demand for rapid product prototyping and development and for fabricating geometrically complex parts. Metal AM technologies can be classified into powder feeding systems, wire feeding systems, and powder bed systems [11]. Compared to powder-based processes, wire-feed AM

© The Author(s), under exclusive license to Springer Nature Switzerland AG 2025

Q. Huang, *Domain-informed Machine Learning for Smart Manufacturing*,

https://doi.org/10.1007/978-3-031-91631-1_7

typically has a larger built area and a higher deposition rate but lower geometric precision [19]. The wire and arc additive manufacturing (WAAM) process makes use of traditional arc welding technology to deposit material layer by layer. Plasma arc welding (PAW), gas metal arc welding (GMAW), and gas tungsten arc welding (GTAW) have been applied to create weld pools and melt the wire feedstock [7]. The feedstock materials include carbon steel, aluminum alloy, nickel alloy, and titanium alloy [3]. Thanks to GMAW's high energy input, high deposition rate, and low-cost equipment [32], metal AM based on GMAW offers substantial benefits over powder bed and powder feed systems for near-net shape production of large-size parts used in the aerospace and petroleum industries [23].

However, WAAM technologies suffer from large shape deviation and high surface roughness. In GMAW, for example, the filler material is melted with an electrical arc between the wire electrode and the top layer surface. During fabrication, layers are repeatedly subjected to re-heating and cooling processes and the generated heat stresses due to the multiple fusion, solidification, and phase change cycles are extremely nonlinear and transient. This results in significant residual stresses as well as shape deviation, which are some of the primary deterrents for widespread adoption of WAAM technologies [18, 33].

To achieve the desired final part geometry and metallurgical properties [18], process characterization of new materials and new geometrical features are regularly carried out to optimize deposition pattern, heat input, and deposition speed, among other variables. This process involves experimentation depositing a single-track, single-layer, multilayer, and certain geometrical features such as crossing, cylinder geometry, or bulk material to identify a desirable deposition process window.

To reduce the number of physical experiments used for process characterization, numerical studies have been devoted to physics-based pattern generation [8], thermal and stress distribution modelling [9, 47], and microstructure and mechanical property evaluation of WAAM fabricated parts [20, 29, 46]. Among others, Mughal et al. [32] examined residual stress deformation of GMAW using finite element modelling and concluded that thermal cycling was the principal source of deformation. Zhao et al. [53] developed a thermal model and demonstrated that the deposition orientation can affect the heat diffusion and the distribution of residuals stresses. In situ process monitoring has also been developed to control the width and height of each layer during the WAAM fabrication process [10, 40]. However, physics-based simulation and prediction of WAAM part deformation not only involves a large number of process variables (e.g., energy input, temperature regime, and material properties) but also the non-smooth surfaces between overlapping bead layers [18], complicated layer interactions, and high residual stresses. Therefore, simulation of large-size parts is computationally prohibitive. Furthermore, the deposited multi-layer geometries are often inconsistent with the design even when the theoretical relationship between printed layer geometry and process parameters is well understood for simple and small geometries [48].

This chapter introduces a domain-informed statistical learning approach to predict the shape deviation of as-printed WAAM parts. From the prediction model, we derive an optimal design compensation plan to modify the CAD design to obtain the desired geometry upon manufacturing. This strategy is based on the prescriptive modeling approach and effect equivalence principle that have been established for other AM processes [25]. It starts with extracting the geometric deviation patterns from point-cloud data through fabrication-aware machine learning [17] and then minimizing shape deviations through the optimal compensation algorithms [12]. For predicting 3D shape deviation, Huang et al. [15] proposed a convolution framework that learns inter-layer interactions and the final stack-up of layers. However, directly implementing this strategy to WAAM processes faces a unique challenge, that is, parts having large shape deformation and high surface roughness at the same time while geometric compensation cannot reduce the roughness patterns. As an example, Fig. 7.1 illustrates the shape deviation profile of a WAAM-built cylindrical wall. The deviation patterns on the flattened outer wall are highly volatile. As a comparison, shape deviations in a stereolithography (SLA) process and a WAAM process are illustrated in Fig. 7.2. Clearly, the total shape deviation of the SLA

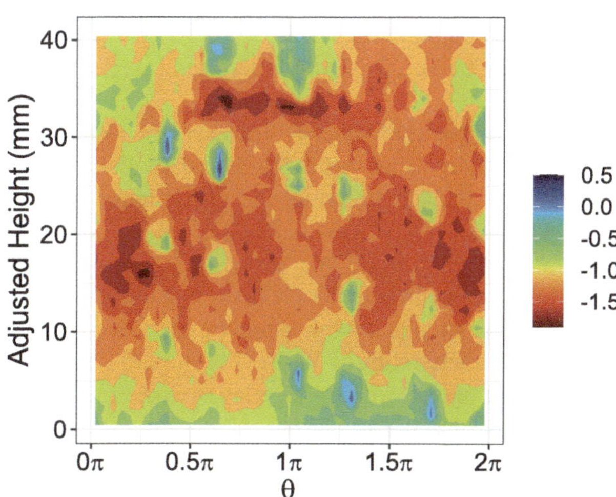

Fig. 7.1 Shape deviation (mm) surface of a 60 mm radius stainless-steel 316L cylindrical wall manufactured using the GMAW process

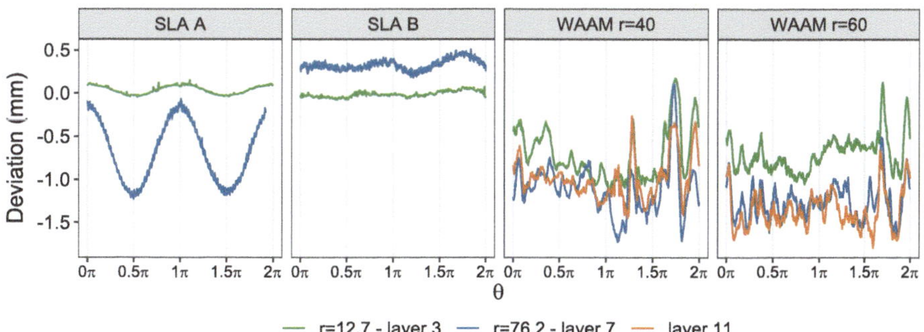

Fig. 7.2 Shape deviation for 2D cylinders printed using an SLA machine before (SLA A) [14] and after repair (SLA B) [36] and the center of layers 3, 7, and 11 of the 40 and 60 mm radius cylindrical walls (WAAM)

printed parts is dominated by uneven material shrinkage with relatively small surface roughness [14]. In contrast, the WAAM manufactured part exhibits changing deviation patterns with large surface roughness varying at different spatial locations such as arc ignitions (e.g., near 1.75π). Accurate separation of local surface roughness from global shape deformation is therefore nontrivial. This has also been demonstrated in a study of laser powder bed fusion (LPBF) processes where part size is significantly smaller [25].

Information extraction from 2D high-dimensional datasets with significant noise is a common problem in many applications such as medical imaging [54] and structural health monitoring [52]. Popular approaches for identifying relevant patterns include robust principal component analysis [4], wavelet decomposition [52], additive tensor decomposition [31], and smooth-sparse decomposition [50], among others [51]. These methods require the data to be sampled on a regular grid and decompose into smooth and sparse components. Although they can handle missing entries, geometric accuracy data is obtained as point-clouds sampled at random locations, and it contains global and local smooth patterns. For such datasets, Kriging methods and generalized additive models (GAMs) [41] offer more flexible alternatives, with the latter being more computationally efficient for large datasets.

To reduce part deformation (not surface roughness), we propose to model part shape deviation using a GAM. These models use penalty terms to control the smoothness of each function and allow main trends significantly smoother than local patterns [44]. We mathematically decompose shape deviation into the main shape deformation trend and surface roughness as a function of local geometrical features in the cylindrical coordinate system. Using tensor-product basis functions, each term is modeled with a univariate basis selected using knowledge of the physical characteristics of WAAM built parts. Then, we use the predicted in-plane deformation pattern to introduce layer-wise optimal geometric compensation in the CAD design of new parts.

7.1.1.1　Experiment of Manufacturing Cylindrical Walls Using WAAM

Two sets of experiments were conducted to validate the proposed shape deviation learning and compensation approach. The first set keeps part design and WAAM settings unchanged to investigate the performance of a GMAW-based WAAM process. Without changing the process parameters, the second set of experiments only modifies the part design to validate the model prediction and derived optimal compensation plan.

In traditional GMAW, small amounts of material are deposited by welding and are constrained by surrounding material, parent material, or base plates. In GMAW-based WAAM, deposited material has more deformation freedom as it is constrained only by the previous layers with increasing freedom as deposition moves away from the base plate. After a certain height, the effect of gravity and material thermal properties become more important to the fabrication quality. Subsequent layer deposition partially remelts the previously deposited layers and acts as a post heat treatment. The thickness of heat-treated material will be highly dependent on the processing parameters. Larger deformation and poor material thermal and mechanical properties can be expected if the part is overheated. Surface finishing deteriorates when accumulated heat in the deposited material changes the bead or increases arc instability.

Part geometries in WAAM can be decomposed into crossings, straight, and curved features, each affecting geometric accuracy in a unique way. The part design for experimentation is chosen to be cylindrical wall with different radii. The continuous and symmetrical shape of cylindrical walls offers a shape deviation pattern representative of common curved structures. However, a perfect cylindrical geometry is one of the most difficult WAAM features to manufacture, as a strategy should be used to avoid arc igniting and extinguishing defects that affect the cylindrical shape, straightness, surface roughness, and flatness of the top surface. The region of the layer adjacent to the arc ignition point has an excessive thickness in contrast to everywhere else in the layer due to the heat sink effect of the base metal in the first few layers [18] and excess material fed due to the high voltage and current (i.e., overshooting) required for arc ignition. In contrast, the layer is thinner at the arc extinguishing region due to the arc pressure on the molten pool. Hence, cylindrical features have routinely been chosen for WAAM process characterization. We investigate the geometrical deformation of cylindrical features by fabricating cylindrical walls of several radii using a GMAW system. The system is comprised of a Panasonic TM-1400WGIII-SAWP 5-axis welding robot, equipped with an additional 2-axis workpiece manipulator (type YA-1RJC62) with a maximum handling capacity of 300 kg, as shown in Fig. 7.3.

Deposition was carried out on a S355 J2 steel base plate ($300 \times 400 \times 30$ mm) that was machined and cleaned using 99.9% industrial ethanol. The consumable solid wire electrode was stainless steel 316L with a diameter of 1.2 mm. The chemical compositions of the wire is given in Table 7.1. Deposition experiments were conducted using a Panasonic Super Active Wire Process (SAWP) welder with

Fig. 7.3 GMAW cell used to fabricate the samples

Table 7.1 Nominal composition of the stainless steel (SS) welding wire

Alloy	Chemical composition (wt%)						
	C	Cr	Ni	Mn	Si	Mo	Fe
SS 316L	0.01	18.5	12.2	1.8	0.8	2.5	Re.

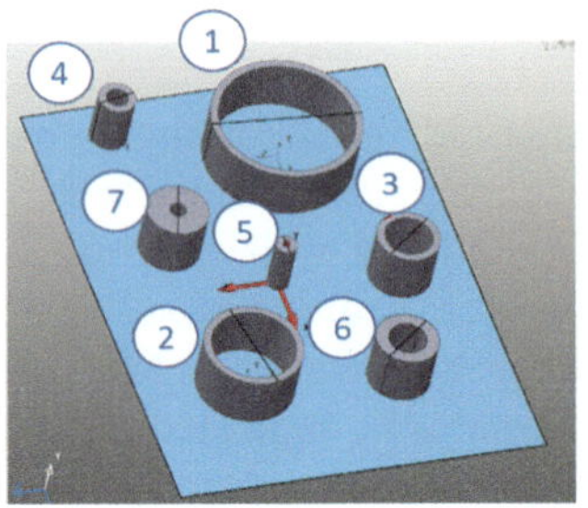

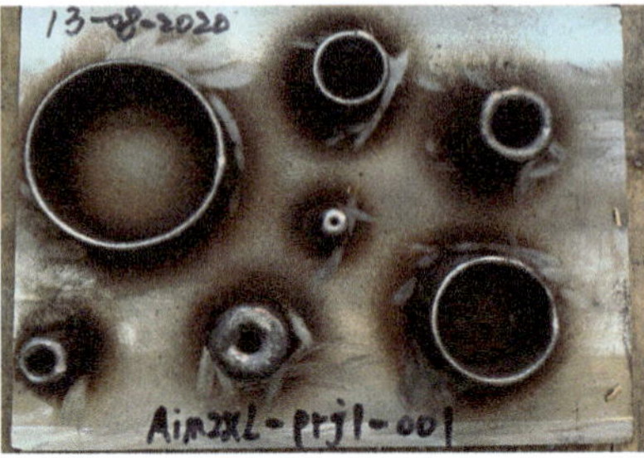

Fig. 7.4 CAD models (left) and manufactured samples (right)

Table 7.2 Main process parameters used to manufacture the cylindrical walls with 45 mm nominal height

	Cylindrical wall						
	1	2	3	4	5	6	7
Radius (mm)	60	40	25	15	10	25	25
Thickness (mm)	6	6	6	6	6	10	18
Average current (A)	140	140	140	140	130	105	142
Average voltage (V)	15.4	15.4	15.4	15.4	15.4	14.1	15.8
Deposition speed (m/min)	0.75	0.75	0.75	0.75	1.00	0.25	0.17
Wire feed speed (m/min)	4.20	4.20	4.20	4.20	3.80	2.50	4.20

shielding gas Inomaxx® plus (35% helium and 2% carbon dioxide in argon) at a gas flow rate of 22 liters per minute. The contact tip was positioned at 15 ± 0.5 mm from the workpiece. Figure 7.4 shows the samples manufactured, and Table 7.2 shows the radius, thickness, and main process parameters for the deposition of various cylindrical walls. Figure 7.5 shows an example of the current and voltage waveforms used by the power supply with current/voltage adjusted to match the average presented on Table 7.2. The acceleration/deceleration of the robot movement was chosen to minimize the effect of kinematic ramping on the arc ignition/extinguishing locations. To minimize arc instability, the maximum current was set to 1.2–1.5 times the nominal current. The specimens were placed with clamps during welding and subsequent cooling process.

Post-fabrication geometrical measurements are collected as point clouds using structured light scanning equipment to assess the geometric accuracy of the manufactured parts. The structured light 3D scanning technique is user-friendly and cost-effective. We confirmed the accuracy of our structured light scanner by using a metal calibration block and observed the accuracy of approximately 50 to 100

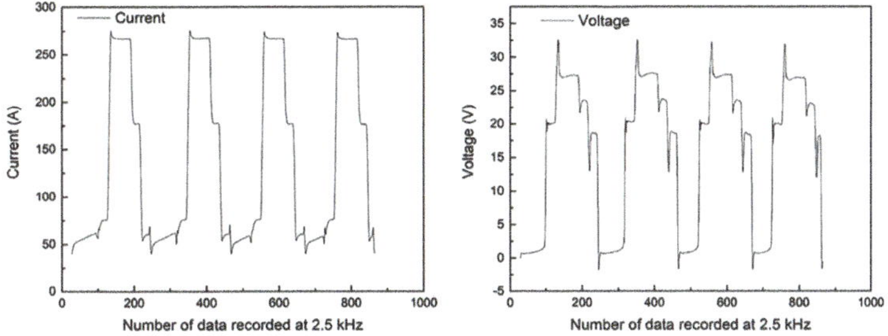

Fig. 7.5 Nominal waveform for current and voltage

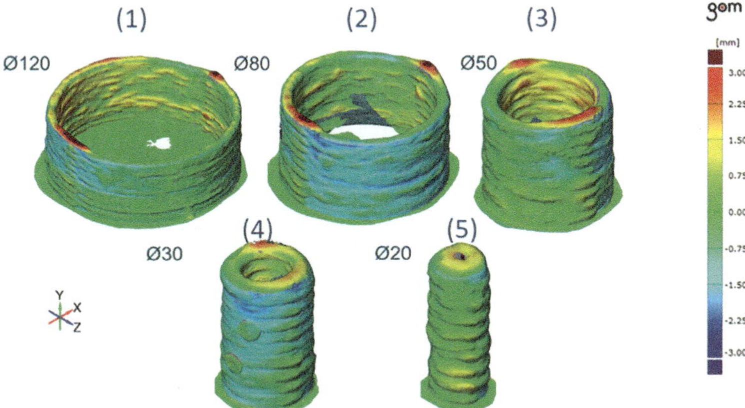

Fig. 7.6 Surface deviation of the 6-mm-thick cylindrical walls

μm, which is sufficient for WAAM fabricated parts. To improve data quality, the measured parts were fitted with reference points, and a thin layer of anti-reflection cover powder (FabConstruct L500 Matteringspray) was applied to the parts. The point-clouds were processed using the GOM software to align the axis of the Cartesian coordinates with the printing direction based on the reference points.

The 6-mm-thick cylindrical walls were selected for analysis, and the shape deviation from their designs is shown in Fig. 7.6. Samples 1 to 5 have a fabricated height of 45.90, 45.66, 45.07, 44.94 and 44.09 mm, respectively. Positive deviation (red) indicates that the mesh surface is higher than the nominal file, whereas negative deviation (blue) indicates that the mesh surface is lower. Deviation ranges from -1 to 3 mm with the 10 mm radius part not meeting print specifications. The instability of the WAAM process (spatter) causes the abundance of extra material on the surface of the parts. The most significant deformation and the largest amount of error are found at the start and end welds.

7.1.2　A Generalized Additive Model for Shape Deviation Modeling and Compensation in WAAM

To improve shape accuracy by compensating the CAD design, we develop a generalized additive model (GAM) to decompose shape deviation profiles into low-order global deformation and high-order surface roughness patterns.

The first step is to convert point-cloud data to an appropriate functional representation. Similar to [17], we transform the point-cloud data to a functional representation of the 2D shape deviation at any given height (in-plane shape deformation). Since all layers have the same radius, we chose the cylindrical coordinates system (CCS) for functional representation. Each point is described by a height z, angle $\theta \in [0, 2\pi]$, and a radius $r(z, \theta)$. Shape deviation is defined as $y(\theta, z) = r(\theta, z) - r_0(\theta, z)$ where r_0 is the nominal (designed) radius and r is the measured radius. Figure 7.1 shows the deviation of the cylindrical wall with 60 mm radius in the CCS.

The shape deviation of WAAM-built parts is modeled as a GAM [42], which links the sum of functions f of the covariates to the expected value of the observed data through a link function. For shape deviation data, the covariates are θ, z, and target radius r_0. We consider the identity link function (i.e., Gaussian distribution of the data). The functions of the covariates capture a global shape deformation pattern f_1 due to geometric approximation and material-related distortion and local surface roughness f_2 induced by process instability. For part k, these effects are represented using basis expansions as follows:

$$
\begin{aligned}
y_k(\theta, z, r_0) &= \sum_{i=1}^{\kappa_1} \psi_i(\theta, z, r_0)\beta_i && \text{deformation } f_1 \\
&+ \alpha(r_0) \sum_{j=1}^{\kappa_2} \phi_j(\theta, z)\gamma_{jk} && \text{roughness } f_2 \\
&+ \epsilon && \text{measurement error}
\end{aligned}
\tag{7.1}
$$

$$
\epsilon \sim \mathcal{N}(0, \sigma^2)
$$

where $r_0(z, \theta)$ is reduced to r_0 for cylindrical walls. ψ and ϕ are basis expansion functions with κ_1 and κ_2 dimensions, respectively. β and γ_k are the vectors of coefficients for the basis expansions. Lastly, σ^2 is the variance of the measurement error.

The model in matrix notation is

$$
\mathbf{y}_k = \Psi_k \boldsymbol{\beta} + \Phi_k \boldsymbol{\gamma}_k + \epsilon,
\tag{7.2}
$$

where $\mathbf{y}_k$ is the vector of observed deviations in part k and Ψ_k and Φ_k are matrices, with each row i being the basis expansion ψ and $\alpha(r_0)\phi$ evaluated at point i

of part k, respectively. These basis expansions are constructed as tensor-products of univariate basis. Parameters are estimated by maximizing the penalized log-likelihood function of the form [42]:

$$\mathcal{L}_p\left(\boldsymbol{\beta}, \boldsymbol{\gamma}_1, \ldots \boldsymbol{\gamma}_K\right) = \mathcal{L}\left(\boldsymbol{\beta}, \boldsymbol{\gamma}_1, \ldots \boldsymbol{\gamma}_K\right) - \sum_{\vartheta \in \Theta^1} \lambda_{1,\vartheta} \boldsymbol{\beta}^T \mathbf{S}_\vartheta^1 \boldsymbol{\beta}$$

$$- \sum_{\vartheta \in \Theta^2} \lambda_{2,\vartheta} \sum_{k=1}^{K} \boldsymbol{\gamma}_k^T \mathbf{S}_{k,\vartheta}^2 \boldsymbol{\gamma}_k \tag{7.3}$$

where K is the number of parts in the dataset, $\mathcal{L}(\cdot)$ is the log-likelihood function, λ are smoothing parameters, and $\Theta^1 = \{\theta, z, r_0\}$, $\Theta^2 = \{\theta, z\}$, and $\mathbf{S}_\vartheta^1$ and $\mathbf{S}_{k,\vartheta}^2$ for $k = 1, \ldots K$ are tensor-product penalty matrices, constructed as in [45]. Restricted maximum likelihood method (REML) is used for parameter estimation and generalized cross-validation for smoothing parameters optimization [43].

As in the study [17], the purpose of modeling the shape deviation is to reduce it by using geometric compensation (GC), which is the process of introducing adjustment $\delta(\theta, z, r_0)$ to the original design and print $r^* = r_0 + \delta$ in order to achieve the desired shape. In [17], GC is done on boundary points of a design. However, deposition of only a few thick layers in WAAM requires compensation to be done layer-wise to reduce in-plane shape deviation of each layer. Moreover, compensation only corrects the deviation due to the shape deformation pattern $f_1(z, (\theta, z, r_0)) = \boldsymbol{\psi}(\theta, z, r_0)^T \boldsymbol{\beta}$, while surface roughness can only be reduced through process optimization and control. Following the strategy in [12], the optimal compensation plan for WAAM fabricated parts can be shown to be

$$\delta(\theta, z, r_0) = -\frac{f_1(\theta, z, r_0)}{1 + f_1'(\theta, z, r_0)} \tag{7.4}$$

where $f_1'(\theta, z, r_0)$ is the partial derivative of f_1 with respect to r_0. We use the mid-point of each layer to determine z.

7.1.2.1   Engineering-Informed Tensor-Product Basis for Cylindrical Walls

Two popular approaches for constructing multivariate basis functions are the use of isotropic basis (i.e., rotation invariant) and the use of tensor products of lower-dimensional basis [42]. Since the patterns of the shape deformation on (θ, z, r_0) are quite different, isotropic basis are ill-suited for our purposes. We model the vector of multivariate basis as a tensor product of univariate basis functions

$$\boldsymbol{\psi}(z, \theta, r_0) = \mathrm{vec}\left(\boldsymbol{\psi}^z(z) \otimes \boldsymbol{\psi}^\theta(\theta) \otimes \boldsymbol{\psi}^{r_0}(r_0)\right), \tag{7.5}$$

where $\mathrm{vec}(\cdot)$ is the vectorization of a tensor, $\otimes$ is the Kronecker product, and $\boldsymbol{\psi}^\cdot$ are the respective univariate basis functions. The same general procedure applies to

ϕ with the appropriate univariate basis and keeping only the basis that model the interactions between z and θ.

The choice of univariate basis has a profound effect on the number of samples required for model training and meaningful interpretability of the results. In particular, they need to reflect the physical characteristics of the WAAM process for each covariate: (1) deviation is the same at $\theta = 0$ and 2π, (2) thick layers induce a cyclical pattern on z with each layer being a cycle, and (3) the part size (r_0) has a non-linear effect on the amount of surface roughness and deformation. Based on these observations, three types of basis functions are chosen in this work [42]:

1. Cubic basis: these are cubic b-splines with equally spaced knots in the range of the covariate. We use these functions to model the effects of z on the roughness pattern.
2. Cyclic cubic basis: these functions have the same value and derivative at the start and end points of its range, which makes them ideal for modeling the effect of θ on both the shape deformation and roughness patterns.
3. Thin-plate regression splines (TPRS): these functions are low-rank approximations of thin-plate basis that are capable of optimally modelling any function. We use these functions to model the effects of z and r_0 on the deformation pattern.

Figure 7.7 shows the first five cubic, four cyclic cubic, and three TPRS basis functions. Note that cubic and cyclic cubic basis are similar around their maximum but have a different behavior on the tails. Figure 7.8 shows the tensor product basis with 5 cyclic cubic basis for r_0, 3 TPRS for z, and 3 TPRS for r_0. The top plot shows the result by fixing $r_0 = 35$ mm and $\theta = \pi$, and the bottom plot shows the result by fixing $r_0 = 35$ mm and $z = 20$. Note that the tensor product produces 36 basis (35 after enforcing identifiability constraints), but only a few basis are observed in both plots due to multiple functions having the same pattern in these ranges. Figure 7.9 presents the surface plots of the three tensor-product basis for multiple radii with a similar pattern in z and θ shifting on its range for each basis. The high-order tensor-product interaction terms that model roughness have a similar behavior with a faster reduction on the basis value. Note that the magnitude of γ_j can be interpreted as the presence of a local deformation due to roughness at the locations around the maximum of ϕ_j.

To assess the effect of roughness over the predicted deformation profile, we construct three increasingly complex models using these basis functions. (1) Fixed effects (FE) serves as a baseline model that considers only the deformation functions ψ. It is constructed using 5 cyclic cubic basis for θ, 3 TPRS for z, and 3 TPRS for r_0. Note that at most. a univariate quadratic relationship between z or r_0 and the deviation profile can be modeled since only a few cylinders of different sizes were manufactured. The resulting basis functions are illustrated in Figs. 7.8 and 7.9. (2) Scaled roughness (SR) is an extension of FE where the roughness is modeled using a high-order tensor-product basis scaled by $\alpha(r_0) = r_0$. The basis functions ϕ are constructed 16 cyclic cubic basis for θ and 32 cubic basis for z such that it absorbs the cyclical vertical patterns associated with thick layers in WAAM. Row i of Φ is $r_0\phi$ evaluated at point i. (3) Gaussian process (GP) is also an extension of FE where

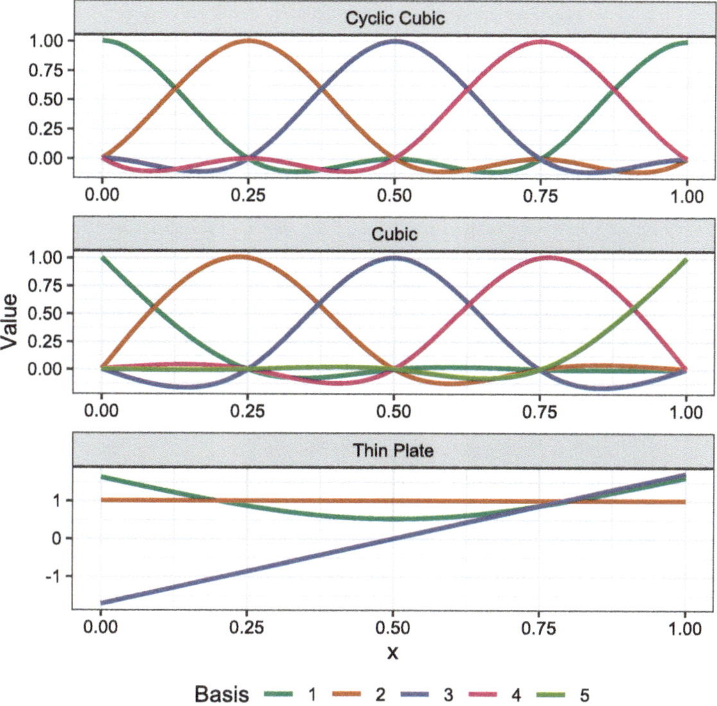

Fig. 7.7 Low-order univariate basis expansions used in this study

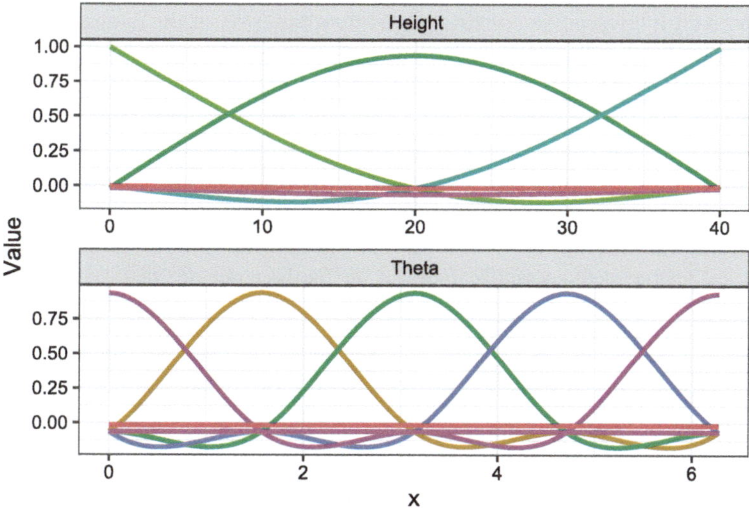

Fig. 7.8 Basis functions values at $r_0 = 35$ mm of a tensor product basis with 5 cyclic cubic basis for θ, 3 TPRS for z, and 3 TPRS for r_0. Top: fixed $\theta = \pi$. Bottom: fixed $z = 20$

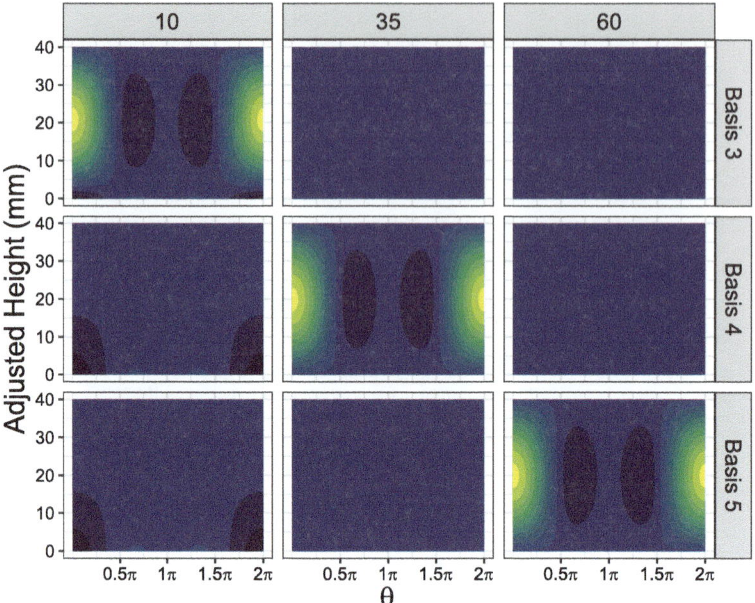

Fig. 7.9 Reparameterized low-order tensor-product basis for different r_0

the roughness is approximated using GP basis [22] for each cylinder, $\alpha(r_0) = 1$, and row i of Φ is ϕ evaluated at point i.

Gaussian processes are a popular choice for non-parametric modeling of smooth functions [35]. For regression, they assume that outputs $\mathbf{y}$ are jointly normally distributed with covariance constructed using a function of the distance between the features $\mathbf{x}$ of the observations. For computational tractability, GP basis are low-rank basis functions built from the approximated covariance between observations. The procedure consists of selecting a subset of representative observations (knots) and computing the covariance matrix $\Omega = \Omega_1 \Omega_2$, where Ω_1 is the covariance between the points in the sample and the knots and Ω_2 is the covariance between knots. ϕ is constructed as a tensor-product of univariate GP smooths using 16 knots with radial [42] covariance function for θ and 32 knots with Matérn covariance function with shape $\nu = 2.5$ for z. In each dimension a, ϕ^a is the rows of Ω^a.

7.1.3 Methodology Validation in WAAM

The shape deviation modeling can provide valuable insights for pre- and post-processing optimization. For the geometrical feature studied here, the models were trained using the point-cloud data of the cylindrical walls with 60 mm, 40 mm, and 15 mm radius (Fig. 7.4). The 25 mm cylindrical wall was used as validation data, and the compensation plans were created for a new 50 mm cylindrical wall. The 10 mm cylinder was excluded from this study because it does not meet the printing quality requirements, the process parameters were different, and most of the deviation patterns are caused by surface roughness. Due to the initial layers melting with the printing base and the excess material on the last layer deposited, only layers 3 to 18 are considered for prediction and compensation. The number of data points in each point-cloud was uniformly down-sampled to 3600 points to avoid overfitting the models to the patterns observed in the larger cylindrical walls. In addition, the height z is adjusted to $z = 0$ at the bottom of layer 3 for all datasets. The deviation profiles used for training the proposed model are shown on Fig. 7.10. All models were fitted using the *mgcv* package in R [42].

We examine the performance of the three shape deviation models to predict the deformation of the parts in the training dataset and to predict the deformation (without roughness) of the 25 mm cylindrical wall. Figure 7.11 presents the prediction of the three models for layers at the bottom (3), middle (11), and top (18) of the cylindrical walls in the training dataset. The highest deviations are visible as peaks at each welding joint. The FE model captures most of the deformation pattern layer-wise. The SR and GP models have a similar performance for the larger cylindrical walls. However, the GP model has a considerably better predictive fit for the smaller cylindrical wall. This suggests that a linear relationship between the roughness pattern and r_0 might be inadequate when r_0 is too small.

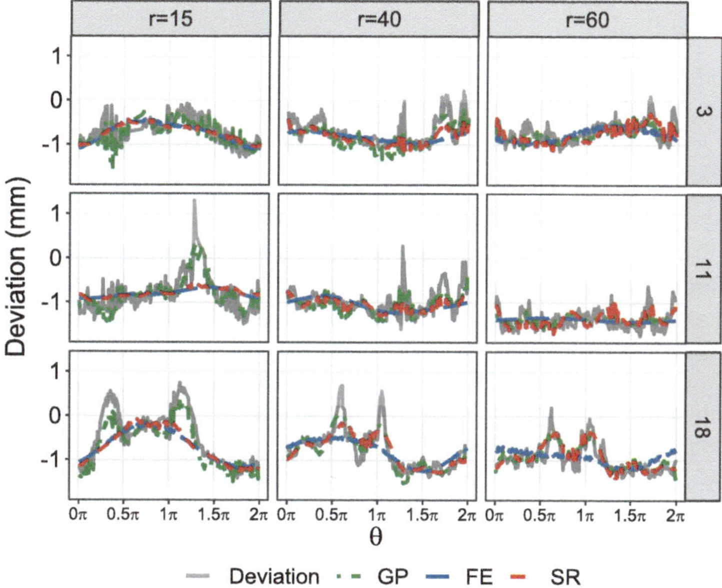

Fig. 7.10 Deviations profiles (mm) of the cylindrical walls for model training

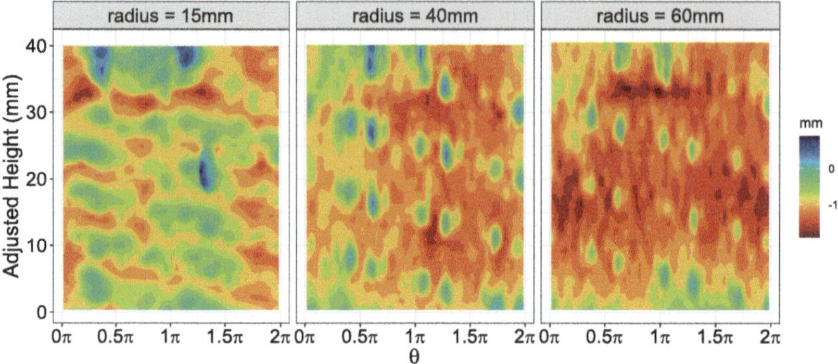

Fig. 7.11 Predicted shape deviation of three layers in all cylindrical walls in the training dataset

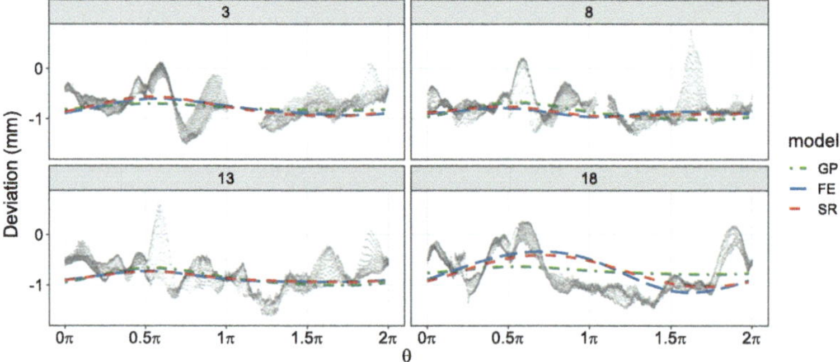

Fig. 7.12 Predicted shape deformation for three layers of the $r_0 = 25$ mm cylinder

Figure 7.12 presents the predicted shape deformation (without roughness pattern) for the cylindrical wall of $r_0 = 25$ mm. For this cylinder, the height z was unavailable, and prediction of the shape deformation is based on the mid-point height of each layer assuming that all layers have the nominal 2.5 mm height. This approximation increases the prediction error. FE and SR seem to perform better than GP because the roughness pattern is too strong in the latter, which reduces its predictive performance for new cylinders. The overall identifiability of the GAM model is measured by the concurvity between functional terms, which is a generalization of co-linearity for additive models [34]. Concurvity measures the proportion of a functional term that can be explained by other functional terms in the model [42]. For the SR model, the observed concurvity of the deformation pattern is 0.02 and 0.20 for the roughness pattern. On the other hand, the observed concurvity of the GP model is over 0.8 for almost all terms, which is expected since it has a functional term for the roughness in each cylinder.

7.1.3.1 Geometric Compensation for Shape Deviation Reduction in WAAM

Geometric compensation is defined as modifications to the printing path that lead to a controlled deformation of the fabricated part and improved accuracy with respect to the target geometry. Similar to [28], overall deviation per unit of area is measured as the absolute change in the external area of the thin walled shapes per unit of area, calculated as follows:

$$\eta = \frac{\Delta S}{S} \tag{7.6}$$

where $S = 2\pi r_0 z^*$ is the nominal area for a cylinder with maximum height z^*; ΔS is the absolute area change approximated as

$$\Delta S = \int_0^{z^*} \int_0^{2\pi} |y(\theta, z)| d\theta dz \approx \sum_{i=1}^{n} \sum_{j=1}^{n} |\bar{y}_{i,j}| w,$$

with the range of θ and z segmented into $n \times n$ cells of area w; and $\bar{y}_{i,j}$ is the mean deviation of points in cell i, j.

To illustrate the expected impact of compensation, Fig. 7.13 shows the predicted shape deviation for the 40 mm cylindrical wall after optimal compensation under the SR model, which is calculated as

$$\hat{y}(\theta, z, r_0, \delta) = f_1(\theta, z, r_0 + \delta) + y(\theta, z, r_0) - f_1(\theta, z, r_0)$$

for each data point. The compensated shape deviation is expected to be centered around zero with over 1 mm of extra material in the arc ignition locations. In the best case, the mean absolute deviation could diminish from 0.96 per mm^2 to 0.22 per mm^2, which implies a 76% reduction.

Experimental compensation is done for a new cylindrical wall with $r_0 = 50$ mm. As discussed above, the main deformation patterns can be predicted for cylindrical walls of reasonable radius. Figure 7.14 shows the predicted deformation and the optimal in-plane compensation plan at any height. GP and SR have a similar deviation profile and compensation. Since FE does not consider surface roughness, the FE predictions and compensation plan have a more complicated pattern than the other two models. The SR model was used to compensate the design of a new $r_0 = 50$ mm cylindrical wall. The layer-wise in-plane compensation plan is shown in Fig. 7.15. Note that compensation starts relatively small, then increases at the middle layers, and lessens again at the top layers.

For the new 50 mm cylindrical wall, the shape deviation pattern is greatly reduced after compensation, as shown in Fig. 7.16. The absolute deviation per mm^2 is 1.51 and 0.80 for the uncompensated and compensated parts, respectively. This translates to a 47.3% reduction in overall shape deviation. Figure 7.17 presents the shape

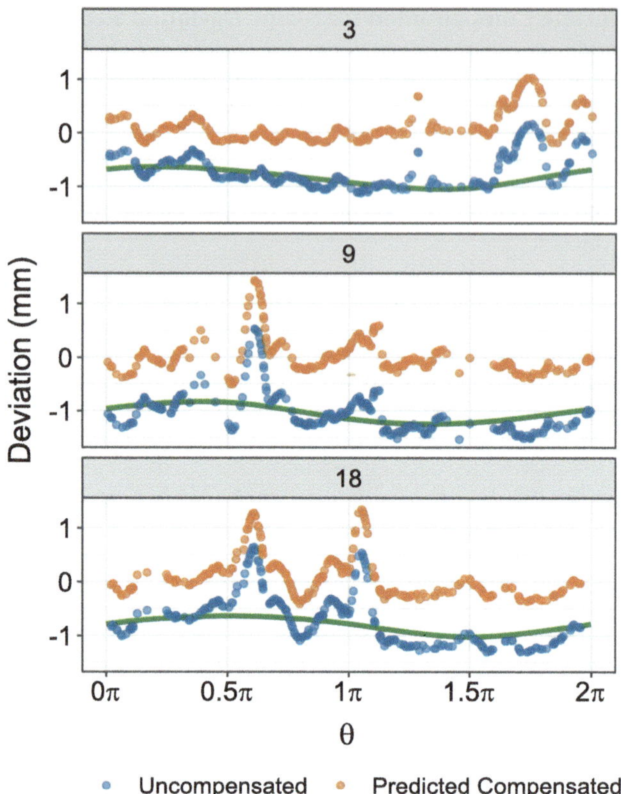

Fig. 7.13 Predicted deformation (green line), shape deviation (blue dots), and predicted shape deviation after compensation (orange dots) at the center of layers 3, 9, and 18 for the 40 mm cylinder under the SR model

deviation patterns and their respective smooth deformations at the center of layers 3, 9, and 18 of the parts. The overall patterns show a similar behavior to those predicted in Fig. 7.13. The smooth trends illustrate the effect of the local roughness and the accumulation of material at the arc ignition locations over the shape deviation. However, the parts present an overall shrinkage not predicted by our model.

Although the proposed prescriptive methodology significantly reduces the overall shape deviation, the dimensional quality of WAAM manufactured components is affected by a number of factors, including environmental conditions, material fluctuations, and changing thermo-mechanical boundary conditions. The deposition parameters utilized in this study caused some significant geometric irregularities. In particular, unforeseen material and environmental changes, such as air flow altering the printing interpass temperature, might have modified the deformation pattern in the second experiment. Consequently, the bumps at the arc ignition locations are more visible in Fig. 7.16 than in Fig. 7.1. Measurement and point cloud data

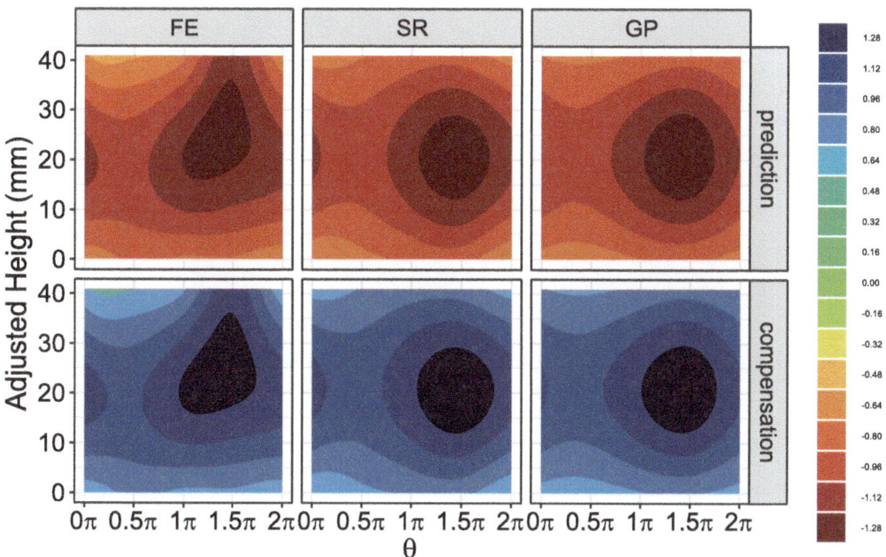

Fig. 7.14 Predicted shape deformation and compensation plan for the $r_0 = 50$ mm cylindrical wall

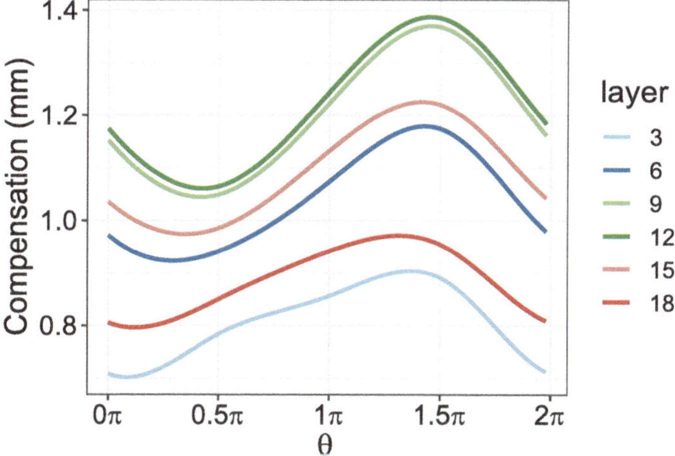

Fig. 7.15 Layer-wise compensation plan for the 50 mm cylindrical wall under the SR model

registration may also have introduced errors. Lastly, the framework in this study could benefit practitioners in the following aspects:

1. Product designers can analyze the predicted deformation patterns of new parts in the initial design stages. Apart from the design freedom provided by WAAM, this ensures that deposition paths that improve geometric accuracy can be achieved early in the design process.

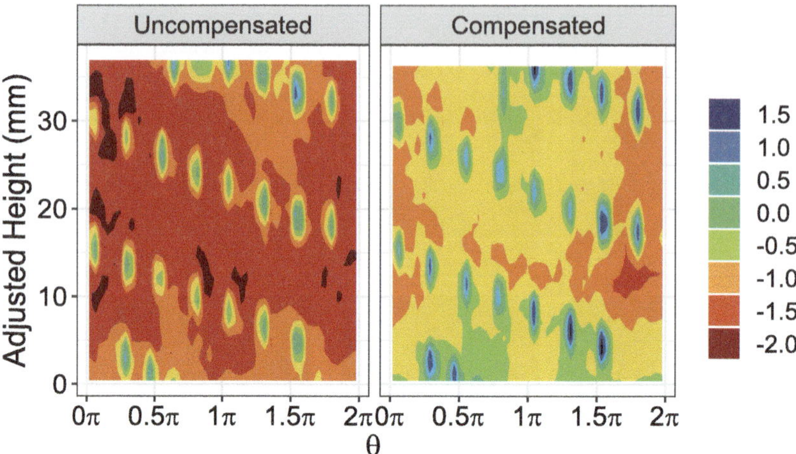

Fig. 7.16 Shape deviation profiles of uncompensated and compensated 50 mm cylindrical walls

2. The usage of this framework by operations engineers minimizes the need for them to deeply understand the complexities of the process, facilitating decision-making, improving part accuracy, and helping automate the WAAM process.

In this section, prescriptive modeling and compensation are introduced to reduce shape deviation in WAAM fabricated products. Shape deviation is more complex in WAAM than other AM processes due to process instability and highly nonlinear stress profiles caused by the high-temperature processes. We model shape deviation as a GAM that decomposes deviation into global deformation and local roughness patterns. Tensor-basis expansions of local geometric features in the cylindrical coordinates system are used to predict shape deviation. Each univariate basis is chosen based on engineering considerations of each geometric feature to obtain meaningful and interpretable results with a few samples. To study the effect of roughness on the deformation prediction, we construct three models with increasing complexity. The FE model considers global deformation, the SR model considers global deformation and local roughness scaled by the part size, and the GP model considers global deformation with Gaussian process basis for the roughness of each individual part.

Experimental validation on cylindrical walls demonstrates the predictive and prescriptive capabilities of the proposed framework. We manufactured four equally tall cylindrical walls with radii 15, 25, 40, and 60 mm using a GMAW system. By considering roughness, the SR and GP models have better deviation prediction performance than the FE model while retaining a complex enough deformation pattern for effective compensation. The optimal geometric compensation of a new 50 mm cylindrical wall results in a 47.3% reduction of absolute deviation per mm^2.

The proposed shape deviation methodology can be applied to a broad category of WAAM as well as AM processes to manufacture smooth geometries. Using

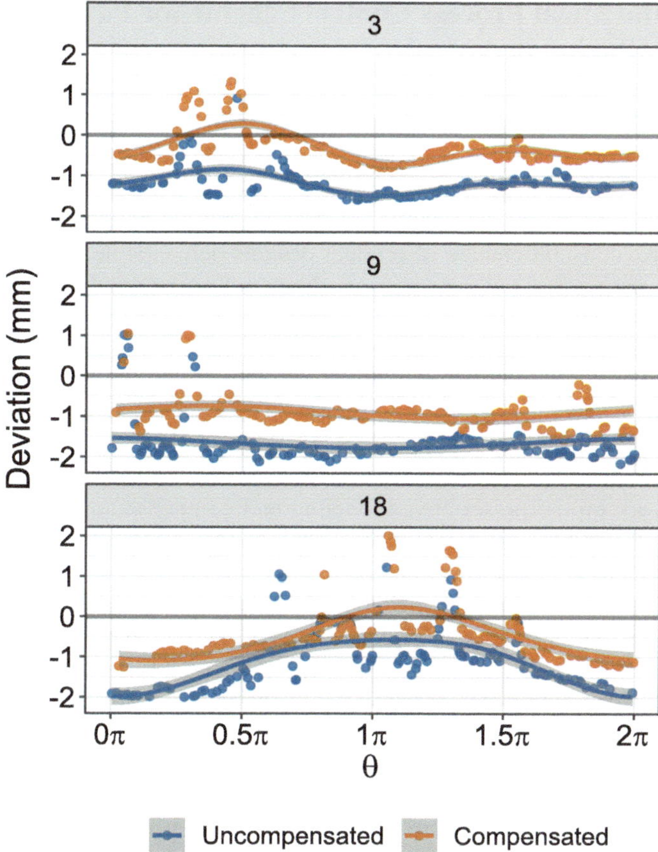

Fig. 7.17 Shape deviation and smoothed shape deformation (using five cyclic cubic basis) at the center of layers 3, 9, and 18 for uncompensated and compensated parts

the equivalence effects principle, model transfer between materials and WAAM processes can be achieved by adding a new term to capture the differences among their shape deviation profiles [36]. Further research is needed to predict the shape deformation patterns of straight structures and crossings between segments withing layers. Such extensions can be achieved by defining an efficient functional representation and univariate basis functions for straight segments and adding extra terms to capture the deviation induced by crossings. Moreover, the effect of relevant process parameters for specific WAAM processes can be incorporated by developing physics-informed basis function as well as adding additive terms for a global process-independent mean deviation pattern and a process or geometry related term. Furthermore, the methodology may potentially be extended to integrate real-time in situ data to further improve shape accuracy.

7.2   A Statistical Process Control Scheme for Part-to-Part Monitoring in AM

Monitoring of AM processes faces major challenge due to the nature of one-of-a-kind manufacturing. In contrast to mass production, AM processes fabricate products with high shape complexity at extremely low volume, which leads to disparate training data with small sample size from one product shape to another. Furthermore, constant change of product designs demands the capability of monitoring the process of building new products not being tried before. Traditional Statistical Process Control (SPC) [30] therefore can hardly be applied directly. A popular solution is to monitor AM process parameters indicative of process conditions. However, process conditions have to be benchmarked by product quality as well. This section puts forth a prescriptive SPC scheme to monitor shape deviation from part to part with different shapes. Only a limited number of test shapes are required to establish control limits. The strategy is extension of our previous work on the prescriptive modeling and compensation of shape deviation to the field of SPC. Experimental investigation using SLA process validates the proposed deviation monitoring approach.

Various strategies have been developed to improve the product accuracy. Among them, online monitoring and feedback control have become a critical area for AM part accuracy enhancement. Most of the efforts focus on monitoring AM process variables such as temperature field and melt pool size as proxy of product quality. To detect abnormality, the prediction of physics-based simulation model is compared with in situ sensing data of process variables [1, 2]. However, high-fidelity physical models are costly to build [24]. Furthermore, majority of these studies are benchmarked by products with relatively simple geometries [38].

Geometric accuracy control in AM involves monitoring of product shapes. Work in [5] models roundness and deviation profiles of cylinders fabricated through traditional subtractive manufacturing. A spatial autoregressive regression model is used to monitor cylinder shape quality. Work in [6] further extends to surface deviation monitoring using Gaussian process models. But the shape modeling and monitoring approach in [5, 6] does not present a clear path to handle shape complexity and limited data faced in AM [14, 16, 21, 27].

Existing process monitoring approaches developed for AM can be viewed as predictive process monitoring in the sense that training samples from the same population are collected to monitor future production of the similar products. AM process monitoring based on physical models intends to predict quality of arbitrary shapes. As mentioned previously, current successful examples are more limited to products with simple geometries. For one-of-a-kind additive manufacturing, we believe that a new category of *prescriptive process monitoring* methods are required to monitor quality of new and untried categories of product shapes.

7.2.1 Prescriptive Monitoring of Shape Deviation from Part to Part

For simplicity of representation, examples of 2D product shapes are used. In real applications, a thin product or a section of a product with small thickness can be approximated with a 2D shape where only in-plane geometric error is of major concern.

Training Data The training data can be illustrated by Fig. 7.18, where three circles with various sizes and two polygons are required to be built and measured.

Assumption The training products are fabricated under the same process settings with the same materials. The process is relatively repeatable and deemed as normal condition. Note that normal condition does not require zero shape deviation.

Objective of Process Monitoring When a new and different shape is built, the monitoring scheme is expected to decide whether the process is in control or out of control.

Challenge Clearly, the challenge of developing a SPC scheme resides in data disparity, limited sample size, and complicated data structure. In addition, decision on new products requires the identification of inherent connection among the data.

Under the general modeling framework of control charts, the critical step is to identify the statistic for monitoring. In AM, training data illustrated in Fig. 7.18 needs a proper transformation in order to derive the control chart model. To illustrate the rationale of the proposed statistics, we will first discuss an ideal AM process without systematic shape deviation patterns. We will then derive the monitoring statistic for realistic AM processes often containing systematic deviation patterns varying from shape to shape.

Suppose an ideal AM process builds a circle and square shape shown in Fig. 7.19. With the presence of only natural process variation, the built products in solid curves should be contained within narrow envelops. Quantities such as roundness or cylindricity are not considered here because these measures tend to be shape-dependent, which limits the ability of quantifying a different shape.

One intuitive measure of shape deviation across different shapes is the percentage of shrinkage/expansion η, which can be defined as

$$\eta = \frac{|\Delta S|_{Actual}}{S_{Nominal}} \tag{7.7}$$

Fig. 7.18 Training data

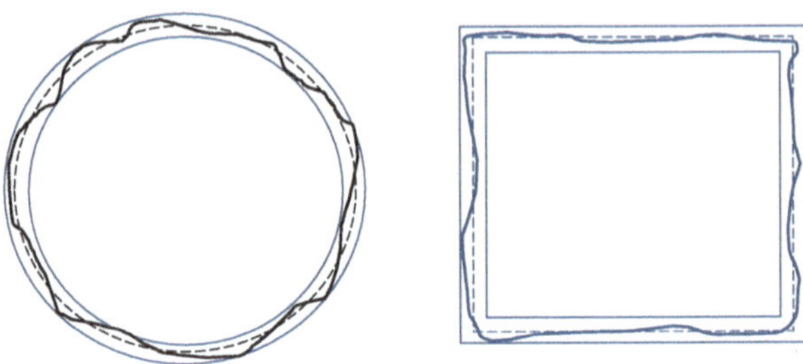

Fig. 7.19 Statistic for AM process monitoring

where $S_{Nominal}$ represents the nominal in-plane surface area of the product and $|\Delta S|_{Actual}$ is the measured, absolute change of surface area from the design. Extension to 3D case naturally follows, which is not discussed here.

Under stable process condition and the same materials, we could expect that products with various shape would have similar percentages of shrinkage. In practice, AM industry often uses it as a benchmark value for different materials. The training data in Fig. 7.18 can be transformed into $|\Delta S|/S$, and individual control charts such as EWMA can be applied for process monitoring.

In practice, however, current AM machines often demonstrate systematic, shape-dependent deviation patterns, even key process variables are within control. In another word, AM machines under normal conditions will have systematic deviation patterns. Directly monitoring $|\Delta S|/S$ could end up with large false alarm rate. Statistics $|\Delta S|/S$ in Eq. (7.7) has to be modified for realistic AM processes.

The modified monitoring statistic has two variations. In the first case, we remove the expected or predicted amount of surface deviation $|\Delta S|_{Predicted}$ from the measured $|\Delta S|_{Actual}$.

$$\eta = \frac{\left| |\Delta S|_{Actual} - |\Delta S|_{Predicted} \right|}{S_{Nominal}} \tag{7.8}$$

The logic justification of this proposal is that the process is deemed in control if the actual amount of deviation is close to the expected one. Since the expected deviation pattern is removed from the measurement, the AM process is equivalent to an ideal process, and the false alarm rate can be reduced.

The first proposal in Eq. (7.8) is passive control strategy in that sense that actual deviation cannot be reduced. To remedy this problem, we propose the second variation of monitoring statistic:

$$\eta = \frac{\left| |\Delta S|_{Actual} - |\Delta S|_{Compensated} \right|}{S_{Nominal}} \tag{7.9}$$

where $|\Delta S|_{Compensated}$ represents the predicted, absolute surface deviation after deviation compensation. If the shape deviation of a new product can be predicted and compensated beforehand, the percentage of shrinkage/expansion will be as small as the ideal AM process under normal condition. The false alarm rate of individual control charts can also be reduced.

The critical element of both proposals, however, is the prescriptive model for predicting the shape deviation of a new product based on training data illustrated in Fig. 7.18. We adopt the prescriptive shape deviation modeling framework developed in [14–16, 27, 39, 49]. To accomplish the prediction goal, we in [16] transform the shape from the Cartesian coordinates system to profiles in the polar coordinates system (PCS). This decouples the geometric shape complexity and makes the proposed prescriptive model free of shape complexity. In-plane shape deviation and optimal compensation plan is established therein. Huang et al. [14] extends the approach from cylindrical shape to polyhedrons by treating an in-plane polygon as being cut from its circumcircle. Luan and Huang [26] further extends the prescriptive modeling and compensation approach from cylinder, polyhedron to arbitrary freeform shapes through a circular approximation with selective cornering strategy.

Denote $\Delta r(\theta, r_0(\theta))$ as the in-plane deviation profile of a product shape in the PCS: $\Delta r(\theta, r_0(\theta)) = r(\theta, r_0(\theta)) - r_0(\theta)$, where $r_0(\theta)$ is the nominal design shape in PCS.

The prescriptive shape deviation model for freeform shape is formulated as:

$$\Delta r(\theta, r_0(\theta)) = f(\theta, r_0(\theta)) + \epsilon(\theta)$$

$$= g_1(\theta, r_0(\theta)) + g_2(\theta, r_0(\theta)) + \epsilon(\theta)$$

where g_1 depicts the generalized cylindrical basis function, g_2 is the cookie-cutter function, and $\epsilon(\theta)$ represents the unmodeled term. With limited number of test shapes, dramatic in-plane accuracy improvement has been achieved through compensation for cylindrical shape (90%), polyhedrons, and freeform shapes (>50%) [14, 16, 21, 26, 37].

With the prescriptive model, $|\Delta S|_{Predicted}$ in Eq. (7.8), according to [12], can be derived as

$$|\Delta S|_{Predicted} = \int_0^{2\pi} r_0(\theta)|f(\theta, r_0(\theta))|d\theta \tag{7.10}$$

$|\Delta S|_{Compensated}$ in Eq. (7.9), according to [12], can be derived as

$$|\Delta S|_{Compensated} = \int_0^{2\pi} r_0(\theta)\Big|f(\theta, r_0(\theta) + x(\theta)) + x(\theta)\Big|d\theta \tag{7.11}$$

where $x(\theta)$ denotes the amount of compensation applied to CAD design to reduce shape deviation.

The absolute shape area deviation with and without compensation can be illustrated by Fig. 6.2.

Under the minimum area deviation criteria, the optimal amount of compensation is [12, 16]

$$x^*(\theta) = -\frac{f(\theta, r_0(\theta))}{1 + f'(\theta, r_0(\theta))}$$

$|\Delta S|_{Compensated}$ in Eq. (7.9) can be simplified as

$$|\Delta S|_{Compensated} = \int_0^{2\pi} r_0(\theta)|\epsilon(\theta)|d\theta \tag{7.12}$$

Clearly, $|\Delta S|_{Compensated}$ in ideal case approaches to zero.

Lastly, $S_{Nominal}$ can be simply computed as

$$S_{Nominal} = \int_0^{2\pi} r(\theta)d\theta$$

7.2.2 Control Chart to Monitor Shape Deviation from Part to Part

With individual measurement of AM-built products and the objective to detect small process shift, we adopt EWMA control chart to monitor the proposed statistic of percentage of shrinkage η defined in Eq. (7.8) or Eq. (7.9).

For a sequence of AM-built products and their η_i numbered in a chronicle order, the EWMA statistics is

$$z_i = \lambda \eta_i + (1 - \lambda)z_{i-1}, \quad z_0 = 0;$$

with the corresponding control limits:

$$UCL = \mu_\eta + L\sigma_\eta \sqrt{\frac{\lambda}{2 - \lambda}[1 - (1 - \lambda)^{2i}]}$$

$$CL = \mu_\eta$$

$$LCL = \mu_\eta - L\sigma_\eta \sqrt{\frac{\lambda}{2 - \lambda}[1 - (1 - \lambda)^{2i}]}$$

where μ_η and σ_η are mean and standard deviation of η_i observed from training data, $0 < \lambda < 1$ is a constant, and L is the width of the control limits.

7.2.2.1 Process capability index C_p for AM processes

The proposed monitoring statistic η enables the adoption of the widely accepted process capability index C_p AM processes. Due to the nature of η, i.e., the percentage of shrinkage/expansion, it is meaningful to only specify a USL (upper specification limit), e.g., 2%. The C_p for AM processes can be defined as

$$C_p = \frac{USL - \mu_\eta}{3\sigma_\eta} \tag{7.13}$$

where μ_η and σ_η represents the mean and standard deviation of statistic η.

7.2.2.2 Estimation of μ_η and σ_η

For training data with n products, e.g., $n = 5$, μ_η can be estimated as the average of $\eta_i, i = 1, 2, \ldots, n.$ σ_η can be estimated through moving ranges $MR_i = |\eta_i - \eta_{i-1}|$, and σ_η is

$$\sigma_\eta = \frac{\overline{MR}}{d_2} \tag{7.14}$$

where $\overline{MR}$ denotes the average of moving ranges and d_2 takes the value of 1.128 since we obtain the range using two consecutive observations.

7.2.3 Monitoring Stereolithography Process: Methodology Demonstration

This section demonstrates the proposed prescriptive SPC scheme to monitor an actual SLA process.

7.2.3.1 Establish Prescriptive Model

In our previous work [13, 14, 16], three circular disks and three polygons are fabricated in a MIP-SLA machine to validate the prescriptive model for cylinder and polyhedron shapes. Design of training sample products are summarized in Table 7.3, and Fig. 7.20 shows their deviation profiles. All test parts have the same height of 0.25 in. To facilitate the identification of the orientation of test parts during or after the building process, a non-symmetric cross with line thickness of 0.02 in is built on the top surface.

The specification of prescriptive deviation model $f(\theta, r_0(\theta))$ is detailed in [26]. As an example, the fitted circular basis model $g_1(\theta, r(\theta))$ is divided into two parts.

Table 7.3 Design of training sample products

Cross-sectional shape	Circumcircle radius
Circle	$r = 0.5''; 1.5''; 3''$
Square	$r = 2''/\sqrt{2}$
Regular pentagon	$r = 1''; 3''$

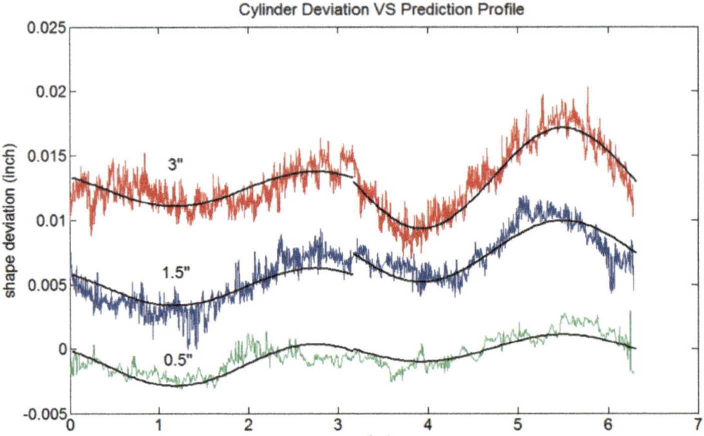

(a) Deviation profiles (red, blue and green lines) and prediction (smooth dark lines) of three circular disks

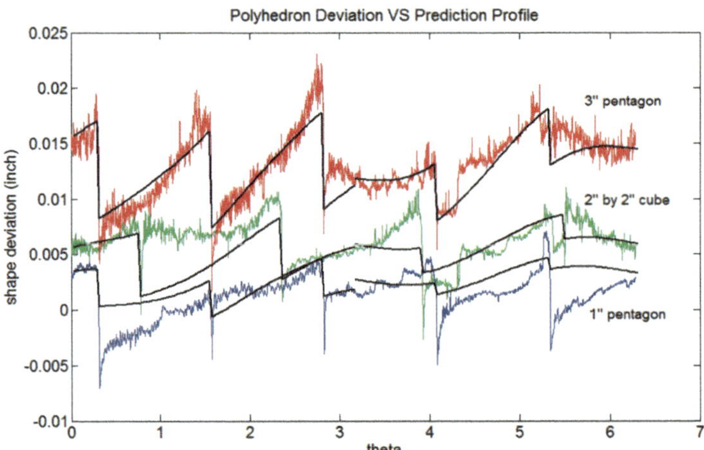

(b) Deviation profiles (red,blue and green lines) and prediction (dark lines) of three polygons

Fig. 7.20 Deviation and prediction profiles of simple trial shapes [26]

For the upper half ($\theta = 0 \sim \pi$) of the product, $g_1(\theta, r(\theta))$ is:

$$x_0^u + \alpha^u (r_0 + x_0^u)^{a^u} + \beta^u (r_0 + x_0^u)^{b^u} \cos(2(\theta + \pi/8))$$

For the lower half ($\theta = \pi \sim 2\pi$) of the product, $g_1(\theta, r(\theta))$ is:

$$x_0^l + \alpha^l (r_0 + x_0^l)^{a^l} + \beta^l (r_0 + x_0^l)^{b^l} (-\sin(2\theta))$$

Note that superscripts u and l indicate parameters corresponding to upper and lower half of the product, respectively.

The cookie-cutter model or $g_2(\theta, r(\theta))$ is first developed in [14] to capture the sharp transitions of polygon shapes.

$$g_2(\theta, r(\theta)) = \beta_2(r_0 + x_0)^\alpha cookie.cutter(\theta - \phi_0)$$

One example of cookie-cutter functions is the square wave model:

$$sign[cos(n(\theta - \phi_0)/2)]$$

where n is the number of sides of a polygon and ϕ_0 is a phase variable to shift the cutting position in the PCS. The sawtooth cookie-cutter model is another alternative, which is adopted in our analysis:

$$saw.tooth(\theta - \phi_0) = (\theta - \phi_0) \text{ MOD } (2\pi/n)$$

where x MOD y = remainder of (x/y).

7.2.3.2 Monitoring Statistic η and Phase I Control Charting

Since all the six training parts in Table 7.3 are fabricated without compensation, Eq. (7.8) is adopted to derive the monitoring statistic η (as shown in Table 7.4)

The Normal Q-Q plot of the sample data η_i is analyzed and shown in Fig. 7.21. It indicates that the normality assumption reasonably holds.

The estimated mean and standard deviation are $\mu_\eta = 3.3563 \times 10^{-4}$ and $\sigma_\eta = 2.4337 \times 10^{-4}$.

For the EWMA control chart with $\lambda = 0.1$ and $L = 2.7$, the average run length (ARL) has $ARL_0 \approx 500$ and $ARL_1 \approx 10.3$ for detecting 1σ mean shift.

Table 7.4 Monitoring statistic η_i of training data

Cross-sectional shape	$|\Delta S|_{Actual}$	$|\Delta S|_{Predicted}$	η ($\times 10^{-4}$)
0.5″ circle	0.00396	0.00318	2.5057
1.5″ circle	0.05988	0.05839	1.5814
3″ circle	0.24503	0.24175	1.7407
2″/$\sqrt{2}$ square	0.04391	0.03830	7.9578
1″ pentagon	0.01092	0.01355	4.8185
3″ pentagon	0.21370	0.21120	1.5328

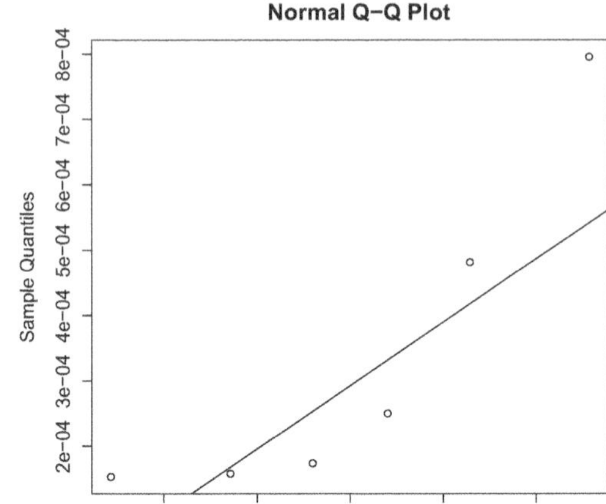

Fig. 7.21 Normal Q-Q plot of sample data η_i

Table 7.5 EWMA control limits

No.	Cross-sectional shape	UCL ($\times 10^{-4}$)	CL ($\times 10^{-4}$)	LCL ($\times 10^{-4}$)
1	$0.5''$ circle	4.0132	3.3562	2.6991
2	$1.5''$ circle	4.2402	3.3562	2.4721
3	$3''$ circle	4.3880	3.3562	2.3243
4	$2''/\sqrt{2}$ square	4.4938	3.3562	2.2185
5	$1''$ pentagon	4.5727	3.3562	2.1396
6	$3''$ pentagon	4.6331	3.3562	2.0792

The corresponding phase I EWMA control chart limits are shown in Table 7.5 and Fig. 7.22.

If the upper specification limit (USL) for the percentage of shrinkage/expansion is 1%, the process capability index following Eq. (7.13) is $C_p = 13.23716$.

As a comparison, if compensation is applied and quality measure in Eq. (7.7) is adopted, the mean and standard deviation of η without compensation are $\mu_\eta = 6.9905 \times 10^{-3}$ and $\sigma_\eta = 5.9972 \times 10^{-3}$. Then the process capability index following Eq. (7.13) is $C_p = 0.1672709$.

The comparison of two C_p indicates that the optimal compensation plan owns the potential to greatly improve the ability of a process to meet the specification.

7.2.3.3 Phase II Control Charting and Validation

In order to test the detection ability of SPC scheme for AM processes, four parts printed under different conditions are employed to simulate the occurrence of process shift.

The first part is a $3''$ dodecagon with compensation. Since this is an untried new shape, the compensation is derived directly from the prescriptive model for

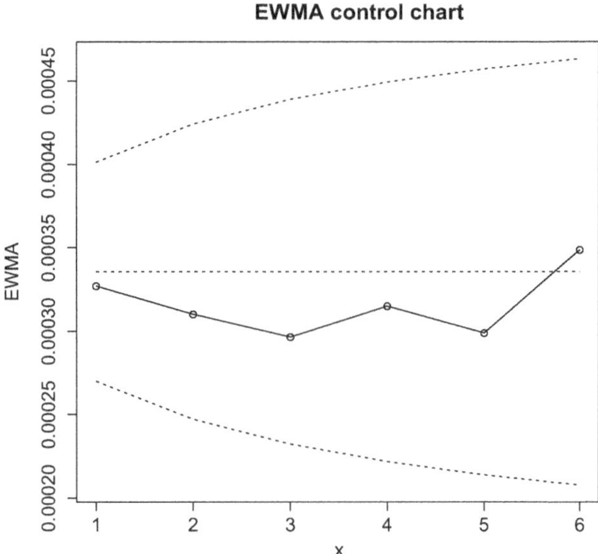

Fig. 7.22 phase I EWMA control chart

Table 7.6 Monitoring statistic η_i of validation data

Cross-sectional shape	$|\Delta S|_{Actual}$	$|\Delta S|_{Predicted}$	$\eta\ (\times 10^{-3})$
$3''$ dodecagon with compensation	0.07282	0	3.9575
$1''$ pentagon with compensation	0.01267	0	2.3245
$1''/\sqrt{2}$ square inner	0.02729	0.00351	6.7521
$2''/\sqrt{2}$ square before repair	0.05461	0.03812	2.3446

polyhedron validated in [14]. Although our compensation has reduced on average 75% deviation [14], there are still some deviations, and patterns remain, indicating that our model could be further improved with this new dataset.

The second part is a $1''$ pentagon with compensation. But the compensation is wrongly added, i.e., the compensation plan is not optimal.

The third part is a 1 by $1''$ square, but this square is the inner boundary of a square cavity. Since the material deviation mechanism is definitely different for inner boundary, its deviation pattern is also different.

The last part is a 2 by $2''$ square printed in a different process condition. As we mentioned in [14], the MIP-SLA machine settings were changed after the experiments in [16]. Therefore, the part produced before machine repair has a different shape of deviation profile.

The monitoring statistic η of these four parts are shown in Table 7.6.

Adding those four data obtained under "abnormal" printing processes, the EWMA control chart is shown in Fig. 7.23. The EWMA chart control correctly detects the changes.

This section presents a SPC scheme to monitor part-to-part changes in AM. Process capability index is also established to quantify the capability of AM processes.

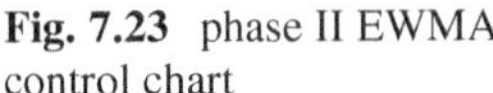

Fig. 7.23 phase II EWMA control chart

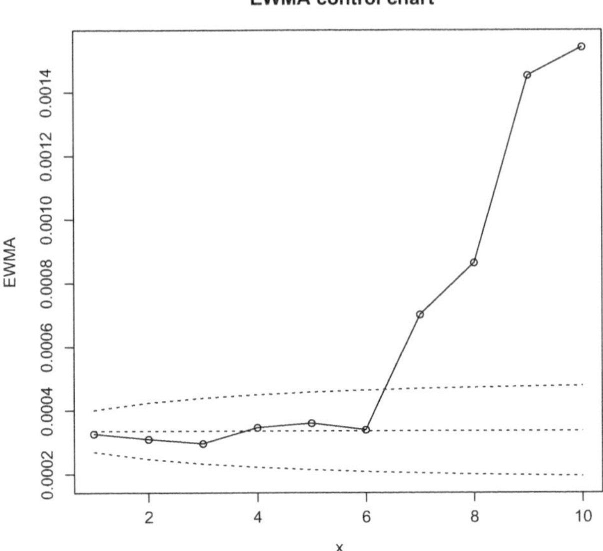

The proposed statistic for monitoring product shape deviation is consistent with industrial practice, i.e., the percentage of shrinkage/expansion. The requirement for training data is minimal, and prescriptive model is adopted to predict the deviation of new and untried product shapes. This monitoring statistic incorporates the predicted deviation, which enables the monitoring of new product in one-of-a-kind manufacturing environments. Real AM process experiments show the promise of applying the proposed SPC scheme. It should be noted that the proposed monitoring statistic only reflects the overall deviation of a product, and there might be other alternatives to catch local deviation conditions. More future work is necessary to establish comprehensive SPC methods for AM.

Acknowledgments The work in this chapter was partially supported by US Oak Ridge National Laboratory with grant # 4000145653 and by University of USC Pratt & Whitney Institute for Collaborative Engineering Board (PWICE) Project.

References

1. Aggarangsi, P., Beuth, J.L., Gill, D.D.: Transient changes in melt pool size in laser additive manufacturing processes. In: Solid Freeform Fabrication Proceedings, University of Texas, Solid Freeform Fabrication Symposium, pp. 2–4, Austin (2004)
2. Birnbaum, A., Aggarangsi, P., Beuth, J.: Process scaling and transient melt pool size control in laser-based additive manufacturing processes. In: Solid Freeform Fabrication Proceedings, pp. 328–339. Proc. 2003 Solid Freeform Fabrication Symposium, Austin (2003)
3. Bourell, D., Kruth, J.P., Leu, M., Levy, G., Rosen, D., Beese, A.M., Clare, A.: Materials for additive manufacturing. CIRP Ann. **66**(2), 659–681 (2017)

4. Candès, E.J., Li, X., Ma, Y., Wright, J.: Robust principal component analysis? J. ACM **58**(3) (2011). https://doi.org/10.1145/1970392.1970395

5. Colosimo, B., Pacella, M., Semeraro, Q.: Statistical process control for geometric specifications: on the monitoring of roundness profiles. J. Quality Technol. **40**(1), 1–18 (2008)

6. Colosimo, B.M., Cicorella, P., Pacella, M., Blaco, M.: From profile to surface monitoring: Spc for cylindrical surfaces via gaussian processes. J. Quality Technol. **46**(2), 95 (2014)

7. Ding, D., Pan, Z., Cuiuri, D., Li, H.: Wire-feed additive manufacturing of metal components: technologies, developments and future interests. Int. J. Adv. Manuf. Technol. **81**(1), 465–481 (2015)

8. Ding, D., Pan, Z., Cuiuri, D., Li, H., Larkin, N., Van Duin, S.: Automatic multi-direction slicing algorithms for wire based additive manufacturing. Robot. Comput.-Integrated Manuf. **37**, 139–150 (2016)

9. Ding, J., Colegrove, P., Mehnen, J., Ganguly, S., Almeida, P.S., Wang, F., Williams, S.: Thermo-mechanical analysis of wire and arc additive layer manufacturing process on large multi-layer parts. Comput. Mater. Sci. **50**(12), 3315–3322 (2011)

10. Garmendia, I., Leunda, J., Pujana, J., Lamikiz, A.: In-process height control during laser metal deposition based on structured light 3d scanning. Procedia CIRP **68**, 375–380 (2018). https://doi.org/https://doi.org/10.1016/j.procir.2017.12.098

11. Gibson, I., Rosen, D., Stucker, B.: Additive Manufacturing Technologies: Rapid Prototyping to Direct Digital Manufacturing. Springer, Berlin (2009)

12. Huang, Q.: An analytical foundation for optimal compensation of three-dimensional shape deformation in additive manufacturing. ASME Trans. J. Manuf. Sci. Eng. **138**(6), 061010 (2016)

13. Huang, Q., Nouri, H., Xu, K., Chen, Y., Sosina, S., Dasgupta, T.: Predictive modeling of geometric deviations of 3d printed products-a unified modeling approach for cylindrical and polygon shapes. In: 2014 IEEE International Conference on Automation Science and Engineering (CASE), pp. 25–30. IEEE, Taipei (2014)

14. Huang, Q., Nouri, H., Xu, K., Chen, Y., Sosina, S., Dasgupta, T.: Statistical predictive modeling and compensation of geometric deviations of three-dimensional printed products. ASME Trans. J. Manuf. Sci. Eng. **136**(6), 061008–061018 (2014)

15. Huang, Q., Wang, Y., Lyu, M., Lin, W.: Shape Deviation Generator-A Convolution Framework for Learning and Predicting 3-D Printing Shape Accuracy. IEEE Trans. Autom. Sci. Eng. **17**(3), 1486–1500 (2020). https://doi.org/10.1109/TASE.2019.2959211

16. Huang, Q., Zhang, J., Sabbaghi, A., Dasgupta, T.: Optimal offline compensation of shape shrinkage for 3d printing processes. IIE Trans. Quality Reliab. **47**(5), 431–441 (2015)

17. Huang, Q., Zhang, J., Sabbaghi, A., Dasgupta, T.: Optimal offline compensation of shape shrinkage for three-dimensional printing processes. IIE Trans. **47**(5), 431–441 (2015)

18. Jafari, D., Vaneker, T.H., Gibson, I.: Wire and arc additive manufacturing: Opportunities and challenges to control the quality and accuracy of manufactured parts. Mater. Design **202**, 109471 (2021)

19. Jafari, D., Wits, W.W., Vaneker, T.H., Demir, A.G., Previtali, B., Geurts, B.J., Gibson, I.: Pulsed mode selective laser melting of porous structures: Structural and thermophysical characterization. Addit. Manuf. **35**, 101263 (2020). https://doi.org/https://doi.org/10.1016/j.addma.2020.101263

20. Jin, W., Zhang, C., Jin, S., Tian, Y., Wellmann, D., Liu, W.: Wire arc additive manufacturing of stainless steels: a review. Appl. Sci. **10**(5), 1563 (2020)

21. Jin, Y., Qin, S., Huang, Q.: Offline predictive control of out-of-plane geometric errors for additive manufacturing. ASME Trans. Manuf. Sci. Eng. **138**(12), 121005(7 pages) (2016)

22. Kammann, E., Wand, M.P.: Geoadditive models. J. Roy. Stat. Soc. C (Appl. Stat.) **52**(1), 1–18 (2003)

23. Karunakaran, K., Suryakumar, S., Pushpa, V., Akula, S.: Low cost integration of additive and subtractive processes for hybrid layered manufacturing. Robot. Comput.-Integrated Manuf. **26**(5), 490–499 (2010)

24. Khairallah, S.A., Anderson, A.T., Rubenchik, A., King, W.E.: Laser powder-bed fusion additive manufacturing: Physics of complex melt flow and formation mechanisms of pores, spatter, and denudation zones. Acta Materialia **108**, 36–45 (2016)
25. Luan, H., Grasso, M., Colosimo, B.M., Huang, Q.: Prescriptive data-analytical modeling of laser powder bed fusion processes for accuracy improvement. J. Manuf. Sci. Eng. **141**(1), 011008 (2019)
26. Luan, H., Huang, Q.: Prescriptive modeling and compensation of in-plane geometric deviations for 3d printed freeform products. IEEE Trans. Autom. Sci. Eng. (2016). https://doi.org/10.1109/TASE.2016.2608955
27. Luan, H., Huang, Q.: Prescriptive modeling and compensation of in-plane shape deformation for 3-D printed freeform products. IEEE Trans. Autom. Sci. Eng. **14**(1), 73–82 (2017)
28. Luan, H., Post, B.K., Huang, Q.: Statistical process control of in-plane shape deformation for additive manufacturing. In: 2017 13th IEEE Conference on Automation Science and Engineering (CASE), pp. 1274–1279 (2017). https://doi.org/10.1109/COASE.2017.8256276
29. Martina, F., Colegrove, P.A., Williams, S.W., Meyer, J.: Microstructure of interpass rolled wire+ arc additive manufacturing ti-6al-4v components. Metallurg. Mater. Trans. A **46**(12), 6103–6118 (2015)
30. Montgomery, D.C.: Statistical Quality Control, vol. 7. Wiley, New York (2009)
31. Mou, S., Wang, A., Zhang, C., Shi, J.: Additive tensor decomposition considering structural data information. IEEE Trans. Autom. Sci. Eng. **19**(4), 2904–2917 (2021). https://doi.org/10.1109/TASE.2021.3096964
32. Mughal, M., Fawad, H., Mufti, R.: Three-dimensional finite-element modelling of deformation in weld-based rapid prototyping. Proc. Inst. Mech. Eng. C: J. Mech. Eng. Sci. **220**(6), 875–885 (2006)
33. Panchagnula, J.S., Simhambhatla, S.: Manufacture of complex thin-walled metallic objects using weld-deposition based additive manufacturing. Robot. Comput.-Integrated Manuf. **49**, 194–203 (2018)
34. Ramsay, T.O., Burnett, R.T., Krewski, D.: The effect of concurvity in generalized additive models linking mortality to ambient particulate matter. Epidemiology **14**(1), 18–23 (2003)
35. Rasmussen, C., Williams, C.: Gaussian Processes for Machine Learning. Adaptive Computation and Machine Learning. MIT Press, Cambridge (2006)
36. Sabbaghi, A., Huang, Q.: Model transfer across additive manufacturing processes via mean effect equivalence of lurking variables. Ann. Appl. Stat. **12**(4), 2409–2429 (2018). https://doi.org/10.1214/18-AOAS1158
37. Song, S., Wang, A., Huang, Q., Tsung, F.: In-plane shape-deviation modeling and compensation for fused deposition modeling processes. IEEE Trans. Autom. Sci. Eng. (2016). https://doi.org/10.1109/TASE.2016.2544941
38. Tapia, G., Elwany, A.: A review on process monitoring and control in metal-based additive manufacturing. J. Manuf. Sci. Eng. **136**(6), 060801 (2014)
39. Wang, A., Song, S., Huang, Q., Tsung, F.: In-plane shape-deviation modeling and compensation for fused deposition modeling processes. IEEE Trans. Autom. Sci. Eng. **14**(2), 968–976 (2017). https://doi.org/10.1109/TASE.2016.2544941
40. Wang, Y., Lu, J., Zhao, Z., Deng, W., Han, J., Bai, L., Yang, X., Yao, J.: Active disturbance rejection control of layer width in wire arc additive manufacturing based on deep learning. J. Manuf. Process. **67**, 364–375 (2021)
41. Wikle, C., Zammit-Mangion, A., Cressie, N.: Spatio-Temporal Statistics with R. Chapman & Hall/CRC The R Series. CRC Press (2019). https://books.google.com/books?id=FD-IDwAAQBAJ
42. Wood, S.: Generalized Additive Models: An Introduction with R, 2nd edn. CRC Press, Boca Raton (2017)
43. Wood, S.N.: Fast stable restricted maximum likelihood and marginal likelihood estimation of semiparametric generalized linear models. J. Roy. Stat. Soc. B **73**(1), 3–36 (2011)

44. Wood, S.N., Pya, N., Säfken, B.: Smoothing parameter and model selection for general smooth models. J. Am. Stat. Assoc. **111**(516), 1548–1563 (2016). https://doi.org/10.1080/01621459.2016.1180986
45. Wood, S.N., Scheipl, F., Faraway, J.J.: Straightforward intermediate rank tensor product smoothing in mixed models. Stat. Comput. **23**(3), 341–360 (2013)
46. Wu, B., Pan, Z., Ding, D., Cuiuri, D., Li, H., Xu, J., Norrish, J.: A review of the wire arc additive manufacturing of metals: properties, defects and quality improvement. J. Manuf. Process. **35**, 127–139 (2018)
47. Xiong, J., Li, R., Lei, Y., Chen, H.: Heat propagation of circular thin-walled parts fabricated in additive manufacturing using gas metal arc welding. J. Mater. Process. Technol. **251**, 12–19 (2018)
48. Xiong, J., Zhang, G., Hu, J., Li, Y.: Forecasting process parameters for gmaw-based rapid manufacturing using closed-loop iteration based on neural network. Int. J. Adv. Manuf. Technol. **69**(1-4), 743–751 (2013)
49. Xu, L., Huang, Q., Sabbaghi, A., Dasgupta, T.: Shape deviation modeling for dimensional quality control in additive manufacturing. In: IMECE2013-66329. Proceedings of the ASME 2013 International Mechanical Engineering Congress & Exposition, San Diego (2013)
50. Yan, H., Paynabar, K., Shi, J.: Anomaly detection in images with smooth background via smooth-sparse decomposition. Technometrics **59**(1), 102–114 (2017). https://doi.org/10.1080/00401706.2015.1102764
51. Yue, X.: Data Decomposition for Analytics of Engineering Systems: Literature Review, Methodology Formulation, and Future Trends (2019). https://doi.org/10.1115/MSEC2019-2945
52. Yue, X., Yan, H., Park, J.G., Liang, Z., Shi, J.: A wavelet-based penalized mixed-effects decomposition for multichannel profile detection of in-line raman spectroscopy. IEEE Trans. Autom. Sci. Eng. **15**(3), 1258–1271 (2017)
53. Zhao, H., Zhang, G., Yin, Z., Wu, L.: A 3d dynamic analysis of thermal behavior during single-pass multi-layer weld-based rapid prototyping. J. Mater. Process. Technol. **211**(3), 488–495 (2011)
54. Zhou, H., Li, L., Zhu, H.: Tensor regression with applications in neuroimaging data analysis. J. Am. Stat. Assoc. **108**(502), 540–552 (2013). https://doi.org/10.1080/01621459.2013.776499

Chapter 8
Transfer Learning Via Effect Equivalence in AM Systems

Recent advances in the industrial Internet of Things and Cyber–Physical Systems have resulted in greater connections and accessibility of smart manufacturing. A particularly exciting consequence is the development of a new paradigm of smart AM systems that seamlessly integrate computing, manufacturing, and services [3, 9, 27]. Each individual AM or 3D printing machine in such a system enables direct manufacturing of complex shapes from CAD models with reduced labor and costs compared to traditional manufacturing methods [4, 10]. The impacts of such systems are not yet fully realized in practice because their constituent processes may yield inconsistent product quality. Furthermore, Individual processes and machines face varying degrees of insufficient data and physical knowledge. A specified process model through machine learning often has a limited scope of application across the vast spectrum of processes in a manufacturing system that are characterized by different settings of process variables, including lurking variables. Knowledge or model transfer among AM processes is therefore essential to smart AM systems.

This chapter introduces a domain-informed transfer learning approach that enables prediction model transfer across processes in an AM system with small training data. Model transfer is achieved through inference on the equivalent effects of lurking variables in terms of an observed factor whose effect has been modeled under a previously learned process. The methodology is demonstrated in shape deviation model transfer in AM processes.

8.1 Challenges of Model Transfer in AM Systems

The common limitation of machine learning models is that the learned models are limited in their scope of application to the particular processes for which they were specified, in the sense that they typically fail to describe and hence control other processes under different settings. Comprehensive control of an AM system based on these methods is then impractical due to the high operating costs incurred by collecting a large amount of data and constructing new models for distinct processes in the system. In addition, the number of test cases that could possibly be manufactured for a particular process is typically in the single digits because of its nature and capability of one-of-a-kind manufacturing [22]. More importantly, modeling deviations separately for every process fails to yield deeper insights on the entire AM system. Therefore, a significant challenge for comprehensive shape deviation control in an AM system is whether the deviation model for one process can be transferred to model deviations under a different process based on a small number of experimental runs. The more general category of engineering problems is transferring a quality model established for a process A to another process B, where B has substantially fewer trials to avoid wasteful and repetitive model building (i.e., the product design set D_B for B is a small subset of the corresponding product design set D_A for A, with a potentially smaller dimension as well).

This challenge is illustrated by consideration of in-plane deviation profiles under two distinct stereolithography processes A and B in the AM system of Fig. 8.1. Let $\mathcal{X}$ denote the set of compensation plans, and D_A and D_B the respective design sets for the two processes. The component of a statistical deviation model that is important for deviation control, and hence constitutes our primary focus, is its expectation. We denote the expected deviation models for these processes by $f_A : \mathcal{X} \times D_A \to \mathcal{P}$ and $f_B : \mathcal{X} \times D_B \to \mathcal{P}$, where $\mathcal{P}$ is the set of in-plane deviation profiles [11]. We consider engineering processes to be distinct if their designs, parameters, or any other factors are different. Needless to say, processes involving completely different AM machines are distinct. Even if the observed factors' settings for A and B are identical, these processes and their respective expected deviation models can be distinguished by a wide spectrum of unobserved factors related to their product designs, materials, parameters, and conditions. Such lurking variables, whose settings are completely unobserved due to infeasibility of measurement or insufficient knowledge [2], are ubiquitous in AM systems, and complicate the task of model transfer.

Figure 8.2a displays the in-plane deviation profiles for four disks of nominal radii $0.5''$, $1''$, $2''$, and $3''$ manufactured under process A. Each point on a disk is identified by its angle under the polar coordinate system [11]. This particular process was studied by [11], and a simple specification of f_A was formulated to enable the construction of compensation plans that reduce in-plane deviation by one order of

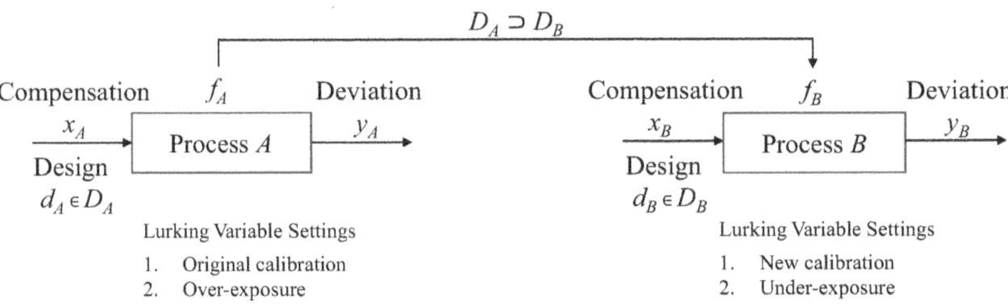

Fig. 8.1 Model transfer across processes in an AM system characterized by different settings of lurking variables

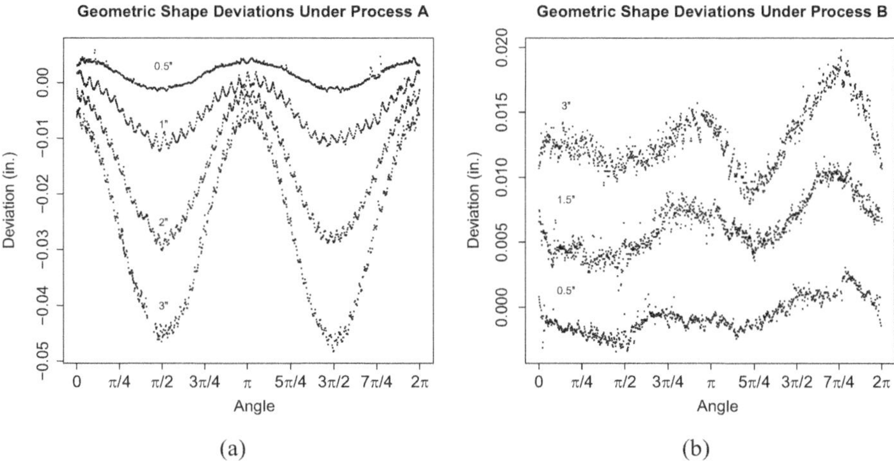

Fig. 8.2 (**a**) Deviation profiles of four disks of nominal radii 0.5″, 1″, 2″, and 3″ manufactured under process A. (**b**) Deviation profiles of three disks of nominal radii 0.5″, 1.5″, and 3″ manufactured under process B, which has a new calibration setting

magnitude. Figure 8.2b displays the different deviation profiles for three disks of nominal radii 0.5″, 1.5″, and 3″ manufactured under a new process B that operates with an unknown and distinct calibration setting. The lurking calibration factor clearly impacts deviation in a complicated manner and f_A fails to directly capture the deviation profiles under B. Resource constraints may prevent the specification of an appropriate model f_B based only on the data from B. It thus becomes of interest to extend the previously learned model f_A to B. In this way, we can leverage all of the data and knowledge across the AM system to acquire a better understanding of the change in the lurking calibration factor for process B.

Current statistical and machine learning methodologies cannot resolve the challenge of deviation model transfer across processes in an AM system in the presence of lurking variables. We address this challenge by incorporating effect equivalence for a lurking variable in terms of an observed factor into the model specified under a previous process. Effect equivalence refers to the common engineering phenomenon in which two factors can generate the same outcomes. Reference [26] conceived of its first quantitative formulation in their study of a machining process. However, they did not consider its application for model transfer across different processes, operating with distinct lurking variables settings, in a system. We develop a new statistical framework and Bayesian method for such model transfer based on effect equivalence.

Under our approach, model transfer proceeds by learning a function that benchmarks a lurking variable's effect on the process mean in terms of a previously studied process and an observed factor's effect on the mean under that process. For the previous AM system, this function is denoted by $T : \mathcal{X} \times D_B \to \mathcal{X}$ and is specified so that $f_B(x_B, d_B) = f_A(T(x_B, d_B), d_B)$ for all $(x_B, d_B) \in \mathcal{X} \times D_B$.

We refer to T as the total equivalent amount of the lurking calibration in terms of compensation and interpret it as returning a compensation plan $T(x_B, d_B) \in X$ for input $(x_B, d_B) \in X \times D_B$ such that the expected deviation profile of the product with input (x_B, d_B) manufactured under process B is equivalent to the expected deviation profile of the product with shape design d_B and compensation plan $T(x_B, d_B)$ manufactured under process A. In this manner, T directly broadens the scope of the previously specified model f_A to the expectation of process B and facilitates one's understanding of B's changed lurking variable setting. Indeed, our analysis on the total equivalent amount of calibration in terms of compensation (summarized in Fig. 8.3a and described in more depth in Sect. 8.3) helps us to understand that the calibration change is essentially an attempted reproduction of the optimum compensation plan of [11], with discrepancies for the $1.5''$ and $3''$ disks that explain their complicated deviations. Also, a shape manufactured under process B can now be thought of as having been equivalently manufactured under A with a compensation plan defined by the inferred total equivalent amount. This enables us to transfer the model f_A for the mean of A to B (Fig. 8.3b).

8.1.1 Previous Considerations of Model Transfer and Lurking Variables

Various statistical methods exist to identify and account for lurking variables in simple models. A prime example for industrial processes is statistical process

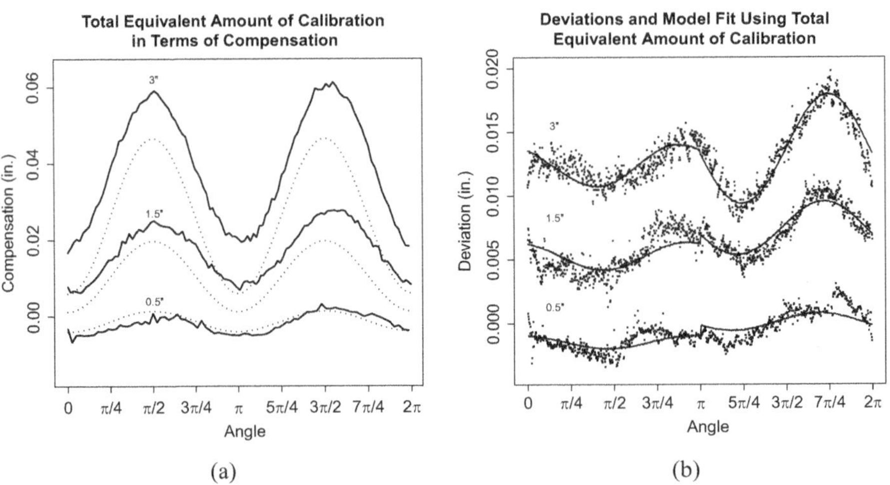

Fig. 8.3 (a) The inferred total equivalent amounts of calibration in terms of compensation (solid lines) for disks under process B, compared to the corresponding optimum compensation plans (dotted lines) from [11]. (b) Posterior predictive means (solid lines) of shape deviations under process B obtained from the transferred deviation model

control [23]. For agricultural settings, [28] discussed how the ANOVA for a group of agronomic experiments should be conducted to account for lurking variables such as soil differences and climate. Reference [13] presented several instances of lurking variables and methods to detect their existence in linear models. Reference [5] described graphical methods for linear models that can illuminate phenomena resulting from lurking variables. However, such methods are ineffective for the nonlinear and complex deviations in AM. Furthermore, the standard practice of conducting follow-up experiments to learn about lurking variables necessarily proceeds on a case-by-case basis, and hence will be expensive, as illustrated by the cryogenic flow meters example in [13].

A broad class of model transfer methods belongs to inductive transfer learning [16]. One such method is TrAdaBoost [6], which predicts the outcomes in a new setting by modifying the relative weights of instances across settings at each step of a boosting algorithm, so as to identify and utilize data from settings most similar to the new setting [17]. These methods enjoy desirable predictive and computational performance and are simple to implement. However, they require prohibitively large samples for AM. In contrast, our framework enables model transfer with limited samples.

A topic related to model transfer is transportability or the extrapolation of experimental findings across domains that differ both in their distributions and in their inherent causal characteristics [1]. Reference [19] and [1] formalized transportability using causal diagrams and the do-calculus of [18]. A contribution of their work is the explication of conditions and transport formulae for causal inference from heterogeneous data. However, they focus on nonparametric inference and effectively assume that the relationships between factors and outcomes are sufficiently well-understood from a large amount of data such that estimation error in probability distributions is not of concern. In contrast, for AM processes, we must address the distinct task of transportability of parametric models across factor settings, and our formulation can handle the case of limited data. In addition, their work may be limited to linear structural equation models in practice, whereas our framework accommodates nonlinear models of interest in AM. Finally, we consider equivalence relations, not probabilistic dependencies, between observed and unobserved factors, so as to transfer models in a functional manner.

Effect equivalence was inspired by the study of a machining process in which observed and unobserved factors could yield identical outcomes. Reference [26] considered the effects of fixture, machine tool, and datum errors on quality, where the latter two are lurking variables. They described the existence of effect equivalence and specified the total equivalent amounts of the lurking variables in terms of fixture error. This approach enabled successful quality control [24, 25]. However, they did not consider model transfer across different processes, characterized by distinct settings of lurking variables.

8.2   Statistical Effect Equivalence Framework for Transfer Learning

For simplicity, we consider in-plane deviations for 3D printed shapes with negligible heights and adopt the functional in-plane deviation representation defined by [11]. Our focus on in-plane deviations is due to two facts. In real applications, a thin product or section of a product is approximated as a 2D shape, with identical top and bottom surface deviations.

We measure the top boundaries of shapes manufactured by our AM processes using a single Micro-Vu Vertex system. The measurement coordinate system is aligned with the design coordinate system based on markers printed on the shapes to ensure consistency and reduce possible measurement errors or misalignment issues. The points $i = 1, \ldots, N$ on the manufactured shapes are then identified by their angles θ_i and observed radii $r^{\mathrm{obs}}(\theta_i)$ under the polar coordinate system.

A CAD model is encoded as a nominal radius function $r^{\mathrm{nom}} : [0, 2\pi] \to \mathbb{R}$. Deviation Y_i for point i on a product with CAD model r^{nom} is defined as

$$Y_i = r^{\mathrm{obs}}(\theta_i) - r^{\mathrm{nom}}(\theta_i).$$

Figures 8.2a and b illustrate this functional deviation representation for several thousands of points on each of the seven disks.

Studies are conducted to model deviation as a function of the continuous treatment factor compensation. This factor is formally defined for each point i as the addition or subtraction of material in the CAD model at that point. The previous notation in the case of compensation becomes

$$Y_i(x_1) = r^{\mathrm{obs}}(\theta_i, x_1) - r^{\mathrm{nom}}(\theta_i),$$

where $r^{\mathrm{obs}}(\theta_i, x_1)$ denotes the observed radius for point i under compensation x_1, and $Y_i(x_1)$ is point i's deviation.

8.2.1   Notation and Assumptions

We denote the K factors for an AM process by F_k and their corresponding set of possible levels by $\mathcal{X}_k$, for $k = 1, \ldots, K$. We consider processes whose factors possess the following two properties. First is that each is either fully observed or unobserved. Second is that the level for any factor is not an outcome of another factor. The first property ensures that each factor is well-defined as either an observed factor or lurking variable, and the second eliminates factor linkages [2]. Factor F_1 will be reserved for compensation, with $\mathcal{X}_1 = \mathbb{R}$.

The covariate vector for point i is denoted by z_i. An example of a covariate vector is $z_i = (\theta_i, r^{\mathrm{nom}}(\theta_i))^{\mathsf{T}}$. As shall be seen in our studies, such covariates serve a useful role for deviation model transfer.

We assume the Stable Unit-Treatment Value Assumption [12] is satisfied. Under this assumption, each point i has a well-defined deviation $Y_i(x_1, \ldots, x_K)$ for $(x_1, \ldots, x_K) \in \prod_{k=1}^{K} \mathcal{X}_k$, and interference does not exist.

To simplify the notations for our definition of effect equivalence, we consider $K = 2$ in this section. We let $p(y \mid z_i, x_1, x_2, \psi)$, $p_1(y \mid z_i, x_1, \psi_1)$, and $p_2(y \mid z_i, x_2, \psi_2)$ denote the probability density functions for $Y_i(x_1, x_2)$, $Y_i(x_1, c_2)$, and $Y_i(c_1, x_2)$, respectively. The domains of p, p_1, and p_2 are $\mathbb{R}$, and ψ, ψ_1, and ψ_2 are parameter vectors with respective parameter spaces Ψ, Ψ_1, and Ψ_2 (examples of which are in Sect. 8.3). Here, $c_1 \in \mathcal{X}_1$ and $c_2 \in \mathcal{X}_2$ represent fixed settings of F_1 and F_2, respectively. These density functions define statistical deviation models for different settings of F_1 and F_2 and will be used to specify likelihood functions for the parameters and total equivalent amounts in our Bayesian method. In terms of the notations in Sect. 8.1, f_A can be thought of as the expected deviation profile model derived from one of the statistical deviation models p_1, and f_B as the expected deviation profile model derived from the other statistical model p_2.

8.2.1.1 Mean Effect Equivalence

Definition 8.1 formalizes effect equivalence with respect to the mean of a deviation model.

Definition 8.1 Factors F_1 and F_2 are equivalent with respect to the mean if, for any point i and $c_1 \in \mathcal{X}_1$, $c_2 \in \mathcal{X}_2$, functions $T_{i,1\to2} : \mathcal{X}_1 \times \mathcal{X}_2 \to \mathcal{X}_2$ and $T_{i,2\to1} : \mathcal{X}_1 \times \mathcal{X}_2 \to \mathcal{X}_1$ exist such that for all $(x_1, x_2) \in \mathcal{X}_1 \times \mathcal{X}_2$:

$$\int_{-\infty}^{\infty} y p(y \mid z_i, x_1, x_2, \psi) dy = \int_{-\infty}^{\infty} y p_2(y \mid z_i, T_{i,1\to2}(x_1, x_2), \psi_2) dy,$$

$$\int_{-\infty}^{\infty} y p(y \mid z_i, x_1, x_2, \psi) dy = \int_{-\infty}^{\infty} y p_1(y \mid z_i, T_{i,2\to1}(x_1, x_2), \psi_1) dy.$$

Function $T_{i,k\to k'}$ is the total equivalent amount of F_k in terms of $F_{k'}$ with respect to the mean for point i. It maps each level combination of F_k and $F_{k'}$ to a level of $F_{k'}$ that generates the equivalent expected outcome under a fixed setting c_k of F_k. Our consideration of fixed settings corresponds to AM systems, in which deviations for a process are generated under fixed levels of the lurking variables. Equivalence of F_k and $F_{k'}$ is denoted by $F_k \overset{e}{\sim} F_{k'}$.

The total equivalent amount in Definition 8.1 enables the transfer of the expectation specified under one process to new processes. For example, if $F_1 \overset{e}{\sim} F_2$ with respect to the mean, where F_1 is observed and F_2 is lurking, then the mean specification under the process in which F_2 is set at c_2 can be transferred to a process with a different setting of F_2 by incorporating $T_{i,2\to1}$ into p_1. In practice, model transfer is facilitated by means of the equivalence of a lurking variable with an observed factor that permits convenient control. Such a factor is referred to as the base factor F_1, and any $T_{i,k \to 1}$ is shortened to $T_{i,k}$. Compensation is a standard base factor for AM processes. As illustrated in our studies, domain knowledge of AM processes can yield equivalencies between their factors and compensation.

8.2.2 Bayesian Inference and Modeling of Total Equivalent Effect

We now address learning of the total equivalent amount of a lurking variable F_2 in terms of compensation F_1 with respect to the mean, which we also refer to as mean effect equivalence. As before, we simplify the exposition by considering two processes characterized by distinct levels c_2 and x_2 of F_2, respectively. We assume that a statistical deviation model, with corresponding probability density function $p_1(y \mid z_i, x_1, \psi_1)$, is specified for the first, and that data are collected on the second, with its model yet to be specified. Let $\psi_1 = (\mu_1, \sigma_1)^\mathsf{T}$, where μ_1 consists of all parameters in the mean under p_1 and σ_1 consists of the remaining parameters in ψ_1. A model for the second process is then $p_1(y \mid z_i, T_{i,2}(x_1, x_2), \mu_1, \sigma_2)$. The $T_{i,2}$ remain to be inferred and modeled to complete the model transfer or transfer learning.

Let $\mathbf{x}_1^{\mathrm{obs}} = (x_{1,1}, \ldots, x_{N,1})^\mathsf{T}$ be the compensations for N points under the second process, and $\mathbf{T}_2(\mathbf{x}_1^{\mathrm{obs}}) = (T_{1,2}(x_{1,1}, x_2), \ldots, T_{N,2}(x_{N,1}, x_2))^\mathsf{T}$ the corresponding vector of their realized total equivalent amounts of the lurking variable in terms of compensation. We infer $\mathbf{T}_2(\mathbf{x}_1^{\mathrm{obs}})$ using the Bayesian calculation of its posterior distribution. The likelihood function for μ_1, σ_2, and $\mathbf{T}_2(\mathbf{x}_1^{\mathrm{obs}})$ follows from p_1, and is denoted by $L(\mu_1, \sigma_2, \mathbf{T}_2(\mathbf{x}_1^{\mathrm{obs}}) \mid \mathbf{z}, \mathbf{y}^{\mathrm{obs}})$, where $\mathbf{y}^{\mathrm{obs}}$ is the vector of outcomes and $\mathbf{z}$ is the matrix of covariates for all the points. For a prior $p(\mu_1, \sigma_2, \mathbf{T}_2(\mathbf{x}_1^{\mathrm{obs}}))$, the joint posterior is

$$p\left(\mu_1, \sigma_2, \mathbf{T}_2\left(\mathbf{x}_1^{\mathrm{obs}}\right) \mid \mathbf{z}, \mathbf{y}^{\mathrm{obs}}\right) \propto L\left(\mu_1, \sigma_2, \mathbf{T}_2\left(\mathbf{x}_1^{\mathrm{obs}}\right) \mid \mathbf{z}, \mathbf{y}^{\mathrm{obs}}\right)$$
$$\times\, p\left(\mu_1, \sigma_2, \mathbf{T}_2\left(\mathbf{x}_1^{\mathrm{obs}}\right)\right).$$

The marginal posterior of $\mathbf{T}_2(\mathbf{x}_1^{\mathrm{obs}})$ then follows by integration. It is important to incorporate data from the first process to improve the precision of our inferences. Letting $\mathbf{D}$ denote the deviations, covariates, and compensations for the first process, the posterior $p(\mathbf{T}_2(\mathbf{x}_1^{\mathrm{obs}}) \mid \mathbf{z}, \mathbf{y}^{\mathrm{obs}}, \mathbf{D})$ is similarly calculated as above. Our Bayesian approach thus enables us to leverage all of the data collected across distinct settings of a lurking variable in a straightforward manner to learn about its total equivalent amounts.

In practice, prior information on μ_1 can be elicited independently of that on σ_2 and $\mathbf{T}_2(\mathbf{x}_1^{\mathrm{obs}})$, because they correspond to distinct processes. Our effect equivalence framework facilitates the elicitation of appropriate priors on $\mathbf{T}_2(\mathbf{x}_1^{\mathrm{obs}})$

that incorporate domain knowledge of AM processes with controlled subjectivity to enable reasonable inferences. Specifically, as compensation is readily interpretable, and experience can be acquired to understand its effect on deviation, an informative prior for $\mathbf{T}_2(\mathbf{x}_1^{\text{obs}})$ can be specified and justified in a straightforward manner. This is demonstrated in Sect. 8.3.

The posterior of $\mathbf{T}_2(\mathbf{x}_1^{\text{obs}})$ can be computed by the blocked Gibbs sampler that draws from the conditional posteriors $p(\mathbf{T}_2(\mathbf{x}_1^{\text{obs}}) \mid \mathbf{z}, \mathbf{y}^{\text{obs}}, \mu_1, \sigma_2)$, $p(\mu_1 \mid \mathbf{z}, \mathbf{y}^{\text{obs}}, \sigma_2, \mathbf{T}_2(\mathbf{x}_1^{\text{obs}}))$, and $p(\sigma_2 \mid \mathbf{z}, \mathbf{y}^{\text{obs}}, \mathbf{T}_2(\mathbf{x}_1^{\text{obs}}), \mu_1)$. If the deviations are conditionally independent given the covariates, total equivalent amounts, and parameters, and if the total equivalent amounts are independent *a priori*, then drawing from $p(\mathbf{T}_2(\mathbf{x}_1^{\text{obs}}) \mid \mathbf{z}, \mathbf{y}^{\text{obs}}, \mu_1, \sigma_2)$ in the Gibbs sampler reduces to independently drawing each entry in $\mathbf{T}_2(\mathbf{x}_1^{\text{obs}})$ from their respective conditional posteriors. Potential computational complexities are thus reduced under this approach, with sampling of the vector of total equivalent amounts simplified into parallel sampling of one-dimensional distributions for the individual entries. This reduction will hold more generally for moderately large N and low-dimensional model parameter vectors. For very large N or high-dimensional parameter vectors associated with more complex processes and products (e.g., the 52-dimensional parameter vector for the hierarchical nonlinear regression model of an irregular polygon in [22]), this approach may not be computationally efficient or feasible.

Such a high-dimensional inferential task can be handled in practice via the construction of a discrepancy measure [14, 20] that uses Bayesian inferences on the parameters from the first process to approximate the full inference for the total equivalent amounts of the second. To illustrate, let $p(\mu_1 \mid \mathbf{D})$ denote the posterior for μ_1 based only on data $\mathbf{D}$ from the first process. Sampling from this distribution is less computationally complex compared to sampling from the full posterior of μ_1 because the former does not involve the unknown realized total equivalent amounts. Define $e : \mathcal{X}_1 \to \mathcal{X}_1$ as

$$e(x_1 \mid z, \mu_1) = \int_{-\infty}^{\infty} y p_1(y \mid z, x_1, \mu_1) dy.$$

For each point i in the second process, we construct the discrepancy measure

$$T_i = \underset{t \in \mathcal{X}_1}{\operatorname{argmin}} \left\{ y_i^{\text{obs}} - e\left(t \mid z_i, \tilde{\mu}_1\right) \right\}^2 \tag{8.1}$$

to infer their realized total equivalent amounts, where $\tilde{\mu}_1 \sim p(\mu_1 \mid \mathbf{D})$. To understand this, note that the calculation of $p(\mathbf{T}_2(\mathbf{x}_1^{\text{obs}}) \mid \mathbf{z}, \mathbf{y}^{\text{obs}}, \mathbf{D})$ by

$$\int p\left(\mathbf{T}_2\left(\mathbf{x}_1^{\text{obs}}\right) \mid \mathbf{z}, \mathbf{y}^{\text{obs}}, \mathbf{D}, \mu_1\right) p\left(\mu_1 \mid \mathbf{z}, \mathbf{y}^{\text{obs}}, \mathbf{D}\right) d\mu_1$$

can be interpreted via sampling, with a posterior draw of $\mathbf{T}_2(\mathbf{x}_1^{\text{obs}})$ obtained by first drawing $\tilde{\mu}_1 \sim p(\mu_1 \mid \mathbf{z}, \mathbf{y}^{\text{obs}}, \mathbf{D})$, and then drawing from $p(\mathbf{T}_2(\mathbf{x}_1^{\text{obs}}) \mid$

$\mathbf{z}, \mathbf{y}^{\text{obs}}, \mathbf{D}, \tilde{\mu}_1)$. The discrepancy measure instead draws $\tilde{\mu}_1 \sim p(\mu_1 \mid \mathbf{D})$, which approximates $p(\mu_1 \mid \mathbf{z}, \mathbf{y}^{\text{obs}}, \mathbf{D})$, and then solves for the unknown realized total equivalent amounts as in Eq. (8.1) given the drawn $\tilde{\mu}_1$. This is less computationally complex than sampling from the conditional posterior of $\mathbf{T}_2(\mathbf{x}_1^{\text{obs}})$ given $\tilde{\mu}_1$. Inferences obtained from the discrepancy measure will be similar to those from the full Bayesian calculation when p_1 is a normal probability density function. We recommend this discrepancy measure for practical model transfer in the case of large N or high-dimensional model parameters.

After the marginal posterior distribution of $\mathbf{T}_2(\mathbf{x}_1^{\text{obs}})$ is obtained, we examine it by means of exploratory data analytic and visualization methods to specify a model $T_2(z_i, x_1; \beta)$ for the total equivalent amounts of points under the second process as a function of their covariates and compensations, with β denoting a parameter vector. Model transfer will then be complete upon incorporating $T_2(z_i, x_1; \beta)$ in p_1.

8.3 Real Case Studies of Shape Deviation Model Transfer in AM

8.3.1 Transfer Learning Formulation to Estimate Lurking Variable Effect of Overexposure

An important case of model transfer in stereolithography involves the lurking variable of overexposure or the unintended expansion of a shape due to the faulty spread of light beams on its boundary. This factor F_2, with its levels consisting of positive real-valued functions on $[0, 2\pi]$, was present in the study of [11] on the disks in Fig. 8.2a. Reference [11] hypothesized the following probability density function for the deviation of a point i on a disk of nominal radius r_i^{nom} under no overexposure:

$$p_1(y \mid z_i, x_1, \mu_1, \sigma_1) = \left(2\pi\sigma_1^2\right)^{-1/2} \exp\left[-\frac{1}{2\sigma_1^2}\left\{ y - x_1 - \alpha_0 \left(r_i^{\text{nom}} + x_1\right)^{a_0} \right.\right.$$
$$\left.\left. - \alpha_1 \left(r_i^{\text{nom}} + x_1\right)^{a_1} \cos(2\theta_i) \right\}^2 \right],$$

$$(8.2)$$

with $z_i = (\theta_i, r_i^{\text{nom}})^{\mathsf{T}}$, $\mu_1 = (\alpha_0, \alpha_1, a_0, a_1)^{\mathsf{T}}$, $\Psi_1 = \mathbb{R}^4 \times \mathbb{R}_{>0}$, and independent outcomes. In terms of the notation in Sect. 8.1, this setting corresponds to a previous process A, and the expected deviation model $f_{i,A} : \mathbb{R} \times \mathbb{R}_{>0} \to \mathbb{R}$ for point i on a disk of nominal radius r_i^{nom} with compensation $x_{i,1}$ manufactured under it is derived from Eq. (8.2) as

$$f_{i,A}\left(x_{i,1}, r_i^{\text{nom}}\right) = x_{i,1} + \alpha_0 \left(r_i^{\text{nom}} + x_{i,1}\right)^{a_0} + \alpha_1 \left(r_i^{\text{nom}} + x_{i,1}\right)^{a_1} \cos(2\theta_i).$$

Also, for N points on a disk of nominal radius r^{nom} with compensation plan $\mathbf{x}_1 = (x_{1,1}, \ldots, x_{N,1})^{\mathsf{T}}$,

$$f_A\left(\mathbf{x}_1, r^{\mathrm{nom}}\right) = \left(f_{1,A}\left(x_{1,1}, r^{\mathrm{nom}}\right), \ldots, f_{N,A}\left(x_{N,1}, r^{\mathrm{nom}}\right)\right)^{\mathsf{T}}.$$

This model performed poorly for the data in Fig. 8.2a and in validation experiments. Reference [11] then identified overexposure as the lurking variable. They effectively provided a physical justification for $F_1 \overset{c}{\sim} F_2$ that pre-specified $T_{i,2}(x_1, x_2(\cdot)) = x_1 + x_2(\theta_i)$ under overexposure $x_2(\cdot) \in \mathcal{X}_2$. The transferred deviation model for this process B then follows from the probability density function $p_1(y \mid z_i, T_{i,2}(x_1, x_2(\cdot)), \mu_1, \sigma_2)$, with a distinct standard deviation σ_2. The remaining task is to learn $T_{i,2}$.

We proceed to infer and model the realized total equivalent amounts of overexposure in terms of compensation with respect to the mean. Following [21] and [11], we assume independent outcomes conditional on covariates and total equivalent amounts. Approximately a thousand equally spaced points were collected from each disk.

The likelihood function for the realized total equivalent amounts is obtained from Eq. (8.2). The following prior is specified:

$$p\left(\mu_1, \sigma_2^2, \mathbf{T}_2\left(\mathbf{x}_1^{\mathrm{obs}}\right)\right) \propto \left(\sigma_2^2\right)^{-4} \exp\left\{-\frac{0.0016}{\sigma_2^2}\right\}$$

$$\times \prod_{i=1}^{N} \mathbb{I}\left\{0 \leq T_{i,2}(0, x_2(\cdot)) \leq 0.015\right\}$$

$$\times \exp\left\{-50\alpha_0^2 - 50\alpha_1^2 - \frac{(a_0 - 1)^2}{8} - \frac{(a_1 - 1)^2}{2}\right\}.$$

This prior is based on previous literature and studies, as we describe further.

First, consider the prior for the total equivalent amounts. We observed from the work of [30] and [29] that a product's size should not impact the spread of light beams on its boundary, and that this impact in terms of the equivalent amount of compensation should be relatively small (in inches). These observations led to our bounds $0 \leq T_{i,2}(0, x_2(\cdot)) \leq 0.015$ for all points, with the upper bound not depending on the nominal radius of the disk on which a point resides. We consider the upper bound 0.015 a conservative prior estimate of the greatest possible total equivalent amount of overexposure in terms of compensation under our process. This value is also justified by noting that we are considering disks of nominal radii $0.5''$ to $3''$, and that overexposure is at most a single-digit percentage of the $0.5''$ disk's radius. After this range of overexposure values was elicited, we then specified the corresponding uniform prior to reflect our uncertainty about which values within it are more likely than others. A different distribution (e.g., a truncated Normal) could also be specified given additional information on the likely overexposure values.

Now consider the prior for $(\alpha_0, \alpha_1, a_0, a_1, \sigma_2^2)^\mathsf{T}$. Our specified prior for these parameters was also informed by domain knowledge of our process. For example, the physical reasoning of [11] suggests that the priors of both a_0 and a_1 should be centered at 1, and we accordingly specified dispersed Normal priors for these parameters. Similarly, the priors for α_0 and α_1 are dispersed normal distributions centered at zero, and the prior for σ_2^2 is an inverse-χ^2 with small degrees of freedom and scale that reflect our prior conception of the level of variation for disk deviations.

It is important to recognize that data do not exist on a process with no overexposure in this case, and so our prior specification must necessarily be informative to prevent identifiability issues. In addition, as the deviation model in Eq. (8.2) is a nonlinear regression, specifying non-informative or improper prior distributions for the total equivalent amounts and model parameters, and verifying that the corresponding posteriors are proper, is difficult. It is simpler in practice to specify proper and informative priors that guarantee a proper posterior and are straightforward to interpret.

Our blocked Gibbs sampler computes the posterior distribution proceeds via three steps. Let $\mu_1^{(t)}$, $\sigma_2^{(t)}$, and $\mathbf{T}_2^{(t)}(\mathbf{x}_1^{\mathrm{obs}})$ denote the draws in iteration t. First, a draw of $\sigma_2^{(t)}$ from $p(\sigma_2 \mid \mathbf{z}, \mathbf{y}^{\mathrm{obs}}, \mathbf{T}_2^{(t-1)}(\mathbf{x}_1^{\mathrm{obs}}), \mu_1^{(t-1)})$ follows as the square root of a scaled inverse-χ^2. Second, a draw of $\mu_1^{(t)}$ from $p(\mu_1 \mid \mathbf{z}, \mathbf{y}^{\mathrm{obs}}, \mathbf{T}_2^{(t-1)}(\mathbf{x}_1^{\mathrm{obs}}), \sigma_2^{(t)})$ is obtained using a Metropolis random walk. Finally, each entry in $\mathbf{T}_2^{(t)}(\mathbf{x}_1^{\mathrm{obs}})$ is independently drawn from their respective posteriors conditional on $\mu_1^{(t)}$ and $\sigma_2^{(t)}$ using a Metropolis random walk. This Gibbs sampler was implemented with 10,000 draws obtained after a burn-in of 200,000. Convergence was verified by trace and autocorrelation plots, and the [8] statistic, of the evaluated log posterior.

We examine the posterior of $\mathbf{T}_2(\mathbf{x}_1^{\mathrm{obs}})$ with a visualization in Fig. 8.4 of the points' posterior means stratified according to their nominal radii. The high-frequency oscillations for the $1''$, $2''$, and $3''$ disks are an artifact of the lower resolution of the process for them. More importantly, we observe that the posterior means lie in a small range. This suggests that the total equivalent amount does not depend on θ_i or r_i^{nom}. It also explains why the corresponding assumption in [11] provided a good fit.

Thus, from this exploratory visualization of the posterior of $\mathbf{T}_2(\mathbf{x}_1^{\mathrm{obs}})$, we model the total equivalent amount as

$$T_2(z_i, x_1; \beta_0) = \beta_0 + x_1, \tag{8.3}$$

with the additive nature of this model pre-specified as before. Parameter β_0 corresponds to x_0 in the model of [11]. The corresponding transferred model is defined by the probability density function

$$p(y \mid z_i, x_1, \mu_2, \sigma_2)$$

$$= (2\pi\sigma_2^2)^{-1/2}\exp\left[-\frac{1}{2\sigma_2^2}\left\{ y - \beta_0 - x_1 - \alpha_0 \left(r_i^{\mathrm{nom}} + \beta_0 + x_1\right)^{a_0} \right. \right.$$

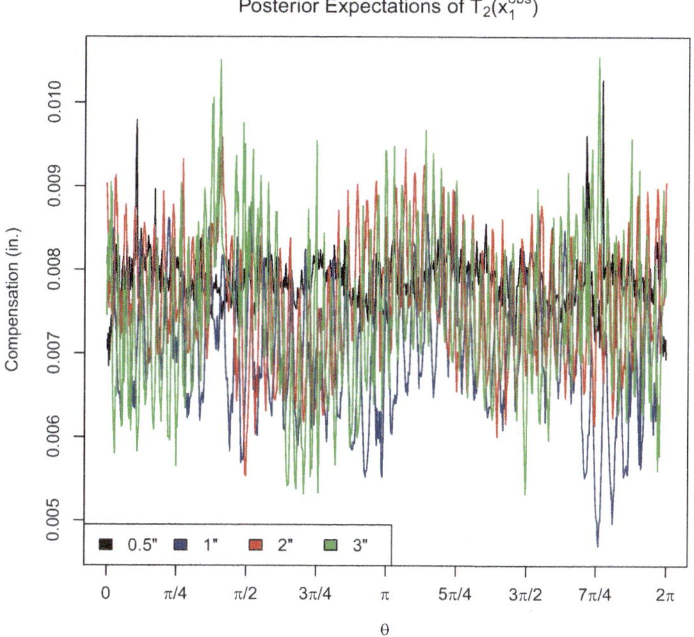

Fig. 8.4 Posterior expectations of the realized total equivalent amounts of overexposure in terms of compensation with respect to the mean

$$
\left. \left. - \alpha_1 \left(r_i^{\text{nom}} + \beta_0 + x_1 \right)^{a_1} \cos(2\theta_i) \right\}^2 \right], \qquad (8.4)
$$

with $\mu_2 = (\alpha_0, \alpha_1, a_0, a_1, \beta_0)^{\mathsf{T}} \in \mathbb{R}^4 \times \mathbb{R}_{>0}$. Note that for points $i = 1, \ldots N$,

$$
f_{i,B}\left(x_{i,1}, r_i^{\text{nom}}\right) = f_{i,A}\left(\beta_0 + x_{i,1}, r_i^{\text{nom}}\right)
$$

$$
= \beta_0 + x_{i,1} + \alpha_0 \left(r_i^{\text{nom}} + \beta_0 + x_{i,1} \right)^{a_0}
$$

$$
+ \alpha_1 \left(r_i^{\text{nom}} + \beta_0 + x_{i,1} \right)^{a_1} \cos\left(2\theta_i\right),
$$

$$
f_B\left(\mathbf{x}_1, r^{\text{nom}}\right) = \left(f_{1,B}\left(x_{1,1}, r^{\text{nom}}\right), \ldots, f_{N,B}\left(x_{N,1}, r^{\text{nom}}\right) \right)^{\mathsf{T}}.
$$

The validation experiments of [11] effectively illustrate the utility of this model for deviation control under overexposure. As $T_{i,2}(x_1, x_2(\cdot)) = x_1 + x_2(\theta_i)$ was pre-specified and x_1 is known, we have explicitly inferred and derived the fixed, unknown overexposure.

8.3.2 *AM Machine Model Calibration Through Transfer Learning*

An important case of model transfer for general types of AM processes involves the lurking variable of calibration, i.e., unknown hardware settings, which is denoted by F_3. In general $F_3 \overset{e}{\sim} F_1$, because a change to the process calibration, which yields a new process B, can be viewed as equivalent to a compensation plan under a previous calibration setting A. This effect equivalence for the processes A and B in Sect. 8.1 can be justified by the physics of stereolithography.

We formally infer and model the total equivalent amount of calibration in terms of compensation with respect to the mean. We incorporate the data and model for process A specified in Eq. (8.4), which operated with overexposure and the original calibration setting, for this inference, and transfer the model to process B. Approximately a thousand equally spaced points were collected from each of the disks under B.

Our prior distribution for $(\mu_2, \sigma_2^2, \sigma_3^2, \mathbf{T}_3(\mathbf{x}_1^{\mathrm{obs}}))^\mathsf{T}$ is

$$
p\left(\mu_2, \sigma_2^2, \sigma_3^2, \mathbf{T}_3\left(\mathbf{x}_1^{\mathrm{obs}}\right)\right) \propto \sigma_2^{-2}\sigma_3^{-2}\beta_0^{-1} \prod_{i=1}^{N} \mathbb{I}\left\{-\frac{r_i^{\mathrm{nom}}}{10} \le T_{i,3}(0, x_3) \le \frac{r_i^{\mathrm{nom}}}{10}\right\}
$$

$$
\times \exp\left\{-\frac{(a_0 - 1)^2}{8} - \frac{(a_1 - 1)^2}{2} - \frac{(\log \beta_0)^2}{2}\right\}.
$$

We specify non-informative priors for α_0, α_1, σ_2^2, and σ_3^2 because our incorporation of data from the previous process eliminates possible identifiability issues. Our priors for the $T_{i,3}(0, x_3)$ now depend on the nominal radius r_i^{nom} of the disk on which a point resides, because our prior knowledge of process calibration does not preclude this dependence. The lower and upper bounds are our prior estimates of their possible magnitudes.

We calculate the posterior of $\mathbf{T}_3(\mathbf{x}_1^{\mathrm{obs}})$ based on all seven disks by blocked Gibbs sampling and summarize it in Fig. 8.5 with visualizations that stratify it according to the nominal radii and halves of the disks. We also calculate the expectations of the discrepancy measure defined in Eq. (8.1), where $\mathbf{D}$ is the data on disks under overexposure, and

$$
e(t \mid z_i, \mu_2) = \beta_0 + t + \alpha_0 \left(r_i^{\mathrm{nom}} + \beta_0 + t\right)^{a_0} + \alpha_1 \left(r_i^{\mathrm{nom}} + \beta_0 + t\right)^{a_1} \cos(2\theta_i).
$$

The posterior $p(\mu_2 \mid \mathbf{D})$ has been calculated from the previous process. The discrepancy measure T_i for point i under the new process is

$$
T_i = \underset{t \in \mathcal{X}_1}{\mathrm{argmin}} \left\{y_i^{\mathrm{obs}} - \tilde{\beta}_0 - t - \tilde{\alpha}_0 \left(r_i^{\mathrm{nom}} + \tilde{\beta}_0 + t\right)^{\tilde{a}_0}\right.
$$

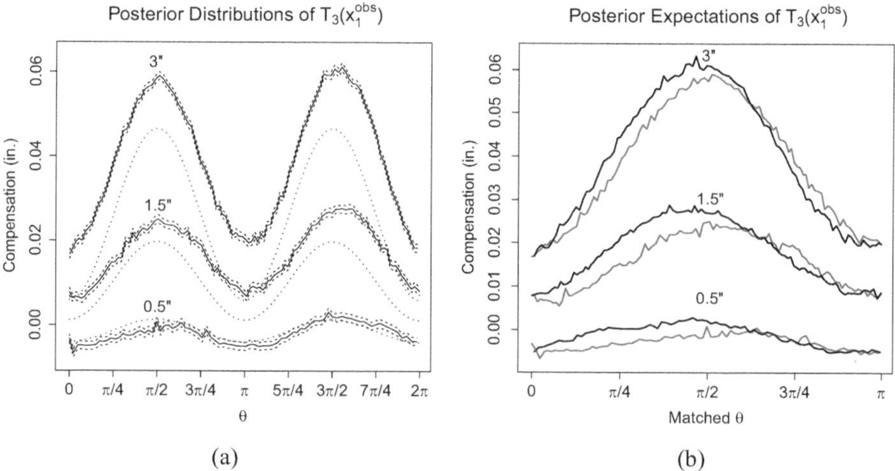

Fig. 8.5 (**a**) Posterior expectations (solid) and 95% central credible intervals (dashed) of the total equivalent amounts of calibration in terms of compensation with respect to the mean. The compensation plans from [11] are dotted lines. (**b**) Posterior expectations of the total equivalent amounts for the upper (gray) and lower (black) disk halves, where the angles on the lower halves are matched to those on the upper halves directly above them

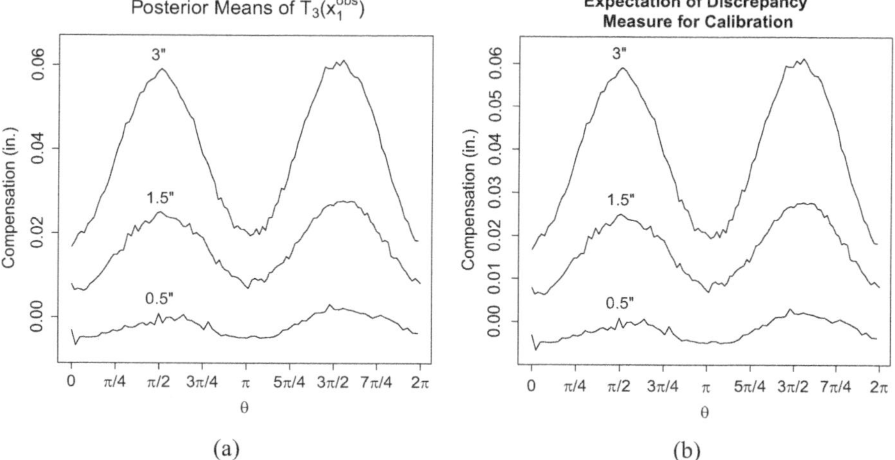

Fig. 8.6 (**a**) Posterior means of the total equivalent amounts of calibration in terms of compensation. (**b**) Expectations of the discrepancy measure

$$
- \tilde{\alpha}_1 \left(r_i^{\mathrm{nom}} + \tilde{\beta}_0 + t \right)^{\tilde{a}_1} \cos(2\theta_i) \Big\}^2 .
$$

Our visual comparison of the results obtained from the blocked Gibbs sampler and the discrepancy measure in Fig. 8.6 demonstrates that they yield identical inferences. In either case, they enable the specification of a total equivalent amount model, as described next.

We observe in Fig. 8.5a that the posterior trends of the realized total equivalent amounts are similar to the optimum compensation plans of [11], but exhibit discrepancies that explain the complicated deviations. For example, these discrepancies explain the apparent under-exposure of B compared to A. Figure 8.5b highlights the asymmetry in the realized total equivalent amounts between the lower and upper halves of the disks, which explains the asymmetrical profiles in Fig. 8.2b.

We incorporate our observations from the exploratory visualizations of the posterior of the total equivalent amounts against angles, nominal radii, and disk halves to model the total equivalent amount of calibration as

$$T_3(z_i, x_1; \beta) = T_2(z_i, x_1; \beta_0)$$

$$+ \mathbb{I}(0 \le \theta < \pi) \left[x_{0,U} + \beta_{0,U} \left(r_i^{\text{nom}} + x_{0,U} \right)^{b_{0,U}} \right.$$

$$\left. + \beta_{1,U} \left(r_i^{\text{nom}} + x_{0,U} \right)^{b_{1,U}} \cos\{2(\theta_i - \psi_U)\} \right]$$

$$+ \{1 - \mathbb{I}(0 \le \theta < \pi)\} \left[x_{0,L} + \beta_{0,L} \left(r_i^{\text{nom}} + x_{0,L} \right)^{b_{0,L}} \right.$$

$$\left. + \beta_{1,L} \left(r_i^{\text{nom}} + x_{0,L} \right)^{b_{1,L}} \cos\{2(\theta_i - \psi_L)\} \right],$$

where $T_2(z_i, x_1; \beta_0)$ is defined in Eq. (8.3) and β is a vector of length 12 containing all parameters in the aforementioned equation. Flat priors are placed on $\beta_{0,U}, \beta_{1,U}, \beta_{0,L}, \beta_{1,L}$, and the priors for the other parameters are

$$b_{0,U}, b_{0,L} \sim \mathrm{N}(1, 2^2), \quad b_{1,U}, b_{1,L} \sim \mathrm{N}(1, 1^2),$$

$$\log\left(\frac{0.5 + x_{0,U}}{0.5 - x_{0,U}}\right), \log\left(\frac{0.5 + x_{0,L}}{0.5 - x_{0,L}}\right) \sim \mathrm{N}(0, 1^2),$$

$$\log\left(\frac{\psi_U/\pi}{1 - \psi_U/\pi}\right), \log\left(\frac{\psi_L/\pi}{1 - \psi_L/\pi}\right) \sim \mathrm{N}(0, (2^{1/2})^2).$$

We assume all parameters in β are mutually independent *a priori*.

The transferred deviation model was fit using Hamiltonian Monte Carlo [7], which is a Markov Chain Monte Carlo algorithm for sampling from a distribution based on Hamiltonian dynamics [15]. Figure 8.7 demonstrates that this transferred model provides a good fit for the new calibration setting. Thus, although the change in calibration introduced strikingly different and complicated deviation profiles, we are able to transfer the model from process A to B in a simple manner.

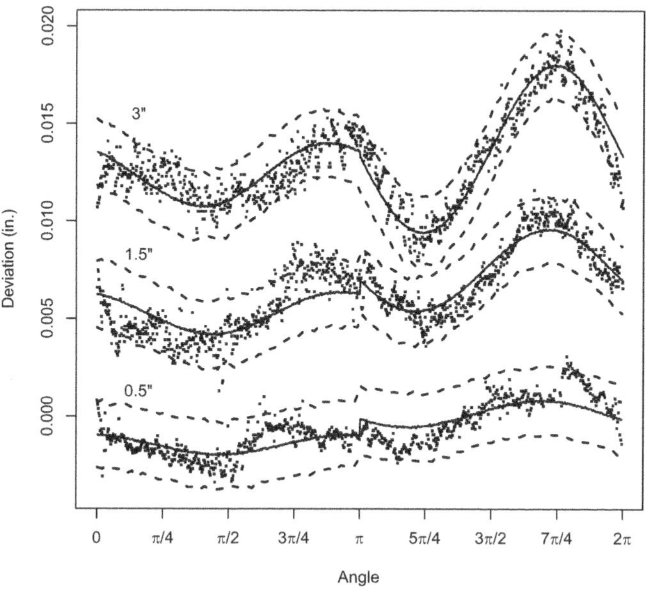

Fig. 8.7 Posterior predictive means (solid) and 95% central posterior predictive intervals (dashed) of shape deviations under process B obtained from the transferred deviation model

This section presents a novel framework for deviation model transfer across distinct AM processes characterized by different settings of lurking variables in an AM system. A key component of our framework is the total equivalent amount of a lurking variable in terms of a base factor with respect to the process mean. We described a Bayesian method for learning the total equivalent amount in terms of compensation for a new process. Inferences on total equivalent amounts facilitate one's understanding of changes in the lurking variables. Once the total equivalent amount is modeled, it can be directly incorporated into a previously learned process's deviation model to effectively transfer it to the new condition. Our two real studies illustrate deviation model transfer across the three stereolithography process conditions of no overexposure, constant overexposure, and the combination of constant overexposure and a new calibration setting. They also demonstrate that, if considerable resources have been expended in the specification of a model for one process, its features can then be transferred in a simple manner, with little further expenditure, to a new process. This addresses the fundamental challenge of deviation modeling in an AM system.

Our general framework can have broader impacts in engineering and statistics. As our framework effectively extends engineering insights across processes, it can permit comprehensive modeling of distinct processes connected in a distributed manufacturing environment. In addition, the construction of a catalog of total equivalent amounts in terms of a base factor can enable insightful statistical conclusions to be made on a system's lurking variables. For example, if it is thought that a system is operating under a new condition, then its inferred total equivalent amount can be compared to the catalog's entries to obtain statistical assessments of the most likely change in its lurking variables. Our framework can also provide a new approach to address the statistical problem of external validity, or the generalizability of empirical findings to new settings, in the presence of lurking variables.

A host of interesting tasks remain for future study. One is relaxing the assumption of fixed lurking variables, to address the case of dynamic lurking variables. Another is to develop automated model transfer algorithms that are applicable to a variety of systems, which can provide further impetus to the growing trend of automation in advanced manufacturing. In line with this is combining our effect equivalence framework with the adaptive Bayesian modeling approach in [22] to enable deviation model building across different shapes and AM processes. When factors can affect multiple model features, an important issue is how the total equivalent amounts with respect to the different features can be learned simultaneously. One important consideration is the effect of factors on the variability of deviation, in which the goal is to identify the levels that minimize it. This and similar considerations are part of our ongoing research on the theory of effect equivalence. Another issue is inference of the total equivalent amounts of multiple lurking variables across processes. For example, in uncontrolled, or observational, studies of AM systems, alignment issues or device precision could constitute additional lurking variables, and their effects would be entangled with those of other lurking variables. From a statistical viewpoint, it would not be possible to disentangle these effects without further information or data. Disentanglement of their effects could perhaps be accomplished by including data from a designed experiment that involved the alignments or device precision. These issues require new research on the design of experiments and observational studies for learning total equivalent amounts.

The work in this chapter was partially supported by US Office of Naval Research with grant # N000141110671 and US National Science Foundation Cyber-Physical Systems Program under Grant CMMI-1544917.

Acknowledgments The work in this chapter was supported by US US National Science Foundation under grant CMMI-1744121.

References

1. Bareinboim, E., Pearl, J.: Causal inference and the data-fusion problem. In: Proceedings of the National Academy of Sciences, pp. 7345–7352 (2016)
2. Box, G.E.: Use and abuse of regression. Technometrics **8**(4), 625–629 (1966)
3. Buckholtz, B., Ragai, I., Wang, L.: Cloud manufacturing: Current trends and future implementations. J. Manuf. Sci. Eng. **137**(4), 040902 (9 pages) (2015)

4. Campbell, T., Williams, C., Ivanova, O., Garrett, B.: Could 3D Printing Change the World? Technologies, Potential, and Implications of Additive Manufacturing. Atlantic Council, Washington, DC (2011)
5. Cook, R.D., Critchley, F.: Identifying regression outliers and mixtures graphically. J. Am. Stat. Assoc. **81**(396), 945–960 (2000)
6. Dai, W., Yang, Q., Xue, G.R., Yu, Y.: Boosting for transfer learning. In: Proceedings of the 24th International Conference on Machine Learning (2007)
7. Duane, S., Kennedy, A., Pendleton, B.J., Roweth, D.: Hybrid Monte Carlo. Phys. Lett. B **195**(2), 216–222 (1987). https://doi.org/10.1016/0370-2693(87)91197-X
8. Gelman, A., Rubin, D.: Inference from iterative simulation using multiple sequences. Stat. Sci. **7**(4), 457–472 (1992)
9. Germany Trade & Invest: Industrie 4.0: Smart manufacturing for the future
10. Gibson, I., Rosen, D., Stucker, B.: Additive Manufacturing Technologies: Rapid Prototyping to Direct Digital Manufacturing. Springer, New York (2009)
11. Huang, Q., Zhang, J., Sabbaghi, A., Dasgupta, T.: Optimal offline compensation of shape shrinkage for 3D printing processes. IIE Trans. Qual. Reliab. **47**(5), 431–441 (2015)
12. Imbens, G., Rubin, D.: Causal Inference for Statistics, Social, and Biomedical Sciences: An Introduction, 1st edn. Cambridge University Press, New York (2015)
13. Joiner, B.: Lurking variables: some examples. Am. Stat. **35**(4), 227–233 (1981)
14. Meng, X.L.: Posterior predictive p-values. Ann. Stat. **22**(3), 1142–1160 (1994)
15. Neal, R.: MCMC using Hamiltonian dynamics. In: Brooks, S., Gelman, A., Jones, G.L., Meng, X.L. (eds.) Handbook of Markov Chain Monte Carlo, pp. 113–162. Chapman & Hall / CRC Press (2010)
16. Pan, S.J., Yang, Q.: A survey on transfer learning. IEEE Trans. Knowl. Data Eng. **22**(10), 1345–1359 (2010)
17. Pardoe, D., Stone, P.: Boosting for regression transfer. In: Proceedings of the 27th International Conference on Machine Learning (2010)
18. Pearl, J.: Causal diagrams for empirical research. Biometrika **82**, 669–710 (1995)
19. Pearl, J., Bareinboim, E.: External validity: from do-calculus to transportability across populations. Stat. Sci. **29**(4), 579–595 (2014)
20. Rubin, D.: Bayesianly justifiable and relevant frequency calculations for the applied statistician. Ann. Stat. **12**(4), 1151–1172 (1984)
21. Sabbaghi, A., Dasgupta, T., Huang, Q., Zhang, J.: Inference for deformation and interference in 3D printing. Ann. Appl. Stat. **8**(3), 1395–1415 (2014)
22. Sabbaghi, A., Huang, Q., Dasgupta, T.: Bayesian model building from small samples of disparate data for capturing in-plane deviation in additive manufacturing. Technometrics **60**(4), 532–544 (2018)
23. Shewhart, W.A.: Economic Control of Quality of Manufacturing Product, 1st edn. Van Nostrand Reinhold, Hoboken, NJ (1931)
24. Wang, H., Huang, Q.: Error cancellation modeling and its application to machining process control. IIE Trans. **38**(4), 355–364 (2006)
25. Wang, H., Huang, Q.: Using error equivalence concept to automatically adjust discrete manufacturing processes for dimensional variation control. ASME J. Manuf. Sci. Eng. **129**(3), 644–652 (2007)
26. Wang, H., Huang, Q., Katz, R.: Multi-operational machining processes modeling for sequential root cause identification and measurement reduction. ASME J. Manuf. Sci. Eng. **127**(3), 512–521 (2005)
27. Wu, D., Rosen, D.W., Wang, L., Schaefer, D.: Cloud-based design and manufacturing: A new paradigm in digital manufacturing and design innovation. Comput. Aided Design **59**, 1–14 (2015)
28. Yates, F., Cochran, W.G.: The analysis of groups of experiments. J. Agricult. Sci. **28**, 556–580 (1938)

29. Zhou, C., Chen, Y.: Additive manufacturing based on optimized mask video projection for improved accuracy and resolution. J. Manuf. Process. **14**, 107–118 (2012)
30. Zhou, C., Chen, Y., Waltz, R.A.: Optimized mask image projection for solid freeform fabrication. ASME Trans. J. Manuf. Sci. Eng. **131**, 061004 (12 pages) (2009)

Chapter 9
Automated Model Generation Via Principled Design of Neural Networks for AM Systems

A significant challenge in quality control of an AM system is the model specification for different computer-aided design products manufactured by constituent AM processes. Current machine learning techniques can require substantial user inputs and efforts to implement in practice.

This chapter presents an automated model generation method based on a class of Bayesian neural networks that enables automated deviation modeling of different shapes and AM processes. The fundamental innovation is the adaptive and principled design of new and connectable neural network structures, which is informed by domain knowledge, specifically, the small-sample learning and transfer learning models established in previous chapters. The power and broad scope of our method are demonstrated with several case studies on both in-plane and out-of-plane deviations for a wide variety of shapes manufactured under different stereolithography processes.

9.1 Need of Automated Shape Deviation Modeling in a Smart AM System

The rapid evolving domain of cyber–physical additive systems demands seamlessly integrate computational models and physical AM processes. In such a system, design inputs and manufacturing process conditions can differ both between and within AM machines across time. The comprehensive control of manufactured products' geometric shape deviations, which inevitably occur due to material phase changes, complicated layer interactions, and process variations that are inherent in AM [38], then becomes a challenging issue for the operation of smart manufacturing in practice.

One general class of geometric accuracy control strategies in AM is based on the use of statistical deviation models, i.e., statistical models in which the

Q. Huang, *Domain-informed Machine Learning for Smart Manufacturing*,
https://doi.org/10.1007/978-3-031-91631-1_9

dependent variable is defined based on shape deviations of additively manufactured products, and the independent variables are defined based on characteristics of the products' CAD models, properties of the materials used in the AM process, and the observed settings of the AM process itself, among other factors. These models enable the derivation of compensation plans or modifications to CAD models that are predicted to reduce the deviations in manufactured products [15, 17]. Achieving comprehensive deviation control in an AM system with this strategy is complicated by four significant issues that result from the nature and capability of AM for one-of-a-kind manufacturing. First is the wide variety of shapes with varying geometric complexities that are of interest for manufacture. Second is the vast spectrum of AM processes or conditions that can yield fundamentally distinct deviations for products manufactured from the same CAD model. Third is the fact that only a small sample of test shapes, typically in the single digits, can possibly be manufactured for any AM process [30]. Finally, the effort that an AM system operator can devote to learn shape deviation models is typically limited. Comprehensive deviation control via compensation plans in an AM system thus requires a method that can leverage previously developed deviation models for different shapes and processes to automate deviation modeling for new shapes and processes using only a small sample of products. Automated modeling of shape deviations here refers to the ability of a methodology to specify a deviation model without requiring extensive efforts on the part of an AM operator to specify new functional forms for the statistical models or to incorporate specialized knowledge of particular AM processes into the statistical models.

Figure 9.1 illustrates this requirement for an AM system with two processes A and B and two shapes 1 and 2. For shape $s \in \{1, 2\}$ manufactured under process $p \in \{A, B\}$, we let $f_{s,p} : \mathcal{X} \to \mathcal{Y}$ denote its deviation model that

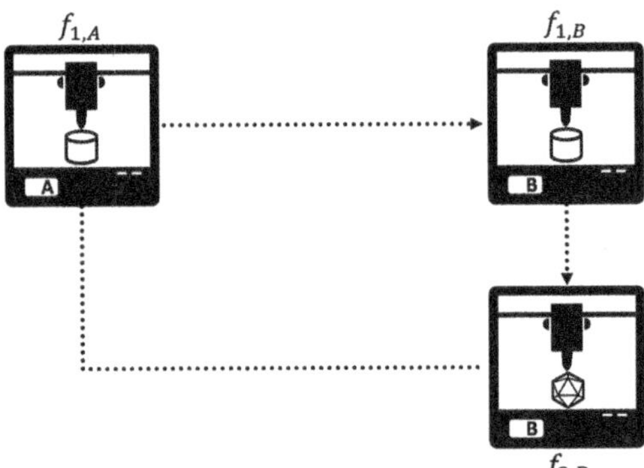

Fig. 9.1 An AM system with two processes A and B and two shapes 1 and 2 of interest for manufacture. The fundamental tasks in this system are to learn the deviation model $f_{1,B}$ for shape 1 under process B using knowledge of the deviation model $f_{1,A}$ for this shape under process A, and to specify the deviation model $f_{2,B}$ of a new shape 2 under process B using knowledge of all of the previous models for shapes that share similar features with shape 2, in an automated manner

returns the expected deviations (with range $\mathcal{Y}$ denoting the set of deviations for an entire manufactured product) under different compensation plans in domain $\mathcal{X}$. First, consider the case in which requests for shape 1 are assigned to both A and B. Suppose $f_{1,A}$ has previously been specified but $f_{1,B}$ has not. Given the system's constraints and limited resources for fulfilling the requests, it then becomes important to reduce the effort in specifying $f_{1,B}$ by automatically adapting $f_{1,A}$ to B based on a small sample of products manufactured under it. Now consider the case in which a request for the new, more complicated shape 2 is assigned to B. Automated learning of the corresponding deviation model $f_{2,B}$ can be performed more effectively by leveraging all of the previously specified models for the different shapes and processes that share similar geometric features with shape 2 under B and using a small sample of new shapes to learn deviation features unique to it.

Current shape deviation modeling techniques cannot address all of the previously described features of AM systems. The methods in [35, 36] specify independent polynomial deviation models for each direction of a shape, and their applications are limited to particular shapes under a single process. Reference [40] uses a transformation-based method to model in-plane deviations. Specifically, this method maps the deviations' sources to different transformations, e.g., scaling and rotation, of the nominal shape. However, it is unable to leverage previously developed models across processes. Reference [17] devised a distinct functional method that decouples geometric shape complexity from deviation modeling, but the focus was on individual shapes, with no consideration paid to specifying models for new shapes or processes in an automated manner. Methods of specifying models for just new sets of shapes based on the combination of the latter approach with the concept of modular deviation features were developed in [15, 22, 30]. Reference [2] devised a transfer learning framework for different shapes by separating shape deviation into two components, with one being independent of the shape and the other being shape-dependent. To address the requirement of deviation modeling across processes, a statistical framework of effect equivalence for model transfer was formulated in [29], and utilized in [18, 28] to specify models across distinct shapes and processes. However, all of these methods can incur a great deal of effort to implement and do not readily enable automated modeling for an AM system, in that they require more manual specifications of functional forms for the deviation models and/or knowledge of the AM process. Also, existing automatic modeling techniques do not address the first two features of AM systems. For example, the approaches in [31, 32] to automate deviation modeling of a surrounding cuboid object based on B-splines and the free-form deformation concept is limited to specified shapes and processes. An additional example is the two-step procedure in [3] that uses Gaussian processes to automate in-plane deviation modeling with the introduction of process parameters, which is limited to a specified shape and requires knowledge and data for the AM process. An automated and efficient methodology for comprehensive shape deviation modeling in an AM system remains to be developed.

This challenge in deviation modeling for AM systems is addressed in this chapter by developing a class of of Bayesian neural networks (NNs), specifically, Bayesian extreme learning machines (ELMs) [13]. This methodology utilizes point-

((a)) Additively manufactured irregular polygon.

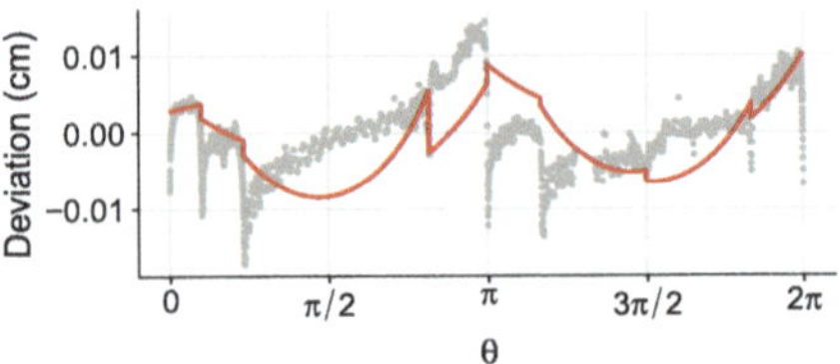

((b)) The model obtained from the method in [15] (RMSE = 5.98×10^{-3}).

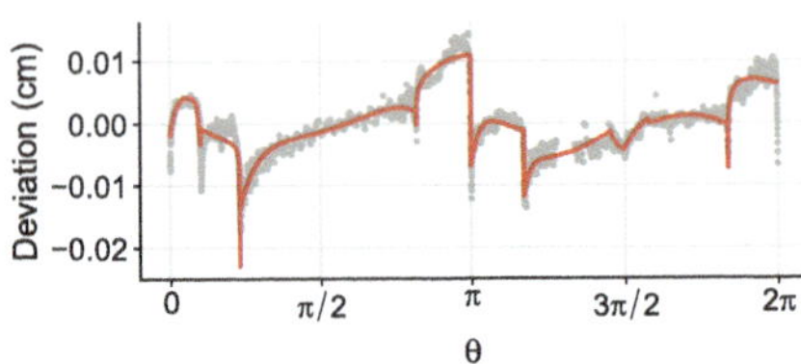

((c)) The model obtained from the method in [30] (RMSE = 1.89×10^{-3}).

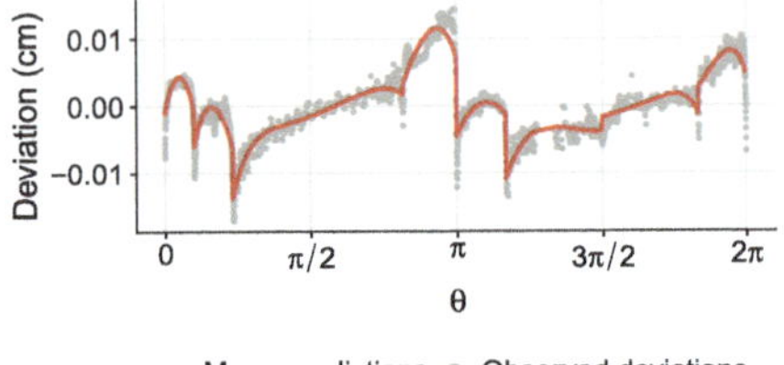

((d)) The model obtained from our Bayesian ELM method (RMSE = 1.87×10^{-3}).

Fig. 9.2 (**a**) An irregular polygon whose in-plane deviations are to be modeled based on the data and models for disks and a single regular pentagon. (**b**) The deviation model obtained from [15]. (**c**) The deviation model obtained from [30]. (**d**) The deviation model obtained from our method

cloud measurement data collected from a small sample of test shapes and does not require detailed knowledge of AM processes. In comparison to the previously described techniques, it can dramatically facilitate automated modeling of both in-plane and out-of-plane deviations for different shapes under distinct processes in an AM system. This advantage is illustrated by the case study of modeling in-plane deviations for the irregular polygon in Fig. 9.2a based on data and models for a small set of disks and a single regular pentagon manufactured using stereolithography (all of which are detailed in Sect. 9.4). Figures 9.2b,c, and d contain the deviation model fits (represented by red lines) obtained respectively from the methods in [15, 30], and our approach. By inspection of the alignment of the red lines with the shape deviations (represented by gray dots), and the root mean squared errors (RMSE) provided in the captions of Fig. 9.2 (which are formally defined in Sect. 9.4), we conclude that our method yields better predictive performance than that of [15] and comparable performance to that of [30]. Furthermore, in contrast to the other two approaches, our method was automated and required fewer user inputs and efforts for its computational implementation. For example, the method in [30] requires the user to specify a complete Bayesian hierarchical model for the irregular polygon, and the computation incurs great efforts.

9.2 Preliminaries

9.2.1 Functional Representation of Shape Deviations and Deviation Compensation

Geometric measurements of an additively manufactured product are collected in the point-cloud format that uses Cartesian coordinates defined with respect to physical axes printed directly on the product. We transform point-cloud data by means of the functional representations of 2D and 3D shape deviations formulated in [17] and [14], respectively. We demonstrate the methodology on in-plane deviations of shapes with negligible heights and out-of-plane deviations of shapes with negligible widths. In both cases, each point on a product is identified by an angle θ. The CAD model for a shape s is defined under the polar coordinate representation by a nominal radius function $r_s^{\text{nom}} : [0, 2\pi] \to \mathbb{R}_{\geq 0}$ with argument θ. As in [30], we assume that each shape s has an associated collection of known parameters γ_s that define r_s^{nom}. For example, γ_s has one entry for a disk (namely, its nominal radius), and is a vector for other shapes. The observed radius for a point θ on product s manufactured under process p is denoted by $r_{s,p}^{\text{obs}}(\theta)$, and its deviation is defined as $\Delta_{s,p}(\theta) = r_{s,p}^{\text{obs}}(\theta) - r_s^{\text{nom}}(\theta)$. An advantage of this representation is that it yields a consistent framework to specify statistical models for in-plane and out-of-plane deviations of different shapes and processes in an AM system. It is important to note that out-of-plane deviations are generated under more complex physical phenomenon (e.g., interlayer bonding effects [18]) than in-plane deviations.

Shape deviations are modeled to derive compensation plans that can enable geometric accuracy control. The compensation factor is defined for each θ as the addition or subtraction of material in the original CAD model at that point (Fig. 9.3). We augment the aforementioned notations from $r_{s,p}^{\text{obs}}(\theta)$ to $r_{s,p}^{\text{obs}}(\theta, x)$ and $\Delta_{s,p}(\theta)$ to $\Delta_{s,p}(\theta, x) = r_{s,p}^{\text{obs}}(\theta, x) - r_s^{\text{nom}}(\theta)$, respectively, to account for θ being given compensation $x \in \mathbb{R}$. Compensation also plays an important role for transferring deviation models across processes, which we describe next.

Fig. 9.3 Illustration of a compensation plan (dashed black line) versus the original CAD model (solid grey line) for an additively manufactured product

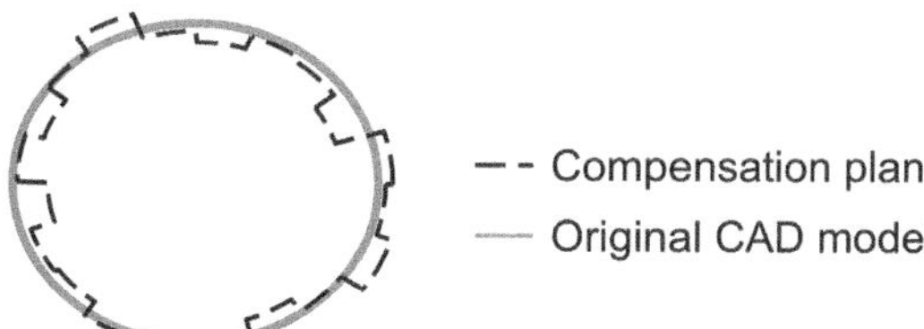

9.2.2 Statistical Effect Equivalence Framework for Model Transfer Across AM Processes

Our methodology incorporates aspects of the statistical effect equivalence framework in [29] to perform model transfer across processes in an AM system. The general effect equivalence concept corresponds to the engineering phenomenon in which the effect of a process condition change can be equivalently generated by changing a particular factor (e.g., compensation) under a fixed process condition and was inspired by the investigation in [37] on a machining process. To illustrate this framework, consider a shape s with nominal radius function r_s^{nom} manufactured under two processes A and B, with in-plane deviations of interest. Suppose that the deviation model under B has yet to be specified, and that the deviation model under A has been specified as

$$\Delta_{s,A}(\theta, x) = \delta_{s,A}(\theta, x) + \epsilon_{s,A}(\theta), \tag{9.1}$$

where $\delta_{s,A}(\theta, x)$ is the systematic (or expected) deviation at θ with compensation x, and the $\epsilon_{s,A}(\theta)$ are random variables representing high-frequency deviation components with expectation $\mathbb{E}\{\epsilon_{s,A}(\theta)\} = 0$ for all θ [17, 30]. We exclude model parameters in Eq. (9.1) to simplify the exposition. In terms of the notation from Sect. 9.1, $\delta_{s,A}(\theta, x)$ also specifies $f_{s,A}$ to model the expected in-plane deviation profile for the entire shape under process A. Then under effect equivalence, $\delta_{s,A}$ is transferred to model deviations for process B via a hypothesized function $T_{s,B \to A}$: $[0, 2\pi] \times \mathbb{R} \to \mathbb{R}$ in the manner

$$\Delta_{s,B}(\theta, x) = \delta_{s,A}(\theta, T_{s,B \to A}(\theta, x)) + \epsilon_{s,B}(\theta). \tag{9.2}$$

In Eq. (9.2), $T_{s,B \to A}$ returns a compensation for each θ such that, in expectation, the deviation of θ with compensation x when the shape is manufactured under B is equivalent to the deviation of θ with compensation $T_{s,B \to A}(\theta, x)$ when the shape is manufactured under A. Function $T_{s,B \to A}$ is referred to as the total equivalent amount (TEA) of B in terms of compensation with respect to the mean of A for shape s. This general concept of the TEA connects new and old processes in AM systems. The task of model transfer across AM processes is thus reduced by effect equivalence to learning the unknown TEA for a new process from a small sample of shapes manufactured under it and the fitted model for a previous process. Once the TEA is learned, it can be entered into $\delta_{s,A}$ as in Eq. (9.2) to perform the model transfer. Bayesian methods for this learning task were developed in [29], but they can require a great deal of effort to implement, and are not automated for applications in AM systems. We describe in Sect. 9.3 how we utilize the effect equivalence framework to design a new and simple NN architecture in our methodology that directly enables automated learning of TEAs, and hence model transfer, based on small samples of products.

9.2.3 Modular Deviation Features for Model Transfer Across Different Shapes

Our methodology also incorporates aspects of the "cookie-cutter" framework in [15] to perform model transfer across different shapes in an AM system based on their modular deviation features. The key idea of the framework is to capture the carving out of a new shape from an old shape. As the new shape is carved out, a new local deviation feature is introduced, with the old shape capturing a more global deviation feature. To illustrate this idea, consider two shapes 1 and 2 with nominal radius functions r_1^{nom} and r_2^{nom}, respectively, that are manufactured under a single process p and whose in-plane deviations are of interest. Suppose that the deviation model for shape 2 has yet to be specified, and that the deviation model for shape 1 has been specified as

$$\Delta_{1,p}(\theta, x) = \delta_{1,p}(\theta, x) + \epsilon_{1,p}(\theta), \tag{9.3}$$

where $\delta_{1,p}(\theta, x)$ and $\epsilon_{1,p}(\theta)$ are defined as in Eq. (9.1). Then a new, hypothesized deviation feature $\delta_{2,p}(\theta, x)$ is introduced under this framework to specify the model for shape 2 as

$$\Delta_{2,p}(\theta, x) = \delta_{1,p}(\theta, x) + \delta_{2,p}(\theta, x) + \epsilon_{2,p}(\theta). \tag{9.4}$$

Feature $\delta_{2,p}$ is referred to as a cookie-cutter basis function, or local deviation feature, for shape 2, and $\delta_{1,p}$ is the shared global deviation feature that connects the shapes [30]. Equation (9.4) can also be viewed as specifying the unified model

$$\Delta_{s,p}(\theta, x) = \delta_{1,p}(\theta, x) + \mathbb{I}(s = 2)\delta_{2,p}(\theta, x) + \epsilon_{s,p}(\theta) \tag{9.5}$$

for these shapes, where $\mathbb{I}(\cdot)$ is the indicator function. Additional details for this framework are in [15, 30]. The task of model transfer across shapes under a single AM process is thus reduced by this framework to learning the unknown $\delta_{2,p}(\theta, x)$ in Eqs. (9.4) and (9.5) from a small sample of new shapes and the fitted model for shape 1. Reference [15] considered pre-specified classes of local deviation features that may not yield successful model transfer for complicated shapes in AM systems. Reference [30] developed an adaptive Bayesian method to learn local deviation features that are applicable to a wide range of shapes. However, the latter method is not automated and can incur significant effort to learn appropriate models. We describe in Sect. 9.3 how we incorporate the concept of modular deviation features in our methodology's NN architecture to address the task of automated model transfer across shapes.

9.2.4 Overview of Neural Networks and Extreme Learning Machines

Our methodology utilizes a new class of single-hidden layer feedforward NNs that we developed to automate and facilitate model transfer across both processes and shapes in an AM system. We briefly review NNs and ELMs here and describe our new class of NNs in Sect. 9.3. To simplify this review, we let $\mathbf{y} = (y_1, \dots, y_N)^\mathsf{T} \in \mathbb{R}^N$ denote the outcomes for N units of analysis, and $\mathbf{z}_i \in \mathbb{R}^K$ the independent variables, or inputs, for unit $i = 1, \dots, N$. In AM, the typical units of analysis are the points θ_i, and $y_i = \Delta_{s,p}(\theta_i, x_i)$ for each point θ_i on a shape s manufactured under process p with compensation x_i. The choice of inputs depends on the product. For example, a useful set for an uncompensated disk with nominal radius r_1^{nom} and negligible height is $\mathbf{z}_i = (\theta_i, r_1^{\text{nom}})^\mathsf{T}$ [17]. Also, a useful set for an uncompensated polygon with nominal radius function r_2^{nom} and negligible height is $\mathbf{z}_i = (\theta_i, r_2^{\text{nom}}(\theta_i), \text{edge}(\theta_i))^\mathsf{T}$, where $\text{edge}(\theta_i)$ denotes the edge containing θ_i [30].

Neural networks enjoy the ability to learn complex relationships across a wide range of domains [4, 20]. Inspired by the brain, NNs involve a composition of "hidden neurons" that are structured via different layers and connections among themselves. Although NNs allow great structural freedom in general, in practice, it is not clear how a structure should be chosen for a particular dataset, and so it is typically specified in an ad-hoc manner. The simplest NN structure is the single-hidden layer feedforward NN with additive hidden neurons and a single activation function $g : \mathbb{R} \to \mathbb{R}$, defined by

$$y_i = \sum_{m=1}^{M} \beta_m \, g\left(\alpha_{m,0} + \mathbf{z}_i^\mathsf{T} \boldsymbol{\alpha}_m\right) + \epsilon_i, \tag{9.6}$$

where the error terms ϵ_i are independent $\mathrm{N}(0, \sigma^2)$ random variables, and the unknown parameters are σ^2, $\boldsymbol{\beta} = (\beta_1, \dots, \beta_M)^\mathsf{T}$, $\alpha_{m,0}$ and $\boldsymbol{\alpha}_m = (\alpha_{m,1}, \dots, \alpha_{m,K})^\mathsf{T}$ for $m = 1, \dots, M$. An example of an activation function is the hyperbolic tangent $g(x) = (e^x - e^{-x})/(e^x + e^{-x})$. These NNs can be considered as universal approximators of nonlinear functions [7] and possess a wide scope of application due to their flexibility and generality. However, one of their limitations that prevents their immediate application for deviation modeling is that they can incur quite some effort to fit to deviations for complex geometries. For example, the traditional backpropagation algorithm for fitting NNs suffers from both slow convergence and the local minimum problem [8]. Other limitations involve non-identifiability issues [26].

Extreme learning machines are a class of single-hidden layer feedforward NNs that were developed in [13] to address the limitations of standard NNs. An ELM does not attempt to infer all of the unknown parameters in Eq. (9.6). Instead, it first randomly sets the parameters in $\alpha_{m,0}$ and $\boldsymbol{\alpha}_m$, and then estimates $\boldsymbol{\beta}$ conditional on

Fig. 9.4 An example of the structure of a standard ELM

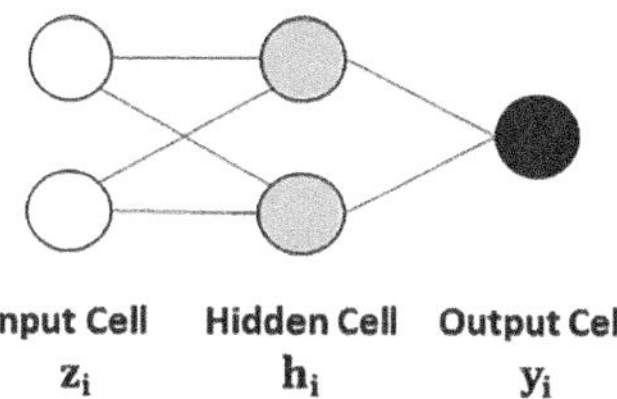

this selection [13]. Thus, an ELM reduces the original NN to a linear regression that simplifies model fitting, maintains the universal approximator property [9–11], and resolves non-identifiability issues. Figure 9.4 illustrates ELMs, with $h_{i,m} = g(\alpha_{m,0} + \mathbf{z}_i^{\mathsf{T}}\boldsymbol{\alpha}_m)$ denoting hidden neuron m for unit i. In practice, little consideration is paid to the proper tuning of the random parameters [12], and they are usually drawn independently from the uniform$(-1, 1)$ distribution [23]. However, the distribution from which the random parameters are drawn impacts the ELM's generalizability and external validity [34]. This impact is exacerbated when only small samples are available. Our methodology refines standard ELMs by making use of new techniques we developed to effectively address the requirements for automated deviation modeling in an AM system given small samples.

9.3 Bayesian Neural Network for Automated Deviation Modeling in an AM System

9.3.1 Outline of Principled Design of Bayesian Neural Network

Our Bayesian ELM methodology for automated deviation modeling proceeds via four steps that are outlined further. Details for each step are provided in the following subsections. Figure 9.5 illustrates our new and connectable ELM structures that enable comprehensive deviation modeling in an AM system.

1: *Model Deviations of Baseline Shape and Process*
   Establish a baseline Bayesian ELM deviation model, $\delta_{s,p}$, for shape s from one set of shapes S manufactured under a fixed process p.
2: *Transfer Baseline Deviation Model to a New Process*
   Transfer $\delta_{s,p}$ to a new process p' by learning a Bayesian ELM for the TEA $T_{s,p'\to p}$ using the fitted baseline model and data from process p'.
3: *Transfer Baseline Deviation Model to a New Shape*
   Transfer $\delta_{s,p}$ to shapes s' from a new set of shapes S' manufactured under p by taking $\delta_{s,p}$ as the global deviation feature and learning a Bayesian ELM for the new local deviation feature $\delta_{s',p}$ using the fitted baseline model and data from shapes in S' manufactured under p.

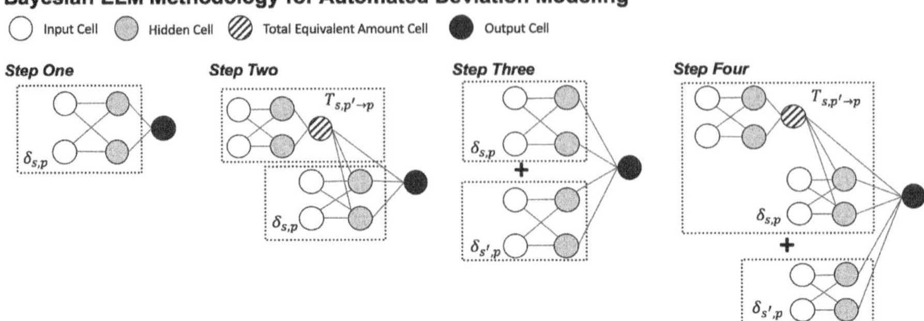

Fig. 9.5 The connectable NN structures in our methodology, where $\delta_{s,p}$ is the baseline deviation model for shape s under process p, $T_{s,p'\to p}$ is the TEA of p' in terms of compensation under p for s, and $\delta_{s',p}$ is the local deviation feature for s' under p

4: *Transfer Baseline Model to a New Shape and Process* Transfer $\delta_{s,p}$ to shapes s' from set S' manufactured under p' by performing the previous two steps and combining their resulting Bayesian ELM models (Fig. 9.5).

9.3.2 Model Deviations of Baseline Shape and Process

The baseline Bayesian ELM model specified in the first step serves as a building block for subsequent deviation modeling. It is learned from a small number of products in set S manufactured under the fixed process p. To describe it, let $N_{S,p}$ denote the total number of points on all of the products manufactured under p, $M_{S,p}$ the chosen number of hidden neurons for the Bayesian ELM model, and g the activation function. For each point $i \in \{1, \ldots, N_{S,p}\}$, let $\mathbf{z}\left(\theta_i, r_{i,S}^{\text{nom}}, x_i\right) \in \mathbb{R}^K$ denote its vector of inputs that is a function of θ_i, the nominal radius function $r_{i,S}^{\text{nom}}$ for the product on which i resides, and compensation $x_i \in \mathbb{R}$ applied to point i. If compensation was not applied, $x_i \equiv 0$. Finally, for drawn hidden neuron parameters $\alpha_{m,k}^{(S,p)}$ ($m = 1, \ldots, M_{S,p}, k = 0, \ldots, K$), let $\mathbf{H}_{S,p}$ be the $N_{S,p} \times M_{S,p}$ matrix whose (i, m) entry is $g\left(\alpha_{m,0}^{(S,p)} + \mathbf{z}\left(\theta_i, r_{i,S}^{\text{nom}}, x_i\right)^{\mathsf{T}} \boldsymbol{\alpha}_m^{(S,p)}\right)$ ($i = 1, \ldots, N_{S,p}, m = 1, \ldots, M_{S,p}$). Then the Bayesian ELM deviation model is

$$\mathbf{y}_{S,p} = \mathbf{x}_{S,p} + \mathbf{H}_{S,p}\boldsymbol{\beta}_{S,p} + \boldsymbol{\epsilon}_{S,p}, \tag{9.7}$$

where $\mathbf{y}_{S,p} \in \mathbb{R}^{N_{S,p}}$ is the vector of deviations, $\mathbf{x}_{S,p} \in \mathbb{R}^{N_{S,p}}$ is the vector of compensations, $\boldsymbol{\beta}_{S,p} \in \mathbb{R}^{M_{S,p}}$ is a vector of unknown parameters, and the error terms in $\epsilon_{S,p}$ are independent and identically distributed $N\left(0, \sigma_{S,p}^2\right)$ random variables with $\sigma_{S,p}^2$ unknown. The addition of compensation in Eq. (9.7) is informed by the physical reasoning in [17]. Our prior probability density function for the parameters is

$$p\left(\boldsymbol{\beta}_{S,p}, \sigma_{S,p}^2 \mid \tau_{S,p}^2\right) \propto \sigma_{S,p}^{-2}\tau_{S,p}^{-M_{S,p}}\exp\left(-0.5\tau_{S,p}^{-2}\boldsymbol{\beta}_{S,p}^{\mathsf{T}}\boldsymbol{\beta}_{S,p}\right),\tag{9.8}$$

and our hyperprior probability density function for $\tau_{S,p}^2$ is the relatively non-informative Inverse-Gamma distribution $p\left(\tau_{S,p}^2\right) \propto \tau_{S,p}^{-6}\exp\left(-0.01\tau_{S,p}^{-2}\right)$. The Gibbs algorithm [6] enables simple and rapid sampling from the posterior distribution of the parameters. Posterior predictions of deviations for S under p immediately follow from these draws and are also used to transfer the baseline model to new processes.

Example 9.1 Consider in-plane deviations for uncompensated disks (shape set 1) manufactured under a process A, with $N_{1,A} = 1000, M_{1,A} = 3$. We set $\mathbf{z}\left(\theta_i, r_{i,1}^{\mathrm{nom}}, x_i\right) = \left(\theta_i, r_{i,1}^{\mathrm{nom}}\right)^{\mathsf{T}}$, and $\mathbf{H}_{1,A}$ is a 1000×3 matrix with entry (i, m) equal to $g\left(\alpha_{m,0}^{(1,A)} + \alpha_{m,1}^{(1,A)}\theta_i + \alpha_{m,2}^{(1,A)}r_{i,1}^{\mathrm{nom}}\right)$. If compensations were applied, we set $\mathbf{z}\left(\theta_i, r_{i,1}^{\mathrm{nom}}, x_i\right) = \left(\theta_i, r_{i,1}^{\mathrm{nom}} + x_i\right)^{\mathsf{T}}$ based on the reasoning in [17], and $\mathbf{H}_{1,A}$ is defined as before.

In contrast to the standard ELM method in [23], we utilize a new mechanism to draw the $\alpha_{m,k}^{(S,p)}$ that we developed to obtain better out-of-sample predictive performance. Specifically, we draw $\alpha_{m,k}^{(S,p)}$ from Uniform $(-a_k, a_k)$ $(m = 1, \ldots, M_{S,p}, k = 1, \ldots, K)$ with the $a_k > 0$ values tuned in an automated manner to avoid saturation of the hidden neurons based on the ratios of the standard deviations of the inputs and knowledge of the non-saturation regions of the activation function. To illustrate, consider Example 9.1 and suppose the hyperbolic tangent is the activation function. The saturation region corresponds to the absolute output of this function being approximately 1 for inputs greater than 3 in absolute value. We set

$$b_\theta = \sqrt{\frac{1}{N_{S,p} - 1}\sum_{i=1}^{N_{S,p}}\left(\theta_i - \bar{\theta}\right)^2},$$

$$b_r = \sqrt{\frac{1}{N_{S,p} - 1}\sum_{i=1}^{N_{S,p}}\left\{r_{i,S}^{\mathrm{nom}}\left(\theta_i\right) - \bar{r}\right\}^2},$$

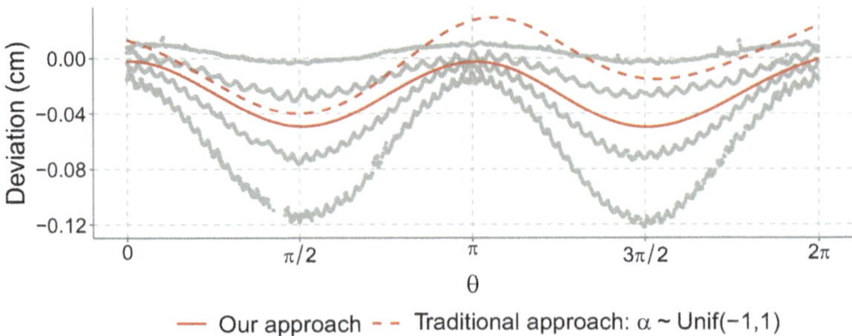

Fig. 9.6 In-plane deviations (gray dots) for four disks manufactured under process A, out-of-sample predictions for a $1.5''(3.81\,\text{cm})$-radius disk obtained using the Uniform$(-1, 1)$ distribution for all $\alpha_{m,k}^{(1,\text{A})}$ (dashed red line), and out-of-sample predictions obtained using our tuned Uniform distributions (solid red line)

$$a_1 = \frac{5b_\theta}{2\left(b_\theta + b_r\right)\max\left\{\theta_i : i = 1, \ldots, N_{S,p}\right\}},$$

$$a_2 = \frac{2 - a_1\max\left\{\theta_i : i = 1, \ldots, N_{S,p}\right\}}{\max\left\{r_{i,S}^{\text{nom}}(\theta_i) : i = 1, \ldots, N_{S,p}\right\}}.$$

with $\bar{\theta}$ and $\bar{r}$ the average of the θ_i and $r_{i,S}^{\text{nom}}(\theta_i)$, respectively. Figure 9.6 illustrates the improved out-of-sample predictions of our mechanism compared to the standard ELM for a $1.5''(3.81\,\text{cm})$ radius disk. In this case, the in-plane deviations of $N_{1,\text{A}} \approx 4000$ points on four disks (which are analyzed in Sect. 9.4) were used to fit the model. The standard ELM yields poor predictions because its use of uniform$(-1, 1)$ saturates its hidden neurons and prevents it from learning the deviation patterns as a function of the inputs. Additional discussions on the critical role of the random assignment of the $\alpha_{m,k}^{(S,p)}$ for the predictive performance of ELM models are in Appendix A.

An important consideration in this step is the choice of the set of shapes and process. Shapes and processes whose deviations are convenient to describe are preferable, as they generally require fewer test shapes to fit and can be more usefully applied in the remaining steps. For example, we observe in Sect. 9.4 that as few as four shapes are sufficient to specify a baseline Bayesian ELM deviation model with both good predictive performance and capability for model transfer.

9.3.3   Transfer Baseline Model to New Processes

We learn the Bayesian ELM for the $T_{s,p'\to p}$ via a discrepancy measure [24, 27] that combines posterior predictions from the baseline model with data from process

p' to extract information on the TEAs. Our measure is defined for each point $i \in \{1, \ldots, N_{S,p'}\}$ on the products under p' as

$$T_{i,S,p'\to p} = \underset{t \geq -r_{i,S}^{\mathrm{nom}}(\theta_i)}{\mathrm{argmin}} \left\{ y_i - t - x_i - \mathbf{h}_i^\mathsf{T} \widetilde{\boldsymbol{\beta}}_{S,p} \right\}^2, \tag{9.9}$$

where y_i is the deviation for point i, $\widetilde{\boldsymbol{\beta}}_{S,p}$ is a random variable distributed according to the posterior of $\boldsymbol{\beta}_{S,p}$ from the previous step, and $\mathbf{h}_i \in \mathbb{R}^{M_{S,p}}$ has entry m equal to $g\left(\alpha_{m,0}^{(S,p)} + \mathbf{z}\left(\theta_i, r_{i,S}^{\mathrm{nom}}, x_i + t\right)^\mathsf{T} \boldsymbol{\alpha}_m^{(S,p)} \right)$. We summarize the discrepancy measure distribution for each i by its expectation $\widehat{T}_{i,S,p'\to p}$, and form the vector $\widehat{\mathbf{T}}_{S,p'\to p} \in \mathbb{R}^{N_{S,p'}}$ containing them. The Bayesian ELM for the TEAs is then specified as

$$\widehat{\mathbf{T}}_{S,p'\to p} = \mathbf{H}_{S,p'\to p} \boldsymbol{\beta}_{S,p'\to p} + \boldsymbol{\epsilon}_{S,p'\to p}, \tag{9.10}$$

where $\mathbf{H}_{S,p'\to p}$ is the $N_{S,p'} \times M_{S,p'\to p}$ matrix whose (i, m) entry is $g\left(\alpha_{m,0}^{(S,p')} + \mathbf{z}\left(\theta_i, r_{i,S}^{\mathrm{nom}}, x_i\right)^\mathsf{T} \boldsymbol{\alpha}_m^{(S,p')} \right)$ for random $\alpha_{m,k}^{(S,p')}$ ($M_{S,p'\to p}$ is the selected number of hidden neurons), $\boldsymbol{\beta}_{S,p'\to p} \in \mathbb{R}^{M_{S,p'\to p}}$ is a vector of unknown parameters, and the error terms in $\boldsymbol{\epsilon}_{S,p'\to p}$ are independent and identically distributed $\mathrm{N}\left(0, \sigma_{S,p'\to p}^2\right)$ random variables with $\sigma_{S,p'\to p}^2$ unknown. Our prior probability density function is

$$p\left(\boldsymbol{\beta}_{S,p'\to p}, \sigma_{S,p'\to p}^2 \mid \tau_{S,p'\to p}^2 \right) \propto \sigma_{S,p'\to p}^{-2} \tau_{S,p'\to p}^{-M_{S,p'\to p}}$$
$$\times \exp\left(-0.5 \tau_{S,p'\to p}^{-2} \boldsymbol{\beta}_{S,p'\to p}^\mathsf{T} \boldsymbol{\beta}_{S,p'\to p} \right), \tag{9.11}$$

and our hyperprior probability density function for $\tau_{S,p'\to p}^2$ is $p\left(\tau_{S,p'\to p}^2\right) \propto \tau_{S,p'\to p}^{-6} \exp\left(-0.01 \tau_{S,p'\to p}^{-2} \right)$. The Gibbs algorithm can again be used to derive the posterior.

As illustrated in Fig. 9.5, the Bayesian ELM TEA model is connected to the baseline Bayesian ELM deviation model in an immediate and simple manner. To formally describe this connection, let $\mathbf{h}_{i,S,p'\to p}$ denote row i of $\mathbf{H}_{S,p'\to p}$, and $\widehat{\boldsymbol{\beta}}_{S,p'\to p}$ contain the posterior modes of each entry in $\boldsymbol{\beta}_{S,p'\to p}$. Also, define $\mathbf{H}_{S,p'}$ as the $N_{S,p'} \times M_{S,p}$ matrix whose (i, m) entry is $g\left(\alpha_{m,0}^{(S,p)} + \mathbf{z}\left(\theta_i, r_{i,S}^{\mathrm{nom}}, x_i + \mathbf{h}_{i,S,p'\to p} \widehat{\boldsymbol{\beta}}_{S,p'\to p}\right)^\mathsf{T} \boldsymbol{\alpha}_m^{(S,p)} \right)$, where $\alpha_{m,0}^{(S,p)}$ and $\boldsymbol{\alpha}_m^{(S,p)}$ are from step one. Then the comprehensive Bayesian ELM deviation model that can be fitted to products from S manufactured under both p and p' is

$$\mathbf{y}_{S,p} = \mathbf{x}_{S,p} + \mathbf{H}_{S,p}\boldsymbol{\beta}_{S,p} + \boldsymbol{\epsilon}_{S,p},$$

$$\mathbf{y}_{S,p'} = \mathbf{x}_{S,p'} + \mathbf{H}_{S,p'\to p}\widehat{\boldsymbol{\beta}}_{S,p'\to p} + \mathbf{H}_{S,p'}\boldsymbol{\beta}_{S,p} + \boldsymbol{\epsilon}_{S,p'}, \qquad (9.12)$$

where the error terms in $\boldsymbol{\epsilon}_{S,p'}$ are independent and identically distributed $N\left(0, \sigma^2_{S,p'}\right)$ random variables with $\sigma^2_{S,p'}$ unknown, and independent of those in $\boldsymbol{\epsilon}_{S,p}$. Note that we utilize posterior modes in Eq. (9.12) to facilitate deviation modeling under p' for general, nonlinear activation functions. This corresponds to concepts in [25] for practical machine learning. Our use of computationally tractable discrepancy measures and TEA parameter estimates distinguishes our method from that in [29], which used intensive Bayesian calculations for all unknowns.

Example 9.2 Consider deviations for disks under a new process B. With $N_{1,\mathrm{B}} = 100$, $M_{1,\mathrm{B}\to\mathrm{A}} = 2$, $\mathbf{H}_{1,\mathrm{B}\to\mathrm{A}}$ is 100×2, and has (i, m) entry $g\left(\alpha^{(1,\mathrm{B})}_{m,0} + \alpha^{(1,\mathrm{B})}_{m,1}\theta_i + \alpha^{(1,\mathrm{B})}_{m,2} r^{\mathrm{nom}}_{i,1}\right)$. Also, $\mathbf{H}_{1,\mathrm{B}}$ has (i, m) entry $g\left(\alpha^{(1,\mathrm{A})}_{m,0} + \alpha^{(1,\mathrm{A})}_{m,1}\theta_i, +\alpha^{(1,\mathrm{A})}_{m,2}\left(r^{\mathrm{nom}}_{i,1} + \mathbf{h}_{i,1,\mathrm{B}\to\mathrm{A}}\widehat{\boldsymbol{\beta}}_{1,\mathrm{B}\to\mathrm{A}}\right)\right)$.

9.3.4   Transfer Baseline Model to New Shapes

The global deviation feature for shapes from a new set S' manufactured under process p is specified according to $\mathbf{H}_{S,p}\boldsymbol{\beta}_{S,p}$ as in Eq. (9.7), and a Bayesian ELM for their local deviation feature $\delta_{s',p}$ is then learned in a principled and automated manner by leveraging the global deviation feature model with data from the new shapes. To formally describe this, let $N_{S',p}$ denote the total number of points on products from S' manufactured under p, and $M_{S',p}$ the chosen number of hidden neurons for the Bayesian ELM of the local deviation feature. For each point $i \in \{1, \ldots, N_{S',p}\}$ on these products, let $\mathbf{w}\left(\theta_i, r^{\mathrm{nom}}_{i,S'}, x_i\right) \in \mathbb{R}^J$ denote its vector of inputs that is a function of θ_i, the nominal radius function $r^{\mathrm{nom}}_{i,S'}$ for the specific product from S' on which i resides, and the applied compensation $x_i \in \mathbb{R}$. Also, let $\mathbf{z}\left(\theta_i, r^{\mathrm{nom}}_{i,S}, x_i\right) \in \mathbb{R}^K$ denote the vector of inputs for the corresponding shape from S whose deviation model captures the global deviation feature for i. We define the $N_{S',p} \times M_{S,p}$ matrix $\mathbf{H}_{S',p,\mathrm{G}}$ whose (i, m) entry is $g\left(\alpha^{(S,p)}_{m,0} + \mathbf{z}\left(\theta_i, r^{\mathrm{nom}}_{i,S}, x_i\right)^{\mathsf{T}}\boldsymbol{\alpha}^{(S,p)}_m\right)$, so that $\mathbf{H}_{S',p,\mathrm{G}}\boldsymbol{\beta}_{S,p}$ captures the global deviation feature of products from S'. To specify the local deviation feature, we draw $\alpha^{(S',p)}_{m,j}$ as independent uniform random variables ($m = 1, \ldots, M_{S',p}, j = 0, \ldots, J$), and define the $N_{S',p} \times M_{S',p}$ matrix $\mathbf{H}_{S',p,\mathrm{L}}$ whose (i, m) entry is $g\left(\alpha^{(S',p)}_{m,0} + \mathbf{w}\left(\theta_i, r^{\mathrm{nom}}_{i,S'}, x_i\right)^{\mathsf{T}}\boldsymbol{\alpha}^{(S',p)}_m\right)$. Then the Bayesian ELM deviation model for S' is specified as

$$\mathbf{y}_{S',p} = \mathbf{x}_{S',p} + \mathbf{H}_{S',p,\mathrm{G}}\boldsymbol{\beta}_{S,p} + \mathbf{H}_{S',p,\mathrm{L}}\boldsymbol{\beta}_{S',p} + \boldsymbol{\epsilon}_{S',p}, \tag{9.13}$$

where $\mathbf{H}_{S',p,\mathrm{L}}\boldsymbol{\beta}_{S',p}$ captures the local deviation feature. As before, $\boldsymbol{\beta}_{S',p} \in \mathbb{R}^{M_{S',p}}$ is a vector of unknown parameters, and the error terms in $\boldsymbol{\epsilon}_{S',p}$ are independent and identically distributed $\mathrm{N}\left(0, \sigma^2_{S',p}\right)$ random variables with $\sigma^2_{S',p}$ unknown. Our prior probability density function for the parameters is

$$p\left(\boldsymbol{\beta}_{S',p}, \sigma^2_{S',p} \mid \tau^2_{S',p}\right) \propto \sigma^{-2}_{S',p} \tau_{S',p}^{-M_{S',p}} \exp\left(-0.5\tau^2_{S',p}\boldsymbol{\beta}_{S',p}^{\mathsf{T}}\boldsymbol{\beta}_{S',p}\right), \tag{9.14}$$

and our hyperprior probability density function for $\tau^2_{S',p}$ is $p\left(\tau^2_{S',p}\right) \propto \tau_{S',p}^{-6}\exp\left(-0.01\tau^2_{S',p}\right)$. We derive the posterior of the parameters and hence predictions of deviations for shapes from S' under process p, via the Gibbs algorithm. By the same reasoning as before, the comprehensive Bayesian ELM deviation model involving Eqs. (9.7) and (9.13) can be fitted to products from both S and S' manufactured under p.

Example 9.3 Consider uncompensated squares under A, which belong to the shape set 2 of polygons. Suppose $N_{2,\mathrm{A}} = 500$. One useful set of inputs that can capture the local deviation feature is $\mathbf{w}\left(\theta_i, r_{i,2}^{\mathrm{nom}}\right) = \left(\theta_i, r_{i,2}^{\mathrm{nom}}(\theta_i), \mathrm{edge}(\theta_i)\right)^{\mathsf{T}}$, where $\mathrm{edge}(\theta_i) \in \{1, 2, 3, 4\}$ indicates the edge containing i. Also, the global deviation feature for a square is captured by the disk with radius $r_{i,1}^{\mathrm{nom}}$ equal to the square's circumradius [15], and so $\mathbf{z}\left(\theta_i, r_{i,1}^{\mathrm{nom}}\right)$ is accordingly defined for each i.

In contrast to the work in [30], our Bayesian ELM method reduces the effort for learning local deviation features, from specifying and fitting an entire nonlinear model for it to the simpler task of selecting the number of hidden neurons. Also, our modular ELM components for the local deviation features are immediately connectable for specifying comprehensive models across sets of shapes. Ultimately, our use of Bayesian statistics in this and the previous steps of our methodology plays a key role in enabling automated deviation modeling across an AM system based on sequential updates of prior deviation models with data from different shapes and processes.

9.4   Real Case Studies of Automated Model Generation in AM

This section presents case studies of the developed Bayesian ELM method for automated deviation modeling of several sets of shapes (Fig. 9.7) and stereolithography

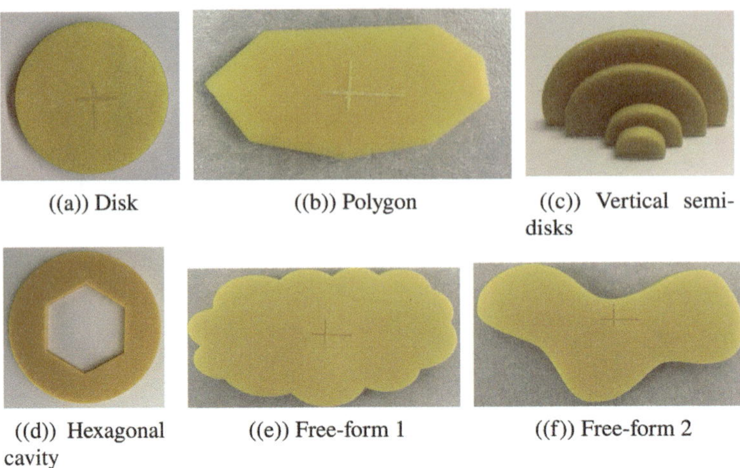

((a)) Disk   ((b)) Polygon   ((c)) Vertical semi-disks

((d)) Hexagonal cavity   ((e)) Free-form 1   ((f)) Free-form 2

Fig. 9.7 The sets of shapes considered in our case studies

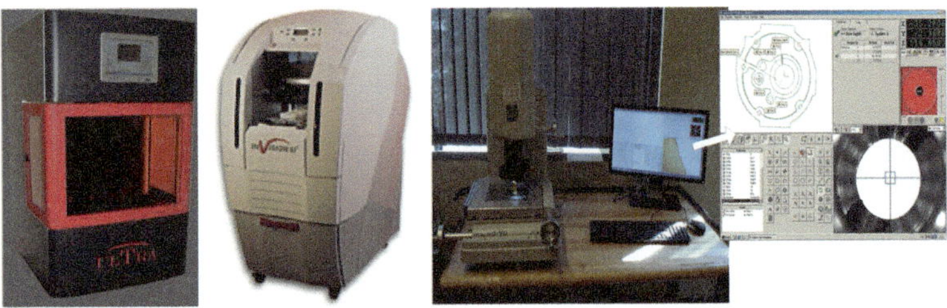

Fig. 9.8 EnvisionTec's ULTRA machine, and the Micro-Vu Vertex system. Photo taken in Dr. Y. Chen's lab at the University of Southern California

conditions. The products were manufactured under different settings of an ULTRA machine (Fig. 9.8), which is a commercial mask image projection stereolithography [39] platform by EnvisionTec. The observed settings for two processes A and B considered throughout are in Table 9.1. All deviations were measured by a Micro-Vu Vertex system (Fig. 9.8). For consistency and comparisons with previous methodologies, the models were fitted for data in inches. To display the results, we converted the data to centimeters (cm) to facilitate understanding for global readers.

The progression of the case studies is in Fig. 9.9. The first step automatically generates the baseline for disks manufactured under process A. The second step generates models for disks manufactured under process B, circular cavities in disks under process C, and out-of-plane deviations for vertical semi-disks under process D. The latter two sets of products were manufactured by the same machine, but their deviations are effectively generated under new processes C and D, respectively, due to the distinct physics involved in deviations of cavities and interlayer bonding effects for vertical products [28], [18]. The third step generates model for the second set of shapes, that is, polygons manufactured under process A. The models from the

Table 9.1 Observed settings
for the ULTRA processes

Variable	Process A	Process B
Product height	0.5″	0.25″
Layer thickness	0.004″	0.00197″
Mask resolution	1920 × 1200	1920 × 1200
Pixel dimension	0.005″	0.005″
Illuminating time/layer	9 s	7 s
Waiting time/layer	15 s	15 s
Resin type	SI500	SI500

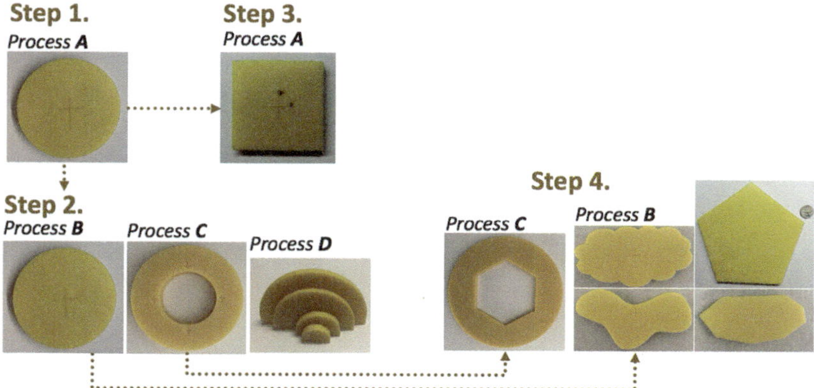

Fig. 9.9 Overview of the case studies

previous three steps are combined in the implementation of the fourth step for new
polygons and 2D freeform products under process B. A discussion on the broader
scope of the results of these case studies is given in the end.

Comparisons between the results obtained from our methodology and those
obtained from previous methodologies are provided in Table 9.2 to further demonstrate the power and advantages of our approach. The results from Ref. [21] are
not comparable to ours, as the authors of [21] propose a prescriptive approach to
model in-plane deviations of free-form shapes. In addition, comparisons between
the results obtained from our methodology and those obtained using an existing
Bayesian ELM method in [33] are summarized in Table 9.3 in the appendix. The
comparisons are based on RMSE, defined as $\left\{ \sum_{i=1}^{n} \left(y_i - \hat{y}_i\right)^2 / n \right\}^{1/2}$ where y_i
and $\hat{y}_i$ are the observed and predicted deviations, respectively, for point i, and n is
the size of the data. In all of these case studies, deviations from approximately 1000
equally spaced points on each individual product were used to form training data
sets for the models, and the activation function was the hyperbolic tangent.

Step 1: Baseline Model for In-Plane Disk Deviations
The baseline Bayesian ELM model for in-plane deviations under process A was
fitted using four uncompensated disks of nominal radii 0.5″(1.27cm), 1″(2.54cm),

Table 9.2 Comparison of RMSE for predictions based on our methodology (under "Our Method") and existing methods (under "Ref.", with the corresponding article under "Ref. Paper"). The RMSEs for each shape, process, and method are given in inches and centimeters, which are separated by |

| Shape | Process | RMSE (in. \| cm) | | Ref. paper |
		Our method	Ref.	
Disk	A	0.0007|0.0019	0.0009|0.0022	[17]
	B	0.0007|0.0017	0.0008|0.0020	[29]
	C	0.0004|0.0011	0.0006|0.0015	[28]
	D	0.0007|0.0017	0.0011|0.0028	[19]
Squares	A	0.0016|0.0041	0.0020|0.0050	[16]
Pentagon	B	0.0017|0.0042	0.0029|0.0074	[16]
			0.0008|0.0019	[30]
Dodecagon	B	0.0007|0.0019	0.0028|0.0071	[16]
Hexagonal cavity	C	0.0008|0.0021	0.0007|0.0017	[28]

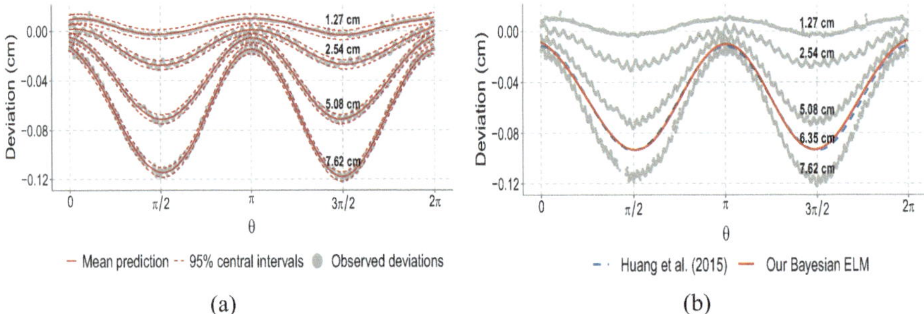

Fig. 9.10 (**a**) In-plane deviations (dots) for four disks under process A, and the posterior predictive means (solid lines) and 95% central posterior predictive intervals (dashed lines) from our Bayesian ELM model. (**b**) Comparison of the posterior predictive means for a 2.5"(6.35cm)-radius disk obtained from our approach (solid) with those obtained from [17] (dashed)

2"(5.08cm), and 3"(7.62cm), respectively. We set $M_{1,A} = 40$ and $\mathbf{z}\left(\theta_i, r_{i,1}^{\text{nom}}\right) = \left(\theta_i, r_{i,1}^{\text{nom}}\right)^{\mathsf{T}}$. The posterior predictions of deviations from our model are summarized in Fig. 9.10a. By inspection, our model provides a good fit to the deviations. Reference [17] previously specified a Bayesian nonlinear regression model for these deviations that was informed by their domain knowledge of the stereolithography process A. In contrast, our Bayesian ELM model was specified without any such knowledge and yields equivalent predictive performances (e.g., as illustrated in Fig. 9.10b for a 2.5"(6.35cm)-radius disk), which serves to illustrate the effective reductions in user efforts and inputs afforded by our method. Another demonstration of the effectiveness of our method compared to an existing Bayesian ELM is in Appendix B. We conclude that the deviation model specified by our method will

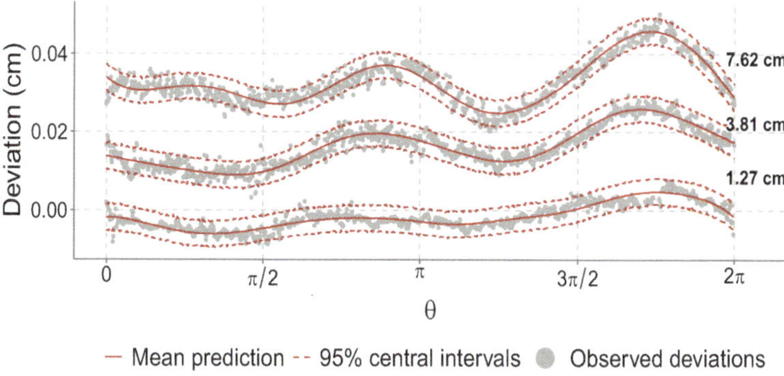

Fig. 9.11 In-plane deviations (dots) for three disks under process B, and the posterior predictive means (solid lines) and 95% central posterior predictive intervals (dashed lines) obtained by the transfer of the baseline deviation model to B

enable the same level of in-plane deviation control for disks under process A as that of the model in [17], which was an order of magnitude reduction in validation experiments.

Step 2: Transfer of Baseline Disk Model to New Processes
Three uncompensated disks of nominal radii $0.5''(1.27\text{cm})$, $1.5''(3.81\text{cm})$, and $3''(7.62\text{cm})$ manufactured under process B are considered in the first case study of the second step of our Bayesian ELM methodology. It is important to note that, besides the observed differences in product height, layer thickness, and illuminating time per layer, process B also differs from A in terms of new lurking variable settings that induce overcompensation [15]. The distinct and complicated nature of process B is clear upon inspection of these disks' deviations in Fig. 9.11. Specifically, in contrast to A, in-plane disk deviations under B increase on average as a function of the nominal radius and are asymmetrical. Our methodology effectively learns these complex features in the transfer of the baseline model to B by means of its model for $T_{1,B\to A}$. It also facilitates model transfer by involving only the single user input of $M_{1,B\to A}$ (which we set to 40) instead of the traditional specification of an entirely new model. The summary of our transferred model's posterior predictions in Fig. 9.11 demonstrates our successful modeling of in-plane disk deviations under B. References [29] and [30] previously specified Bayesian nonlinear regression models for these products' deviations. However, our Bayesian ELM model is preferable to theirs because it is fitted in a much simpler manner using the Gibbs algorithm compared to their computationally demanding Hamiltonian Monte Carlo [5] implementation, with no loss of predictive performance. Our comprehensive Bayesian ELM model for processes A and B is also preferable to fitting a standard NN or ELM model just on the data from B because our method yields smaller predictive uncertainties due to its incorporation of more data.

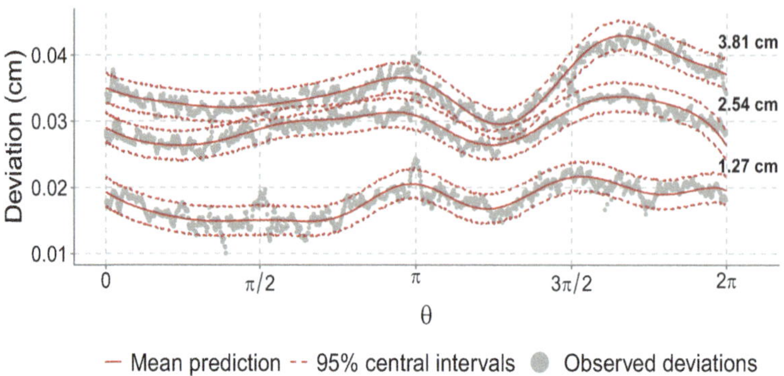

Fig. 9.12 In-plane deviations (dots) for three circular cavities, and the posterior predictive means (solid lines) and 95% central posterior predictive intervals (dashed lines) obtained by our transfer of the baseline deviation model

The second step of our method can also accommodate new deviation processes that arise in complex geometries. Three circular cavities of nominal radii $0.5''(1.27\text{cm})$, $1''(2.54\text{cm})$, and $1.5''(3.81\text{cm})$ contained in disks of nominal radii $1''(2.54\text{cm})$, $2''(5.08\text{cm})$, and $3''(7.62\text{cm})$, respectively, are considered to demonstrate this fact. Although the disks with circular cavities were manufactured under the same machine settings as process B, we consider the deviations for the cavities to have been generated under a different process C, as the physics for deviations of cavities differ from that for in-plane deviations on the outer boundary of a product. The posterior predictions obtained from our model for these products with $M_{1,\text{C}\to\text{A}} = 40$ (Fig. 9.12) indicate that our transferred model performs well in fitting this data. The broader consequence of this case is that our Bayesian ELM methodology can enable comprehensive and automated deviation modeling for both cavity and boundary components in geometrically complex products. These products were also modeled in [28], but our method is more advantageous because it greatly reduces the effort in learning the TEA and fitting the transferred model.

The final case study here involves specifying out-of-plane deviation models for semi-vertical disks manufactured under the same machine settings as process B. This is an especially challenging task because interlayer bonding effects yield complicated vertical deviations [18]. As the physics for out-of-plane deviations differ greatly from that for in-plane deviations, we consider these products to have been manufactured under a new process D, even though the machine settings were the same as those for process B. Four vertical semi-disks of nominal radii $0.5''(1.27\text{cm})$, $0.8''(2.032\text{cm})$, $1.5''(3.81\text{cm})$, and $2''$ (5.08cm) are considered to demonstrate how the second step effectively addresses this challenge. Figure 9.13 summarizes the posterior predictions from our transferred model with $M_{1,\text{D}\to\text{A}} = 40$. Reference [18] previously modeled out-of-plane deviations using nonlinear regression. Again, our approach is preferable to that in [18] because it automates the

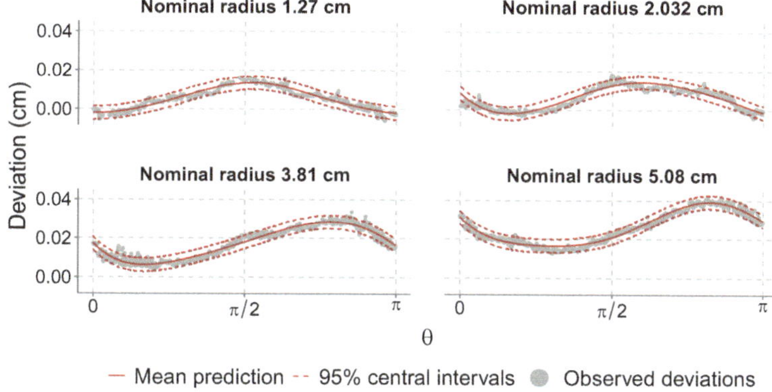

Fig. 9.13 Out-of-plane deviations (dots) for four vertical semi-disks, and the posterior predictive means (solid lines) and 95% central posterior predictive intervals (dashed lines) obtained by our transfer of the baseline deviation model

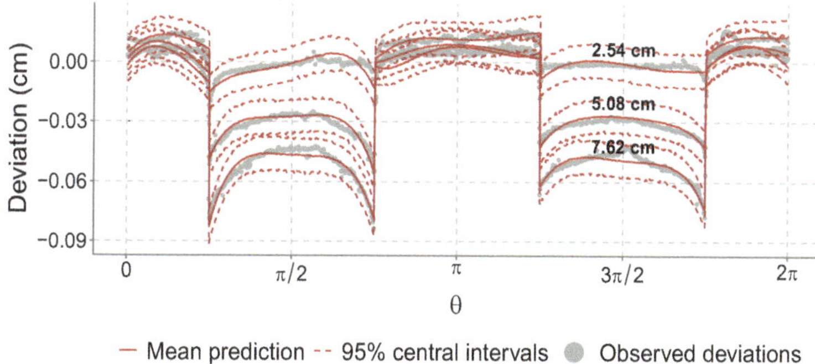

Fig. 9.14 In-plane deviations (dots) for three squares, and the posterior predictive means (solid lines) and 95% central posterior predictive intervals (dashed lines) obtained by our transfer of the disk deviation model to these polygons

specification of a deviation model with good predictive performance and reduces the user's efforts in leveraging both in-plane and out-of-plane deviation data to perform the model transfer. The advantages of our model compared to existing ELM models are established in Appendix B.

Step 3: Transfer of Baseline Disk Model to a New Shape
Three uncompensated squares of circumradius $1''(2.54\text{cm})$, $2''(5.08\text{cm})$, and $3''(7.62\text{cm})$ manufactured under A are considered in this case study of the transfer of the baseline deviation model to the new shape set 2 of polygons via the third step of our methodology. Additional, more complicated products from this set are

in the next subsection. The straight edges and sharp corners in these polygons introduce complex local deviation features that can be difficult to model (Fig. 9.14). The third step effectively learns the complex local deviation features with the previously specified $\delta_{1,A}$ as the global deviation feature. We set $M_{2,A} = 50$,

$$\mathbf{w}\left(\theta_i, r_{i,2}^{\text{nom}}\right) = \left(\theta_i, r_{i,2}^{\text{nom}}(\theta_i), \text{edge}(\theta_i)\right)^{\mathsf{T}}, \text{ and } \mathbf{z}\left(\theta_i, r_{i,1}^{\text{nom}}\right) = \left(\theta_i, r_{i,1}^{\text{nom}}\right)^{\mathsf{T}} \text{ for}$$

each point i on a square, and fit the comprehensive Bayesian ELM model to the in-plane deviations of both disks and squares under A. The summary of our fit for squares in Fig. 9.14 indicates the success in our model transfer. A comparison of our model's posterior predictions for these products with those obtained by the approach in [15] further highlights the high predictive performance that can result from the application of our methodology compared to other methods that pre-specify local deviation features.

Step 4: Transfer of Baseline Model to New Shapes and Processes
An uncompensated regular pentagon of circumradius $3''(7.62\text{cm})$, regular dodecagon of circumradius $3''(7.62\text{cm})$, and irregular polygon with smallest bounding circle of radius $1''(2.54\text{cm})$ (Fig. 9.2) manufactured under B are considered in this first case study of the fourth step of our method. The in-plane deviations of these products were modeled in [15] and [30] by complicated nonlinear regression models that took quite some effort to specify and fit. In contrast, the simpler combination of our connectable ELM structures for $T_{1,B\to A}$ and the polygon local deviation feature, which we learned in the previous steps, enables us to specify and fit deviation models for these products in a more automated and effortless manner. The sole user input is the number of hidden neurons for the various ELM structures, which is clearly simpler than learning entirely new models as under current deviation modeling methods. Our models' posterior predictions (Figs. 9.15a,b, and 9.2c) demonstrate their high predictive performance compared to the more complicated models in previous work. The RMSE for the fitted model of the pentagon in [30] is smaller than that of our model because the former did not consider points near the vertices of the shape, which exhibit more complicated trends than the other points, whereas our model does.

A hexagonal cavity of circumradius $1.8''(4.572\text{cm})$ contained in a $3''(7.62\text{cm})$-radius disk is considered as another case to further demonstrate how the fourth step can accommodate polygons under a new process. The deviations for this product are generated under process C, and we immediately connect $T_{1,C\to A}$ with the local deviation feature for polygons to transfer the baseline model to this product. The fit of the transferred model is summarized in Fig. 9.15c. Reference [28] modeled this product's deviations using Bayesian nonlinear regression, but our approach is preferable because it yields a model with comparable predictive performance, as seen in Table 9.2, that requires less effort to specify and fit.

We conclude with two free-form shapes under B (corresponding to the free-form products in Fig. 9.7). Note that free-form 2 ($s = 7$, Fig. 9.7f) requires the definition of two centers due to multiple correspondences on one polar

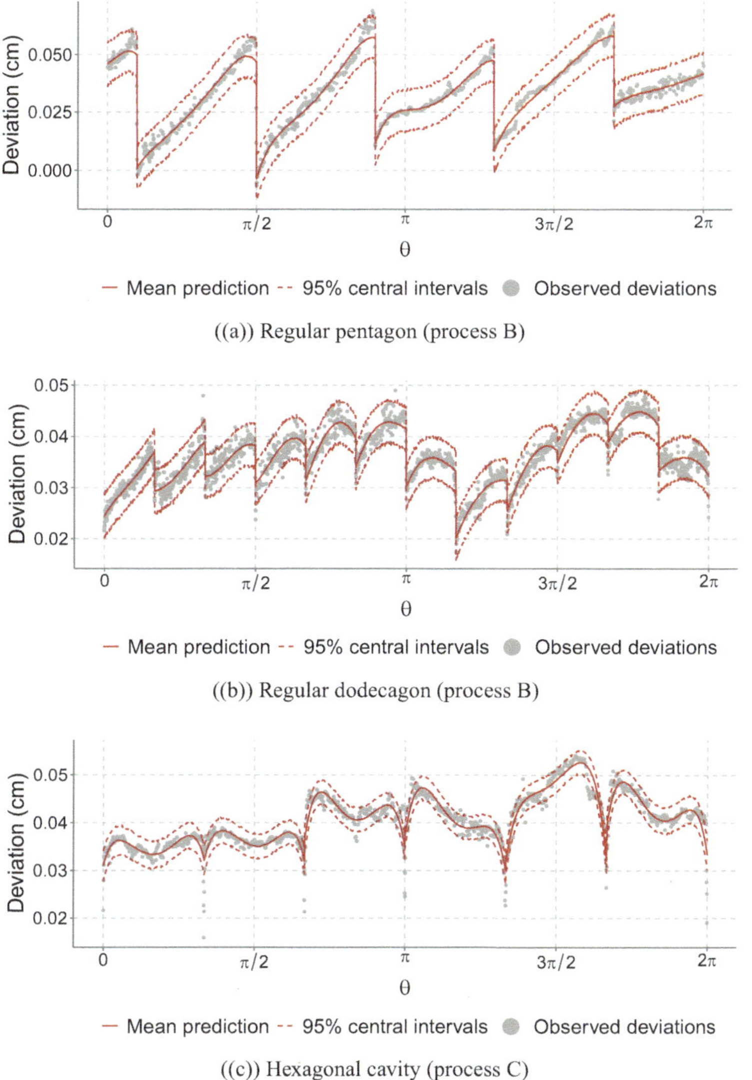

((a)) Regular pentagon (process B)

((b)) Regular dodecagon (process B)

((c)) Hexagonal cavity (process C)

Fig. 9.15 In-plane deviations (dots) for polygons under different processes, and the posterior predictive means (solid lines) and 95% central posterior predictive intervals (dashed lines)

angle (Fig. 9.17). We set $\mathbf{w}\left(\theta_i, r_{i,6}^{\mathrm{nom}}\right) = \left(\theta_i, r_{i,6}^{\mathrm{nom}}\left(\theta_i\right)\right)^{\mathsf{T}}$, and $\mathbf{w}\left(\theta_i, r_{i,7}^{\mathrm{nom}}\right) = \left(\theta_i, r_{i,7}^{\mathrm{nom}}\left(\theta_i\right), \mathrm{center}\left(\theta_i\right)\right)^{\mathsf{T}}$ for each point i on this shape, where $\mathrm{center}\left(\theta_i\right) \in \{1, 2\}$ indicates the center used to obtain $r_{i,7}^{\mathrm{nom}}\left(\theta_i\right)$. The nominal radius functions for both free-form shapes are expressed in point cloud data formats. Our connectable

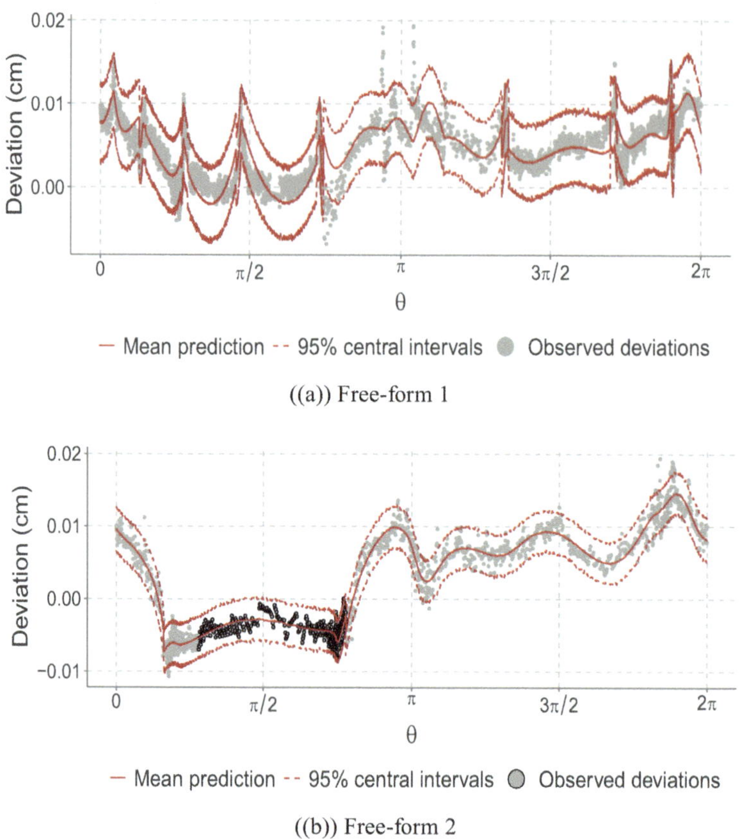

((a)) Free-form 1

((b)) Free-form 2

Fig. 9.16 In-plane deviations (dots) for two free-form shapes under B, and the posterior predictive means (solid lines) and 95% central posterior predictive intervals (dashed lines)

ELM structures possess a sufficiently broad scope so as to account for the new, complicated deviation features that arise in the complex geometries of these products. Figure 9.16 summarizes the posterior predictions of the transferred models obtained from the fourth step. The ability of our method to automate modeling of these free-form shapes is a significant demonstration of its effectiveness for AM systems, especially as current deviation modeling methods cannot accommodate free-form shapes with the same ease as our method.

Three broad results about general properties of our methodology can be drawn based on our wide-ranging case studies on solid and hollow products, regular and free-form shapes, and in-plane and out-of-plane deviations. First, in contrast to existing methods, our use of ELMs eliminates identifiability issues in fitting a baseline deviation model. Second, our structured approach to leveraging data and models across different processes and shapes eliminates identifiability issues in

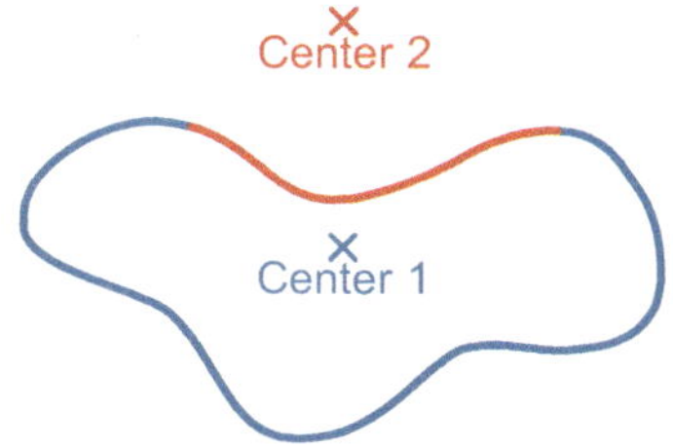

Fig. 9.17 The two centers used to define shape deviations for free-form 2. The correspondences between the centers and the points on the free-form shape are indicated by colors

learning TEAs and local deviation features. Third, our new random mechanism for ELMs effectively prevents saturation of hidden neurons. Thus, we can conclude that our Bayesian ELM methodology can yield deviation models with good predictive performance in an automated manner, involving negligible effort, for a wide variety of shapes and processes in AM systems. It is important to note that no specific domain knowledge for the stereolithography machine involved in these case studies was employed in our method. This fact further demonstrates the general scope of our method.

The use of polar coordinates to define shape deviations can introduce limitations in our methodology. For example, there can be multiple distinct points that correspond to one polar angle for concave shapes (such as in free-form 2 in Fig. 9.17). Following the work in [22], we address such issues in practice by defining multiple centers across different parts of a shape so that angles and points are in one-to-one correspondence.

Additive manufacturing systems possess great potential to fundamentally transform the manner in which people interact with manufacturing. Indeed, by reducing fabrication complexity and liberating product design for online users, AM systems can inaugurate an exciting new era of cybermanufacturing with capabilities that far exceed those of current manufacturing systems. However, comprehensive deviation modeling for geometric accuracy control of the vast variety of shapes manufactured under distinct processes in an AM system is a significant issue that must be addressed to realize their potential. We developed an automated Bayesian ELM deviation modeling methodology that can effectively address this challenge. Our method sequentially leverages different sets of data and models via four simple steps to automate deviation modeling of new shapes and AM processes. Statistical innovations in our method include principled and connectable NN structures that facilitate deviation modeling in these steps and a new random mechanism for the hidden neurons that yields improved predictive performances for the Bayesian ELM models. The use of the Bayesian paradigm in our method is important because it provides a straightforward framework for updating and transferring prior models and using all data collected from an AM system to improve predictions and reduce uncertainties. Our extensive case studies demonstrate the power and broad scope of our method to produce effective deviation models in a simple and efficient manner, without requiring specialized domain knowledge on specific processes. The significance of this is that our method can abstract from particular shapes and processes to underpin cross-cutting deviation modeling in AM systems. Accordingly, our modeling methodology can enable smarter geometric accuracy control in AM systems and thereby help to advance their future growth and adoption.

Appendix A: Preventing Saturation of Hidden Neurons in Bayesian ELM Deviation Models

Saturation of a hidden neuron refers to the activation function returning a relatively constant value for a range of inputs. An illustration for the hyperbolic tangent activation function is in Fig. 9.18. To illustrate saturation in the context of deviation modeling via ELMs, consider modeling in-plane deviations of uncompensated disks (shape set 1) manufactured under a fixed process p as in the case studies. Let the inputs for each point i be $\mathbf{z}\left(\theta_i, r_{i,1}^{\mathrm{nom}}\right) = \left(\theta_i, r_{i,1}^{\mathrm{nom}}\right)^{\mathsf{T}}$, and set $h_{i,m} = g\left(\alpha_{m,0}^{(1,p)} + \alpha_{m,1}^{(1,p)}\theta_i + \alpha_{m,2}^{(1,p)} r_{i,1}^{\mathrm{nom}}\right)$ as the neuron in entry (i, m) of $\mathbf{H}_{1,p}$ for random $\alpha_{m,0}^{(1,p)}, \alpha_{m,1}^{(1,p)}, \alpha_{m,2}^{(1,p)}$. Suppose $\alpha_{m,1}^{(1,p)} = 0.9, \alpha_{m,0}^{(1,p)}, \alpha_{m,2}^{(1,p)} \geq 0$, and $\theta_i = 2$. Then $h_{i,m} = g\left(\alpha_{m,0}^{(1,p)} + 1.8 + \alpha_{m,2}^{(1,p)} r_{i,1}^{\mathrm{nom}}\right) \geq 0.94$ as $r_{i,1}^{\mathrm{nom}} \geq 0.5$. Thus $h_{i,m} \approx 1$, and so is effectively an "activated neuron" irrespective of $r_{i,1}^{\mathrm{nom}}$. The broader, practical lesson to be drawn is that if many neurons are saturated, then the relationships between deviation and inputs will not be effectively learned.

To illustrate how our new random assignment mechanism addresses the saturation issue in a principled manner, consider again the previous setting. Suppose $(-2.5, 2.5)$ is taken as the non-saturation region for the activation function. In our mechanism, we choose a_1, a_2 such that $0 \leq |a_1\theta_i + a_2 r_{i,1}^{\mathrm{nom}}| \leq 2.5$ across most of the θ_i and $r_{i,1}^{\mathrm{nom}}$. Specifically, we allocate a proportion λ of the range $(0, 2.5)$ to input θ that depends on the standard deviations b_θ and b_r, with $\lambda = b_\theta/(b_\theta + b_r)$, and define $a_1 = 5\lambda/(2\theta_{\mathrm{max}})$, $a_2 = (2 - a_1\theta_{\mathrm{max}})/r_{\mathrm{max}}$, where θ_{max} and r_{max} are the largest angle and nominal radius, respectively. Note that although a_1 and a_2 involve the maximum values of the corresponding inputs, our mechanism is not equivalent to rescaling the respective inputs to lie in the range $(-1, 1)$. The $\alpha_{m,0}^{(1,p)}$ are still drawn from Uniform$(-1, 1)$ independently. Also, this new mechanism does not force all

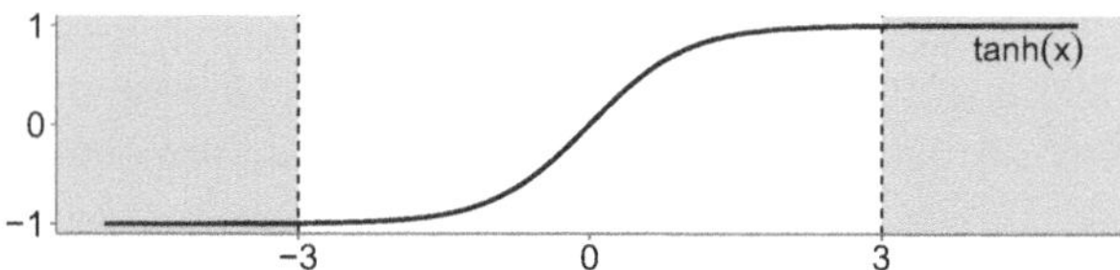

Fig. 9.18 Saturation region (shaded) for the hyperbolic tangent

hidden neurons to be non-saturated. The general case involving more inputs follows in a similarly straightforward fashion by allocating different portions of (0, 2.5) to them.

Standard tuning approaches to avoid saturation in ELMs (and thereby enhance their predictive performances) involve input normalization and/or rescaling, and the selection of different activation functions (e.g., the rectified linear unit). However, if no changes are made to the random mechanism, these methods typically fail to yield satisfactory models, and are particularly ineffective in the AM context. We accordingly developed this principled random mechanism for our Bayesian ELM methodology so as to reduce the likelihood of randomly selecting a large number of saturated hidden neurons.

Appendix B: Bayesian ELM Model Comparisons

Reference [33] proposed a Bayesian framework for ELMs, referred to as BELMs, in which the parameters are optimized iteratively using the ML-II method in [1]. The authors did not modify the usual random mechanism for the hidden neuron parameters. We compare the deviation models obtained from our methodology to those obtained from BELMs for the case studies in Sect. 9.4 that have at least four products. For each, one of the products is taken as the test data and the others are the training data. For example, in the case study of Section, the deviations for the $2''$(5.08cm)-radius disk were taken as the test data, and the other disks were used to fit the models. This was done because at least three shapes are needed to learn the relationships between the inputs and deviation. The methodologies are compared graphically and via root mean squared error (RMSE), defined as $\left\{ \sum_{i=1}^{n} \left(y_i - \hat{y}_i \right)^2 / n \right\}^{1/2}$ where y_i and $\hat{y}_i$ are the observed and predicted deviations, respectively, for point i, and n is the size of the test data. In all of these comparisons, the inputs were not scaled to lie within $(-1, 1)$, as that yields worse results.

First, consider in-plane disk deviations under stereolithography process A. Deviations from the $0.5''$(1.27cm), $1''$(2.54cm), and $3''$(7.62cm)-radii disks formed the training data, and those from the $2''$(5.08cm)-radius disk were the test data. We set $M_{1,A} = 40$ for both methods. We conclude from the comparisons in Fig. 9.19a and Table 9.3(a) that our method yields better out-of-sample predictions.

Second, consider out-of-plane deviations of vertical semi-disks under stereolithography process B. Three vertical semi-disks of nominal radii $0.5''$(1.27cm), $0.8''$(1.232cm), and $3''$(7.62cm) formed the training data, and the $1.5''$(3.675cm)-radius vertical semi-disk formed the test data. We set the number of hidden neurons for both methods to 40. Figure 9.19b and Table 9.3(b) summarizes the comparison of the predictions from these two methods and leads again to the conclusion that our method is better.

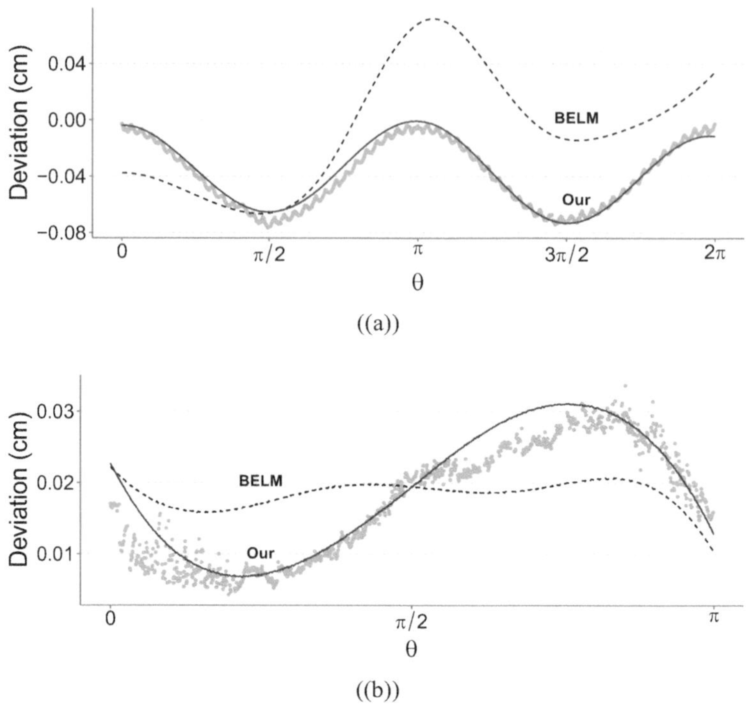

((a))

((b))

Fig. 9.19 Posterior predictive mean trends obtained from our methodology (solid line), and those obtained from the standard BELM method (dashed lines). (**a**) In-plane deviations (dots) for the test disk under process A. (**b**) Out-of-plane deviations (dots) for the vertical semi-disk under B

Table 9.3 Comparison of the posterior summaries for RMSE (specifically, mean and standard deviation) under our methodology and the standard BELM. (a) In-plane deviations of test disk under process A. (b) Out-of-plane deviations of test semi-disk under process B

Model	RMSE | SD (in.)	RMSE | SD (cm)
Our	0.066 | 0.003	0.167 | 0.009
BELM	0.605 | 0.020	1.53 | 0.048

(a)

Model	RMSE | SD (in.)	RMSE | SD (cm)
Our	0.041 | 0.002	0.105 | 0.004
BELM	0.094 | 0.003	0.238 | 0.007

(b)

Acknowledgments The work in this chapter was supported by US National Science Foundation Cyber-Physical Systems Program under grant CMMI-1544917.

References

1. Berger, J.O.: Statistical Decision Theory and Bayesian Analysis. Springer, New York (1985)
2. Cheng, L., Tsung, F., Wang, A.: A statistical transfer learning perspective for modeling shape deviations in additive manufacturing. IEEE Robot. Autom. Lett. **2**(4), 1988–1993 (2017)

3. Cheng, L., Wang, A., Tsung, F.: A prediction and compensation scheme for in-plane shape deviation of additive manufacturing with information on process parameters. IISE Trans. **50**(5), 394–406 (2018)
4. Deng, L., Yu, D., et al.: Deep learning: methods and applications. Found. Trends® Signal Process. **7**(3–4), 197–387 (2014)
5. Duane, S., Kennedy, A., Pendleton, B.J., Roweth, D.: Hybrid Monte Carlo. Phys. Lett. B **195**(2), 216–222 (1987). https://doi.org/10.1016/0370-2693(87)91197-X
6. Geman, S., Geman, D.: Stochastic relaxation, Gibbs distributions, and the Bayesian restoration of images. IEEE Trans. Pattern Anal. Mach. Intell. **6**(6), 721–741 (1984)
7. Hornik, K., Stinchcombe, M., White, H.: Multilayer feedforward networks are universal approximators. Neural Networks **2**(5), 359–366 (1989)
8. Huang, G., Huang, G.B., Song, S., You, K.: Trends in extreme learning machines: A review. Neural Networks **61**, 32–48 (2015)
9. Huang, G.B., Chen, L.: Convex incremental extreme learning machine. Neurocomputing **70**(16), 3056–3062 (2007)
10. Huang, G.B., Chen, L.: Enhanced random search based incremental extreme learning machine. Neurocomputing **71**(16), 3460–3468 (2008)
11. Huang, G.B., Chen, L., Siew, C.K.: Universal approximation using incremental constructive feedforward networks with random hidden nodes. IEEE Trans. Neural Networks **17**(4), 879–892 (2006)
12. Huang, G.B., Wang, D., Lan, Y.: Extreme learning machines: a survey. Int. J. Mach. Learn. Cybern. **2**(2), 107–122 (2011)
13. Huang, G.B., Zhu, Q.Y., Siew, C.K.: Extreme learning machine: a new learning scheme of feedforward neural networks. In: Proceedings. 2004 IEEE International Joint Conference on Neural Networks, 2004, vol. 2, pp. 985–990. IEEE (2004)
14. Huang, Q.: An analytical foundation for optimal compensation of three-dimensional shape deviations in additive manufacturing. J. Manuf. Sci. Eng. **138**(6), 061010 (2016)
15. Huang, Q., Nouri, H., Xu, K., Chen, Y., Sosina, S., Dasgupta, T.: Statistical predictive modeling and compensation of geometric deviations of 3D printed products. ASME Trans. J. Manuf. Sci. Eng. **136**(6), 061008 (10 pages) (2014)
16. Huang, Q., Nouri, H., Xu, K., Chen, Y., Sosina, S., Dasgupta, T.: Statistical predictive modeling and compensation of geometric deviations of three-dimensional printed products. ASME Trans. J. Manuf. Sci. Eng. **136**(6), 061008–061018 (2014)
17. Huang, Q., Zhang, J., Sabbaghi, A., Dasgupta, T.: Optimal offline compensation of shape shrinkage for 3D printing processes. IIE Trans. Quality Reliab. **47**(5), 431–441 (2015)
18. Jin, Y., Qin, S., Huang, Q.: Offline predictive control of out-of-plane geometric errors for additive manufacturing. ASME Trans. Manuf. Sci. Eng. **138**(12), 121005 (7 pages) (2016)
19. Jin, Y., Qin, S.J., Huang, Q.: Prescriptive analytics for understanding of out-of-plane deformation in additive manufacturing. In: 2016 IEEE International Conference on Automation Science and Engineering (CASE), pp. 786–791. IEEE (2016)
20. Libbrecht, M.W., Noble, W.S.: Machine learning applications in genetics and genomics. Nature Rev. Genet. **16**(6), 321 (2015)
21. Luan, H., Huang, Q.: Predictive modeling of in-plane geometric deviation for 3d printed freeform products. 2015 IEEE International Conference on Automation Science and Engineering (CASE 2015), Gothenberg, Sweden (2015)
22. Luan, H., Huang, Q.: Prescriptive modeling and compensation of in-plane geometric deviations for 3D printed freeform products. IEEE Trans. Autom. Sci. Eng. **14**(1), 73–82 (2017)
23. McDonnell, M.D., Tissera, M.D., Vladusich, T., van Schaik, A., Tapson, J.: Fast, simple and accurate handwritten digit classification by training shallow neural network classifiers with the "extreme learning machine" algorithm. PLOS ONE **10**(8), e0134254 (2015)
24. Meng, X.L.: Posterior predictive p-values. Ann. Stat. **22**(3), 1142–1160 (1994)
25. Meng, X.L.: Machine learning with human intelligence: Principled corner cutting (pc 2). In: Plenary Invited Talk, Annual Conference on Neural Information Processing Systems (NIPS), vol. 163 (2010)

26. Prieto, A., Prieto, B., Ortigosa, E.M., Ros, E., Pelayo, F., Ortega, J., Rojas, I.: Neural networks: An overview of early research, current frameworks and new challenges. Neurocomputing **214**, 242–268 (2016)
27. Rubin, D.: Bayesianly justifiable and relevant frequency calculations for the applied statistician. Ann. Stat. **12**(4), 1151–1172 (1984)
28. Sabbaghi, A., Huang, Q.: Predictive model building across different process conditions and shapes in 3D printing. 2016 IEEE International Conference on Automation Science and Engineering (CASE 2016), Dallas, Texas (2016)
29. Sabbaghi, A., Huang, Q.: Model transfer across additive manufacturing processes via mean effect equivalence of lurking variables. Ann. Appl. Stat. **12**(4), 2409–2429 (2018)
30. Sabbaghi, A., Huang, Q., Dasgupta, T.: Bayesian model building from small samples of disparate data for capturing in-plane deviation in additive manufacturing. Technometrics **60**(4), 532–544 (2018)
31. Schmutzler, C., Boeker, C., Zaeh, M.F.: Investigation of deviations caused by powder compaction during 3D printing. Procedia CIRP **57**, 698–703 (2016)
32. Schmutzler, C., Zimmermann, A., Zaeh, M.F.: Compensating warpage of 3D printed parts using free-form deformation. Procedia CIRP **41**, 1017–1022 (2016)
33. Soria-Olivas, E., Gomez-Sanchis, J., Martin, J.D., Vila-Frances, J., Martinez, M., Magdalena, J.R., Serrano, A.J.: BELM: Bayesian extreme learning machine. IEEE Trans. Neural Networks **22**(3), 505–509 (2011)
34. Tao, X., Zhou, X., He, Y.L., Ashfaq, R.A.R.: Impact of variances of random weights and biases on extreme learning machine. JSW **11**(5), 440–454 (2016)
35. Tong, K., Joshi, S., Lehtihet, E.: Error compensation for fused deposition modeling (fdm) machine by correcting slice files. Rapid Prototyping J. **14**(1), 4–14 (2008)
36. Tong, K., Lehtihet, E., Joshi, S.: Parametric error modeling and software error compensation for rapid prototyping. Rapid Prototyping J. **9**(5), 301–313 (2003)
37. Wang, H., Katz, R., Huang, Q.: Multi-operational machining processes modeling for sequential root cause identification and measurement reduction. J. Manuf. Sci. Eng. **127**(3), 512–521 (2005)
38. Wang, W., Cheah, C., Fuh, J., Lu, L.: Influence of process parameters on stereolithography part shrinkage. Mater. Design **17**(4), 205–213 (1996)
39. Zhou, C., Chen, Y.: Additive manufacturing based on optimized mask video projection for improved accuracy and resolution. J. Manuf. Process. **14**(2), 107–118 (2012)
40. Zhu, Z., Anwer, N., Mathieu, L.: Shape transformation perspective for geometric deviation modeling in additive manufacturing. Procedia CIRP **75**, 75–80 (2018)

Part V

Domain-Informed Machine Learning for Nanomanufacturing of Nanostructures

Chapter 10
Scale-Up Modeling for Nanomanufacturing

Nanomanufacturing represents the future of US manufacturing. Nanostructured materials and processes have been estimated to increase their market impact to about $340 billion per year in the next 10 years [28]. In the past decade, tremendous efforts have been devoted to basic nanoscience discovery, novel process development, and concept proof of nano devices. Yet much less research activities have been undertaken in nanomanufacturing to duplicate the success of transforming quality and productivity performance of traditional manufacturing. High cost of processing has been a major barrier of transferring the fast-developing nanotechnology from laboratories to industry applications [1]. The process yield of current nano devices is typically 10% or less [21, 22]. Hence, there is an imperative need of process improvement methodologies for nanomanufacturing.

10.1 Nanomanufacturing Scale-Up

To keep up with the increasing societal and economical need of nanomanufacturing (NM), research on scale-up NM has been a important field of study, which entails scale-up process research and scale-up methodology research. The scale-up methodology research includes establishing modeling, simulation, and control methodologies that enable and support economical production at commercial scale.

Two categories of scale-up NM research are defined:

C1. Scale-up process research: Identifying and developing NM processes and processing techniques with the potential of economical production at commercial scale, and

C2. Scale-up methodology research: Establishing modeling, simulation, and control methodologies that enable and support economical production at commercial scale.

Q. Huang, *Domain-informed Machine Learning for Smart Manufacturing*,
https://doi.org/10.1007/978-3-031-91631-1_10

The first category focuses on the process-level issues, while the second devotes its attention to system-level methodological issues such as yield and quality improvement. We further classify the scale-up methodology research (C2) into four subcategories based on their research objectives, nature of the problems, measurement of outcomes, and methodology domains (Table 10.1). For instance, we classify the objectives of this scale-up methodology as: (i) improving process repeatability (i.e., increasing **quantity**), (ii) scaling up **size**, (iii) increasing production **throughput**, and (iv) reduce defects (i.e., increasing **yield**). Quantity, size, throughput, and yield consist of the four pillars of a scale-up methodology (Fig. 10.1), where cost is considered as the implied outcome from the four pillars. Although this classification and comparison in Table 10.1 may be arguable, particularly considering that fact that limited research has been done, our intention is to promote more thinking and efforts to establish the scale-up methodologies for nanomanufacturing.

The scale-up methodology research spans areas of robust design, process monitoring and control, metrology, and reliability. Since NM process modeling provides the basis for process monitoring and control, guided inspection and sensing strategy, and more efficient experimental design strategy for robust synthesis of nanomaterials [19], we focus on domain-informed machine learning of nanostructure growth models for scalable NM.

The existing literature on nanostructure growth process modeling generally falls into three categories: (i) physical modeling, (ii) statistical modeling, and (iii) physical–statistical modeling approaches. These three nanostructure modeling strategies have their own suitable domains of application. With sufficient physical

Table 10.1 Subcategories of scale-up methodology and comparison

	Quantity	Size	Throughput	Yield
	1 →N	S → L	S → F	L → H
Objective	From one to many	From small to large size or area	From slow to fast processing rate	From low to high quality
Nature of problem	Improve process repeatability	Scale size up	Increase production rate	Reduce defects
Measure of outcome	Large number of units with low variations	Full-scale geometry at full-scale system	Short production cycle	Low fraction nonconforming, low defect rate
Methods	variation reduction	Dimensional analysis	Process design	Process & quality control

Fig. 10.1 Four pillars of scale-up methodology

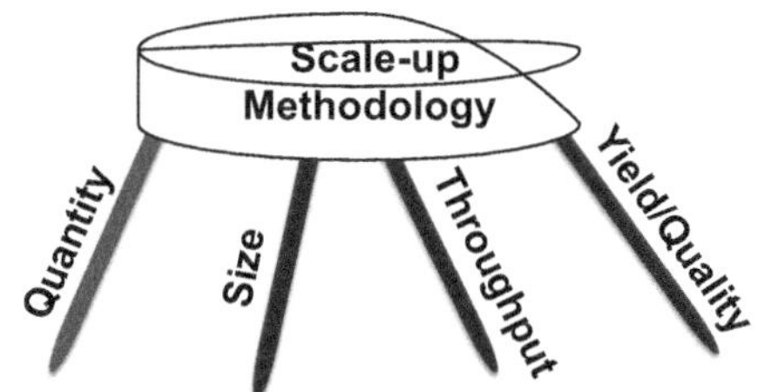

knowledge, physical modeling is readily a choice even when experimental data are less accessible. When growth mechanisms are debatable, uncertainties in the first principles will invalidate the purely physical modeling approaches.

On the other hand, statistical modeling faces issue of limited data in NM due to costly growth experiment and structure characterization. Moreover, the collected data tend to have large variations because of the poor *in situ* control of process variables such as growth temperature gradient. As a result, a large pool of candidate models can statistically fit the data. The data requirement for statistical modeling alone is rarely satisfied [35].

Physical–statistical modeling comes into play when physical knowledge or data alone are insufficient, but the combined information makes it feasible to improve the understanding of one growth mechanism or select among a few candidate mechanisms. When the uncertainties in physical knowledge and data continue to increase and growth mechanisms still require to be derived or investigated, new modeling strategies have to be developed to accommodate large uncertainties in two domains. Wang and Huang [35] made the first attempt to devise the cross-domain model building and validation (CDMV) approach. Both strategies fall into domain-informed machine learning.

10.2 Scale-Up Modeling for Manufacturing Nanoparticles Using Microfluidic T-Junction

As a zero-dimensional nanostructures, nanoparticles have great potentials to revolutionize the industry and improve our life in various fields such as energy, security, medicine, food, and environmental science. Droplet-based microfluidic reactors serve as an important tool to facilitate monodisperse nanoparticles with a high yield. Depending on process settings, droplet formation in a typical microfluidic T-junction is explained by different mechanisms, squeezing, dripping or squeezing-to-dripping. Therefore, the manufacturing process can potentially operate under multiple physical domains due to uncertainties. Although mechanistic models have been developed for individual domains, a modeling approach for the scale-up manufacturing of droplet formation across multiple domains does not exist. Establishing an integrated and scalable droplet formation model, which is vital for scaling up microfluidic reactors for large-scale production, faces two critical challenges: high dimensionality of modeling space and ambiguity among boundaries of physical domains. This work establishes a novel and generic formulation for the scale-up of multiple-domain manufacturing processes and provides a scalable modeling approach for the quality control of products, which enables and supports the scale-up of manufacturing processes that can potentially operate under multiple physical domains due to uncertainties.

Synthesis of solid particles, liquid droplets, and gas bubbles is critical for pharmaceutical and chemical engineering applications [8, 37, 41]. Droplet-based microfluidic devices manipulate immiscible fluids in channels of micrometer size

[2, 5, 6, 11, 14, 24, 29, 31, 43]. The high surface-area-to-volume ratio within microchannels guarantees uniform temperature throughout the reaction volume, and convective mixing within droplets ensures rapid homogenization. These properties make droplet microfluidic reactors a viable technology for the scalable synthesis of high-quality metal nanoparticles.

In order to produce particles on an industrial scale with low cost, microfluidic reactors need to be scaled up for high throughput and high yield with the foremost control over droplet size [6, 23, 25, 30]. Scale-up modeling, which refers to the process modeling approaches that enable and support economical production at commercial scale, is thus crucial for the quality control of nanoparticles[40]. Scale-up modeling of the droplet formation in the microfluidic channels face several key challenges:

- *High dimensionality*. Description of droplet formation in microfluidic channel involves a large number of physical parameters or quantities. Even by conducting dimensional analysis using scaling law [7, 14, 15], the number of obtained dimensionless numbers can be large, which gives rise to a high-dimensional problem. For instance, droplet size after scaling is proposed to be a function of more than five dimensionless numbers in squeezing-to-dripping domain with each being a combination of multiple physical parameters [15, 16]. This poses both experimental and modeling challenges to understand the response surface in a high-dimensional space. Currently, the droplet formation experiment in practice uses the one-factor-at-a-time approach, i.e., testing one dimensionless number at a time and fixing the rest [6, 7, 12, 14, 15, 34, 38, 42], which only guarantees the understanding in a projected low-dimensional space.

- *Multiple physical domains*. The droplet formation in a microfluidic channel is multiple-domain [6, 12, 34, 38, 42], and a microfluidic T-junction can produce droplets in squeezing, dripping, or squeezing-to-dripping domain (Due to our focus on producing monodisperse droplets, jetting domain is not herein discussed). Since different domains are dominated by different mechanisms, model structures depicting droplet formation vary with physical domains, which significantly increases the complexity in experimentation and modeling. Current practice is to discover individual domains through experimentation and then establish individual models within each domain. For instance, the domains are classified based on the capillary number Ca: squeezing with $Ca < 0.002$, squeezing-to-dripping with $0.002 < Ca < 0.01$, dripping with $0.01 < Ca < 0.3$, and jetting domain with $Ca > 0.3$ [39]. In the squeezing domain, the droplet formation process is explained by the pressure drop across the droplet during formation; while in the dripping domain, the process is interpreted by the balance between shear force and interfacial force. However, there are inconsistencies and ambiguity in defining the boundaries of physical domains [7], which hinders the application of these models in full-scale manufacturing.

This chapter introduces a unified scale-up model across multiple domains under uncertainties.

10.3 Scalable Modeling Methodology for Multiple-domain Manufacturing Process

In order to scale up manufacturing processes, we need to first address the so-called scale-up issue: how to translate the understanding of a process from lab scale to commercial scale such that the optimal properties can be determined a priori for future operations. The science base of achieving the scalability of engineering models is the scaling law, which captures the scale-invariant characteristics of an engineering system [3, 44]. Dimensional analysis serves as an important tool to transform the parameter space to the so-called Π-space spanned of dimensionless numbers, which achieves dimension reduction and ensures the scalability of models. Note that although there is no unique selection of dimensionless numbers, the dimension of Π-space is uniquely determined by dimensional analysis.

Droplet formation in microfluidic channels represents a class of "multiple-domain" scale-up modeling problems, which involve different physical phenomena and mechanisms [20, 36]. Since the boundaries between different physical domains can sometimes be ambiguous, it is often not known a priori which physical domain the process belongs to. Our objective is to establish a unified scale-up model across multiple domains under uncertainties to facilitate full-scale production. Our methodology is illustrated in Fig. 10.2. We first conduct dimensional analysis to formulate the problem in the Π-space. Then, we identify the *primary factor* and *secondary factor* for further dimension reduction.

Definition 10.1 The **primary factor** consists of dimensionless numbers that characterize the physical domains of engineering systems; the rest of dimensionless numbers form the **secondary factor**. Each physical mechanism, which is likely to dominate a certain domain, is then characterized by a model structure that **only** depends on the primary factor, termed as a **basis function**. Basis functions are assumed to be linearly independent.

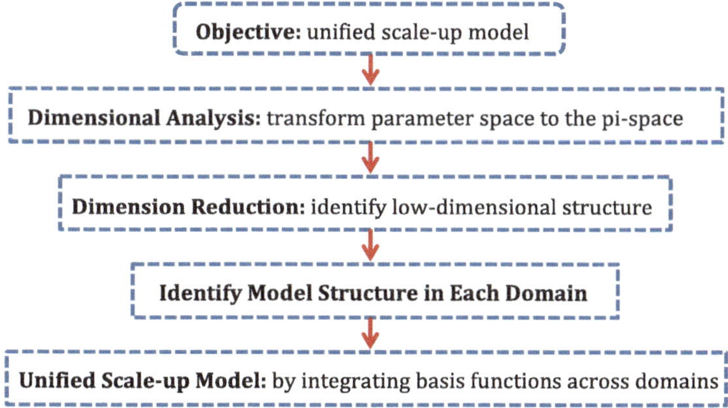

Fig. 10.2 Methodology to construct the scale-up model for a multivariate physical system

The rationale of this definition is based on the observation that physical domains of engineering systems are often be classified by one or only a very few dimensionless numbers, which form the *primary factor* henceforth. For example, the Reynolds number is widely used to predict flow patterns in the case of a bounding surface. Laminar flow occurs at low Reynolds numbers where viscous forces dominate, and turbulent flow occurs at high Reynolds numbers where the flow is dominated by inertial forces [27]. In the case of microfluidic droplet formation, four domains are differentiated by the capillary number defined in Eq. (10.4). Since the multiple-domain property of the process is fulled captured by the primary factor, the original scale-up modeling problem can be divided into several subproblems in lower-dimensional subspaces. Each physical domain is dominated by one mechanism, and the secondary factor contributes to the weights of each mechanism. For instance, the effect of flow rate ratio on the dimensionless size of droplets appears to be linear across different domains [12, 14, 34, 38, 42], and can thus be treated as a weighting factor of domains.

Based on this rationale, we are now in the position to formulate the high-dimensional multiple-domain scale-up modeling problem. Let z denote the response of interest, and let $\mathbf{x}^*$, $\mathbf{x}^\circ$ denote the *primary factor (PF)* and the *secondary factor (SF)*, respectively. In the high-dimensional Π-space, there exists the functional relation in Eq. (10.1), where ϵ and $\boldsymbol{\theta}$ denote the random noise and model parameters, respectively.

$$z = \Phi(\mathbf{x}^*, \mathbf{x}^\circ) + \epsilon \tag{10.1}$$

Projecting the response surface $\Phi(\mathbf{x}^*, \mathbf{x}^\circ)$ onto the lower-dimensional space spanned by the primary factor $\mathbf{x}^*$, we obtain $\Phi(\mathbf{x}^*|\mathbf{x}^\circ)$, i.e., the response conditioning on given settings of the secondary factor $\mathbf{x}^\circ$. Since the model structure of $\Phi(\mathbf{x}^*|\mathbf{x}^\circ)$ in each domain is dictated by $\mathbf{x}^*$ only, we adopt a set of *linearly independent* basis functions $S = \{f_i(\mathbf{x}^*)\}_{i=1}^{K}$ to represent conditional response function $\Phi(\mathbf{x}^*|\mathbf{x}^\circ)$ in Eq. (10.2), where $f_i(\mathbf{x}^*)$ is the ith basis function used to characterize a model structure, and its coefficient β_i characterizes the effect of secondary factor, i.e., a scaling factor for modeling structure $f_i(\mathbf{x}^*)$. Note that the definition of primary and secondary factor guarantees the existence of model decomposition in Eq. (10.2).

$$\Phi(\mathbf{x}^*|\mathbf{x}^\circ) = \sum_{i=1}^{K} \beta_i f_i(\mathbf{x}^*) \tag{10.2}$$

where $f_i(\mathbf{x}^*)$ is the ith basis function used to characterize a model structure, and its coefficient β_i characterizes the effect of secondary factor, i.e., a scaling factor for modeling structure $f_i(\mathbf{x}^*)$. In dripping domain with high capillary number, for instance, experimental studies show that droplet size is proportional to $Ca^{-0.25}$ [12, 33], i.e., a candidate basis function in the dripping domain is $f_i(Ca) = Ca^{-0.25}$. It is important to note that the basis functions may vary across domains.

Based on the conditional response model in Eq. (10.2), we deduce the full model in Eq. (10.3). Since the measured response is finite, there exists a reference frame such that the response is always non-negative, and Tonelli's theorem holds. The exchange between sum and integration is thus valid.

$$
\begin{aligned}
\Phi(\mathbf{x}^*, \mathbf{x}^\circ; \boldsymbol{\theta}) &= \int_{\mathbf{x}^\circ} \Phi(\mathbf{x}^*|\mathbf{x}^\circ)d\mathbf{x}^\circ = \int_{\mathbf{x}^\circ} \sum_{i=1}^{K} \beta_i(\mathbf{x}^\circ) f_i(\mathbf{x}^*)d\mathbf{x}^\circ \\
&= \sum_{i=1}^{K} \left(\int_{\mathbf{x}^\circ} \beta_i(\mathbf{x}^\circ)d\mathbf{x}^\circ \right) f_i(\mathbf{x}^*) \\
&= \sum_{i=1}^{K} g_i(\mathbf{x}^\circ) f_i(\mathbf{x}^*)
\end{aligned}
\tag{10.3}
$$

where $g_i(\mathbf{x}^\circ)$ can be interpreted as a weight function for $f_i(\mathbf{x}^*)$ given the settings of secondary factor $\mathbf{x}^\circ$, noting that $g_i(\mathbf{x}^\circ)'s$ share the function form with different parameters.

Remark The model formulation in (10.3) essentially suggests the statistical additive model framework [4, 10, 18] for the high-dimensional multiple-domain scale-up modeling problem. The formulation enables the application of modeling techniques in statistics for model building and estimation.

Although experimental literature assists to identify candidate basis functions, there exist discrepancies due to incomplete physical understanding and variations in experimental conditions and facilities. Furthermore, disagreement and ambiguity in defining the boundaries of physical domains will increase complexity of model building as well. To accommodate these uncertainties, we will investigate the following framework.

Assumption Let $\mathbb{S}_d$ denote the subset of basis functions characterizing the model structure of $\Phi(\mathbf{x}^* \mid \mathbf{x}^\circ)$ in the dth domain, $d = 1, 2, \ldots, D$. We assume $\mathbb{S}_d = \mathbb{S}$ for all d.

Under this framework, the equality $f_i(\mathbf{x}^*) = 0$ holds only for countable settings of $\mathbf{x}^*$. So the basis functions of $\Phi(\mathbf{x}^* \mid \mathbf{x}^\circ)$ do not degenerate in any continuous domain. Physically, the assumption means that all physical mechanisms, which are characterized by the complete set of basis functions, coexist in all physical domains with different weights.

Remark Note that the setup of this framework is able to avoid the issue of defining the transition points or boundaries between different physical domains upfront. However, implicitly the model $\Phi(\mathbf{x}^*, \mathbf{x}^\circ)$ selects proper basis function(s) to characterize different physical domains by varying the weight $g_i(\mathbf{x}^\circ)$ for each basis function $f_i(\mathbf{x}^*)$. The influence of a particular mechanism is described by some subset of $\{g_i(\mathbf{x}^\circ)f_i(\mathbf{x}^*), i = 1, \ldots, K\}$. It follows that both domains and domain transitions can be obtained from the model $\Phi(\mathbf{x}^*, \mathbf{x}^\circ; \boldsymbol{\theta})$: (i) A domain is identified to be dominated by a certain mechanism if this mechanism contributes to the majority of response in this domain; (ii) If none of the mechanisms contributes to the majority of response in certain domain, this domain is a transitional domain.

10.4   Scale-up Modeling of Droplet Formation in a Coated Microfluidic T-Junction

This section presents the detailed solution procedure for the high-dimensional scale-up modeling problem formulated in Sect. 10.3.

10.4.1   Dimensional Analysis and Dimension Reduction

Before conducting dimensional analysis to obtain the transformed Π-space, we first introduce the droplet formation process in a microfluidic T-junction to identify relevant physical quantities in the parameter space. As shown in Fig. 10.3, the carrier oil (continuous phase) is injected via inlet 1, while the reagent streams are introduced via inlets 2 and 4. A stream injected via inlet 3 is used to prevent diffusive mixing between reagent streams before droplet formation. The immiscible fluid of droplets is called "dispersed phase."

Notations of physical quantities used in the rest of the paper are listed as follows.

Notation	Physical quantity
$w_c(w_d)$	Width of channel into which continuous (dispersed) phase flows
h	Depth of channel
$\mu_c(\mu_d)$	Dynamic viscosity of continuous (dispersed) phase
$\rho_c(\rho_d)$	Density of continuous (dispersed) phase
$Q_c(Q_d)$	Volumetric flow rate of continuous (dispersed) phase
σ	Interfacial tension

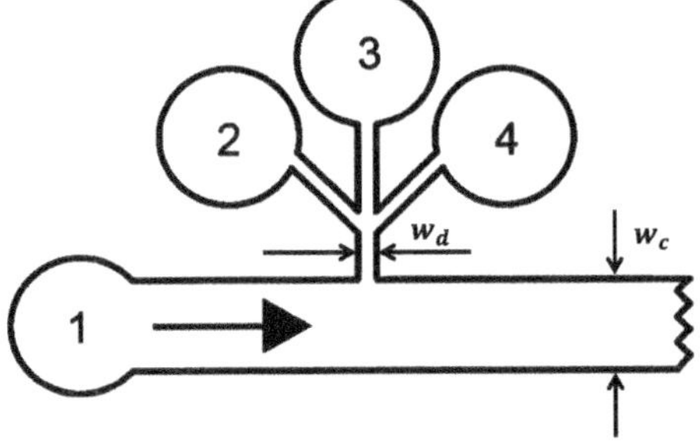

Fig. 10.3 Structure of the two-phase microfluidic T-junction [23]

Table 10.2 Relevant list of physical quantities

	Quantity	Dimension
Geometry	w_c, w_d	L
	h	L
Material	μ_c, μ_d	$ML^{-1}T^{-1}$
	ρ_c, ρ_d	ML^{-3}
Mechanics	Q_c, Q_d	$L^3 T^{-1}$
	σ	MT^{-2}

The parameter space consists of parameters that characterize the geometric structure, properties of the materials, and mechanical control variables, which are listed in the relevance list (Table 10.2). Dimensional analysis is then conducted to generate dimensionless π numbers in Eq. (10.4) that span the transformed Π-space [7, 14, 15].

$$
\pi_0 = \bar{L} = \frac{L}{w_c}, \quad \pi_1 = Ca = \frac{\mu_c Q_c}{\sigma w_c h}, \quad \pi_2 = \lambda = \frac{\mu_d}{\mu_c},
$$
$$
\pi_3 = Q = \frac{Q_d}{Q_c}, \quad \pi_4 = W_w = \frac{w_d}{w_c}, \quad \pi_5 = W_h = \frac{h}{w_c},
$$
$$
\pi_6 = \rho = \frac{\rho_d}{\rho_c}, \quad \pi_7 = Re = \frac{\rho_c Q_c}{\mu_c w_c}. \tag{10.4}
$$

Corresponding to the formulation in Sect. 10.3, we choose the response to be the dimensionless droplet length $z = \pi_0 = \bar{L}$ due to our interest in droplet size. The scale-up modeling problem in the transformed Π-space is then formulated by the π numbers in the form $\pi_0 = \Phi(\pi_1, \pi_2, \pi_3, \pi_4, \pi_5, \pi_6, \pi_7)$, i.e. $\bar{L} = \Phi(Ca, \lambda, Q, W_w, W_h, Re, \rho)$, with $(\mathbf{x}^*, \mathbf{x}^\circ) = (Ca, \lambda, Q, W_w, W_h, Re, \rho)$.

To reduce dimensionality, the capillary number is selected as the primary factor, i.e., $x^* = Ca$, whereas the remaining dimensionless numbers are identified as secondary factor, which will be explained in Sect. 10.4.3. For typical microchannel flows, the Reynolds number Re is very small. Once the microfluidic T-junction design and fluidic materials are determined, the only remaining controllable dimensionless number other than Ca is the flow rate ratio, the effect of which has been investigated in each domain. We also qualitatively investigate the effect of W_h by comparing droplet formation in a microfluidic T-junction with $W_h = 1$ and $W_h = 2$. Therefore, Q and W_h form the two-dimensional secondary factor $\mathbf{x}^\circ = (Q, W_h)$. Despite the demonstration in the study of droplet formation in a coated microfluidic T-junction, the strategy to reduce the dimension of a high-dimensional scale-up problem by identifying low-dimensional structures can be applied to a generic high-dimensional scale-up problem.

10.4.2 Experimental Setup and Data Collection

In order to systematically characterize droplet formation across multiple domains, we first select a reference geometry and keep the fluid pair fixed. We used two geometries that maintained a $w_d : w_c$ ratio of $1 : 4$, with $w_c = 200\,or\,400\,\mu$m. Microfluidic devices were coated with a low-surface-energy fluoropolymer coating using initiated chemical vapor deposition (iCVD) as described previously [23, 25]. The continuous phase was a polychlorotrifluoroethene (PCTFE) oil (trade name/vendor: Halocarbon oil), with $\mu_c = 100\,$mPa $\cdot$ s, and the dispersed phase was de-ionized water. Dimensional parameters for the reference system are given in Table 10.3. The width ratio W_w is set to be 0.25, while the depth–width ratio $W_h = 1, 2$. Droplet length was measured as a function of capillary number Ca and flow rate ratio Q for the reference system by selecting five different flow rate ratios $Q = 0.05, 0.25, 0.5, 1, 2$. For each fixed flow rate ratio, the capillary number is varied from $Ca = 0.00193$ to 0.15432. To keep the flow rate ratio fixed, both Q_c and Q_d must vary as Ca varies.

As reported in [25], for $Ca < 0.05$ with $w_c = 200\,\mu$m or $Ca < 0.01$ with $w_c = 400\,\mu$m respectively, droplet size increased with increasing flow rate ratio at each fixed capillary number explored in the experiment and decreased with increasing Ca at each fixed flow rate ratio (see Fig. 10.4). Unlike droplet formation in uncoated microfluidic channels, droplet formation in the coated device remained in the dripping regime above the threshold of $Q = 0.05$, and the droplet size appeared to either plateau or increase according to different flow rate ratios.

10.4.3 Model Structures and Basis Functions

Candidates of basis functions that characterize model structures for droplet formation in uncoated microfluidic T-junction in each domain can be acquired from experimental literature. In squeezing domain ($Ca \lesssim 0.01$ according to [7], $Ca < 0.002$ in [39]), the scaling law is given in the form $\bar{L} = \sigma + \omega Q$ [17], where

Table 10.3 Experimental settings of physical quantities

Fluid System		Viscosity [mPa $\cdot$ s]
Continuous phase	Halocarbon oil	100
Dispersed phase	De-ionized water	1
Device geometry		Dimension [μm]
Channel width	w_c	200–400
	w_d	50–100
Channel depth	h	400
Flow rate control		Dimension μL/h
Continuous phase	Q_c	250–200,000
Dispersed phase	Q_d	12.5–200,000

σ and ω are parameters determined by channel geometry. In dripping domain with high Ca where droplets are unconfined, the dimensionless droplet length is proportional to $Ca^{-0.25}$ approximately [12, 32]. In the squeezing-to-dripping domain, according to the correlations discussed in existing studies [6, 12, 34, 38, 42], the scale-up model regarding the capillary number with all the other dimensionless numbers fixed is either given in the general form $\bar{L}(Ca) \propto Ca^{-\alpha}$ with $0 < \alpha < 0.25$ or by the dimensionless droplet volume $\bar{V}(Ca)$ being linear combination of $Ca^{-\alpha_1}$ and $Ca^{-\alpha_2}$ based on approximation models. According to the assumption in Sect. 10.3, droplet formation in the squeezing-to-dripping domain exhibits intermediate phenomenon. This explains why $\bar{L}(Ca) \propto Ca^{-\alpha}$ with $0 < \alpha < 0.25$ in squeezing-to-dripping domain.

We therefore choose the basis functions $f_1(Ca) = 1$, $f_2(Ca) = Ca^{-\alpha}$ to characterize the decrease of droplet size in squeezing and dripping domain respectively at relatively low capillary numbers, where α is a positive parameter to be determined for coated devices.

In addition to the decrease in droplet size as the capillary number increases at relatively low Ca, we observed either a plateau or an increase of droplet size at higher capillary numbers before jetting occurred. We attribute this to the difference between our system and most in the literature using PDMS channels. The water contact angle of PDMS channels is $112°$ while the low surface energy coating applied to our system renders the channels more hydrophobic, resulting a water contact angle larger than $120°$ [25]. Wall effects by coating become significant when the droplet length is larger than the channel geometry. Therefore, an additional basis function $f_3(Ca) = Ca^{\gamma}$ is proposed to characterize the increase in droplet size at higher capillary numbers with $\gamma > 0$. The total number of basis functions is $K = 3$.

Besides the primary factor $x^* = Ca$, two secondary factor $x = (Q, W_h)$ are considered to explore the form of $g_i(x)$ in Eq. (10.3). From observation and existing literature, it is known that (i) the droplet size is linear to flow rate ratio with remaining π numbers fixed at high viscosity contrast [12, 14, 34, 38, 42]; (ii) droplet formation with regard to Ca does not depend on the geometry of microfluidic T-junction in squeezing [17], squeezing-to-dripping [6] and in dripping domain [12], indicating the independence of α and γ in W_h. The weight function $g_i(Q, W_h)$ is proposed to be

$$g_i(Q, W_h) = \beta_{i,0}(W_h) + \beta_{i,1}(W_h)Q. \tag{10.5}$$

where the intercept and slope of the linear model $\beta_{i,0}$, $\beta_{i,1}$ are functions of the other secondary factor W_h.

We analyze the effect of W_h qualitatively by regarding it as a treatment factor due to the following reasons: (i) The effect of geometry has not been well investigated across domains, i.e., concrete forms of β_{i0} and β_{i1} are not available for some i; (ii) W_h is a two-level factor in our experiments, which can be represented by $j = W_h$ with $j = 1, 2$. Similar to [26], an equivalent expression of Eq. (10.5) can be given in Eq. (10.6), using a dummy variable T defined such that $T = 0$ for $j = 1$ and $T = 1$ for $j = 2$.

$$g_i(T, Q) = \eta_i + \tau_i T + \beta_i Q + w_i T Q, \quad i = 1, \ldots, K. \tag{10.6}$$

Remark This is essentially an ANCOVA (Analysis of Covariance) model setup in statistical design of experiments, which is a generalized linear model blending ANOVA (Analysis of Variance) with regression. $g_i(T, Q)$ is the response for a given value of covariate Q and a selection of treatment T, τ_i refers to the difference in $\beta_{i,0}(W_h)$ due to the treatment effect, β_i is the regression coefficient for the covariate Q, and w_i is the contrast in $\beta_{i,1}(W_h)$ due to an interaction between the treatment factor T and the covariate Q.

Based on the assumption formulated in Sect. 10.3 and discussions earlier, the basis functions of $\Phi(Ca|Q)$ are obtained as $\mathbb{S} = \{1, Ca^{-\alpha}, Ca^{\gamma}\}$. According to Eq. (10.6), the unified scale-up model under the assumption is given in the form of Eq. (10.7), with $K = 3$ and $T = 0, 1$ denoting $W_h = 1, 2$ respectively.

$$\Phi(Ca, Q, T) = \sum_{i=1}^{K} \{\eta_i + \tau_i T + \beta_i Q + w_i T Q\} f_i(Ca) \tag{10.7}$$

10.4.4   Model Estimation

The nls() function in R language was applied for Nonlinear Least Squares (NLS) estimation [9, Appendix], using a form of Gauss–Newton iteration that employs numerically approximated derivatives. The full model in Eq. (10.7) is further reduced to the model in Eq. (10.8) by eliminating redundant terms to minimize the residual standard error, estimation given in Table 10.4.

$$\Phi(Ca, Q, T) = (\eta_2 + \tau_2 T)Ca^{-\alpha} + [\eta_3 + \beta_3 Q + w_3 T Q]Ca^{\gamma} \tag{10.8}$$

Table 10.4 Reduced model estimation

Parameter	Estimate	Std.error	P-value
α	0.39953	0.04079	3.12e$-$15
γ	0.24503	0.01758	< 2e$-$16
η_2	0.09094	0.02596	0.000764
τ_2	0.05991	0.01521	0.000177
η_3	0.83829	0.11187	8.94e$-$11
β_3	1.70621	0.11252	< 2e$-$16
w_3	0.84682	0.08789	6.36e$-$15

10.5 Physical Insights and Domain Boundaries

10.5.1 Physical Insights Through High-Dimensional Modeling

Compared to one-factor-at-a-time approach, our generic methodology allows investigation into the overall model structure in the high-dimensional space, providing additional insights into the droplet formation process in coated microfluidic devices.

1. *Flexibility for full-scale production*
   As observed from Fig. 10.4, the concise model in Eq. (10.8) captures well the influence of Ca, Q and W_h simultaneously, providing an opportunity to optimize multiple parameters simultaneously for full-scale production.
2. *Interpretation of multiple physical domains*
   Domains are identified by dominant physical mechanisms characterized by corresponding basis functions. As observed from Eq. (10.8), dripping mechanism characterized by $f_2(Ca)$ dominates the decrease in droplet size before $f_3(Ca)$

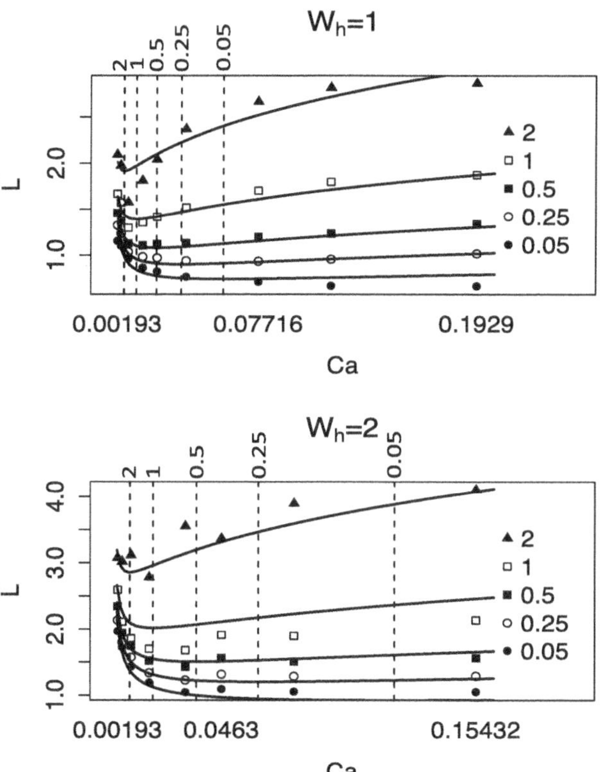

Fig. 10.4 Change in dimensionless droplet length $\bar{L}$ with respect to the capillary number Ca, flow rate ratio Q (from 0.05 to 2), and depth–width ratio W_h

starts to dominate the increase in droplet size. The transition between domains depends on the secondary factor.

3. *Effect of coating*
   Coating enables a wider range of producing stable droplets, allowing higher flow rate ratios and capillary numbers. Although physical domains can still be classified according to different dominant mechanisms characterized by corresponding basis functions of the capillary numbers, the range of each physical domain could be very different from that of droplet formation in uncoated microfluidic devices, e.g., squeezing mechanism characterized by $f_1(Ca)$ is not significant within our explored range, whereas in a coated T-junction, squeezing dominates droplet formation for $Ca < 0.002$ and is competitive with dripping mechanism for $0.002 < Ca < 0.01$.

4. *Effect of secondary factor*
   Transitions between physical domains are not independent of secondary factor (Q, W_h discussed in our case). This will be further explained in the discussion of identifying physical domains.

10.5.2   *Identification of Physical Domains and Boundaries*

The scale-up model in Eq. (10.8) not only provides some insights into the droplet formation process in a coated microfluidic T-junction but also demonstrates the possibilities of detecting physical domains dominated by different mechanisms without ambiguity in boundaries that are characterized by values of primary factor in literature. Physical domains in this scale-up droplet formation process are detected by identifying dominant mechanisms.

1. *Identification of relevant mechanisms*
   As observed in model reduction from Eqs. (10.7) to (10.8), $f_1(Ca)$ is eliminated from the final model since it is not significant across investigated domains. This implies that squeezing mechanism characterized by $f_1(Ca)$ in our coated devices is not significant within the explored range, i.e., coating leads to a decrease in the lower bound of dripping domain with respect to the primary factor Ca as a benefit of low surface energy.

   Remark As observed from Fig. 10.4, there exists a slight lack of fit under $W_h = 2$. This can be attributed to the unconspicuous postponement of dripping domain compared to the case when $W_h = 1$. In other words, when $W_h = 2$, droplet formation near the lower bound of explored Ca may actually fall in the squeezing-to-dripping domain, although the basis function $f_1(Ca)$ is still recognized as insignificant due to the narrow range of the squeezing-to-dripping domain even if there exists.

2. *Identification of dominant mechanisms*
   Dripping mechanism characterized by $f_2(Ca)$ dominates the decrease of droplet size at low capillary numbers within our explored range, while droplet size starts to increase and then tends to reach a plateau as Ca increases to a certain level. The domain in which the size of stable droplets increases is identified as "dripping-to-jetting" domain due to the coexistence of partial characteristics: (i) stable droplets are produced in this domain with no observations of jets; (ii) droplet size tends to reach a plateau near the explored upper bound of Ca, consistent with the scale-up model in jetting domain [13], which merely depends on Ca.
3. *Identification of boundaries*
   Noting that $f_2(Ca) = Ca^\alpha)$ and $f_3(Ca) = Ca^\gamma$ are both monotonic with respect to Ca, a transition point between dripping and dripping-to-jetting domain can be defined as the solution to $\partial\Phi(Ca, Q, T)/\partial Ca = 0$ given in Eq. (10.9).

$$Ca_1(Q, T) = \left[\frac{\alpha(\eta_2 + \tau_2 T)}{\gamma[\eta_3 + (\beta_3 + w_3 T)Q]}\right]^{\frac{1}{\alpha+\gamma}}. \tag{10.9}$$

When $Ca < Ca_1(Q, T)$, droplet size decreases, i.e., dripping mechanism depicted by $Ca^{-\alpha}$ explains a larger portion of change in droplet size; when $Ca > Ca_1(Q, T)$, droplet size starts to increase, i.e., dripping-to-jetting mechanism depicted by Ca^β contributes to a larger percentage of size change.
As shown in Table 10.5 and Fig. 10.4 (dashed lines), the value of $Ca_1(Q, T)$ is smaller at higher flow rate ratio Q, and the transition is postponed due to extra confinement with $W_h = 2$ ($T = 1$) compared to the case with $W_h = 1$ ($T = 0$).
4. *Effect of secondary factor*
   To further demonstrate the effect of secondary factor on boundaries between physical domains, we define $Ca_2(Q, T)$ as the solution to $\partial^2\Phi(Ca, Q, T)/\partial Ca^2 = 0$ given in Eq. (10.10) and Table 10.6, noting that $Ca_2(Q, T) > Ca_1(Q, T)$.

$$Ca_2(Q, T) = \left[\frac{\alpha(1 + \alpha)(\eta_2 + \tau_2 T)}{\gamma(1 - \gamma)[\eta_3 + (\beta_3 + w_3 T)Q]}\right]^{\frac{1}{\alpha+\gamma}}. \tag{10.10}$$

In dripping domain with $Ca < Ca_1(Q, T) < Ca_2(Q, T)$, $\partial^2\Phi(Ca, Q, T)/\partial Ca^2$ is positive with $Ca_2(Q, T)$ decreasing in Q, which explains the sharper decrease at lower Q when Ca is close to the lower bound of explored range. In dripping-to-jetting domain with $Ca > Ca_1(Q, T)$, the increase in droplet size

Table 10.5 $Ca_1(Q, T)$

	Q				
	0.05	0.25	0.5	1	2
$T = 0$	0.0055	0.012	0.023	0.036	0.059
$T = 1$	0.0071	0.017	0.036	0.062	0.12

Table 10.6 $Ca_2(Q, T)$

	Q				
	0.05	0.25	0.5	1	2
$T = 0$	0.014	0.032	0.060	0.094	0.15
$T = 1$	0.019	0.044	0.093	0.16	0.31

with respect to Ca first accelerates in the range of $Ca < Ca_2(Q, T)$ and then decelerates to reach a plateau at higher $Ca > Ca_2(Q, T)$.

This chapter introduces a generic multiple-domain scale-up problem and a scalable modeling approach to predict manufacturing processes that can potentially operate under multiple physical domains due to uncertainties. The approach addresses two critical challenges in scale-up modeling for multiple-domain manufacturing processes: high dimensionality and multiple physical domains. The challenge of high dimensionality is addressed by identifying low-dimensional model structures through dimensional analysis and adoption of primary factor. The ambiguity in boundaries between physical domains is addressed by interpreting multiple-domain manufacturing processes as an outcome of coexisting mechanisms. Physical domains are identified by identifying dominant mechanisms.

The formulation and approach have been applied and demonstrated to investigate the scale-up droplet formation process in the coated microfluidic T-junction. The unified scale-up model across multiple domains not only captures well the joint effect of capillary number, flow rate ratio, and depth width ratio but also leads to some physical insights into the process: (i) Droplet formation in the coated device can be explained by coexisting mechanisms with weights varying across domains. (ii) Variables besides the capillary number influence the transition from dripping to dripping-to-jetting domain, which is postponed at lower flow rate ratio and higher depth–width ratio. (iii) The low-surface-energy fluoropolymer coating has proved to highly extend the range of domain for producing stable droplets either in dripping or dripping-to-jetting domain.

Acknowledgments The work in this chapter was partially supported by US National Science Foundation under CAREER grant number CMMI-1055394.

References

1. 2005 NCMS Survey of Nanotechnology in the U.S. Manufacturing Industry: National Center for Manufacturing Science, Ann Arbor, MI (2006)
2. Anna, S.L., Bontoux, N., Stone, H.A.: Formation of dispersions using "flow focusing" in microchannels. Appl. Phys. Lett. **82**(3), 364–366 (2003)
3. Barenblatt, G.I.: Scaling. Cambridge Texts in Applied Mathematics (2003)
4. Buja, A., Hastie, T., Tibshirani, R.: Linear smoothers and additive models. Ann. Stat. **17**(2), 453–510 (1989). www.summon.com
5. Christopher, G.F., Anna, S.L.: Microfluidic methods for generating continuous droplet streams. J. Phys. D Appl. Phys. **40**(19), R319 (2007)

6. Christopher, G.F., Noharuddin, N.N., Taylor, J.A., Anna, S.L.: Experimental observations of the squeezing-to-dripping transition in t-shaped microfluidic junctions. Physical Rev. E Stat. Nonlinear Soft Matter Phys. **78**(3 Pt 2), 036317 (2008). www.summon.com

7. De Menech, M., Garstecki, P., Jousse, F., Stone, H.A.: Transition from squeezing to dripping in a microfluidic t-shaped junction. J. Fluid Mech. **595**, 141–161 (2008). www.summon.com

8. deMello, A.J.: Control and detection of chemical reactions in microfluidic systems. Nature **442**(7101), 394–402 (2006). www.summon.com

9. Fox, J., Weisberg, S.: An R Companion to Applied Regression. Sage (2010)

10. Friedman, J.H., Stuetzle, W.: Projection pursuit regression. J. Am. Stat. Assoc. **76**(376), 817–823 (1981). www.summon.com

11. Fu, T., Ma, Y., Funfschilling, D., Li, H.Z.: Bubble formation and breakup mechanism in a microfluidic flow-focusing device. Chem. Eng. Sci. **64**(10), 2392–2400 (2009). www.summon.com

12. Fu, T., Ma, Y., Funfschilling, D., Zhu, C., Li, H.Z.: Squeezing-to-dripping transition for bubble formation in a microfluidic t-junction. Chem. Eng. Sci. **65**(12), 3739–3748 (2010). www.summon.com

13. Fu, T., Wu, Y., Ma, Y., Li, H.Z.: Droplet formation and breakup dynamics in microfluidic flow-focusing devices: from dripping to jetting. Chem. Eng. Sci. **84**, 207–217 (2012)

14. Garstecki, P., Fuerstman, M.J., Stone, H.A., Whitesides, G.M.: Formation of droplets and bubbles in a microfluidic t-junction-scaling and mechanism of break-up. Lab Chip **6**(3), 437–446 (2006). www.summon.com

15. Glawdel, T., Ren, C.L.: Droplet formation in microfluidic t-junction generators operating in the transitional regime. iii. dynamic surfactant effects. Phys. Rev. E Stat. Nonlinear Soft Matter Phys. **86**(2 Pt 2), 026308 (2012). www.summon.com

16. Gupta, A., Kumar, R.: Effect of geometry on droplet formation in the squeezing regime in a microfluidic t-junction. Microfluid. Nanofluid. **8**(6), 799–812 (2010). www.summon.com

17. Gupta, A., Kumar, R.: Effect of geometry on droplet formation in the squeezing regime in a microfluidic t-junction. Microfluid. Nanofluid. **8**(6), 799–812 (2010)

18. Hastie, T., Tibshirani, R.: Generalized Additive Models, vol. 43. Chapman and Hall, New York (1990). www.summon.com

19. Huang, Q.: Integrated nanomanufacturing and nanoinformatics for quality improvement (invited). In: 44th CIRP International Conference on Manufacturing Systems (2011)

20. Huang, Q., Wang, L., Dasgupta, T., Zhu, L., Sekhar, P.K., Bhansali, S., An, Y.: Statistical weight kinetics modeling and estimation for silica nanowire growth catalyzed by pd thin film. IEEE Trans. Autom. Sci. Eng. **8**(2), 303–310 (2011)

21. Implications of Emerging Micro- and Nanotechnologies: The National Academies Press, Washington DC (2002)

22. Kuo, W.: Challenges related to reliability in nano electronics. IEEE Trans. Reliab. **55**(1), 169–169 (2006)

23. Lazarus, L.L., Riche, C.T., Marin, B.C., Gupta, M., Malmstadt, N., Brutchey, R.L.: Two-phase microfluidic droplet flows of ionic liquids for the synthesis of gold and silver nanoparticles. ACS Appl. Mater. Interfaces **4**(6), 3077–3083 (2012). www.summon.com

24. Link, D., Anna, S.L., Weitz, D., Stone, H.: Geometrically mediated breakup of drops in microfluidic devices. Phys. Rev. Lett. **92**(5), 054503 (2004)

25. Riche, C.T., Zhang, C., Gupta, M., Malmstadt, N.: Fluoropolymer surface coatings to control droplets in microfluidic devices. Lab Chip **14**(11), 1834–1841 (2014). www.summon.com

26. Rogosa, D.: Comparing nonparallel regression lines. Psychol. Bull. **88**(2), 307 (1980)

27. Schlichting, H., Gersten, K.: Boundary-Layer Theory. Springer, New York (2000)

28. Societal Implications of Nanoscience and Nanotechnology: National Science Foundation, Arlington, VA (2001)

29. Tan, J., Li, S.W., Wang, K., Luo, G.S.: Gas–liquid flow in t-junction microfluidic devices with a new perpendicular rupturing flow route. Chem. Eng. J. **146**(3), 428–433 (2009). www.summon.com

30. Teh, S.Y., Lin, R., Hung, L.H., Lee, A.P.: Droplet microfluidics. Lab Chip **8**, 198–220 (2008). http://doi.org/10.1039/B715524G
31. Thorsen, T., Roberts, R.W., Arnold, F.H., Quake, S.R.: Dynamic pattern formation in a vesicle-generating microfluidic device. Phys. Rev. Lett. **86**(18), 4163–4166 (2001). www.summon.com
32. Van der Graaf, S., Nisisako, T., Schroen, C., Van Der Sman, R., Boom, R.: Lattice boltzmann simulations of droplet formation in a t-shaped microchannel. Langmuir **22**(9), 4144–4152 (2006)
33. van der Zwan, E., van der Sman, R., Schroën, K., Boom, R.: Lattice boltzmann simulations of droplet formation during microchannel emulsification. J. Colloid Interface Sci. **335**(1), 112–122 (2009). www.summon.com
34. Wang, K., Lu, Y.C., Xu, J.H., Luo, G.S.: Determination of dynamic interfacial tension and its effect on droplet formation in the t-shaped microdispersion process. Langmuir **25**(4), 2153–2158 (2009). www.summon.com
35. Wang, L., Huang, Q.: Cross-domain model building and validation (cdmv): A new modeling strategy to reinforce understanding of nanomanufacturing processes. IEEE Trans. Autom. Sci. Eng. **10**(3), 571–578 (2013)
36. Wang, L., Huang, Q.: Cross-domain model building and validation (cdmv): A new modeling strategy to reinforce understanding of nanomanufacturing processes. IEEE Trans. Autom. Sci. Eng. **10**(3), 571–578 (2013)
37. Wang, S.S., Jiao, Z.J., Huang, X.Y., Yang, C., Nguyen, N.T.: Acoustically induced bubbles in a microfluidic channel for mixing enhancement. Microfluid. Nanofluid. **6**(6), 847–852 (2009). www.summon.com
38. Xu, J.H., Li, S.W., Tan, J., Luo, G.S.: Correlations of droplet formation in t-junction microfluidic devices: from squeezing to dripping. Microfluid. Nanofluid. **5**(6), 711–717 (2008). www.summon.com
39. Xu, J.H., Li, S.W., Tan, J., Luo, G.S.: Correlations of droplet formation in t-junction microfluidic devices: from squeezing to dripping. Microfluid. Nanofluid. **5**(6), 711–717 (2008). www.summon.com
40. Xu, L., Wang, L., Huang, Q.: Growth process modeling of semiconductor nanowires for scale-up of nanomanufacturing: A review. IIE Trans. **47**(3), 274–284 (2015)
41. Xu, Q., Hashimoto, M., Dang, T.T., Hoare, T., Kohane, D.S., Whitesides, G.M., Langer, R., Anderson, D.G.: Preparation of monodisperse biodegradable polymer microparticles using a microfluidic flow-focusing device for controlled drug delivery. Small (Weinheim an der Bergstrasse, Germany) **5**(13), 1575–1581 (2009). www.summon.com
42. Zhang, Y., Wang, L.: Experimental investigation of bubble formation in a microfluidic t-shaped junction. Nanoscale Microscale Thermophys. Eng. **13**(4), 228–242 (2009). www.summon.com
43. Zhao, C.X., Middelberg, A.P.J.: Two-phase microfluidic flows. Chem. Eng. Sci. **66**(7), 1394–1411 (2011). www.summon.com
44. Zlokarnik, M.: Scale-Up in Chemical Engineering. Wiley, New York (2006)

Chapter 11
Domain-Informed Bayesian Hierarchical Modeling of Nanowire Growth at Multiple Scales

This chapter introduces domain-informed modeling of nanowire (NW) growth process at multiple scales of interest for prediction. The main idea is to integrate available data and physical knowledge through a Bayesian hierarchical framework with consideration of scale effects. At each scale, the NW growth model describes the time-space evolution of NWs across different sites on a substrate. The model consists of two major components: NW morphology and local variability. The morphology component represents the overall trend characterized by growth kinetics. The area-specific variability is less understood in nanophysics due to complex interactions among neighboring NWs. The local variability is therefore modeled by an intrinsic Gaussian Markov random field (IGMRF) so as to separate itself from the growth kinetics in the morphology component. Case studies are provided to illustrate the NW growth process model at the coarse and fine scales, respectively.

11.1 Prediction of Nanowire Growth

Understanding the first principles of nanostructure synthesis is certainly critical to improve process yield. Yet the physical laws are not completely understood at nanoscale. Current growth kinetics models are generally deterministic [7, 12, 27, 33], involving a large number of constants to be estimated to a high level of accuracy. These models provide understanding of growth behavior at coarse scale, lack of description of local variability across different sites on a substrate. Due to processing uncertainties, existing models are unable to predict realistic nanomanufacturing processes [20].

In traditional manufacturing, statistical quality control and engineering-driven statistical analysis of manufacturing process have achieved great success in yield and productivity improvement [21, 29, 34]. Yet the scale effects bring new challenges upon quality control in nanomanufacturing. First, manufacturing of quality-

Q. Huang, *Domain-informed Machine Learning for Smart Manufacturing*,
https://doi.org/10.1007/978-3-031-91631-1_11

engineered nanostructures demands prediction, monitoring, and control of process variations at multiple scales, e.g., the overall NW growth rate vs. the growth rates at different sites of a substrate. Second, nanostructure growth kinetics discovered in nanophysics provides valuable insights for process improvement. However, deterministic kinetics models fail to address process uncertainties. Third, there is a lack of *in situ* observation of most properties at the nanoscale during processing. The offline SEM (scanning electron microscope) or TEM (transmission electron microscopy) inspection is time-consuming and costly. These challenges call for a systematic methodology to model and control nanomanufacturing processes.

Efforts have been undertaken to develop robust design methods to synthesize desired nanostructures [6] and to study the reliability of nano electronics [1, 18, 19] (see review in [9, 16]).

11.2   Nanowire Growth Process and Modeling Strategy

Nanomaterials such as nanotubes or NWs are expected to have wide applications in nanoelectronics and optoelectronic devices [17]. Yet the process of synthesizing nanomaterials is complex. As shown in Fig. 11.1, metal catalyst such as gold is first deposited on {111} surface of a Si substrate. The substrate is then heated above eutectic temperature to form liquid Au-Si alloy. Through a chemical vapor deposition (CVD) process, the liquid droplet becomes the preferred site for absorbing Si atoms from vapor, causing the liquid to be supersaturated with Si. The deposit particles travel to the liquid–solid interface and vertically grow layer-by-layer mediated by nucleation. The driving force for this the so-called vapor–liquid–solid (VLS) growth mechanism is the supersaturation of semiconductor atoms (Si) in the liquid alloy [33].

In addition, nanostructure growth normally involves multiple correlated processing steps, e.g., catalyst deposition and VLS growth. Processing uncertainties during deposition could cause variations in the shape and diameter of metal catalyst

Fig. 11.1 Vapor–liquid–solid mechanism of NW growth [33]

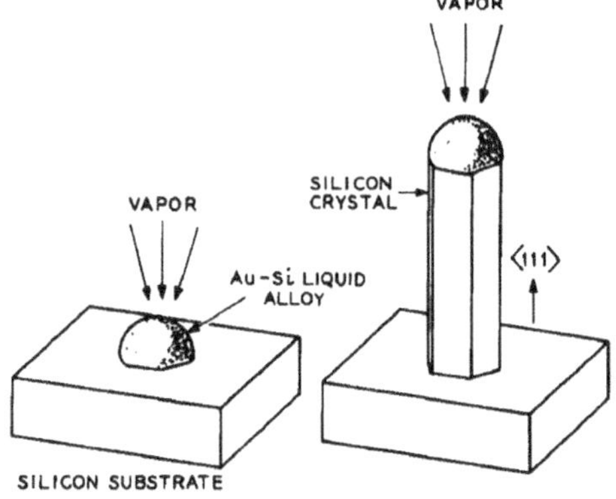

Fig. 11.2 Growth variability within a silicon substrate

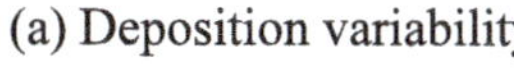

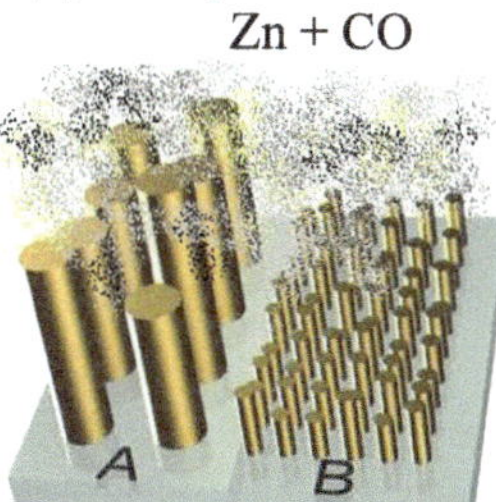

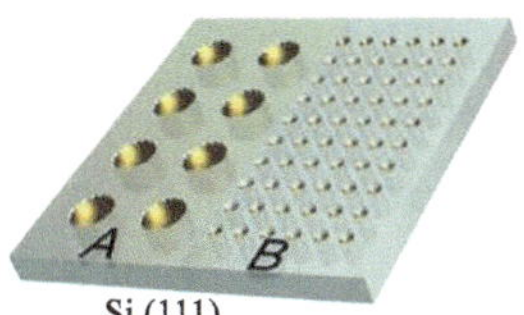

from sites to sites, as illustrated by regions A and B in Fig. 11.2a. During VLS growth process, the NWs in region A might absorb more species from the vapor and therefore inhibit the growth in region B. This phenomenon could aggregate the growth variability on a substrate (see Fig. 11.2b).

Effective control of such a complex NW growth process demands a process model integrated with nanophysics. The challenge is the knowledge disparity at different scales, i.e., limited understanding of growth kinetics at fine scales (greater uncertainties) vs. relatively established work at coarse scales (less uncertainties). The following three principles are essential when developing our modeling framework:

- *Openness*: In light of the rapid advance in nanoscience, the modeling framework should be open enough to integrate the forthcoming knowledge at fine scales without dramatic changes in model structures.
- *Separation*: The model components with different degrees of uncertainties should be separable so that during model estimation, the model structures with less confidence in physics will have no or little impact on those with better physical knowledge.
- *Structured simplicity*: Due to limited measurement data, it is preferred to have a simple process model (containing less unknown parameters) with a physical structure embedded with growth knowledge.

Specifically, the existing kinetics models depict the NW growth at coarse scale [7, 12, 27], and their model structure will be adopted to describe NW morphology. Based on the separation principle, the local variability component, where no physical models are available to describe growth variability across sites, is constructed to be invariant to morphology. This morphology-local decomposition is placed under a Bayesian hierarchical model. As shown in Fig. 11.3, at a fine scale, the quality features $X(s, t)$ of NWs on a substrate consist of morphology $\eta_{k-1}(s, t)$, local variability $\phi(s)$, and noise ε, where t and s denote time and the collection of sites on a substrate, respectively. Model parameters ξ of $X(s, t)$ are determined by process variables from growth kinetics. The classical Bayesian hierarchical modeling [25], illustrated at the right panel of Fig. 11.3, has long been applied to integrate available knowledge for model building (see engineering applications in [3, 11, 23]).

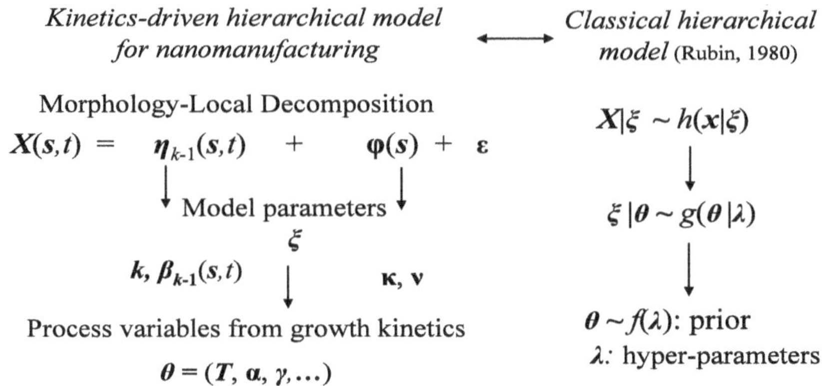

Fig. 11.3 Physics-driven hierarchical modeling of growth process at micro/nano scale

This strategy provides a nanomanufacturing process modeling methodology, which addresses the issue of knowledge disparity at different scales through morphology-local decomposition under a Bayesian hierarchical modeling framework.

11.3   Hierarchical Modeling of Nanowire Growth Process

The nanostructure growth is a dynamic process, which is to be characterized by a space–time random field model. The random field model has the hierarchical structure defined in Fig. 11.3.

11.3.1   Space–Time Random Field Representation of Nanostructures

Let X represent the quality features of NWs on a substrate. At a certain scale and time t, the observed feature at site s_i of the substrate is denoted as $X(s_i, t)$, $s_i \in \mathbb{R}^p$, $p \in \{1, 2, 3\}$, $i = 1, 2, \ldots, n$. For simplicity of presentation, we hereafter only show the case with $p = 1$. As an example, $X(s_1, t)$ could be 150 nanometer (nm) representing the average length of all NWs at site A of a substrate shown in Fig. 11.2b. All the features at sites $s = (s_1, s_2, \ldots, s_n)^{\mathrm{T}}$ are represented as $X(s, t)$:

$$X(s, t) = [X(s_1, t),\ X(s_2, t),\ \ldots,\ X(s_i, t),\ \ldots,\ X(s_n, t)]^{\mathrm{T}} \tag{11.1}$$

which is a space–time random field on lattice s. A real example is shown in Fig. 11.4 [4] where NW lengths vary from site to site, and random field is a better characterization of nanostructures. Similar idea has been developed in stochastic modeling of microstructure [30]. But it should be noted that those works mainly

Fig. 11.4 Vertically aligned
NWs with a mean length of
400 nm [4]

Fig. 11.5 Morphology-local
decomposition of NW length
field

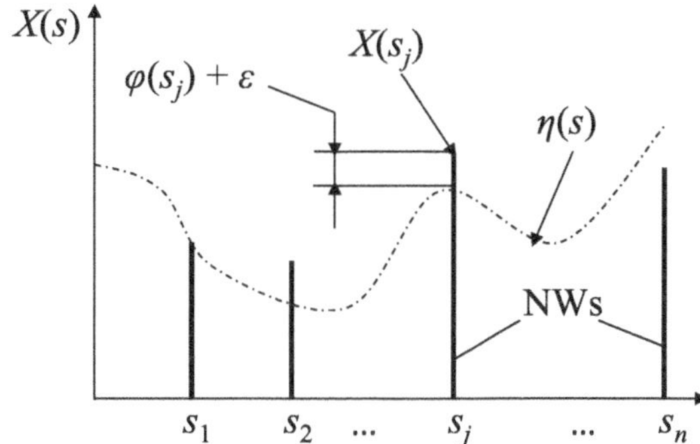

focus on (offline) mathematical characterization of material structures. We go
beyond that by linking the mathematical model with process physics to set the
foundation for *in situ* process control.

11.3.2 Morphology–Local Decomposition of Nanostructure Random Field

We propose to decompose the nanostructure random field $X(s, t)$ into morphology
or profile $\eta_{k-1}(s, t)$, local variation $\phi(s)$, and noise ε,

$$X(s, t) = \eta_{k-1}(s, t) + \phi(s) + \varepsilon \tag{11.2}$$

where $\eta_{k-1}(s, t)$ could be a plane or a surface with relative complex profile
evolving over the growth process. The subscript $k - 1$ represents the order of a
polynomial, which will be introduced later in Eq. (11.3). $\phi(s)$ represents local
fluctuation riding on the morphology. We suggest an intrinsic Gaussian Markov
random field (IGMRF) model for $\phi(s)$. Figure 11.5 illustrates the idea in a 2D graph
for simplicity.

The rationale of the morphology-local decomposition is as follows. Currently,
there is better understanding of global behaviors of the growth kinetics. But limited

physical knowledge is available for area-specific variability because of latent and unobserved factors. The decomposition therefore aims at engaging growth kinetics through $\eta_{k-1}(s, t)$ and local variation modeling through $\phi(s)$.

Due to the lack of physical studies, we only consider the case that local variability at site s_i is mainly determined by its interaction with neighbors and remains stable over time. At the initial stage of modeling work, this time-invariant assumption could simply the model structure with the hope that the dynamics will be captured by morphology $\eta_{k-1}(s, t)$. The model in Eq. (11.2), however, could be extended to consider $\phi(s, t)$.

Remark An alternative method of modeling the nanostructure is to use the kriging model in spatial statistics [5, 24]. $X(s_i, t)$ would be expressed as $X(s_i, t) = \mu(s_i, t) + Z(s_i)$, where $\mu(s_i, t)$ is a mean function and $Z(s_i)$ is weak stationary Gaussian field (see a kriging example in [10]). The kriging model also captures the trend and local variability, and actually, there is a connection between Gaussian random field and Gaussian Markov random field [26]. We prefer a Markovian property for local variability component and a noise term ε in Eq. (11.2) for three reasons:

- First, we intend to separate the modeling error and noise (see a general discussion in [11]). The modeling error in Eq. (11.2) dominates in $\phi(s)$ due to the lack of understanding at fine scales. However, molecular dynamics simulation [22] often uses, e.g., the Lennard–Jones potential [13] to describe the interactions among NWs or nanotubes. The separation gives the opportunity to model the structure of $\phi(s)$ at nanoscale.
- Second, GMRF has shown the flexibility to specifying the neighboring structure on a lattice and efficiency in computation [26]. Material properties such as anisotropy are easier to be modeled. It is a viable strategy to model the nanostructure growth process with effective integration of nanphysics.
- Third, in the hierarchical model estimation through Markov chain Monte Carlo Simulation (MCMC), GMRF such as conditional autoregressive (CAR) model has shown better computational efficiency [2]. Our simulation studies also support the claim.

11.3.2.1   Modeling of Nanowire Morphology

The morphology is devised to have order $k - 1$ polynomial representation on s,

$$
\eta_{k-1}(s, t) = \begin{pmatrix} \eta(s_1, t) \\ \eta(s_2, t) \\ \vdots \\ \eta(s_n, t) \end{pmatrix} = \begin{pmatrix} 1 & s_1 & \cdots & \frac{1}{(k-1)!} s_1^{k-1} \\ 1 & s_2 & \cdots & \frac{1}{(k-1)!} s_2^{k-1} \\ \vdots & \vdots & \cdots & \vdots \\ 1 & s_n & \cdots & \frac{1}{(k-1)!} s_n^{k-1} \end{pmatrix} \begin{pmatrix} \beta_0(t) \\ \beta_1 \\ \vdots \\ \beta_{k-1} \end{pmatrix} = S_{k-1} \beta_{k-1}
$$

$$\tag{11.3}$$

where $\boldsymbol{\beta}_{k-1} = [\beta_0(t), \beta_1, \cdots, \beta_{k-1}]^{\mathrm{T}}$ are coefficients with specific model structures. $\boldsymbol{\beta}_{k-1}$ is in the middle layer of the hierarchical framework given in Fig. 11.3. The detailed structure of $\boldsymbol{\beta}_{k-1}$ is derived as follows to connect with process variables $\boldsymbol{\theta}$.

The literature on growth kinetics did not report an integrated model relating the nanowire length with time and temperature [7, 12, 27]. We propose the intercept $\beta_0(t)$ to be:

$$\beta_0(t) = \alpha_1 \exp\left[-\frac{E_a}{k_B T_0} - \frac{\alpha_2}{t}\right], \quad \text{if } \beta_0(t) < L_f, \text{ or } t < t_0, \tag{11.4a}$$

$$\beta_0(t) = \alpha_3 \exp\left[-\frac{E_a}{k_B T_0}\right] t + b, \quad \text{if } \beta_0(t) \geq L_f, \text{ or } t \geq t_0. \tag{11.4b}$$

where t_0 is unknown transition point; E_a, k_B, T_0, and L_f are activation energy (kJ/mol), Boltzmann constant (J/K), average temperature (K), and diffusion length, respectively. Definition of these quantities can be found in [7, 12, 27]. Continuity constraint is imposed at transition point t_0, i.e., $\alpha_1 \exp\left[-\dfrac{E_a}{k_B T_0} - \dfrac{\alpha_2}{t_0}\right] = \alpha_3 \exp\left[-\dfrac{E_a}{k_B T_0}\right] t_0 + b.$

The unknown coefficients α_1, α_2, and α_3 are adopted to combine a set of physical constants such as degree of supersaturation, atomic volume, diffusion coefficient, and adatom concentration. Combining multiple unknowns into one would require much less data in model estimation. The drawback is that the model has to be estimated again if any of those physical constants are subject to changes. Since our ultimate goal is to control and monitor growth conditions, data under the same process condition will be collected anyway, and the drawback is therefore not a concern.

The structures of the rest coefficients in $\boldsymbol{\beta}_{k-1}$ or $[\beta_1, \cdots, \beta_{k-1}]^{\mathrm{T}}$ are determined by properties of the NW random field. Little work has reported the kinetics of NW morphology except for the overall trend or the intercept term $\beta_0(t)$. By the definition of $\boldsymbol{\eta}_{k-1}(s, t)$ in Eq. (11.3), we could easily show that

$$\eta(s_i) - \eta(s_j)$$

$$= \left[s_i - s_j, \frac{1}{2}(s_i^2 - s_j^2), \cdots, \frac{1}{(k-1)!}(s_i^{k-1} - s_j^{k-1})\right] (\beta_1, \ldots, \beta_{k-1})^{\mathrm{T}} \tag{11.5}$$

Equation (11.4) suggests that if the temperature has a space–time profile, i.e., being $T(s, t)$ instead of a constant T_0, the growth rate variability may lead to a complex morphology. By defining $\boldsymbol{q} \overset{\text{def}}{=} \left[s_i - s_j, \frac{1}{2}(s_i^2 - s_j^2), \cdots, \frac{1}{(k-1)!}(s_i^{k-1} - \right.$

$$\left. s_j^{k-1}) \right]^{\mathrm{T}}, \text{ the following model is postulated for the structure of } [\beta_1, \ldots, \beta_{k-1}]^{\mathrm{T}}:$$

$$\eta(s_i) - \eta(s_j) = \boldsymbol{q}^{\mathrm{T}}[\beta_1, \ldots, \beta_{k-1}]^{\mathrm{T}} = \alpha_4 \exp\left[-\frac{E_a}{k_B T(\boldsymbol{q}, t)} \right] \tag{11.6}$$

where $T(\boldsymbol{q}, t)$ represents the temperature gradient between two sites at time t. If the gradient does not vary with time, we could drop the time index in the model. Since temperature is probably one of the key growth variables, here we choose it to illustrate the modeling strategy. If other key process variables should be considered in Eq. (11.6), those variables may modify the constant α_4 by imposing a certain structure. To further consider the uncertainty of estimating $\alpha_1 \sim \alpha_4$ with limited data, we could introduce prior distributions for $\alpha_1 \sim \alpha_4$ and update them with new observation.

11.3.2.2   Modeling of Nanowire Area-Specific Variability

As defined in Eq. (11.2), the area-specific variability $\boldsymbol{\phi}(s)$ represents local fluctuation riding on the morphology. In addition, $\boldsymbol{\phi}(s)$ is expected to capture the dependence/interaction among neighboring sites. Based on the remark made previously, we adopt an IGMRF model with rank deficiency k for local variability $\boldsymbol{\phi}(s) = [\phi(s_1), \phi(s_2), \ldots, \phi(s_i), \ldots, \phi(s_{k-1})]^{\mathrm{T}}$. It has density function [26]

$$f[\boldsymbol{\phi}(s)] = (2\pi)^{-\frac{n-k}{2}} |\mathbf{Q}|^{*\frac{1}{2}} \exp\left[\frac{1}{2}[\boldsymbol{\phi}(s) - \boldsymbol{\mu}^*]^{\mathrm{T}} \mathbf{Q}[\boldsymbol{\phi}(s) - \boldsymbol{\mu}^*] \right] \tag{11.7}$$

where $\mathbf{Q}$ is precision matrix with rank $n - k$ and $| \circ |^*$ denote the generalized determinant. Note that the rank deficiency k of precision matrix $\mathbf{Q}$ corresponds to an order $k - 1$ polynomial in Eq. (11.3). For an IGMRF of first order ($k = 1$), the conditional mean of $\phi(s_i)$ is simply a weighted average of its neighbors, not involving an overall mean, i.e., $E[\phi(s_i)|\boldsymbol{\phi}(-s_i)] = -\dfrac{1}{Q_{ii}} \sum_{j:j\sim i} \phi(s_j)$ [26]. Notation $j : j \sim i$ denotes sites s_i's that are neighbors of s_i.

The unique property of IGMRF of order k is that it is invariant to an order $k - 1$ polynomial (see detail in [26]). That is another major reason of adopting IGMRF in this study. This property facilitates the separation of morphology from local variability, which gives the freedom to integrate physical laws into morphology modeling without compromising the flexibility of statistical modeling of area-specific variability and dependence.

Hence, the local variability $\boldsymbol{\phi}(s)$ specified by Eq. (11.7) will mainly be determined by structure of precision matrix $\mathbf{Q}$. One key factor is the order of IGMRF or the rank deficiency of $\mathbf{Q}$. A low order $k(\leq 5)$ is preferred in our model because of two reasons. First, large k would result in a high-order polynomial η_{k-1} with more fluctuation in morphology. That may cause the difficulty of distinguishing morphology and local variability. Second, it is hard to properly connect the higher-order morphology with process variables because it demands a better understanding of growth kinetics at finer scales. With the similar level of goodness of fit, we tend to choose the model with lower order. Hence the morphology-local decomposition is a compromise between the limited process knowledge at fine scales and the desire to model and control local variability.

It should be mentioned that the morphology-local decomposition model could take slightly different form from Eq. (11.2), for instance,

$$X(s, t) = \mu(s, t) + \varepsilon \tag{11.8a}$$

$$\log[\mu(s, t)] = \log[\eta_{k-1}(s, t)] + \phi(s) \tag{11.8b}$$

The logarithm transformation of mean $\mu(s, t)$, not $X(s, t)$, is based on the observation that the growth kinetics models usually take exponential form (Eqs. 11.4 and 11.6). Terms $\eta_{k-1}(s, t)$ and $\phi(s)$ follow the same model structures defined in Eqs. (11.4)–(11.7). The transformation enables the morphology-local mode to accommodate a wide range of growth conditions and provides more flexibility in model fitting.

11.3.2.3 Nanostructure Growth Model Estimation

The nanostructure growth model, structured by Eqs. (11.2)–(11.7), integrates the growth kinetics with statistical modeling of uncertainties through the Bayesian hierarchical framework shown in Fig. 11.3. The model inputs for estimation and prediction are from two sources, i.e., (limited) process/product data and prior physical knowledge for hyper-parameters λ (Fig. 11.3). The process/product data to be collected include, for instance, the growth temperature and its temporal–spatial distribution in the CVD process, carrying gas flow rate, nanowire length, and distribution on a substrate.

At fine scales, a number of issues make the model estimation process challenging. The first is a "missing data" problem. Since the nanoscale *in situ* sensing technology is still under development [8], nanoscale measurement of process/product during processing is almost unavailable. Offline SEM or TEM inspection is extremely time-consuming and may only provide information on a few tiny areas on a substrate at the time of realtime decision. Therefore, *in situ* process/product information and offline product data at nanoscale are either missing or incomplete. Second, the number of unknown parameters is still relatively large even though we consolidated the unknown physical constants (Eq. 11.4). Third, there is normally no analytical solution for the hierarchical model set up by Eqs. (11.2)–(11.7). While Maximum likelihood estimation (MLE) is generally hard to be applied, Markov chain Monte Carlo simulation (MCMC) [15, 32] is frequently used to address the aforementioned issues in model estimation.

Taking the NW growth as an example, let NW length field $X = (X_{obs}, X_{miss})$, where X_{obs} is the observed NW lengths at some sites and X_{miss} at the other sites is unobservable due to resource/time constraints. If two or more replicates are available, sites with missing measurement could vary from substrate to substrate. Given the *in situ* measurement of temperature T and prior distributions $f_0(\theta)$ for process variables θ such as $\alpha_1 \sim \alpha_3$ in Eq. (11.4), conditional distribution of θ under available observation can be obtained. By drawing samples via MCMC method, we could estimate θ and ξ in the hierarchical model (Fig. 11.3). More details about MCMC can be found in [15]. In the case studies, we will implement Gibbs sampling using WinBUGS software [31] for model estimation.

11.4   Case Studies

The case studies will demonstrate the procedure of hierarchical modeling and estimation of the NW growth process under uncertainties. Figure 11.6 shows how the NW length varies with temperature and time [12]. The data therein were collected over time ($t = 15s$, $30s$, $180s$, $900s$) under six growth conditions ($T = 365°C$, $380°C$, $400°C$, $420°C$, $430°C$, $440°C$). The first four conditions are used for model building. We did not consider the two high-temperature conditions because there were no observation at time 180s and 900s. The growth in [12] may not last that long for higher-temperature settings.

In the first example, we will focus on modeling and Bayesian estimation of the NW growth process at the coarsest scale, i.e., the overall growth across entire substrate. Comparing to the existing NW kinetics models, we not only consider the uncertainties but also estimate the transition times under different process conditions with missing values. In the second example, we will simulate local variation based on the data in [12]. Since we have not identified the work that studies the growth variations over substrate due to temperature distribution, the second example only considers the first-order IGMRF model for describing the local variations. It should

Fig. 11.6 Growth rate vs. time, temperature [12]

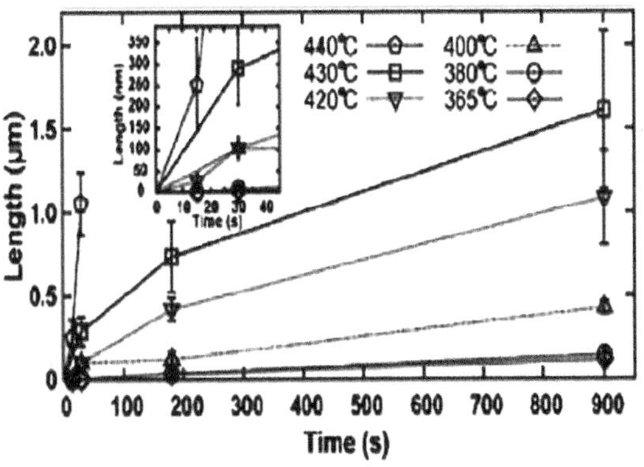

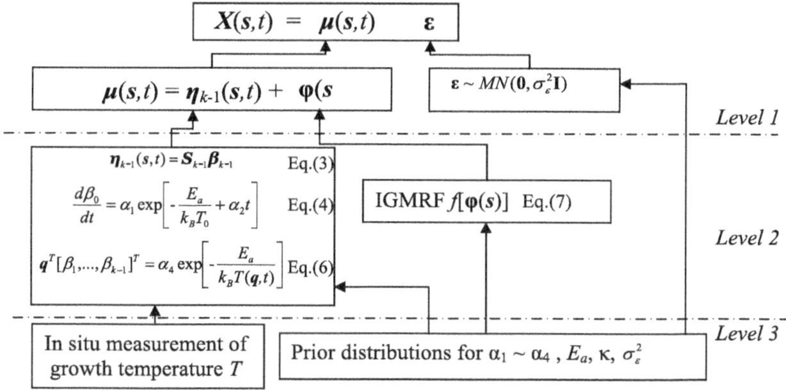

Fig. 11.7 The hierarchical modeling procedure

be noted from Fig. 11.6 that the amount of data is quite limited comparing to the numbers of model parameters to be estimated. Results suggest that the MCMC method works very well in this "missing value" scenario.

The three-level hierarchical modeling procedure is summarized in Fig. 11.7, which will be detailed in the two cases.

11.4.1 Dynamic Nanowire Length Model at a Coarse Scale

At the coarsest scale, we treat a whole substrate as one site. NW length $X(s, t) = X(t)$ with $s = 0$. The hierarchical model for this growth process can be set up as follows:

$Level\ 1:$
$$X(t) \sim N(\mu(t), \sigma_\varepsilon^2)$$

$Level\ 2:$
$$\mu(t) = \beta_0(t)$$

$$= I_{t<t_0}\left[\alpha_1 \exp\left(-\frac{E_a}{k_B T} - \frac{\alpha_2}{t}\right)\right] + I_{t \geq t_0}\left[\alpha_3 \exp\left(-\frac{E_a}{k_B T}\right)t + b\right]$$

$Level\ 3:$
$$Non\text{-}informative\ priors:$$

$$\sigma_\varepsilon \sim uniform(0, 20), \quad t_0(T) \sim uniform(1, 40)$$

$$\alpha_2(T) \sim uniform(1, 20), \quad \alpha_1, \alpha_3 \sim uniform(0, 10^{30})$$

where $b = \alpha_1 \exp\left(-\frac{E_a}{k_B T_0} - \frac{\alpha_2}{t_0}\right) - \alpha_3 \exp\left(-\frac{E_a}{k_B T_0}\right) t_0$, and $I_A(t) = 1,\ if\ t \in A$; and 0, otherwise.

Since model parameters such as diffusion length L_f, transition time t_0, and α_3 defined in Eq. (11.4) vary with temperature T, four prior distributions will be assigned to each parameter under four temperature conditions. For simplicity of presentation, we denote the parameter as a function of temperature when assigning priors. The activation energy Ea is 230kJ/mol [12]. If needed, this parameter can easily be introduced as a prior distribution under the current framework.

The computational issue for this model setup is that the MCMC procedure would often crash because of sampling from such as wide ranges, particularly the priors of α_1 and α_3. Using a piece-wise nonlinear regression [28], we could have a rough estimate for α_1 and α_3, which are at the order of 10^{19} and 10^{17}, respectively. To avoid the problem of sampling large numbers, we re-parameterize $\mu(t)$ and yield the following model:

$$Level\ 1: \qquad\qquad X(t) \sim N(\mu(t), \sigma_\varepsilon^2)$$

$$Level\ 2: \mu(t) = I_{t<t_0}\left[a_1 \exp\left(-\frac{\alpha_2}{t}\right)\right] + I_{t \geq t_0}\left[a_3 t + a_1 \exp\left(-\frac{\alpha_2}{t_0}\right) - a_3 t_0\right]$$

$$Level\ 3: \qquad\qquad \sigma_\varepsilon \sim uniform(0, 20),$$

$$t_0(T) \sim uniform(1, 40), \quad \alpha_2(T) \sim uniform(1, 20)$$

$$a_1 \sim N(\mu_{a_1}, \sigma_{a_1}^2), \quad a_3 \sim N(\mu_{a_3}, \sigma_{a_3}^2)$$

$$\log(\mu_{a_1}) = \log(\alpha_1) - \frac{E_a}{k_B T}, \quad \log(\mu_{a_3}) = \log(\alpha_3) - \frac{E_a}{k_B T}$$

$$\log(\alpha_1) \sim N(0, 0.001),$$

$$\log(\alpha_3) \sim N(0, 0.001), \quad \sigma_{a_1}, \ \sigma_{a_3} \sim uniform(0, 10)$$

During WinBUGS simulation, we ran three Markov chains at the same time with initial values far apart from each other. The first 50,000 iterations constitute burn-in period, while another 50,000 iterations were used for computing posterior statistics. The computation time is at the order of a few minutes for 50,000 iterations using a 64-bit desktop workstation. Gelman–Rubin statistics provided by WinBUGS are around 1 and confirmed the convergency.

The numerical results are summarized in Table 11.1, and the fitted model based on posterior means is shown in Fig. 11.8. The model seems to fit the data well. The 95% confidence interval for the parameters is generally wide due to the limited amount of data. The findings from the results are listed as follows.

- As defined in Eq. (11.4), $\alpha_1 \exp[-\frac{E_a}{k_B T_0}]$ change the ÒscaleÓ of the curve, while α_2 fine-tunes the cure shape. It makes sense because higher temperature may elevate the growth process, but the growth dynamics reflected by the shape of length curves is expected to be the same. As confirmed by the experimental data in Fig. 11.6, the analysis to some degree justified the proposed structure for $\beta_0(t)$. The ÒshapeÓ parameter α_2 can be useful to fine-tune the growth curve.

- The most sensitive parameter is α_1 or its intermediate parameter $a_1(T)$. It is correlated with α_2 by the nature of the model structure. When more data are

Table 11.1 Bayesian Estimates via MCMC for Case One

Parameter	Mean	Std. Dev.	Median	95% C.I.
α_1	1.32E19	6.8E13	4.44E19	[0, 2.72E46]
$a_1(365°C)$	0.59	7.76	0.33	[−15.17, 16.62]
$a_1(380°C)$	1.63	7.85	1.08	[−14.19, 17.95]
$a_1(400°C)$	9.20	11.47	7.93	[−10.9, 32.28]
$a_1(420°C)$	16.49	15.59	17.53	[−9.10, 43.92]
$\alpha_2(365°C)$	15.76	8.31	15.88	[1.82, 29.29]
$\alpha_2(380°C)$	15.65	8.34	15.70	[1.77, 29.29]
$\alpha_2(400°C)$	10.89	8.72	8.09	[1.15, 28.73]
$\alpha_2(420°C)$	7.34	8.60	2.30	[1.03, 28.24]
α_3	2.83E17	7.29E13	2.74E17	[0, 3.27E44]
$a_3(365°C)$	0.11	0.02	0.11	[0.07, 0.16]
$a_3(380°C)$	0.15	0.02	0.15	[0.10, 0.19]
$a_3(400°C)$	0.51	0.02	0.51	[0.47, 0.56]
$a_3(420°C)$	1.25	0.03	1.25	[1.20, 1.30]
σ_ε	19.81	0.19	19.86	[19.3, 19.99]
$t_0(365°C)$	20.14	11.29	19.96	[1.89, 39.01]
$t_0(380°C)$	19.52	11.24	18.99	[1.86, 38.86]
$t_0(400°C)$	10.20	8.15	7.88	[1.25, 32.37]
$t_0(420°C)$	4.79	3.19	4.07	[1.09, 12.56]

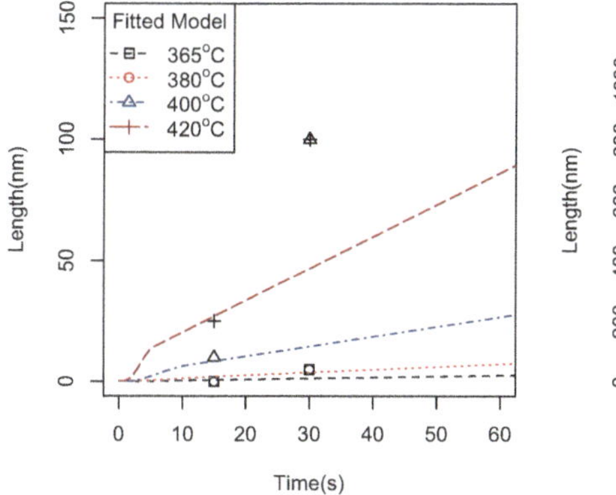

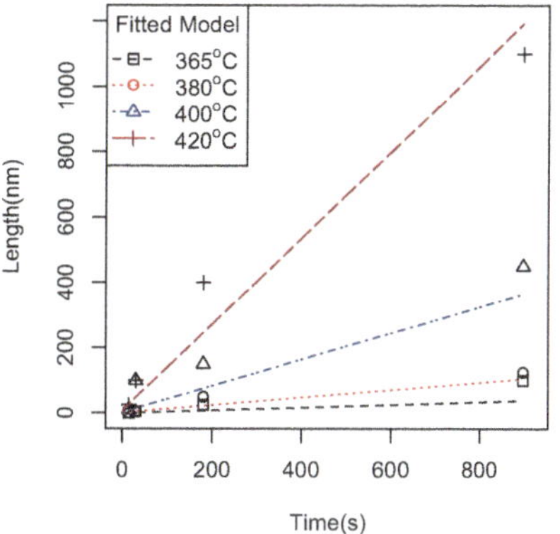

Fig. 11.8 Fitted model $\hat{\beta}_0(t)$ in case one

available, it will be helpful by introducing a correlation structure between α_1 and α_2.

- The parameter α_3 has a wide interval. But the slope of the linear phase or $a_3(T)$ is more robust, as suggested by its intervals. The main reason is that the linear structure requires relatively less data comparing to the exponential phase. Therefore, from the data collection point of view, it demands more data during exponential growth period when high confidence for α_1 is needed.

- It naturally brings up the issue of the exponential-linear transition point $t_0(T)$. The posterior estimates of $t_0(T)$ are surprisingly well considering the fact that there were missing observation before the transition time under $T = 400°C$ and $420°C$. The estimation is in agreement with physics that transition time would decrease with temperature and the balancing point will be reached faster with higher energy. This agreement can be attributed to that we embedded the physical model structure initially. The embedded 'prior' knowledge to certain degree compensates the lack of data.

11.4.2 Dynamic Nanowire Length Model at a Fine Scale

Suppose at a finer scale, we are interested to investigate both the overall growth and the final growth variability across the substrate. For the simplicity of presentation, we study three equally divided grids aligned on on the substrate. When the morphology is relatively uniform, we could use the first-order IGMRF to approximate the local variability, i.e., the conditional mean of $\phi(s_i)$ is simply a weighted average of its neighbors. $X(s, t) = [X(1, t), X(2, t), X(3, t)]^{\mathrm{T}}$, where $s = [1, 2, 3]^{\mathrm{T}}$. Using the morphology-local model or $X(s, t) = \eta_0(s, t) + \phi(s) + \varepsilon$, we have $\eta_0(s, t) = S_0\beta_0 = (1, 1, 1)^{\mathrm{T}}\beta_0(t)$. The density and precision matrix for $\phi(s)$ are [26]

$$f[\phi(s)] \propto \kappa^{(n-3)/2} \exp\left[-\frac{1}{2}\phi(s)^{\mathrm{T}}Q\phi(s)\right], \text{ with } Q_{ij} = \kappa \begin{cases} n_i, & \text{if } i = j; \\ -1, & \text{if } i \sim j; \\ 0, & \text{o/w.} \end{cases}$$

$$(11.9)$$

where n_i represents the number of neighbors of site i, and $i \sim j$ means i and j are neighbors.

For each temperature setting, we simulated the three-grid IGMRF with $\kappa = 1/50$. The simulated data will be added to the growth data at $t = 900s$ in [12]. The hierarchical model in case one is extended as follows:

$$Level\ 1: \qquad X(t, s) \sim N(\mu(t, s), \sigma_\varepsilon^2)$$

$$Level\ 2: \qquad \mu(t, s) = I_{t<t_0}\left[a_1 \exp\left(-\frac{\alpha_2}{t}\right)\right]$$

$$+ I_{t \geq t_0}\left[a_3 t + a_1 \exp\left(-\frac{\alpha_2}{t_0}\right) - a_3 t_0\right] + I_{t=900s}\phi(s)$$

$$Level\ 3: \qquad \phi(s) \sim CAR(\kappa), \quad \kappa^{-2} \sim uniform(0, 20)$$

Other non-informative priors same as those in case one

Table 11.2 Bayesian Estimates via MCMC for Case Two

Parameter	Mean	Std. Dev.	Median	95% C.I.
α_1	1.01E19	5.62E13	0.87E19	[0, 8.20E45]
$\alpha_2(365°C)$	15.75	8.32	15.86	[1.78, 29.31]
$\alpha_2(380°C)$	15.66	8.33	15.74	[1.79, 29.28]
$\alpha_2(400°C)$	10.63	8.65	7.65	[1.15, 28.64]
$\alpha_2(420°C)$	6.96	8.40	2.18	[1.03, 28.1]
α_3	2.52E17	4.69E13	2.09E17	[0, 2.20E44]
κ^{-2}	12.71	4.15	12.83	[4.76, 19.56]
$t_0(365°C)$	20.21	11.29	20.09	[1.90, 38.99]
$t_0(380°C)$	19.51	11.26	18.97	[1.85, 38.85]
$t_0(400°C)$	10.57	8.36	8.25	[1.26, 33.32]
$t_0(420°C)$	5.04	3.31	4.34	[1.10, 13.11]
σ_ε	19.81	0.19	19.87	[19.31, 20.0]

where CAR represents the conditional autoregressive representation of GMRF. WinBUGS defined function *car.normal* for IGMRF, which was used in this MCMC simulation.

The numerical results are given in Table 11.2. In general, the precision of parameter estimation is slightly better (α_1, α_3, $a_3(T)$ omitted) or similar ($\alpha_2(T)$, t_0, σ_ε, $a_1(T)$ omitted) in comparison with case one. The posterior means of the model parameters are very close to case one. The estimated $\hat{\kappa}^{-2}$ is larger that the true value. It can be attributed to the small number of grids prescribed in the simulation. Larger grids would provide more data and confidence to assess the correlation among them.

Clearly, the procedure is directly applicable to finer scales with measurement at each observation time, e.g., $t = 15s$, $30s$, $180s$, $900s$, if data are available. A more challenging issue is to correctly estimate the order of the local variability $\phi(s)$. We have done initial work in [14]. Due to page limitation, details will not be discussed herein.

The chapter introduces a domain-informed modeling methodology to enable the prediction of nanowire growth process at each scale. To take advantage of existing growth kinetics models, morphology-local decomposition of nanostructure field is conducted, in which the extended growth kinetics model is incorporated to reflect the morphology and local variability is represented by an intrinsic Gaussian Markov random field. The morphology model describes the general time–space evolution of the nanostructure growth specified by key process variables. To take into consideration the large uncertainty due to limited measurement and physical knowledge at fine scale, we put the morphology-local model under a Bayesian hierarchical framework. Two case studies are provided to demonstrate the modeling and estimation procedure at the coarse and fine scales.

Acknowledgments The work in this chapter was partially supported by US National Science Foundation under grant CMMI-1002580.

References

1. Bae, S.J., Kim, S.J., Kuo, W., Kvam: Statistical models for hot electron degradation in nano-scaled mosfet devices. IEEE Trans. Reliab. **56**(3), 392–400 (2007)
2. Banerjee, S., Carlin, B.P., Gelfand, A.E.: Hierarchical Modeling and Analysis for Spatial Data. Chapman & Hall /CRC, New York (2004)
3. Bayarri, M.J., Berger, J.O., Paulo, R., Sacks, J., Cafeo, J., Cavendish, J., Lin, C.H., Tu, J.: A framework for validation of computer models. Technometrics **49**, 138–154 (2007)
4. Chik, H., Liang, J., Kouklin, N., Xu, J.: Highly ordered zno nanorod array by 2nd order self-assembly for optoelectronic devices. Appl. Phys. Lett. **84**, 3376–3378 (2004)
5. Cressie, N.: Statistics for Spatial Data, revised edition edn. Wiley, New York (1993)
6. Dasgupta, T., Ma, C., Joseph, V.R., Wang, Z.L., Wu, C.F.J.: Statistical modeling and analysis for robust synthesis of nanostructures. J. Am. Stat. Assoc. **103**, 594–603 (2008)
7. Dubrovskii, V.G., Sibirev, N.V., Girlin, G.E., Harmand, J.C., Ustinov, V.M.: Theoretical analysis of the vapor-liquid-solid mechanism of nanowire growth during molecular beam epitaxy. Phys. Rev. E **73**, 021603 (2006)
8. Howe, J., Mori, H., Wang, Z.: In situ high-resolution transmission electron microscopy in the study of nanomaterials and properties. MRS Bull. **33**, 115–121 (2008)
9. Jeng, S.L., Lu, J.C., Wang, K.: A review of reliability research on nanotechnology. IEEE Trans. Reliab. **56**(3), 401–410 (2007)
10. Joseph, V.R., Hung, Y., Sudjianto, A.: Blind kriging: A blind kriging: A new method for developing metamodels. ASME J. Mech. Des. **130**, 031102–1–8 (2008)
11. Kennedy, M.C., O'Hagan, A.: Bayesian calibration of computer models (with discussion). J. Roy. Stat. Soc. B **63**, 425–464 (2001)
12. Kikkawa, J., Ohno, Y., Takeda, S.: Growth rate of silicon nanowires. Appl. Phys. Lett. **4**, 123109 (2005)
13. Lennard-Jones, J.E.: On the determination of molecular fields. ii. from the equation of state of a gas. Proc. Roy. Soc. Lond. A **106**(738), 463–477 (1924)
14. Liu, G.: Nanostructure Morphology Variation Modeling and Estimation for Nanomanufacturing Process Yield Improvement. Master Thesis, University of South Florida (2009)
15. Liu, J.S.: Monte Carlo Strategies in Scientific Computing. Springer, New York (2001)
16. Lu, J.C., Jeng, S.L., Wang, K.: A review of statistical methods for quality improvement and control in nanotechnology. J. Qual. Technol. **41**(2), 148–164 (2009)
17. Lu, W., Lieber, C.: Semiconductor nanowires. J. Phys. D Appl. Phys. **39**, R387–R404 (2006)
18. Luo, W., Kuo, Y., Kuo, W.: Dielectric relaxation and breakdown detection of doped tantalum oxide high-k thin films. IEEE Trans. Device Mater. Reliab. **4**(3), 488–494 (2004)
19. Luo, W., Yuan, T., Kuo, Y., Lu, J., Yan, J., Kuo, W.: Breakdown phenomena of zirconium-doped hafnium oxide high-k stack with an inserted interface layer. Appl. Phys. Lett. **89**(1), 072901 (2006)
20. Manufacturing at the Nanoscale: Report of the National Nanotechnology Initiative Workshop (2002–2004)
21. Montgomery, D.: Design and Analysis of Experiments, 7th edn. Wiley, New York (2009)
22. Rafii-Tabar, H.: Computatioal Physics of Carbon Nanotubes. Cambridge University Press, Cambridge (2008)
23. Reese, C.S., Wilson, A.G., Hamada, M., Martz, H.F., Ryan, K.J.: Integrated analysis of computer and physical experiments. Technometrics **46**, 153–164 (2004)
24. Ripley, B.: Spatial Statistics. Wiley, New York (1981)
25. Rubin, D.: Using empirical bayes techniques in the law school validity studies. J. Am. Stat. Assoc. **75**, 801–816 (1980)
26. Rue, H., Held, L.: Gaussian Markov Random Fields: Theory and Applications. Chapman & Hall /CRC, New York (2005)
27. Ruth, V., Hirth, J.P.: Kinetics of diffusion-controlled whisker growth. J. Chem. Phys. **41**(10), 3139–3149 (1964)

28. Seber, G.A.F., Wild, C.J.: Nonlinear Regression. Wiley, New York (1989)
29. Shi, J.: Stream of Variation Modeling and Analysis for Multistage Manufacturing Processes. CRC Press, New York (2006)
30. Sobczyk, K., Kirkner, D.: Stochastic Modeling of Microstructures. Birkhauser, Boston (2001)
31. Spiegelhalter, D., Thomas, A., Best, N., Lunn, D.: WinBUGS User Manual. http://www.mrcbsu.cam.ac.uk/bugs
32. Tanner, M., Wong, W.: The calculation of posterior distributions by data augmentation. J. Am. Stat. Assoc. **82**, 528–540 (1987)
33. Wagner, R.S., Ellis, W.C.: Vapor-liquid-solid mechanism of single crystal growth. Appl. Phys. Lett. **4**, 89–90 (1964)
34. Wu, C., Hamada, M.: Experiments: Planning, Analysis, and Parameter Design Optimization. Wiley, New York (2000)

Chapter 12
Cross-Domain Model Building and Validation for Nanomanufacturing Processes

Understanding the nanostructure growth faces issues of limited data, lack of physical knowledge, and large process uncertainties. These issues result in modeling difficulty because a large pool of candidate models almost fit the data equally well. Through the domain-informed machine learning strategy, we derive the process models from physical and statistical domains, respectively, and reinforce the understanding of growth processes by identifying the common model structure across two domains. This cross-domain model building strategy essentially validates models by domain knowledge rather than by (unavailable) data. It not only increases modeling confidence under large uncertainties, but also enables insightful physical understanding of the growth kinetics. We present this method by studying the weight growth kinetics of silica nanowire under two temperature conditions. The derived nanowire growth model is able to provide physical insights for prediction and control under uncertainties.

12.1 Modeling Nanomanufacturing Processes Under Large Uncertainties

Nanostructures with a length scale from 1 to 100 nm have attracted much research interests in recent decades due to their superior electrical, mechanical, optical, and biological properties [3, 13, 33, 35]. To achieve the full potential of nanotechnology, nanomanufacturing has to reach to the level of economical production at commercial scale. Despite our rapid growing knowledge in nanoscience and nanotechnology, credible prediction of nanomanufacturing processes has been a challenging task, which hinders the subsequent control activities for quality and yield improvement.

Prediction and control of nanomanufacturing processes often demand the understanding of nanostructure formation and growth. Such understanding has been traditionally achieved through theoretical or physical modeling based on the

Q. Huang, *Domain-informed Machine Learning for Smart Manufacturing*,
https://doi.org/10.1007/978-3-031-91631-1_12

first principles and validation from experimental observations. Nanomanufacturing, however, faces large uncertainties, not only from nanoscience regarding the understanding of nanostructure growth mechanisms but also from variations and noise in measurement data. The limitation of purely physical modeling approach becomes evident because competing mechanisms can exist for the same growth process. Taking nanowire growth as an example, we review three categories of modeling strategies: (a) physical modeling, (b) statistical modeling, and (c) physical–statistical modeling.

In physical modeling of nanowire growth, two major growth mechanisms have been extensively studied: diffusion and absorption. The diffusion mechanism studies the concentration gradient of growth materials that drives the growth. The growth models are presented in the form of diffusion equations with appropriate boundary conditions. Applications can be found in VLS (vapor–liquid–solid) growth modeling of nanowire [8, 26] as well as selective area metal–organic vapor phase epitaxy (SA-MOVPE) growth processes [4, 23]. The absorption-based growth models assume that the growth is driven by materials absorbed and incorporated by the nanowire [14]. There are also works indicating that absorption and diffusion mechanisms coexist in the process [7, 30], which will be hard to confirm under large measurement uncertainties.

In empirical modeling of nanowire growth, statistics is playing an ever-increasing role in experiment design, process modeling, and nanostructure characterization due to large variabilities observed. Xu et al. [37] designed sequential experiments for ZnO nanowire growth using pick-the-winner rule as well as one-pair-at-a-time effect analysis. Dasgupta et al. [5] identified robust process parameters for given morphologies using design of experiment and GLM. Wang et al. [34] proposed methods based on sequential sampling and imputation to characterize length, diameters, orientations, and densities of ZnO nanowires. Xu and Huang [36] modeled and estimated interactions (i.e., spatial correlations) among neighboring nanostructures to provide a benchmark metric of nanostructure variability and quality.

In physical–statistical modeling of nanowire growth, Huang [16] developed a physics-driven hierarchical modeling approach to integrate available data and growth kinetics with consideration of scale effects. At each scale, the nanowire growth model describes the time–space evolution of general morphology and local variability riding on the trend. Huang et al. [17] proposed an exponential–linear model structure based on diffusion-based growth mechanism to describe the weight growth kinetics of silica nanowire growth. Dasgupta et al. [6] developed a model connecting density of nanowires with thickness of polymer films where different thicknesses correspond to two physical mechanisms: pore-dominated and diffusion-dominated growth.

These three nanostructure modeling strategies have their own suitable domains of application. As illustrated in Fig. 12.1, with sufficient physical knowledge, physical modeling is readily a choice even when experimental data are less accessible. When growth mechanisms are debatable, uncertainties in the first principles will invalidate the purely physical modeling approaches. In the case of MOVPE growth

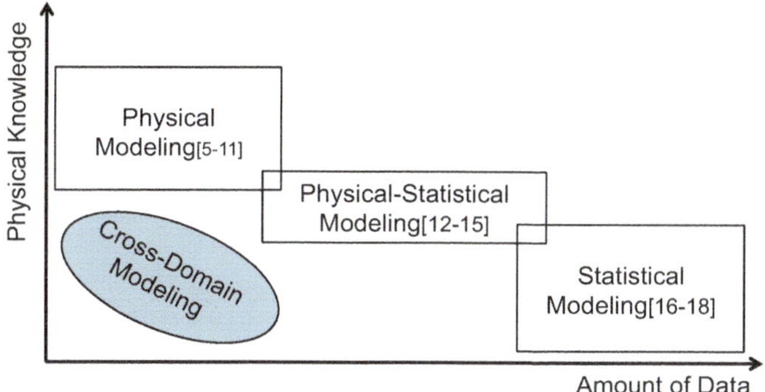

Fig. 12.1 Modeling strategies comparison

of GaAs nanowires, multiple physical models have been proposed based on different mechanism such as diffusion-induced growth [23] and absorption-induced growth [18]. But no conclusive evidences assist to determine either of these mechanisms.

On the other hand, statistical modeling faces issue of limited data in nanomanufacturing due to costly growth experiment and structure characterization. Moreover, the collected data tend to have large variations because of the poor *in situ* control of process variables such as growth temperature gradient. As a result, a large pool of candidate models can statistically fit the data. The data requirement for purely statistical modeling is rarely satisfied.

Physical–statistical modeling comes into play when physical knowledge or data alone are insufficient, but the combined information makes it feasible to improve the understanding of one growth mechanism [16] or select among a few candidate mechanisms [6]. When the uncertainties in physical knowledge and data continue to increase and growth mechanisms still require to be derived or investigated, a new modeling scheme, or the cross-domain model building and validation (CDMV) approach, shall be developed. The CDMV derives models from both physical and statistical domains and cross-validates the model from two domains against one another (Fig. 12.2), which is challenging as it requires understanding in both domains.

To illustrate the CDMV approach, we will investigate the selective growth of silica nanowires on silicon substrate with Pd thin film as catalyst [28, 29]. The weight change of the substrate is a macroscopic measure of nanowire growth and is measured at different times under two temperature settings, namely 1100°C and 1050°C. It can serve as a good index in bio-sensing where the overall number of nanowires, instead of the uniformity of nano structures, directly influences the performance. Since there is no theoretical investigation of weight kinetics for nanowire growth, large uncertainties in both physical and statistical domains justify the development of CDMV approach.

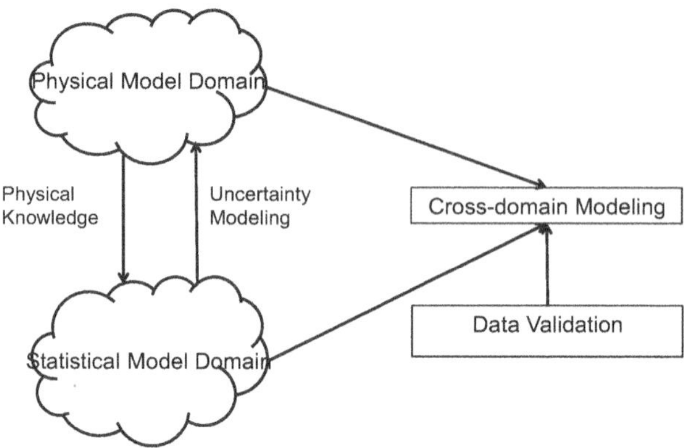

Fig. 12.2 Cross-domain model building and validation (CDMV)

12.2 Related Physical and Statistical Models and Uncertainties

12.2.1 Candidate Physical Models

There is great uncertainty regarding the understanding of growth mechanisms in silica nanowire growth. According to [29], the mechanism of nanowire growth was VLS growth with silicide phase acting as catalyst. The mechanism was experimentally validated by the selection of metal film thickness, growth temperature, and gas flow. However, [28] claimed evidences of oxide-assisted growth. At present, the existences and relative importances of those two different growth mechanisms are still open for discussion.

In general, existing modeling work for nanowire growth aims to understand length growth based on VLS mechanism. Since it was first introduced by Wagner and Ellis [32], both adsorption-induced VLS models [12] and diffusion-induced VLS models [26] have been developed. Ruth and Hirth [26] proposed a NW growth rate model:

$$\frac{dL}{dt} = \frac{2\Omega L(N_\infty - N_0)}{\tau R}, \text{ if NW length } L \leq L_f$$

$$\frac{dL}{dt} = \frac{2\Omega L_f(J - \frac{N_0}{\tau})}{R}, \text{ if } L > L_f \tag{12.1}$$

where D is diffusion coefficient, τ is mean life-time of the adatom on the NW sidewall, Ω is atomic volume of Si, J is impingement flux, R is NW radius, N_0 is adatom concentration at the liquid alloy, and N_∞ is adatom concentration at the base substrate. $L_f = \sqrt{2D\tau}$ is diffusion length. Dubrovskii et al. [8] developed a more complicated model to unify the two types of VLS models in molecular beam epitaxy:

$$\frac{dL}{dt} = V_0\Big[(1 + \frac{R_1}{R\cosh(\lambda)})(\Phi + 1)$$

$$- (1 + \frac{R_2}{R}\tanh(\lambda))(\zeta + 1)$$

$$- \frac{1}{V_*}\frac{dR}{dt}\Big] - (1 - \epsilon)V$$

where R_1 and R_2 describing diffusion-induced contribution, R is the droplet radius, ζ denotes supersaturation in nucleation, Φ denotes supersaturation of gaseous phase, V is deposition rate, λ is the ratio of whisker length L to the adatom diffusion length on the side surface, V_* is a kinetic parameter relates to liquid- and solid-phase volume ratio while ϵ is the relative difference between the deposition rate and growth rate of the substrate surface.

As we discussed earlier, the growth can also be attributed to oxide-assisted growth mechanism. But there is no quantitative understanding due to the difficulty of measuring the curly and bundled nanowires grown under this mechanism.

The complicated model formulation for VLS growth, coupled with the lack of quantitative models for oxide-assisted growth, makes the findings through pure physical modeling approaches inconclusive.

12.2.2 Candidate Statistical Models

Growth process modeling has been under statistical investigation as well [2, 27]. Examples include growth of population [24], novel technology adaptation [9], energy demand [1] as well as growth of tumors [21]. The plethora of statistical models in this area can be categorized into two major approaches: polynomial fitting with on consideration of subject knowledge, and sigmoid function fitting based on assumptions of growth processes [27]. We only consider the second approach in this chapter as polynomial fitting does not provide any physical insight.

The sigmoid growth models, however, have difficulties in model selection. To illustrate this point, we'll use two popular sigmoid growth models, namely, logistic and Gompertz growth model to fit our data and show the difficulty in choosing the appropriate model. Note that the nanowire weight change data in our study has S-shape as well.

The logistic growth model assumes that the growth rate is proportional to the current size and remaining growth potential, and the growth is driven by autocatalytic reaction [2]. Logistic growth has a increasing growth rate at first and declining growth rate with saturation afterward. It provides a simple interpretation when the growth rate increases with a growing population but is limited by available resources. The parametrization we used here is due to Fletcher [10]:

$$W(t) = \frac{W^*}{1 + \frac{W^*-W(0)}{W(0)}\exp(\frac{-4r_M t}{W^*})} \tag{12.2}$$

in which W^* is maximum growth weight, $W(0)$ is initial weight, and r_M is the maximum growth rate. We should note that for logistic growth, a nonzero initial weight is required to start growth as the growth rate is proportional to the existing growth. In our case, we assume that the initial weight represents initial nucleation.

The Gompertz growth model has been applied to tumor [20] and population growth [19] studies with the assumption that the relative growth rate decreases with logarithm of growth size. It has relatively slow growth at the beginning:

$$W(t) = W^*\exp(-\exp(-k(t - t_i))) \tag{12.3}$$

where W^* is maximum growth weight, k is a rate constant, and t_i is inflection point.

We fit those two models using nls() function in R with the silica nanowire data (Fig. 12.3). The residual sum of squares of both models under two conditions are summarized in Table 12.1. As we can see, it is difficult to identify which model is superior simply from the fitting results. Note that the two models have different assumptions regarding the growth mechanisms. Uncertainties in nanomanufacturing data make it difficulty to select among candidate models for purely statistical models.

The analysis of existing physical and statistical modeling approaches shows the need of a new modeling approach to handle uncertainties in both physical and statistical domains for nanomanufacturing.

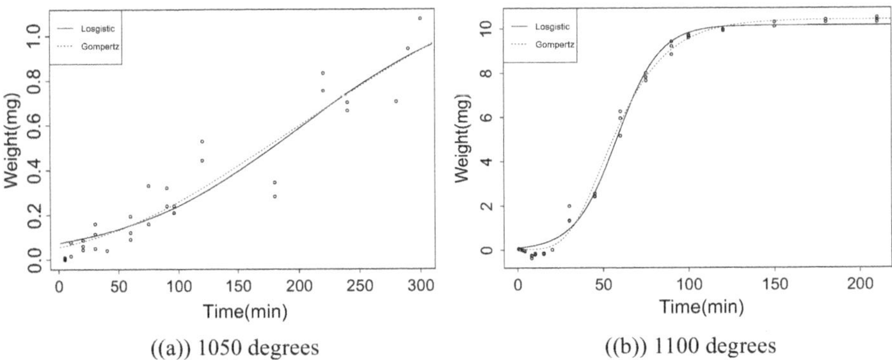

Fig. 12.3 Logistic and Gompertz model under different temperatures: solid line for logistic model and dotted line for Gompertz model

Table 12.1 Residual sum of squares of logistic and Gompertz model

Temperatures	Logistic	Gompertz
1050	0.3326	0.3079
1100	6.1487	4.6114

12.3 Cross-Domain Model Building and Validation Approach to Understand Nanomanufacturing Processes

Cross-domain model building and validation are a strategy of Integrated Nanomanufacturing and Nanoinformatics [15]. We derive the process models from physical and statistical domains, respectively, and reinforce the understanding of growth processes by identifying the common model structure across two domains. This cross-domain model building strategy essentially validates models by domain knowledge rather than by (unavailable) data.

We demonstrate the approach using the silica nanowire growth process studied in [17]. Based on the existing physical understandings, the nanowire growth is largely attributed to material absorption into nanowire surface. As shown in Fig. 12.4, there are two sources contributing to the nanowire growth: the direct impingement of silicon vapor to the Si-Pd droplet at the top of the nanowire and source supply from nanowire side surface. We will analyze the weight change of a single nanowire and then extends the model to the whole substrate through uniformity assumption. Thus, the behavior of a typical nanowire's growth in the middle of the substrate is assumed to represent the overall weight growth on the substrate. This assumption, although restrictive and idealized, still captures the main behaviors of the growth process as we will show later.

We will first derive the growth models for top impingement and side growth separately based on the available physical understanding.

12.3.1 Growth Attributed to Direct Top Impingement

Because of direct impingement, the vapor Si particles are absorbed by the droplet formed of Pd catalyst and Si, then nucleate and contribute to the nanowire growth [25, 32]. While the nanowire grows, the Pd catalyst in the droplet, however, gradually diffuses into the silica nanowire. As a result, the growth is fast at first

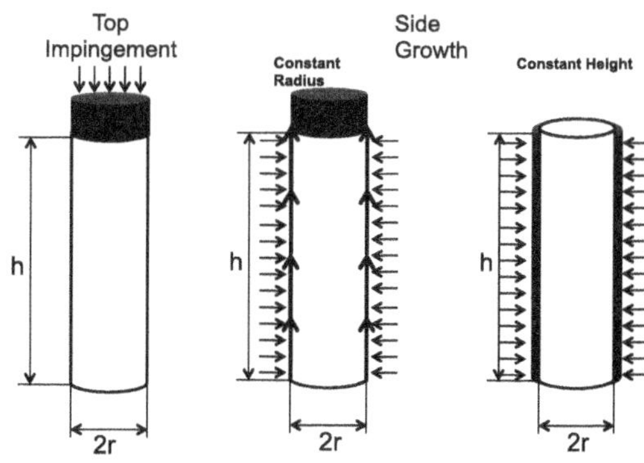

Fig. 12.4 Schematic of growth from two sources

when the catalyst is abundant and will slow down as the Pd catalyst is consumed. When the Pd is exhausted, the growth stops. We postulate the rate of weight change as

$$\frac{dW}{dt} \propto (W^* - W) \tag{12.4}$$

W is the growth weight, t is the growth time, and W^* is the maximum possible weight.

The proposed model bases on assumptions that the growth rate is proportional to the catalyst amount left. Furthermore, W^* is a constant determined by the amount of Pd catalyst and should not depend on growth time t nor weight W. We also assume that the Pd catalyst consumption is proportional to the weight growth. The formulation results in desirable growth behavior observed in experiments: large growth rate at first and steady decreasing growth rate with final stop. More importantly, with current data, it would be impossible to fit more advanced physical models as described in [25] due to multiple unknown physical coefficients and constants. The simple formulation has the advantages of both estimability and easy intepretability.

Equation (12.4) perfectly matches the confined exponential growth model in statistical literature [2]:

$$\frac{dW}{dt} = a(W^* - W) \tag{12.5}$$

where a is a rate coefficient.

Remark Model structures from both physical and statistical domains agree with each and enhance our belief and understanding of nanowire growth driven by top impingement of materials. On the other hand, we gain better insight of the underlying mcchanism for statistical confined exponential model in the context of nanomanufacturing.

12.3.2 Growth Attributed to Side Absorption

The nanowire growth attributed to the absorption from side surface is harder to characterize as there are more factors involved, namely, nanowire side area and space/resource competition among different nanowires. The relationship between nanowire side area and growth rate depends on the geometric shape of the nanowire and we consider two growth scenarios: (1) constant diameter nanowire and (2) constant height nanowires.

When nanowires do not taper and keep a constant diameter during the growth process, all Si absorbed on the side face diffuses to the top and contributes to the length growth. With assumed constant concentration of Si vapor around the nanowires, the growth is proportional to the size of absorbing area, thus we can deduce:

$$\frac{dW}{dt} \propto S = 2\pi r h \tag{12.6}$$

$$W = \rho \pi r^2 h \tag{12.7}$$

In (12.6) and (12.7), S is the side surface area of nanowire, r is the radius of nanowire, h is the height, and ρ is the density. As r is constant in this case,

$$\frac{dW}{dt} \propto \frac{2W}{\rho r} \propto W \tag{12.8}$$

On the other extreme, the nanowires do not grow in height but grow laterally over the whole nanowire uniformly. Based on (12.6) and (12.7), we hold h as a constant:

$$\frac{dW}{dt} \propto S = 2\pi r h = \frac{2\sqrt{W\pi h \rho}}{\rho} \propto \sqrt{W} \tag{12.9}$$

The derivation is based on the assumption that only r changes in this case and other parameters can be treated as constants over time.

The real growth is somewhat between the two scenarios, and we generalize the model as:

$$\frac{dW}{dt} \propto W^\alpha, 0.5 \leq \alpha \leq 1 \tag{12.10}$$

The growth from side surface, however, is also limited by the space for the nanowires and we assume that the space/resource competition can be modeled as growth rate is proportional to remaining space/resource:

$$\frac{dW}{dt} \propto W^\alpha(W' - W) \tag{12.11}$$

Equation (12.11) can describe the initial growth rate increase with growth side surface area as well as the limiting behavior when the competition of space/resource for growth among nanowires makes further growth impossible. W' is used to denote the maximum weight possible due to the space/resource limitation.

Remark The uncertainty regarding the side growth is clear in the physical domain. But the logistic growth model in statistical domain is quite close to Eq. (12.11) with $\alpha = 1$. The similarity in the model structures from two domains greatly assists us to choose the S-shape logistic growth model over the Gompertz growth model.

12.3.3 Cross-Domain Modeling of Nanowire Growth

Due to the complexity of nanowire growth under various conditions, the true growth model can be (i) dominated by either top impingement or side absorption or (ii) driven by both growth mechanisms with varying percentage of contributions. We therefore propose a general model for overall nanowire weight growth, which combines Eqs. (12.11) and (12.5):

$$\frac{dW}{dt} = a^*(W^* - W) + aW^\alpha(W' - W) \tag{12.12}$$

In this model, coefficients a^* and a can be viewed as weights for top impingement and side absorption mechanisms, respectively. The weights can be determined through statistical model selection procedure for specific growth conditions. The corresponding growth mechanisms can thus be uncovered within this modeling framework.

The model (12.12), even with $\alpha = 1$, however, poses challenges for parameter estimation as there are four unknown parameters in a quadratic polynomial of W on the right hand side of the differential equation. As the quadratic expression can be uniquely determined by three parameters, one of the four unknown parameters becomes inestimable. Furthermore, our physical knowledge of W^* and W' is very limited.

The model structure given in (12.12) reminds us the combined model of confined exponential and logistic model discussed in statistical literature [2]:

$$\frac{dW}{dt} = a^*(W^* - W) + aW(1 - \frac{W}{W^*}) \tag{12.13}$$

$$= -\frac{aW^2}{W^*} + (a - a^*)W + a^*W^* \tag{12.14}$$

This formulation requires the consolidation of two parameters W' with W^* in (12.12) into the same parameter W^*. What it means is that for the same growth process, the limiting weight, contributed either from top impingement or from side absorption, should be the same. We take this assumption, and it turns out that numerical result of estimation seems to support it.

To obtain a closed formed solution for the differential equation is possible. Let $k(t) = \exp((a + a^*)(t + C))$, then:

$$W = \frac{W^*(-a^* + ak(t))}{a(1 + k(t))} \tag{12.15}$$

C is the constant associated with initial value W_0.

$$W_0 = \frac{W^*(a \exp[(a + a^*)C] - a^*)}{a(1 + \exp[(a + a^*)C])} \tag{12.16}$$

Or we can write:

$$C = \frac{1}{a + a^*} \ln\left(\frac{a^* W^* + a W_0}{a W^* - a W_0}\right) \tag{12.17}$$

The growth rate function would be:

$$\frac{\mathrm{d}W}{\mathrm{d}t} = \frac{W^*(a + a^*)^2 k(t)}{a[1 + k(t)]^2} \tag{12.18}$$

Based on the growth rate function, the inflection point would be:

$$t_i = -C, \ W_i = \frac{(a - a^*)W^*}{2a} \tag{12.19}$$

Considering the error of weight measurement, we assume there is a random error term ϵ, which follows a normal distribution $N(0, \sigma^2)$ with σ^2 unknown.

12.3.4 Uncover Growth Kinetics for Silica Nanowire Growth

In this section, we will apply the cross-domain modeling result to silica nanowire growth and analyze the growth kinetics under two growth temperatures: 1100°C and 1050°C.

Nanowire Model Estimation for Two Growth Conditions Detailed descriptions of the process conditions and data for 1100°C (57 data points observed) can be found in our previous work [17]. To estimate the parameters $\theta = (a^*, a, W^*, C)$ in (12.15), the nonlinear regression is performed by using PORT library [11] implemented in nls() function in R language. As the algorithm does not guarantee convergence to a global minimum due to the fact that $W(t)$ is not convex, we start from five widely separated sets of initial values shown in Table 12.2.

Five different sets of initial values give same estimation for $\theta = (a^*, a, W^*, C)$ for 1100°C:

$$\theta_{1100} = (0.0024 \pm 0.0005, 0.068 \pm 0.004,$$

$$10.24 \pm 0.08, -57.04 \pm 0.73)$$

Table 12.2 Initial parameter values for 1100°C

Run Number	a^*	a	W^*	C
1	0.2	0.01	8	−80
2	0.1	0.005	6	−60
3	0.05	0.002	4	−40
4	0.02	0.001	2	−20
5	0.01	0.0005	1	−10

As one may notice from Fig. 12.5, the result of curve fitting is good considering the inherent variability in the data.

The model estimation under 1050°C, however, faces tremendous challenge. Although we have 31 data points, all those data are collected before growth saturation due to safety concerns. Since we have only partial information, there are huge uncertainties in the data regarding the maximum growth weight and the inflection point (Fig. 12.6, please notice that we don't have any data after 310 min).

In order to achieve best understanding with incomplete information, we propose to use a Bayesian estimation scheme to characterize the uncertainty involved. The idea is that each parameter has its own distribution, and we can update the prior knowledge of those parameters by observation. The prior distributions for parameters are assumed to be normal distributions as follows:

Fig. 12.5 Estimation under 1100°C

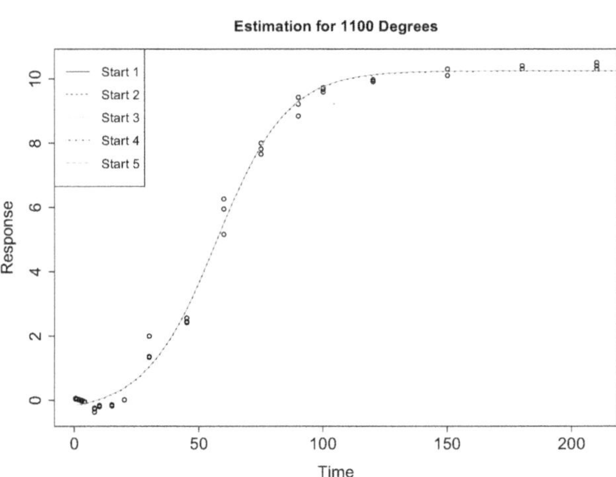

Fig. 12.6 Uncertainty under 1050°C

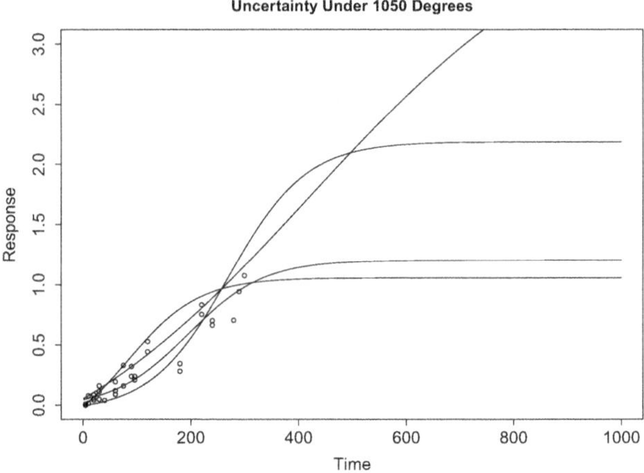

$$a^* \sim N(0.068, 0.316^2), a \sim N(0.002, 0.141^2),$$

$$W^* \sim N(6.5, 2^2), C \sim N(-150, 316^2),$$

$$\sigma \sim U(0.05, 1)$$

For rate parameter a^* and a, as we don't have much knowledge about how they would change with temperature, the prior distribution is centered at estimated value under 1100°C with large variability. For inflection point C, we noticed that if we use a logistic model for growth under 1050°C, the estimation is $C \approx -150$ and we use it as the center of our prior distribution, and we acknowledged the fact that the information is limited by using a large variance. The maximum weight W^*'s prior distribution is based on possible maximum and minimum values from the data for two temperature settings. The random noise is assumed to have a standard variation from 0.05 to 1 as observed under two temperature settings.

With the prior distributions, we use Markov chain Monte Carlo simulation [22] with WinBUGS [31] to draw posterior samples. Three widely separated chains are used, and first half of 100,000 runs is discarded as burn-in. The estimation result is shown in Table 12.3. One of the interesting result is that $-C$ is much larger than 300, and we didn't have any data points after growth saturation.

We checked the histograms and trace plots(omitted here), and all of them affirmed that convergence had been achieved.

Uncover Growth Kinetics With models established, we set to discover physical insights for the two growth conditions. First, we aim to connect two models by recognizing that model parameters are temperature-dependent. For instance, as rate coefficient, a^* and a's dependence on temperature should follow Arrhenius equation:

$$a^* \propto \exp(-\frac{E_{a^*}}{kT}) \tag{12.20}$$

$$a \propto \exp(-\frac{E_a}{kT}) \tag{12.21}$$

E_{a^*} and E_a are corresponding growth processes' activation energy, k is the Boltzmann constant, and T is temperature.

Table 12.3 Bayesian Estimation for 1050°C

Parameter	Mean	S.D	2.5%	97.5%
a^*	3.43E−4	1.25E−4	1.01E−4	5.75E−4
a	3.18E−3	1.67E−3	8.13E−4	6.82E−3
W^*	6.15	0.73	4.74	7.58
C	−586.5	89.31	−743.5	−356.9
σ	0.099	0.013	0.077	0.13

The other two parameters W^* and C are also related to temperature. The dependence, however, is more complicated, and we have no corresponding quantitative models. Taking W^* as an example, as a parameter associated with space/resource and catalyst limitation, it is depended on temperature as the Si evaporation process and Pd diffusion process are temperature-dependent. Another factor is that with different temperatures, the number of initial Pd-Si droplets formulated may be different. Due to our limited knowledge of those factors, W^* and C's temperature dependence will be left for future investigation.

From Table 12.4, it can be observed that a^* decreases slower than a when temperature is lowered from 1100°C to 1050°C. As a^* is associated with catalyzed growth from the top and a is associated with growth from side, it can be assumed that the activation energy E_{a^*} is smaller than E_a, which is consistent with our observation. More specifically, we can calculate from the Arrhenius equation that the ratio $\frac{E_{a^*}}{E_a} = 0.635$.

If we compare the estimation of logistic growth model (Table 12.5, using parametrization in (12.2)) with our model, and computed corresponding rate parameter a, we would notice that while logistic growth model is a good approximation of our model under 1100°C, the rate and maximum growth parameters under 1050°C are dramatically different from our model. This can be explained as side growth dominated the process under 1100°C. With lower temperature, however, top impingement becomes more important due to smaller activation energy. In this new circumstance, logistic growth model based on side growth alone is no longer adequate, and our cross-domain model should be used.

Establishing quantitative nanomanufacturing process models faces challenges of large uncertainties in process physics and experimental observations. Current three modeling strategies, i.e., physical, statistical, and physical–statistical modeling, while being successful in their specific domains of application, may not be sufficient to address those challenges In order to utilize all available information and achieve greater confidence in process modeling, this chapter introduces a cross-domain modeling and validation approach to achieve domain-informed machine learning in nanomanufacturing.

We demonstrate this new modeling strategy for silica nanowire growth process where both physical growth mechanisms and measurement data contain large

Table 12.4 Estimation Comparison under 1100 and 1050°C

	a^*	a	W^*	C
1100	0.0024	0.068	10.24	−57.04
1050	3.43E−4	3.18E−3	6.15	−586.5
Rate Ratio(1100/1050)	6.997	21.38	−	−

Table 12.5 Logistic Growth Estimation under 1100°C and 1050°C

	$W(0)$	W^*	r_M	a
1100	0.0940	10.15	0.204	0.0788
1050	0.07403	1.185	0.00398	0.0132

uncertainties. By deriving physical models and cross-validating with growth models in statistical domains, we are able to establish a generic formulation of growth models suitable for statistical selection of growth mechanisms under various conditions. Using both nonlinear regression and Bayesian approach, models under two growth conditions are characterized satisfactorily. More importantly, we are able to uncover the dominant physical growth mechanisms for each growth condition. The developed approach provides a powerful alternative for nanomanufacturing process analysis and characterization.

Acknowledgments The work in this chapter was supported by US National Science Foundation under grant CMMI-1002580.

References

1. Ang, B., Ng, T.: The use of growth curves in energy studies. Energy **17**(1), 25–36 (1992)
2. Banks, R.: Growth and Diffusion Phenomena: Mathematical Frameworks and Applications, vol. 14. Springer, New York (1994)
3. Cao, L., Fan, P., Vasudev, A., White, J., Yu, Z., Cai, W., Schuller, J., Fan, S., Brongersma, M.: Semiconductor nanowire optical antenna solar absorbers. Nano Lett. **10**(2), 439–445 (2010)
4. Chu, H.J., Yeh, T.W., Stewart, L., Dapkus, P.D.: Wurtzite InP nanowire arrays grown by selective area MOCVD. Physica Status Solidi (C) **7**(10), 2494–2497 (2010)
5. Dasgupta, T., Ma, C., Joseph, V., Wang, Z., Wu, C.: Statistical modeling and analysis for robust synthesis of nanostructures. J. Am. Stat. Assoc. **103**(482), 594–603 (2008)
6. Dasgupta, T., Weintraub, B., Joseph, V.: A physical–statistical model for density control of nanowires. IIE Trans. **43**(4), 233–241 (2011)
7. Dayeh, S., Soci, C., Bao, X., Wang, D.: Advances in the synthesis of InAs and GaAs nanowires for electronic applications. Nano Today **4**(4), 347–358 (2009)
8. Dubrovskii, V.G., Sibirev, N.V., Girlin, G.E., Harmand, J.C., Ustinov, V.M.: Theoretical analysis of the vapor-liquid-solid mechanism of nanowire growth during molecular beam epitaxy. Phys. Rev. E **73**, 021603 (2006)
9. Fisher, J., Pry, R.: A simple substitution model of technological change. Technol. Forecast. Soc. Change **3**, 75–88 (1972)
10. Fletcher, R.: The quadric law of damped exponential growth. Biometrics, 111–124 (1974)
11. Fox, P., Hall, A., Schryer, N.: The PORT mathematical subroutine library. ACM Trans. Math. Software (TOMS) **4**(2), 104–126 (1978)
12. Givargizov, E.: Fundamental aspects of VLS growth. J. Crystal Growth **31**, 20–30 (1975)
13. Gudiksen, M., Lauhon, L., Wang, J., Smith, D., Lieber, C.: Growth of nanowire superlattice structures for nanoscale photonics and electronics. Nature **415**(6872), 617–620 (2002)
14. Hamano, T., Hirayama, H., Aoyagi, Y.: New technique for fabrication of two-dimensional photonic bandgap crystals by selective epitaxy. Jpn. J. Appl. Phys. **36**(3A), L286–L288 (1997)
15. Huang, Q.: Integrated nanomanufacturing and nanoinformatics for quality improvement (invited). In: 44th CIRP International Conference on Manufacturing Systems (2011)
16. Huang, Q.: Physics-driven bayesian hierarchical modeling of nanowire growth process at each scale. IIE Trans. **43**, 1–11 (2011)
17. Huang, Q., Wang, L., Dasgupta, T., Zhu, L., Sekhar, P.K., Bhansali, S., An, Y.: Statistical weight kinetics modeling and estimation for silica nanowire growth catalyzed by Pd thin film. IEEE Trans. Autom. Sci. Eng. **8**, 303–310 (2011)

18. Ikejiri, K., Sato, T., Yoshida, H., Hiruma, K., Motohisa, J., Hara, S., Fukui, T.: Growth characteristics of GaAs nanowires obtained by selective area metal-organic vapour-phase epitaxy. Nanotechnology **19**(26), 265604 (2008)
19. Kozusko, F., Bajzer, Z.: Combining Gompertzian growth and cell population dynamics. Math. Biosci. **185**(2), 153–167 (2003)
20. Laird, A.: Dynamics of tumour growth. Br. J. Cancer **18**(3), 490 (1964)
21. Laird, A.: Dynamics of tumour growth: comparison of growth rates and extrapolation of growth curve to one cell. Br. J. Cancer **19**(2), 278 (1965)
22. Liu, J.S.: Monte Carlo Strategies in Scientific Computing. Springer, New York (2001)
23. Noborisaka, J., Motohisa, J., Fukui, T.: Catalyst-free growth of GaAs nanowires by selective-area metalorganic vapor-phase epitaxy. Appl. Phys. Lett. **86**(21), 213102 (2005)
24. Pearl, R., Reed, L.: On the rate of growth of the population of the united states since 1790 and its mathematical representation. Proc. Natl. Acad. Sci. USA **6**(6), 275 (1920)
25. Roper, S., Davis, S., Norris, S., Golovin, A., Voorhees, P., Weiss, M.: Steady growth of nanowires via the vapor-liquid-solid method. J. Appl. Phys. **102**(3), 034304–034304 (2007)
26. Ruth, V., Hirth, J.P.: Kinetics of diffusion-controlled whisker growth. J. Chem. Phys. **41**(10), 3139–3149 (1964)
27. Seber, G., Wild, C.: Nonlinear Regression, vol. 503. Wiley-IEEE, New York (2003)
28. Sekhar, P., Bhansali, S.: Manufacturing aspects of oxide nanowires. Mater. Lett. **64**(6), 729–732 (2010)
29. Sekhar, P., Sambandam, S., Sood, D., Bhansali, S.: Selective growth of silica nanowires in silicon catalysed by Pt thin film. Nanotechnology **17**(18), 4606–4613 (2006)
30. Soci, C., Bao, X., Aplin, D., Wang, D.: A systematic study on the growth of gaas nanowires by metal-organic chemical vapor deposition. Nano Lett. **8**(12), 4275–4282 (2008)
31. Spiegelhalter, D., Thomas, A., Best, N., Lunn, D.: WinBUGS User Manual. http://www.mrcbsu.cam.ac.uk/bugs
32. Wagner, R.S., Ellis, W.C.: Vapor-liquid-solid mechanism of single crystal growth. Appl. Phys. Lett. **4**, 89–90 (1964)
33. Wanekaya, A., Chen, W., Myung, N., Mulchandani, A.: Nanowire-based electrochemical biosensors. Electroanalysis **18**(6), 533–550 (2006)
34. Wang, F., Hwang, Y., Qian, P., Wang, X.: A statistics-guided approach to precise characterization of nanowire morphology. ACS Nano **4**(2), 855–862 (2010)
35. Wu, B., Heidelberg, A., Boland, J.: Mechanical properties of ultrahigh-strength gold nanowires. Nature Mater. **4**(7), 525–529 (2005)
36. Xu, L., Huang, Q.: Modeling the interactions among neighboring nanostructures for local feature characterization and defects detection. IEEE Trans. Autom. Sci. Eng. **Accepted** (2012). https://doi.org/10.1109/TASE.2012.2209417
37. Xu, S., Adiga, N., Ba, S., Dasgupta, T., Wu, C., Wang, Z.: Optimizing and improving the growth quality of ZnO nanowire arrays guided by statistical design of experiments. ACS Nano **3**(7), 1803–1812 (2009)

Chapter 13
Physical–Statistical Modeling of Graphene Growth Processes

As a zero-band semiconductor, graphene is an attractive material for a wide variety of applications such as optoelectronics. Among various techniques developed for graphene synthesis, chemical vapor deposition (CVD) on copper foils shows high potential for producing few-layer and large-area graphene. Since fabrication of high-quality graphene sheets requires the understanding of growth mechanisms, and methods of characterization and control of grain size of graphene flakes, analytical modeling of graphene growth process is therefore essential to controlled fabrication. The graphene growth process starts with randomly nucleated islands that gradually develop into complex shapes, grow in size, and eventually connect together to cover the copper foil. To model this complex process, this chapter introduces a physical–statistical approach under the assumption of self-similarity during graphene growth. The growth kinetics is uncovered by separating island shapes from area growth rate.

13.1 Learning Graphene Growth Kinetics

The unique zero-band electronic structure of graphene [6] leads to exciting room-temperature properties such as the quantum Hall effect [21] and high electron mobility [4]. With its extraodinary properties, graphene shows high potential in many applications, such as optoelectronics, chemical and biosensing [12, 23, 24].

Since obtaining graphene through micromechanical cleavage [20] is not amenable to commercial applications, various synthesis techniques have been developed, e.g., thermal decomposition of SiC by the surface segregation of C dissolved in metal substrates [18, 25, 29] and by decomposition of hydrocarbons on metals [17, 19]. Among them, chemical vapor deposition (CVD) on Ni and Cu has attracted most of the attention. While graphene growth on Ni seems to be limited by small grain size, presence of multilayers at the grain boundaries, and the high solubility of carbon [14], growth on Cu foils has shown great promise for a

Q. Huang, *Domain-informed Machine Learning for Smart Manufacturing*,
https://doi.org/10.1007/978-3-031-91631-1_13

large area, monolayer graphene. The low solubility of carbon on Cu is believed to self-limit graphene to monolayer.

As the fabrication of high-quality graphene sheets for great applications requires the understanding of growth mechanism, and methods of characterization and control of grain size of graphene flakes, analytical modeling of graphene growth process is essential to controlled fabrication [13]. Currently, the graphene growth is explained by two main mechanisms: absorption and diffusion. For the diffusion model, it has been proposed that CVD growth of graphene on Ni is due to a C segregation or precipitation process and that a fast cooling rate in conjunction with thin films is needed to suppress the formation of multiple graphene layers [29]. On the contrary, the graphene growth on Cu has been shown to be driven by surface absorption [14]. Yet existing works have been mainly focusing on identifying optimal recipes such as flow rate, pressure and temperature to obtain large-scale monolayer graphene on the metal surface. Little research has been reported on growth kinetics modeling that is able to explain the growth rate for graphene with complex shapes.

Although graphene growth modeling is fairly limited, methods have been developed for nanowire growth modeling [8, 9]. This line of research aims to describe process kinetics under uncertainties for prediction and control. Comparing the 1D structure of nanowire, graphene growth over 2D copper foils imposes greater challenge, where both the growth velocity of the covered area and the shape of the graphene islands have to be modeled.

To address this modeling challenge, this chapter introduces an analytical framework for the graphene growth processes, where we separate the area growth/coverage velocity and the geometric shapes of graphene islands. Though we use four-lobed islands as an example, the modeling approach can be applied to other island shapes such as square, hexagon, even flowers in literature [26, 28, 30].

13.2 A Physical–Statistical Framework for Graphene Growth Process Modeling

We develop in this section a fairly generic framework to analyze the graphene growth process. We aim to describe the foil coverage by various shapes of graphene islands over time and to uncover the kinetics model driving the graphene growth.

Though widely applicable, the methodology is illustrated with the process of graphene growth on Cu foil. According to [16, 27], four-lobed symmetric islands nucleate heterogeneously at bunches of Cu steps, pinning sites, and other surface imperfections. Initially compact in shape, the graphene islands become increasingly ramified over time and develop a lobed structure. The long axes of the four graphene lobes are oriented approximately the same from island to island and are usually close to Cu (001) in-plane directions. Despite their ramified shape, the graphene islands eventually combine to form a complete polycrystalline film. However, the spaces

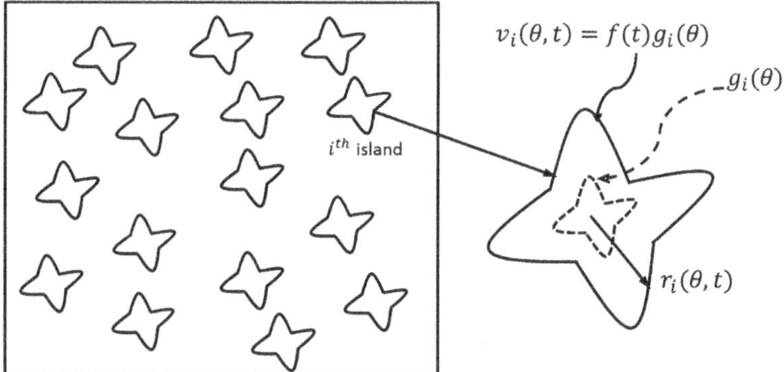

Fig. 13.1 Schematic of graphene growth

Fig. 13.2 SEM images of graphene islands on copper foil [16]

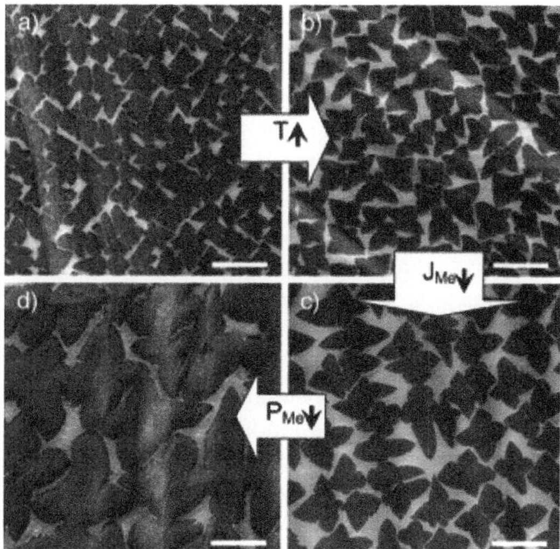

between islands fill slowly after island impingement occurs. Figure 13.1 illustrates the schematics of the graphene growth process to be modeled, and Fig. 13.2 shows SEM snap-shot images of graphene islands under different growth conditions by varying temperature from (a) to (b), varying methane flow rate from (b) to (c), varying methane partial pressure from (c) to (d) [16].

Note that there is another line of research on morphological stability in crystal growth area [5], where the physical principles are derived to explain crystal morphologies. For instance, Ivantsov model [10] or Hunt model [1] aims at establishing the dendritic growth models based on the first principles, e.g., on the diffusion-controlled transport around a parabolic solid–liquid interface. On the contrary, our objective is to describe the process from the perspective of controlled fabrication. As can be seen from Fig. 13.2, the islands have variability in their shapes caused by process uncertainties, which are normally not captured by deterministic dendritic growth models.

To model the growth of graphene islands, we start by describing the growth velocity $v_i(\theta, t)$ of the *ith* island at angle θ and time t. By characterizing the statistical behavior of all graphene islands, we will uncover the underlying kinetics model $f(t)$ that drives the growth process. Following this strategy, the model for the angular dependent growth velocity $v_i(\theta, t)$ is proposed to be

$$v_i(\theta, t) = f(t)g_i(\theta) \tag{13.1}$$

under the assumption that graphene islands with similar geometric shape $g_i(\theta)$ are driven by the same growth kinetics $f(t)$. The growth kinetics $f(t)$ for a given process condition represents the underlying driving force that pushes islands with shape $g_i(\theta)$ growing in size over time (Fig. 13.1). $g_i(\theta)$ is normalized to have unit area, and it can take any shape of graphene islands. The dimension of $f(t)$ is *Length* $\times$ *Time*$^{-1}$, and $g(\theta)$ is dimensionless.

Remark Model (13.1) has two different, but equivalent interpretations. First, on average, graphene islands under the same process condition increase their sizes at the same rate $f(t)$. Within an island, the growth speed varies from angle to angle, and the difference is carved out by its shape function $g_i(\theta)$. Second, we can view that each island starts with shape $g_i(\theta)$, and it is scaled up in size at the rate of $f(t)$. This is essentially the *self-similarity assumption* for graphene growth. In either case, uncovering kinetics $f(t)$ is crucial to understand the growth process.

Remark Although it is reasonable to suspect that island shape $g(\theta)$ might change with time as well, we argue that model (13.1) is a good approximation. First, the shapes of graphene islands are similar, and it is reasonable to separate the growth kinetics $f(t)$ from shape component in Eq. (13.1). Second, the graphene growth involves the island formation and island growth, while the former (nucleation) is much faster than the later (growth). Model (13.1) is mainly used to approximate the growth process, where the island shape is more stable. For a system with fast-low dynamics, ignoring the fast dynamics (nucleation) will only miss the characterization of the very short initial period. It will generally not distort the description of the slow dynamics (growth). Third, model (13.1) could also accommodate time-varying property by further relaxing the strict assumption of $g(\theta)$ to $g_i(\theta) = g(\theta) + \zeta_i$.

Among the limited graphene process modeling literature, the angular dependent growth velocity was proposed in [27] to explain fourfold shape, where the growth velocity v of one fold of the graphene island has the form of $v(\theta) = (1 + s) + (s - 1)\cos(2\theta)$, with θ being the angle of the edge normal relative to the slow-growth direction, and s being the ratio of the velocities in the slow and the fast directions, respectively. No clear definition of process kinetics model was provided therein.

Since the total area of covered foil and shape of graphene islands are measurable, we will relate $v_i(\theta, t)$ to the total area in order to derive $f(t)$. We consider two cases, where in the first case, the total number of the graphene islands remains constant after nucleation, and in the second case, the number of the graphene islands increases with the time.

13.2.1 Constant Number of Graphene Islands After Nucleation

Denote by $S(t)$ and $v_{area}(t)$ the total covered area on Cu foil and its area growth velocity, respectively. Now we can build the relations between $v_{area}(t)$ and $f(t)$ given constant N islands on the copper foil at time t.

$$v_{area}(t) = \frac{dS(t)}{dt} = \sum_{i=1}^{N} \frac{dS_i(t)}{dt}$$

where $S_i(t)$ denotes the area of the ith graphene island at time t. As $S_i(t) = \int_0^{2\pi} \frac{r_i^2(\theta,t)}{2} d\theta$, $\frac{d}{dt} r_i(\theta,t)^2 = 2r_i(\theta,t)\frac{d}{dt}r_i(\theta,t)$ and $r_i(\theta,t) = \int_0^t v_i(\theta,s)ds$, we get

$$v_{area}(t) = \sum_{i=1}^{N} \frac{d}{dt} \int_0^{2\pi} \frac{r_i^2(\theta,t)}{2} d\theta$$

$$= \sum_{i=1}^{N} \int_0^{2\pi} v_i(\theta,t) \left(\int_0^t v_i(\theta,s)ds \right) d\theta$$

$$= \sum_{i=1}^{N} \left[\int_0^{2\pi} g_i(\theta)^2 d\theta \right] f(t) \int_0^t f(s)ds \qquad (13.2)$$

where $r_i(\theta,t)$ is the distance from the origin to the boundary of the ith island at angle θ.

Furthermore, we assume that the shapes of graphene islands follow the same probability distribution with mean $g(\theta)$, which acknowledges the fact that there are shape variations among islands $g_i(\theta)$. As the number of islands N is often quite large, according to the law of large numbers, we have:

$$\frac{\sum_{i=1}^{N} \left[\int_0^{2\pi} g_i(\theta)^2 d\theta \right]}{N} \to E\left[\int_0^{2\pi} g_i(\theta)^2 d\theta \right] \qquad (13.3)$$

as N tends to ∞. Taking (13.3) into consideration, we can rewrite (13.2) as:

$$NE\left[\int_0^{2\pi} g_i(\theta)^2 d\theta \right] f(t) \int_0^t f(s)ds \to v_{area}(t) \quad \text{a.s.} \qquad (13.4)$$

Now we let $C = E\left[\int_0^{2\pi} g_i(\theta)^2 d\theta\right]$, then

$$v_{area}(t) = NCf(t)\int_0^t f(s)ds \tag{13.5}$$

Notice that the area of the shape $g(\theta)$ equals to $\int_0^{2\pi} \frac{1}{2} g(\theta)^2 d\theta$. So $\frac{C}{2}$ is the average area of the shape $g_i(\theta)$. As $g_i(\theta)$ is used to model the shape regardless of its orientation and size, we can let $C = 2$ for simplicity. It means the mean area of $g_i(\theta)$ is normalized to be unit area. If we further define $F(t) = \int_0^t f(s)ds$, we will get a differential equation:

$$F(t)dF(t) = \frac{v_{area}(t)}{2N}dt \tag{13.6}$$

Then the solution $F(t)^2 = \frac{1}{N} S(t)$, where $S(t) = \int_0^t v_{area}(s)ds$. We have

$$F(t) = \sqrt{\frac{S(t)}{N}} \tag{13.7}$$

By taking derivative of model (13.7), we can get the kinetics model $f(t)$

$$f(t) = \frac{v_{area}(t)}{2\sqrt{NS(t)}} \tag{13.8}$$

For a given geometric shape $g_i(\theta)$, the growth velocity of the ith island at angle θ and time t can thus be derived from model (13.1),

$$v_i(\theta, t) = \frac{v_{area}(t)}{2\sqrt{NS(t)}} g_i(\theta) \tag{13.9}$$

Remark on $F(t)$ Model (13.7) and $F(t)^2 = \frac{1}{N} S(t)$ indicate that if the total covered area $S(t)$ is divided into N grids, each grid has area of $F(t)^2$. In another word, we can transform a graphene island with any shape $g(\theta)$ to an equivalent square shape with side length $F(t)$ as seen in Fig. 13.3. $F(t)$ now has a clear meaning, i.e., the length of the side of the transformed square.

Remark on $f(t)$ Based on the understanding of $F(t)$ in Fig. 13.3, the kinetics model $f(t)$ derived in (13.8) can be conveniently interpreted as the velocity of side growth of the equivalent square shape. This theoretical understanding enables us to investigate and compare the growth kinetics of graphene islands with various shapes via the equivalent transformation.

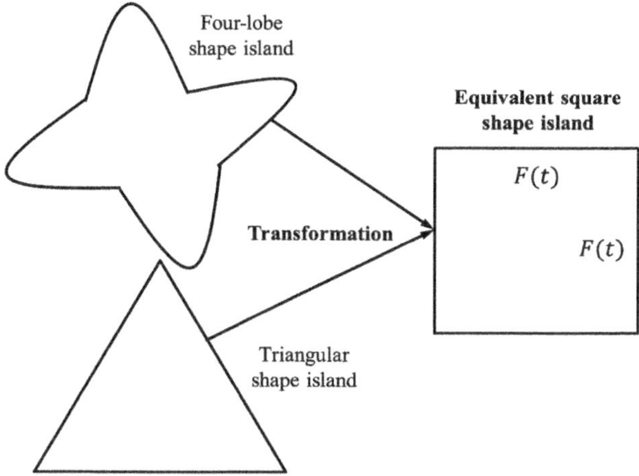

Fig. 13.3 The illustration of equivalent square

13.2.2   Varying Number of Graphene Islands over Time

To consider the case that the number of graphene islands increases over time, we denote by $N(t)$ the number of islands on the Cu foil at time t. Since the number of nucleated islands $N(t)$ at time t is random, we can take expectation of model (13.5) from both sides:

$$E[v_{area}(t)] = E[N(t)]Cf(t)F(t) \tag{13.10}$$

Similarly, if we take $C = 2$, we can get

$$F(t)dF(t) = \frac{E[v_{area}(t)]}{2E[N(t)]}dt \tag{13.11}$$

Let $S'(t) = \int_{s=0}^{t} \frac{E[v_{area}(s)]}{E[N(s)]}ds$, then

$$F(t) = \sqrt{S'(t)} \tag{13.12}$$

Taking derivative, we have the kinetics model with varying $N(t)$ as

$$f(t) = \frac{E[v_{area}(t)]}{2E[N(t)]\sqrt{S'(t)}} \tag{13.13}$$

Remark For varying $N(t)$, the computation of expectation is more involved. Fortunately, as argued in [27], the number of islands doesn't increase very much along the growing process. Case I provides a reasonable description for most cases.

Further study will be necessary for cases where the total number of island change with time.

13.3 Kinetics of Graphene Growth on Copper Foil

The analytical framework developed in Sect. 13.2 is applicable to graphene growth with various shapes. In this section, we use four-lope graphene islands as an example to uncover the kinetics of graphene growth on Cu foil.

Model (13.8) or (13.13) suggests that we can derive $f(t)$ if analytical model of $S(t)$ or $v_{area}(t)$ is obtained. Since $S(t)$ is relatively easier to be derived in comparison to $f(t)$, graphene area growth velocity $v_{area}(t)$ will first be derived and a parametric model will be developed to describe the island shapes $g_i(\theta)$. From model (13.1), we get the complete model to describe the growth velocity of each island at a specific angle and time.

13.3.1 Graphene Area Growth Velocity Modeling

In this subsection, we focus on deriving the analytical model for area growth velocity on Cu foil, i.e., the analytical expression of $v_{area}(t)$ or $S(t)$ equivalently.

To derive this model, we first need to understand the graphene growth mechanism. In the experiment of graphene growth on Cu [15], the carbon isotope labeling in conjunction with Raman spectroscopic mapping was used to track carbon during the growth process. Figure 13.4 presents schematically the possible distributions of ^{12}C and ^{13}C in graphene films based on absorption when ^{12}CH$_4$ or ^{13}CH$_4$ are introduced sequentially. As the graphene grown with the sequential dosing of ^{12}CH$_4$ and ^{13}CH$_4$ grows by surface adsorption, the isotope distribution in the local graphene regions will reflect the dosing sequence employed. So from the physical aspect, the main mechanism for graphene growth on Cu is due to absorption. Furthermore, we can see clearly when the Cu foil is covered by the ^{12}CH$_4$ or ^{13}CH$_4$, it cannot absorb any more because of the low solubility of C on Cu foil.

Fig. 13.4 The illustration of surface absorption [15]

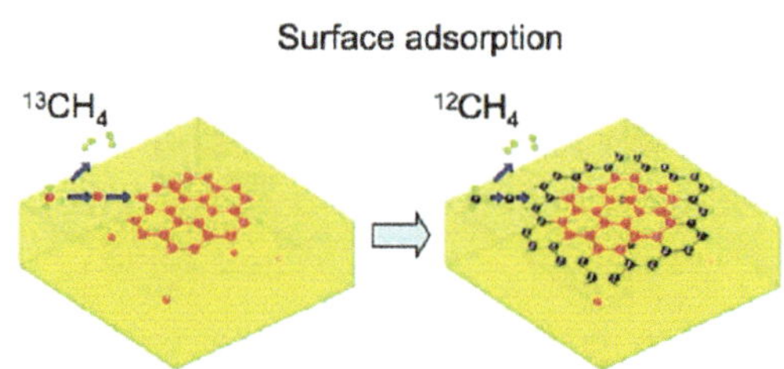

Given the absorption mechanism of graphene growth on Cu foil, the area coverage rate $v_{area}(t)$ should be proportional to the uncovered area on foil as the C cannot be absorbed to the covered area anymore. We therefore postulate the rate of area coverage as

$$v_{area}(t) = \frac{dS(t)}{dt} \propto S^* - S(t) \tag{13.14}$$

where S^* is the maximum area to be covered on a given Cu foil. For some parameter $a > 0$ as the rate coefficient, we have

$$\frac{dS(t)}{dt} = a(S^* - S(t)) \tag{13.15}$$

Solving this ordinary differential equation, we obtain the percentage of coverage $S(t)/S^*$ and the graphene area growth velocity model $v_{area}(t)$ as

$$\frac{S(t)}{S^*} = 1 - e^{-at} \tag{13.16}$$

$$v_{area}(t) = aS^* e^{-at} \tag{13.17}$$

The model suggests that Cu foil is covered relatively fast initially and will gradually slow down as the remaining uncovered area decreases.

Remark Model (13.16) is actually the so-called confined exponential growth model in the statistical literature [2], where the model has been widely used to describe growth phenomena such as technology transfer [22], diffusion theories, and models in geography [7], social science [11], and agriculture [3]. Generally speaking, when the growth rate is proportional to the remaining resources and there is no competition for resources, the confined exponential model would nicely describe the growth processes. This understanding is consistent with the description of absorption mechanism in [15], which reinforces our modeling confidence.

Deriving Kinetics Model for Graphene Growth on Cu foil Once we have $v_{area}(t)$ from (13.17), we finally can derive the growth kinetics model $f(t)$ based on model (13.8) or (13.13). Specifically, from (13.8), we derive the kinetics model as

$$f(t) = \frac{aS^* e^{-at}}{\sqrt{2N(S^* - S^* e^{-at})}} \tag{13.18}$$

$$= a\sqrt{\frac{S^*}{2N}} \frac{cosh(at) - sinh(at)}{\sqrt{(1 - cosh(at) + sinh(at))}}$$

And $F(t)$ or the length of each side of the equivalent square shape island is

$$F(t) = \sqrt{\frac{S^*}{N}(1 - e^{-at})} \qquad (13.19)$$

$$= \sqrt{\frac{S^*}{N}(1 - cosh(at) + sinh(at))}$$

Case Studies We further validate our model using two experiments in [16]. Data in Figs. 13.5 and 13.6 show the percentage of graphene coverage on Cu foils under two different process conditions, i.e., the ratio of covered area $S(t)$ over the actual total Cu area A_{Cu} at time t. In Fig. 13.5 the Cu surface was exposed to methane at a pressure of 285 mTorr, and the surface was fully covered with graphene after about 1.5 min. In Fig. 13.6, the methane pressure was 160 mTorr, and the Cu surface coverage reached only 90

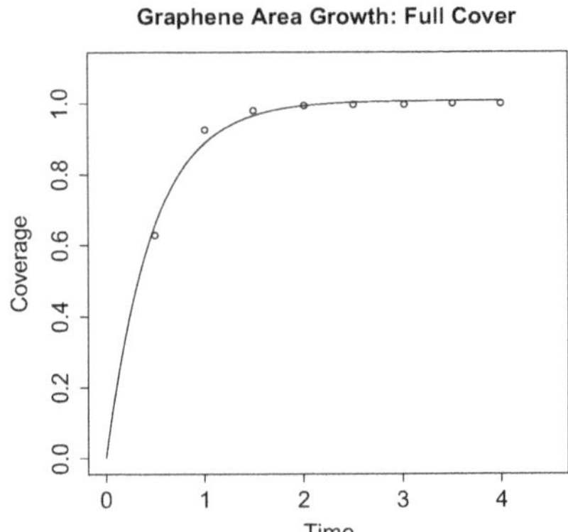

Fig. 13.5 Growth at pressure 285 mTorr

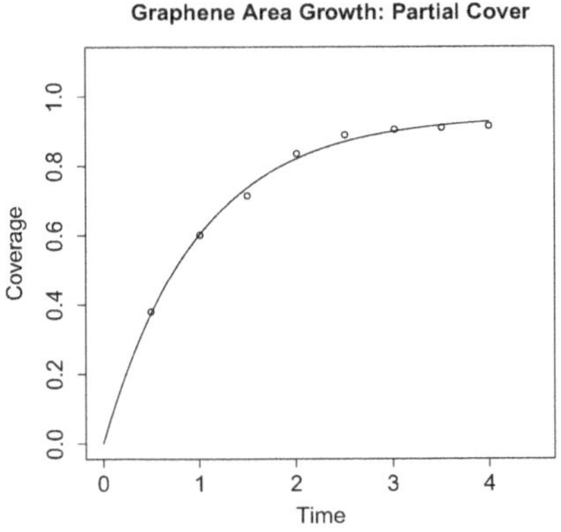

Fig. 13.6 Growth at pressure 160 mTorr

Although the exact total copper area A_{Cu} was not provided in [16], we can simply modify model (13.16) as

$$\frac{S(t)}{A_{Cu}} = \frac{S^*}{A_{Cu}}(1 - e^{-at}) \tag{13.20}$$

To fit the parameters S^*/A_{Cu} and a in this model, we use nonlinear regression method implemented by nls() in R software. The following tables show the estimated parameters and their standard errors. The curves in Figs. 13.5 and 13.6 represent the least square fitting of the modified confined exponential model.

Remark The theoretical limit of covered area S^* can be larger, equal to, or smaller than the total area of Cu foil A_{Cu}. As can been seen from Table 13.1, the theoretical limit S^* in the experiment of Fig. 13.5 is close to Cu foil area A_{Cu} with $S^*/A_{Cu} = 1.007424$, while in the experiment of Fig. 13.6, the process can only reach the limit of $S^*/A_{Cu} = 0.94348$. To design a new graphene growth process, S^* should be far greater than A_{Cu} if 100% coverage of Cu foils in a short time period is required.

We derive the kinetics model for the experiment of Fig. 13.5 given that there are N_1 islands

$$f(t) = \frac{1.5062\sqrt{A_{Cu}/N_1}\,e^{-2.122291t}}{\sqrt{1 - e^{-2.122291t}}} \tag{13.21}$$

$$F(t) = 1.0037\sqrt{A_{Cu}/N_1}\sqrt{1 - e^{-2.122291t}}$$

For the experiment of Fig. 13.6 with N_2 islands, we have

$$f(t) = \frac{0.6724\sqrt{A_{Cu}/N_2}\,e^{-1.01570t}}{\sqrt{(1 - e^{-1.01570t})}} \tag{13.22}$$

$$F(t) = 0.9713\sqrt{A_{Cu}/N_2}\sqrt{1 - e^{-1.01570t}}$$

Note that N/A_{Cu} is defined in [16] as the domain density, i.e., the average number of graphene islands per unit area.

Remark If the domain densities are both 10 $(\mu m^2)^{-1}$ in the two experiments, we can get the kinetic from the fitted results illustrated in Fig. 13.7, it can be seen that at

Table 13.1 Estimation of parameters in the confined exponential model for two cases

Estimation results		Estimate	Std. error	t value	$Pr(> |t|)$
Fully coverage	S^*/A_{Cu}	1.007424	0.009513	105.90	$4.78 * 10^{-11}$
	a	2.122291	0.113784	18.65	$1.53 * 10^{-6}$
Partial coverage	S^*/A_{Cu}	0.94348	0.01085	86.96	$1.56 * 10^{-10}$
	a	1.01570	0.03975	25.55	$2.37 * 10^{-7}$

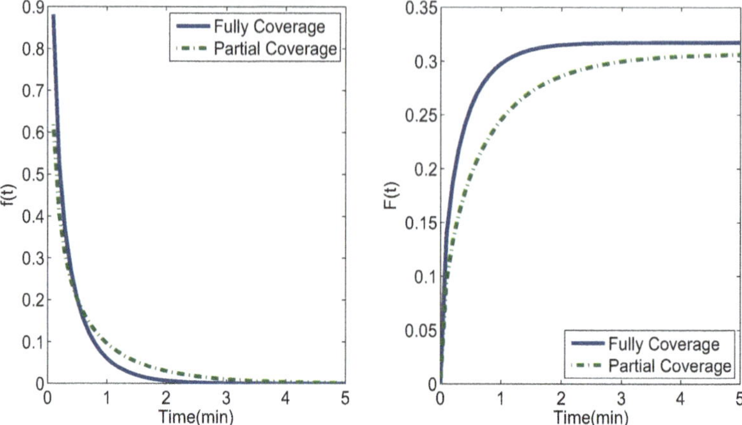

Fig. 13.7 Growth kinetics of two experiments

a higher gas pressure (285 mTorr), graphene grew much faster at the beginning and slowed down more quickly after most of the area was covered. It is also noted that the asymptotic side length $F(t)$ of the equivalent square shape island is almost the same for both experiments (0.3174 vs. 0.3072), which suggests that the final size or area of an individual graphene island *cannot be substantially increased by varying gas pressures* (3.3% increase by area from 160 mTorr to 285 mTorr). But the rate of reaching the theoretical limit will be higher by increasing the gas pressure for graphene growth under the experimental conditions in [16].

13.3.2 Graphene Island Shape Modeling

To establish the angular dependent growth velocity $v_i(\theta, t)$ model in (13.1), the final step is to model the shape of graphene islands or $g_i(\theta)$. The following model is proposed

$$g_i(\theta) = g(\theta) + \zeta_i \quad and \quad \zeta_i \sim N(0, \sigma^2). \tag{13.23}$$

Note that we could also accommodate the time-varying shape by assuming $g_i(\theta) = g(\theta) + \zeta_i(t)$ and conduct statistical testing on the time-varying property if data are available.

Specifically, for four-lope shape islands, we measured 20 islands and present the data (light blue) in the polar coordinate system (Fig. 13.8).

It can be seen that each island can be divided into eight segments and each segment can be approximated by an exponential function. We therefore propose a parametric shape model for $g(\theta)$ as

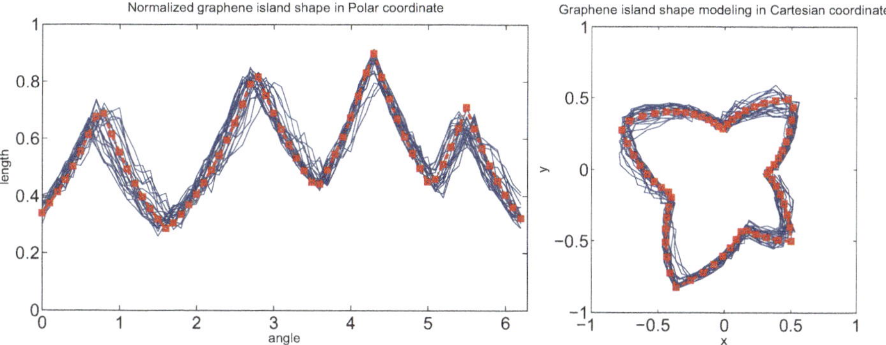

Fig. 13.8 Fitting result of parametric graphene island shape model

Table 13.2 Estimation of parameters in the parametric shape model

	c_1	c_2	c_3	c_4	c_5	c_6	c_7	c_8
Estimation	0.3394	1.6491	0.0602	8.9700	0.0099	62.8768	0.0018	342.5271
Std. error	0.0231	0.1213	0.0042	0.7620	0.0007	5.4614	0.0002	26.1146
	k_1	k_2	k_3	k_4	k_5	k_6	k_7	k_8
Estimation	0.9855	-1.0955	0.9543	-0.8558	1.0538	-0.9875	1.0899	-1.1234
Std. error	0.0826	0.1024	0.0716	0.0924	0.0912	0.0646	0.1018	0.0928

$$
g(\theta) = \begin{cases}
c_1 e^{k_1 \theta} & (0 \le \theta < \frac{\log c_1 - \log c_2}{k_2 - k_1}) & \text{(13.24a)} \\[2ex]
c_i e^{k_i \theta} & (\frac{\log c_{i-1} - \log c_i}{k_i - k_{i-1}} \le \theta < \frac{\log c_i - \log c_{i+1}}{k_{i+1} - k_i}, 2 \le i \le 7) & \text{(13.24b)} \\[2ex]
c_8 e^{k_8 \theta} & (\frac{\log c_7 - \log c_8}{k_8 - k_7} \le \theta < 2\pi) & \text{(13.24c)}
\end{cases}
$$

where $c_i, k_i, 1 \le i \le 8$ are the parameters to be estimated. Recalling the discussion in Sect. 13.2, we need to normalize $g(\theta)$ such that $C = \int_0^{2\pi} g(\theta)^2 d\theta = 2$.

The parameters of the model are estimated by maximizing the log-likelihood function as follows:

$$
L(\theta; \sigma^2) = -\frac{1}{2\sigma^2} \sum_{i=1}^{20} \sum_{j=1}^{n} (g_i(\theta_j) - g(\theta_j))^2 - 10n\log(\sigma^2) \tag{13.25}
$$

Using *fmincon* function in MATLAB, we obtain model parameters listed in Table 13.2. The standard errors of each estimate are derived form the corresponding diagonal elements of the inverse of the Fisher information matrix. The fitted model is shown in Fig. 13.8 (diamond marked curve). We can see that the proposed model catches the key characterization of graphene island shape.

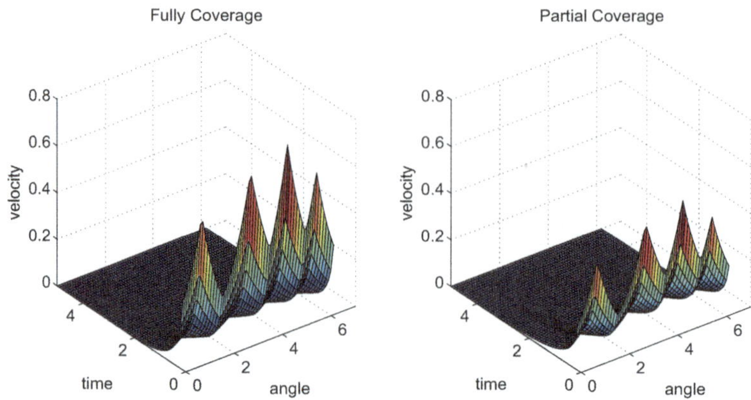

Fig. 13.9 Graphene growth rate $v_i(\theta, t)$

Connecting the growth rate models with the shape analysis, we can have an overall view of the graphene growth process on the copper foil. From (13.1), (13.18) and (13.23), we get the mean angular dependent velocity

$$v(\theta, t) = a\sqrt{\frac{S^*}{2N}}\frac{ae^{-at}}{\sqrt{(1 - e^{-at})}}g(\theta)$$

$$= a\sqrt{\frac{S^*}{2N}}\frac{cosh(at) - sinh(at)}{\sqrt{(1 - cosh(at) + sinh(at))}}g(\theta) \qquad (13.26)$$

and Fig. 13.9 illustrates $v(\theta, t)$ for the two cases in Figs. 13.5 and 13.6. The model is certainly helpful to conduct simulation study of graphene growth with any shape.

13.4 Discussion on Model Applications

The developed graphene growth model can be applied in many ways other than process simulation. Figure 13.7 already demonstrated the comparison study of process parameters, i.e., gas pressure and their impact on growth kinetics and the size of graphene islands. Our analysis suggests that the final size or area of an individual graphene island cannot be substantially increased by varying gas pressures. But the rate of reaching the theoretical limit will be higher by increasing the gas pressure for graphene growth under the experimental conditions in [16]. In this section, we further discuss some process insights derived from the graphene process model, where the derived information can help the control and optimization of graphene fabrication.

The first scenario is related to process design and selection. Suppose the specification for growth process is to achieve 90% coverage at $t = 1$ min, and we need to select a capable process. Based on model $S(t)/A_{\text{Cu}} = S^*/A_{\text{Cu}}(1 - e^{-at})$, we can derive $S(t)/A_{\text{Cu}} = S^*/A_{\text{Cu}}(1 - e^{-2.3026t})$ from the process specification. From model (13.18), we can derive the required process with kinetics model as $f(t) = \dfrac{1.6282\sqrt{A_{\text{Cu}}/N}e^{-2.3026t}}{\sqrt{1-e^{-2.3026t}}}$ and a domain density of N/A_{Cu}.

The second scenario is to explore the process requirement for reducing the domain density. Since high domain density will increase the in-plane electric resistance due to grain boundaries, growing single crystalline graphene sheet is desirable for electronic applications. With the derived growth kinetics model, we could investigate the change of the process kinetic $f(t)$ with varying domain densities N/A_{Cu} (Fig. 13.10). If we establish the knowledge base of connecting between $f(t)$ with physical processes, the study will assist to select capable processes meeting the requirement for domain densities.

This chapter introduces a physical–statistical learning framework to describe the graphene growth processes. The graphene islands on the same substrate are assumed to increase their sizes at the same rate. Within each island, the growth speed varies from angle to angle, and the difference is carved out by its shape function. By separating the growth kinetics from geometric shapes, we transform a graphene island with any shape to an equivalent square shape. The derived kinetics model can be conveniently interpreted as the velocity of side growth of the equivalent square shape. This theoretical understanding enables us to investigate and compare the growth kinetics of graphene islands with various shapes.

Through discussions, we demonstrated the applications of the developed analytical framework, for instance, the selection of Cu foil area, the investigation of the limiting coverage area and its connection with growth parameters, and process design. Although we demonstrate the methodology for graphene on the copper foil, the framework is generic and applicable to other graphene processes as well. Our

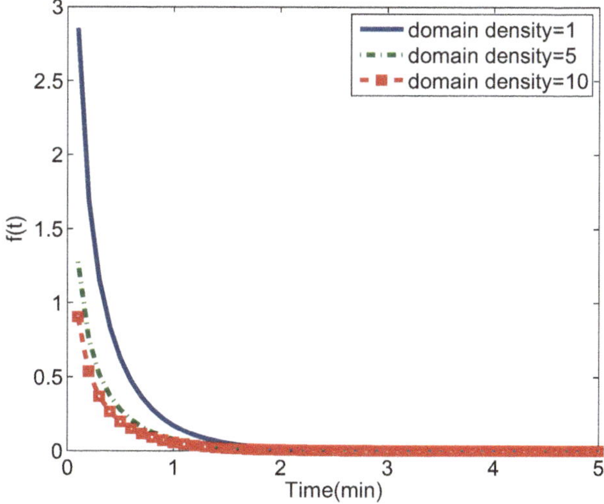

Fig. 13.10 Process kinetics with varying domain densities

future study will directly relate the derived kinetics model with process parameters, which assists to provide more insights on the control of graphene manufacturing

Acknowledgments The work in this chapter was supported by US National Science Foundation under CAREER grant number CMMI-1055394.

References

1. Allen, D., Hunt, J.: Diffusion in the semi-solid region during dendritic growth. Metallurg. Mater. Trans. A **10**(9), 1389–1397 (1979)
2. Banks, R.: Growth and Diffusion Phenomena: Mathematical Frameworks and Applications, vol. 14. Springer, New York (1994)
3. Batschelet, E.: Introduction to Mathematics for Life Scientists, vol. 2. Springer, New York (1979)
4. Bolotin, K., Sikes, K., Hone, J., Stormer, H., Kim, P.: Temperature-dependent transport in suspended graphene. Phys. Rev. Lett. **101**(9), 96802 (2008)
5. Feigelson, R.S.: 50 Years Progress in Crystal Growth: A Reprint Collection. Elsevier B.V., New York (2004)
6. Geim, A., Novoselov, K.: The rise of graphene. Nature Mater. **6**(3), 183–191 (2007)
7. Griffith, D., Lea, A.: Evolving Geographical Structures: Mathematical Models and Theories for Space-Time Processes, vol. 15. Kluwer Academic, Dordrecht (1983)
8. Huang, Q.: Physics-driven bayesian hierarchical modeling of the nanowire growth process at each scale. IIE Trans. **43**(1), 1–11 (2010)
9. Huang, Q., Wang, L., Dasgupta, T., Zhu, L., Sekhar, P., Bhansali, S., An, Y.: Statistical weight kinetics modeling and estimation for silica nanowire growth catalyzed by pd thin film. IEEE Trans. Autom. Sci. Eng. (99) **8**(2), 303–310 (2011)
10. Ivantsov, G.: The temperature field around a spherical, cylindrical, or pointed crystal growing in a cooling solution. Dokl. Akad. Nauk SSSR **58**, 567–569 (1947)
11. Kemeny, J., Snell, J.: Mathematical Models in the Social Sciences. MIT Press, Cambridge, MA (1978)
12. Kim, K., Zhao, Y., Jang, H., Lee, S., Kim, J., Kim, K., Ahn, J., Kim, P., Choi, J., Hong, B.: Large-scale pattern growth of graphene films for stretchable transparent electrodes. Nature **457**(7230), 706–710 (2009)
13. Kumar, A., Zhou, C.: The race to replace tin-doped indium oxide: which material will win? ACS Nano **4**(1), 11–14 (2010)
14. Li, X., Cai, W., An, J., Kim, S., Nah, J., Yang, D., Piner, R., Velamakanni, A., Jung, I., Tutuc, E., Banerjee, S.K., Colombo, L., Ruoff, R.S.: Large-area synthesis of high-quality and uniform graphene films on copper foils. Science **324**, 1312 (2009)
15. Li, X., Cai, W., Colombo, L., Ruoff, R.: Evolution of graphene growth on ni and cu by carbon isotope labeling. Nano Lett. **9**(12), 4268–4272 (2009)
16. Li, X., Magnuson, C., Venugopal, A., An, J., Suk, J., Han, B., Borysiak, M., Cai, W., Velamakanni, A., Zhu, Y., et al.: Graphene films with large domain size by a two-step chemical vapor deposition process. Nano Lett. **10**(11), 4328–4334 (2010)
17. Loginova, E., Bartelt, N., Feibelman, P., McCarty, K.: Evidence for graphene growth by c cluster attachment. New J. Phys. **10**, 093026 (2008)
18. Loginova, E., Bartelt, N., Feibelman, P., McCarty, K.: Factors influencing graphene growth on metal surfaces. New J. Phys. **11**, 063046 (2009)
19. N'Diaye, A., Bleikamp, S., Feibelman, P., Michely, T.: Two-dimensional ir cluster lattice on a graphene moire on ir (111). Phys. Rev. Lett. **97**(21), 215501 (2006)

20. Novoselov, K., Geim, A., Morozov, S., Jiang, D., Zhang, Y., Dubonos, S., Grigorieva, I., Firsov, A.: Electric field effect in atomically thin carbon films. Science **306**(5696), 666–669 (2004)
21. Shafiei, S., Nourbakhsh, A., Ganjipour, B., Zahedifar, M., Vakili-Nezhaad, G.: Diameter optimization of vls-synthesized zno nanowires, using statistical design of diameter optimization of vls-synthesized zno nanowires, using statistical design of experiment. Nanotechnology **18**, 355708 (2007)
22. Sharif, N., for Technology Transfer (India), R.C., Economic, U.N., for Asia, S.C., the Pacific: Management of Technology Transfer and Development. UNESCAP (1983)
23. Stankovich, S., Dikin, D., Dommett, G., Kohlhaas, K., Zimney, E., Stach, E., Piner, R., Nguyen, S., Ruoff, R.: Graphene-based composite materials. Nature **442**(7100), 282–286 (2006)
24. Stoller, M., Park, S., Zhu, Y., An, J., Ruoff, R.: Graphene-based ultracapacitors. Nano Lett. **8**(10), 3498–3502 (2008)
25. Sutter, P., Flege, J., Sutter, E.: Epitaxial graphene on ruthenium. Nature Mater. **7**(5), 406–411 (2008)
26. Warner, J., Rümmeli, M., Gemming, T., Büchner, B., Briggs, G.: Direct imaging of rotational stacking faults in few layer graphene. Nano Lett. **9**(1), 102–106 (2008)
27. Wofford, J., Nie, S., McCarty, K., Bartelt, N., Dubon, O.: Graphene islands on cu foils: the interplay between shape, orientation, and defects. Nano Lett. **10**(12), 4890–4896 (2010)
28. Wu, B., Geng, D., Guo, Y., Huang, L., Xue, Y., Zheng, J., Chen, J., Yu, G., Liu, Y., Jiang, L., et al.: Equiangular hexagon-shape-controlled synthesis of graphene on copper surface. Adv. Mater. **23**(31), 3522–3525 (2011)
29. Yu, Q., Lian, J., Siriponglert, S., Li, H., Chen, Y., Pei, S.: Graphene segregated on ni surfaces and transferred to insulators. Appl. Phys. Lett. **93**, 113103 (2008)
30. Zhang, Y., Zhang, L., Kim, P., Ge, M., Li, Z., Zhou, C.: Vapor trapping growth of single-crystalline graphene flowers: Synthesis, morphology, and electronic properties. Nano Lett. **12**(6), 2810–2816 (2012)

Chapter 14
Stochastic Modeling of Graphene Growth Kinetics

Graphene is an emerging nanomaterial for a wide variety of novel applications. Controlled synthesis of high-quality graphene sheets requires analytical understanding of graphene growth kinetics. The graphene growth via chemical vapor deposition starts with randomly nucleated islands that gradually develop into complex shapes, grow in size, and eventually connect together to form a graphene sheet. Models proposed for this stochastic process do not, in general, permit assessment of uncertainty. This chapter introduces a stochastic modeling framework for the growth process and Bayesian inferential models. The modeling approach accounts for the data collection mechanism and allows for uncertainty analyses, for learning about the kinetics from experimental data. Furthermore, we link the growth kinetics with controllable experimental factors, thus providing a framework for statistical design and analysis of future experiments.

14.1 Uncertainties in Graphene Growth Process Modeling

Graphene is a flat monolayer of carbon atoms tightly packed into a two-dimensional (2D) honeycomb lattice and is a basic building block for graphitic materials of all other dimensionalities [5]. In the realm of materials science and condensed matter physics, graphene represents an exciting new class of materials that continually offers new inroads into low-dimensional physics [5] and a broad spectrum of applications including optoelectronics, chemical, and biosensing [9, 18, 19]. The unique room-temperature properties of graphene, such as the quantum hall effect [17] and high electron mobility [3], make its high yield production particularly desirable.

Although several methods currently exist for synthesizing graphene, recent interest has been focused on synthesizing graphene through the chemical decomposition of hydrocarbons on metals due to relatively low production costs. Specifically,

Q. Huang, *Domain-informed Machine Learning for Smart Manufacturing*,
https://doi.org/10.1007/978-3-031-91631-1_14

chemical vapor deposition of hydrocarbons on copper has proved particularly promising in producing high-quality large-area monolayer graphene [10] and is thus our focus.

The synthesis of large-area high-quality monolayer graphene requires some understanding of its growth mechanism and the impact of underlying process factors (e.g., temperature and pressure), since faster growth kinetics will lead to larger-area graphene. The mechanisms underlying graphene growth and their relationship to the process conditions have recently received much interest [2, 4, 8, 20, 23], albeit through deterministic models. Chapter 13 [22] introduced a physical–statistical modeling approach to deterministically model the growth kinetics of the graphene flakes, using the well-studied confined exponential model. The model, henceforth referred to as the WH-model [22], neither permits quantification of uncertainty associated with the growth kinetics nor relates the kinetics to the process conditions. This chapter introduces a stochastic growth model, which shares some mathematical properties with the WH-model, links it to the growth kinetics, and then proposes inferential models for learning about the kinetics from experimental data. Furthermore, we propose links between controllable experimental factors and the kinetics and consequently provide a framework for the statistical design and analysis of future experiments.

14.1.1 Brief Overview of Graphene Growth and the WH-Model

Reference [7] clearly explain the processes involved in the formation of graphene on copper. A typical experimental run consists of passing methane over a copper foil (or tray), at some predetermined value for the process conditions (for example, [7] investigated the growth of graphene on a copper surface from $720\,°C$ to $1050\,°C$).

As stated by [7] and shown in Fig. 14.1, the breakdown of methane on the copper surface increases the concentration (C_{cu}) of the active carbon atoms (or adatoms), until it reaches a critical supersaturation level ($C_{nuc} \geq 3.82 \times 10^{15}\,\text{cm}^{-2}$), where formation of stable graphene nuclei (islands) takes place (Fig. 14.1 stage (i)). This formation is also referred to as *nucleation*. Reference [7] note that impurities, surface roughness, grain boundary grooves, and stepped terraces on the copper foil often play important roles in graphene nucleation. In particular, they demonstrate that sites exhibiting those properties are energetically favorable sites for nucleation and that nucleation occurs almost exclusively in those regions.

As the nucleation and growth of the graphene islands deplete the adsorbed carbon atoms surrounding them, the C_{cu} is quickly reduced to a level where the nucleation rate is negligible. The growth of the nuclei continues until the supersaturated amount of surface carbon atoms above the equilibrium level C_{eq} is consumed and an equilibrium between graphene, surface carbon, and methane is reached (Fig. 14.1 stage (ii)). This equilibrium is dependent on the twin balancing acts of adsorption/chemisorption and desorption, as well as attachment and detachment of carbon atoms to each other. Depending on the available carbon, graphene nuclei

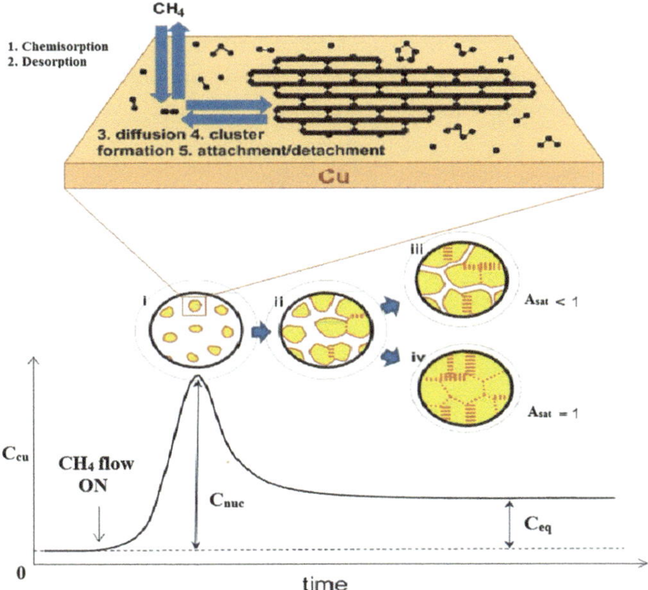

Fig. 14.1 Overall illustration of the nucleation and growth mechanism of graphene on copper [7]. Graphene growth proceeds from stage (i) to stage (ii), at which point it continues to either stage (iii) or stage (iv)

either coalesce to form eventually a saturated continuous film (Fig. 14.1 stage (iii)) or stop growing to reach a final incomplete coverage (Fig. 14.1 stage (iv)), where A_{sat} denotes the proportion of saturation on the copper surface.

The WH-model [22] in Chap. 13 seeks to uncover the kinetics driving the growth of graphene during its production process. Two fundamental assumptions of the model are: (a) single growth mechanism: individual graphene islands under the same process conditions are driven by the same kinetics, and (b) self-similarity: the growth kinetics and island shapes have separable effects on the area growth velocity of each graphene island. Based on the aforementioned assumptions, the following multiplicative model was proposed for the angular dependent area growth velocity, $v_i(\theta, t)$, of the ith graphene island at angle θ and time t:

$$v_i(\theta, t) = g_i(\theta)q(t) , \tag{14.1}$$

where $g_i(\theta)$ describes the shape of the ith graphene island, and $q(t)$ is the *deterministic* kinetics driving the growth at time t. It is important to note at this point that we describe the shape, $g_i(\theta)$, in polar coordinates as is natural when dealing with 2D closed profiles. Another way to interpret the model is to imagine that each island starts with shape $g_i(\theta)$ and is scaled up by the kinetic function [22].

In general, it is very difficult to observe or measure $v_i(\theta, t)$, and thus, this model is instead linked to the more measurable (and thus more easily modeled) total area covered on the copper foil/tray at time t, $S(t)$. In Fig. 14.1 for instance, $S(t)$ represents the sum total of the shaded areas within each circle in stages (i)–(iv). Denoting the constant number of graphene islands on the tray at any time t by N, [22] derived the following model for the growth kinetics:

$$q(t) = \frac{\frac{d}{dt} S(t)}{2\sqrt{N S(t)}}, \tag{14.2}$$

based on the assumption that $g_i(\theta)$ follows the model:

$$g_i(\theta) = g(\theta) + \epsilon_i, \quad \text{where} \quad \epsilon_i \sim \mathcal{N}(0, \sigma_\epsilon^2). \tag{14.3}$$

The aforementioned model appears to have several limitations. First, it does not allow for modeling and evaluation of uncertainty, because (14.2) contradicts the assumption that $q(t)$ is deterministic. This is due to the fact that all the $g_i(\theta)$, and consequently $S(t)$, are random. Second, the assumption of N remaining constant over time is clearly unrealistic and overly restrictive, especially for the first few instants of time. Third, the model does not account for the time-varying random behavior (attachment/detachment) of carbon atoms along the profile of each graphene island. Fourth, it does not suggest a way to link different process conditions to the growth kinetics. Finally, it does not allow us to understand what type of experimental data are needed to estimate the model parameters and how to analyze such data with the objective of estimating model parameters and making predictions from the fitted model.

Some of the models proposed in this chapter retain the fundamental structure (14.1) of the WH-model, but make several generalizations. First, the component shape $g_i(\theta)$ is assumed to be random, but restricted to have area, A_{nuc}. As [21] note, a simple estimate for the critical size of surviving carbon cluster is about 140–320 atoms. Combining this with the lower bound for C_{nuc} (described earlier) gives the estimate $A_{nuc} \approx 6.02 \times 10^{-14}\,\mathrm{cm}^2$. We note here that $A_{nuc} = 1$ was proposed by [22], and this is appropriate only if the goal is to compare the kinetics for different copper trays. In our proposed model, the kinetics are reformulated in terms of $E(S(t))$ (where $E(X)$ stands for expectation of a random variable X), through acknowledgment of the effect of random carbon atom attachment and detachment along the profile of each island. Another major change is the use of the random process $N(t)$ instead of a constant (N) number of graphene islands to make the model much more realistic. Next, we consider the possibility of having multiple trays of graphene islands, each tray representing a single experimental run at a specific value of the process conditions. Such a representation facilitates the task of linking the growth kinetics with the process conditions. Finally, we present a parametric approach for estimating $q(t)$, along with a procedure for quantifying the uncertainty in the estimation as well. In particular, we focus on Bayesian methods as they provide a natural framework for specifying the interdependence of the various components of the parametric model.

14.2   A Generalized Stochastic Graphene Growth Kinetics Model

Let $N(t)$ denote the random number of islands on the tray at time t. Denoting by $v_i(\theta, t)$, the angular velocity of the ith island at time t and angle θ, and $\tilde{v}_i(\theta, t)$ its expectation over the random microscale attachment/detachment of carbon atoms, we can generalize (14.1) to

$$\tilde{v}_i(\theta, t) = E\left[v_i(\theta, t)\right] = g_i(\theta)q(t), \qquad \{i = 1, \ldots, N(t)\}, \qquad (14.4)$$

where $g_i(\theta)$ denotes the fixed area random shape of island i and $q(t)$ denotes the kinetics driving the growth time t. Our objective is to estimate $q(t)$ as a function of time t and to eventually compare $q(t)$ for multiple process conditions, since each kinetics curve reflects the effects of the process conditions on the growth rate.

Let $t_{0i} \geq 0$ denote the random time the nucleation process begins for the ith island, and let t_{0i}^* (where $t_{0i}^* > t_{0i}$) denote the random time of nucleation for that island. Note that $t_{0i}^* - t_{0i} \approx 0$ due to the relatively small value of A_{nuc}, and for mathematical simplicity, we will assume $t_{0i}^* - t_{0i}$ to be infinitesimally small.

Now, let $S_i(t_{0i}, t)$ denote the area covered by the ith island by time t, and let $\tilde{S}_i(t_{0i}, t)$ denote the equivalent area covered when the random microscale behavior of the carbon atoms is averaged out. Two hypothetical curves $S_i(t_{0i}, t)$, one with $t_{0i} = 0$, i.e., instant nucleation (broken-line curve) and the other with an arbitrary $t_{0i} > 0$ (solid-line curve), are shown in Fig. 14.2.

We mention at this point (and show in chapter Appendix A1) that $S_i(t_{0i}, t)$ only depends on the random shape, $g_i(\theta)$, through its non-random area, A_{nuc}. Therefore, any randomness in $S_i(t_{0i}, t)$ is completely determined by the distribution of the nucleation times, t_{0i}, as well as the random attachment/detachment of carbon atoms

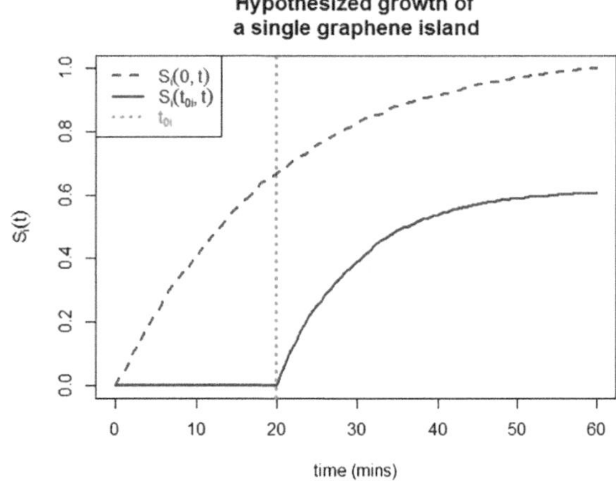

Fig. 14.2 Hypothesized growth curves for a single island under assumptions of instant ($S_i(0, t)$), and random delayed ($S_i(t_{0i}, t)$) nucleation.

to the edges of each island. Consequently, any randomness in $\tilde{S}_i(t_{0i}, t)$ is completely explained by the distribution of the nucleation times, t_{0i}.

The total area $S.(t)$ covered by the graphene islands on the tray by time t can be expressed as

$$S.(t) = \sum_{i=1}^{N(t)} S_i(t_{0i}, t) = \sum_{i=1}^{N(\infty)} S_i(t_{0i}, t) \cdot 1\{t_{0i}^* \le t\} = \sum_{i=1}^{N(\infty)} S_i(t_{0i}, t) \cdot 1\{t_{0i} < t\},$$

(14.5)

and equivalently, under the process where the random behavior of the carbon atoms is averaged out, the total area $\tilde{S}.(t)$ covered by the graphene islands on the tray by time t can be expressed as

$$\tilde{S}.(t) = \sum_{i=1}^{N(t)} \tilde{S}_i(t_{0i}, t) = \sum_{i=1}^{N(\infty)} \tilde{S}_i(t_{0i}, t) \cdot 1\{t_{0i}^* \le t\} = \sum_{i=1}^{N(\infty)} \tilde{S}_i(t_{0i}, t) \cdot 1\{t_{0i} < t\}.$$

(14.6)

We will now make use of a simple example to show the effects of the nucleation times and random attachment/detachment of carbon atoms on a hypothesized tray.

Example: Individual Island Growth

Suppose, for a hypothetical growth process for graphene, we have that $q(t) = 100e^{-t}$. Further suppose that we observe three different graphene islands, each with a distinct shape ($g_i : i = 1, 2, 3$) and nucleation time ($t_{0i} : i = 1, 2, 3$). We let g_1, g_2 and g_3 represent specific circular, three-lobed, and four-lobed shapes, respectively, with A_{nuc} set to 10^{-5} cm^2. Specifically, we first select coefficients $\beta_{i,k}$ ($i = 1, 2, 3; k = 0, 1, \ldots, 8$) for unnormalized shapes h_1, h_2 and h_3 in polar coordinates such that

$$\log(h_i(\theta)) = \beta_{i,0} + \sum_{m=1}^{4} \beta_{i,2m-1} \cos(m\theta) + \beta_{i,2m} \sin(m\theta),$$

and then we normalize appropriately by defining g_1, g_2 and g_3 through

$$g_i(\theta) = \sqrt{A_{nuc}} \cdot \frac{h_i(\theta)}{\sqrt{\frac{1}{2} \int_0^{2\pi} h_i^2(\theta)d\theta}}.$$

For the circlular shape, g_1, we set $\beta_{1,0} = 1$, and set $\beta_{1,k} = 0$ for all $k \ne 0$, so that $g_1(\theta)$ is constant, as expected of circles. For the three-lobed shape, g_2, we set $\beta_{2,k} = 1$ for $k \in \{0, 5, 6\}$, and set $\beta_{2,k} = 0$ otherwise. Finally, for the four-lobed shape, g_3, we set $\beta_{3,k} = 1$ for $k \in \{0, 7, 8\}$, and set $\beta_{3,k} = 0$ otherwise.

Next, we set nucleation times, $t_{01} = 0$ mins, $t_{02} = 0$ mins and $t_{03} = 1$ min, and select observation times $t = 0$, 0.2, 1, or 4.5 mins. In order to represent the effect of random carbon attachment and detachment from the profile edges, we add some noise to (14.4). Specifically, we set $v_i(\theta, t) = \tilde{v}_i(\theta, t) \cdot e_i(\theta, t)$, where $e_i(\theta, t)$ is a log-normal random variable with parameters $\mu = -0.005$ and $\sigma = 0.1$, i.e., $e_i(\theta, t) \sim \mathcal{LN}(-0.005, 0.1^2)$. Note that $E(e_i(\theta, t)) = 1$ always. Figure 14.3a shows the underlying growth process on the tray, when the random carbon attachment is averaged out, while Fig. 14.3b shows the corresponding growth process as observed.

From Fig. 14.3a, we can see that average growth along any direction is proportional to the initial shape at nucleation, and that the total area covered on the tray is highly dependent on the nucleation time for each graphene island. Taking the circular island as an example under this average growth process, the radius and area at time $t = t_{01}$ are given by 0 cm and 0 cm^2, respectively. At time $t = t_{01}^* = t_{01} + \epsilon$ (for some $\epsilon > 0$, an infinitesimally small number by assumption), the radius and area are $\sqrt{A_{nuc}/\pi}$ cm and A_{nuc} cm^2, respectively. More generally, at some general time t mins (where $t > t_{01}^*$), the radius and area become $\sqrt{A_{nuc}/\pi} \cdot 100(1 - e^{-t})$ and $A_{nuc} \cdot \left[100(1 - e^{-t})\right]^2$ respectively. Note that the assumption of infinitesimally small $t_{0i}^* - t_{0i}$ implies that the interval (t_{0i}, t_{0i}^*) will contain a general observation time t with zero probability and allows us to use the same distribution for t_{0i} and t_{0i}^*. Furthermore, according to (14.6), the total area covered on the tray, by all three graphene islands in our example, at time $t = 4.5$ mins is therefore $\tilde{S}(4.5) = \tilde{S}_1(0, 4.5) + \tilde{S}_2(0, 4.5) + \tilde{S}_3(1, 4.5)$, where $\tilde{S}_1(0, 4.5)$, $\tilde{S}_2(0, 4.5)$, and $\tilde{S}_3(1, 4.5)$ are the areas of the three shapes observed in the bottom right panel of Fig. 14.3a.

Figure 14.3b shows the lack of smoothness along the profile of each island, resulting from the random behavior of carbon atoms along the edge of each island. Since we will henceforth be concerned only with aggregated measures, such as the total area covered on each tray, such detailed modeling of the observed carbon atom behavior along the edge of each island will be unnecessary, and we will only need to model the noisy behavior at the aggregated level.

We now state a result (proof in chapter Appendix A1), which provides us with a closed-form expression for the kinetics, $q(t)$.

1 Assuming the multiplicative model (14.4), the kinetics $q(t)$ driving the growth on a specific tray has the following form:

$$q(t) = \frac{1}{\sqrt{A_{nuc}}} \left[\frac{\xi'(t) - \varpi'(t)}{2\sqrt{\xi(t) - \varpi(t)}} + \varphi'(t) \right] , \qquad (14.7)$$

for $\xi(t) > \varpi(t)$ at all times ($t > 0$), where

$$\xi(t) = E\left(\tilde{S}_i(t_{0i}, t) | t_{0i} \leq t \right) ,$$

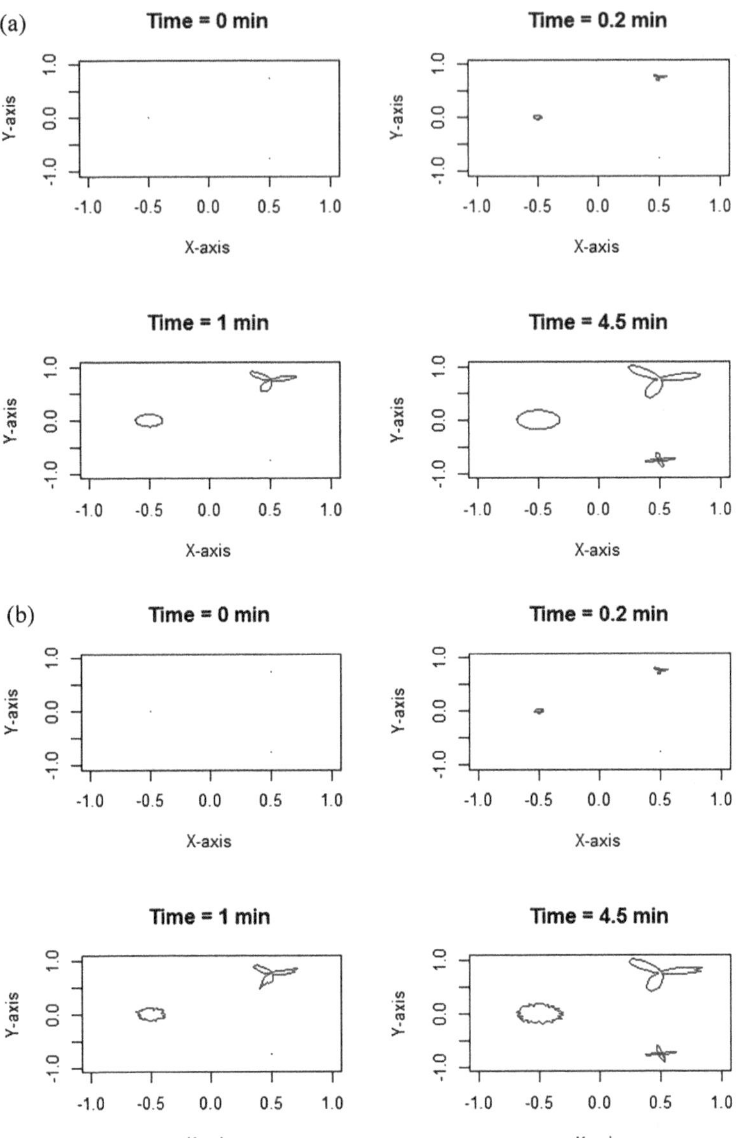

Fig. 14.3 (a) The growth of three graphene islands averaged over random carbon attachment. (b) The actual growth of three graphene islands. The rough profiles show the effect of random attachment of carbon atoms

$$\varphi(t) = E\left(\sqrt{\tilde{S}_i(0, t_{0i})}|t_{0i} \leq t\right), \text{ and}$$

$$\varpi(t) = Var\left(\sqrt{\tilde{S}_i(0, t_{0i})}|t_{0i} \leq t\right).$$

Remarks

(1) Note that the area forfeited at any time t by an island with nucleation time t_{0i} is given by the difference $S_i(0, t) - S_i(t_{0i}, t)$. Thus, the quantity $\tilde{S}_i(0, t_{0i})$ under the square-root sign in the expressions for $\varphi(t)$ and $\varpi(t)$ can be interpreted as the forfeited area of the ith island at its nucleation time t_{0i}, when the random behavior of the edge carbon atoms is averaged out. This is because $S_i(t_{0i}, t_{0i}) = 0$ by definition.

(2) The expectations in the expressions for $q(t)$ and $\varphi(t)$ and the variance in the expression for $\varpi(t)$ are with respect to the distribution of the nucleation times, which we determine in Sect. 14.2.

(3) Under instant nucleation, $\tilde{S}_i(0, t)$ is the area covered by the ith island, when the random carbon atom behavior at the edge of the island is averaged out. In a departure from our example, $\tilde{S}_i(0, t)$ need not be implicitly defined through $\tilde{v}_i(\theta, t)$ and the distribution of carbon atom attachment/detachment; we can derive analytical expressions for $\tilde{S}_i(0, t)$ by first assuming suitable models for $E(S_i(0, t))$ and the nucleation time distribution $f(t_{0i})$, and then setting $\tilde{S}_i(0, t) = E(S_i(0, t))$. Note that in general, we use $\tilde{S}_i(t_{0i}, t) = E(S_i(t_{0i}, t)|t_{0i})$, so that we can rewrite $\xi(t)$ from the aforementioned theorem as

$$\xi(t) = E\left(\tilde{S}_i(t_{0i}, t)|t_{0i} \leq t\right) = E\left(S_i(t_{0i}, t)|t_{0i} \leq t\right).$$

(4) From Theorem 1 and the previous three remarks, it follows that in order to estimate $q(t)$, we need first to decide on suitable models for $E(S_i(0, t))$, $f(t_{0i})$ (the nucleation time distribution), and $E(S_i(t_{0i}, t))$ (the expected graphene area for ith island at time at t, with random nucleation time, $t_{0i} \sim f(t_{0i})$). These models are derived in Sects. 14.1–14.2. While developing these models, we make use of the models already proposed by [22] whenever appropriate. Multiple model modifications that make uncertainty analyses possible and that reflect more realistic conditions have additionally been proposed.

A Model for $E(S_i(0, t))$

Let $S^0(t)$ denote the graphene area on the tray under the assumption of zero nucleation times for all islands on the tray ($t_{0i} = 0$ for all i). From (14.5), we can write by substituting $t_{0i} = 0$

$$S^0_{\cdot}(t) = \sum_{i=1}^{N(\infty)} S_i(0, t) \,.$$

Applying the law of iterated expectations, we have that

$$E(S^0_{\cdot}(t)) = E(N(\infty)) \cdot E(S_i(0, t)) \,,$$

or

$$\tilde{S}_i(0, t) = E(S_i(0, t)) = \frac{E(S^0_{\cdot}(t))}{E(N(\infty))} \,. \tag{14.8}$$

In developing a model for $E(S^0_{\cdot}(t))$, we revisit the model proposed by [22], who noted that the area growth over the copper tray at time t (which we denote by $S^0(t)$) is well described by a confined exponential growth model, which is essentially the following first-order differential equation (ODE):

$$\frac{d}{dt}\{S^0(t)\} = \alpha(S^* - S^0(t)) \,.$$

Solving the ODE under the assumption of deterministic $S^0(t)$ then leads to the following expression for the area

$$S^0(t) = S^*(1 - e^{-\alpha t}) \,. \tag{14.9}$$

Here, $\alpha > 0$ is the parameter governing the expansion/growth on the tray, and S^* is the area of the copper tray (the maximum achievable area).

By assuming a constant number of islands, [22] implicitly assume zero nucleation times for all the islands. It is thus natural to develop a model for $E(S_i(0, t))$ from this confined exponential model. Furthermore, note that the limiting (at infinite time) area according to (14.9) is S^*, which indicates complete coverage of the tray. We relax this assumption and make an adjustment for incomplete coverage. Denoting the graphene area on the tray under the assumption of zero nucleation times for all islands on the tray as $S^0(t)$, now consider the following model for the expected area

$$E(S^0_{\cdot}(t)) = E(S^0_{\cdot}(\infty)) \cdot (1 - e^{-\alpha t}) \,, \tag{14.10}$$

where $\alpha > 0$ is the parameter governing the expansion/growth on the tray, and $E(S^0_{\cdot}(\infty))$ is the expected limiting area on that tray. By making the assumption that higher-valued growth parameters lead to more tray coverage, we further propose the following

$$E(S_.^0(\infty)) = (1 - e^{-\alpha\omega})S^* ,$$ (14.11)

where $\omega > 0$ is a parameter that governs the relationship between α and coverage. Consequently, from (14.10) and (14.11),

$$E(S_.^0(t)) = S^*(1 - e^{-\alpha\omega})(1 - e^{-\alpha t}) .$$ (14.12)

Substituting the expression for $E(S_.^0(t))$ from (14.12) into (14.8), the final model for $E(S_i(0, t))$ is thus given by

$$\tilde{S}_i(0, t) = E(S_i(0, t)) = \frac{(1 - e^{-\alpha\omega})S^*}{E(N(\infty))}(1 - e^{-\alpha t}) .$$ (14.13)

In the next section, we develop a distribution for the nucleation times, $f(t_{0i})$ and subsequently obtain $E(N(\infty))$.

A Model for $f(t_{0i})$, the Nucleation Time Distribution

One model for $f(t_{0i})$ stems from recognizing $N(t) = \sum_{i=1}^{\infty} 1_{\{t_{0i} \leq t\}}$ as a constrained birth process. By proposing the Poisson process $N(t) \sim PP(\lambda_j(t))$ (where $\lambda(t)$ is the underlying non-homogeneous intensity function for the tray), we have that:

$$f(t_{0i}) = \frac{\lambda(t_{0i})}{\int_0^{\infty} \lambda(s)ds} .$$ (14.14)

Any choice of the intensity function will need to take the following consideration into account. The appearance of a new graphene island depends on the proportion of area left untouched on the tray at time t. The less the available area, the lower the number of islands that can appear. We thus propose the following intensity function

$$\lambda(t) = \beta S^* \left(\frac{E(S_.^0(\infty)) - E(S_.^0(t))}{E(S_.^0(\infty))} \right) ,$$

where β is some positively valued parameter, S^* is the limiting area on the tray according to (14.9), and $E(S^0(t))$ is given by (14.10). After substitution of our model for $E(S_.^0(t))$ from (14.10), we have that

$$\lambda(t) = \beta S^* e^{-\alpha t} .$$

We can now obtain the nucleation time distribution as

$$f(t_{0i}) = \alpha e^{-\alpha t_{0i}} \quad \text{which implies} \quad t_{0i} \sim Expo(\alpha) , \tag{14.15}$$

where $Expo(a)$ denotes the exponential distribution with parameter a.

Subsequently, we also obtain that

$$E(N(t)) = \frac{\beta S^*}{\alpha}(1 - e^{-\alpha t}) \quad \text{and} \quad E(N(\infty)) = \frac{\beta S^*}{\alpha} . \tag{14.16}$$

This inverse relation between the average number of islands at infinite time and the growth parameter matches the intuition that faster growth leaves less room for new islands to form, thus resulting in a lower number of islands at infinite time.

A Model for $E(S_i(t_{0i}, t))$

Based on the models derived in the last two sections, we can now develop a model for the expected graphene area for a single graphene island, with random nucleation time, $t_{0i} \sim Expo(\alpha)$. In order to achieve this, we first propose the following link between the expected graphene area on a tray assumed to have instantaneous nucleation ($E(S^0(t))$), and the expected graphene area on a tray where such an assumption is absent ($E(S_.(t))$)

$$E(S_.(t)) = E(S^0(t)) \cdot Pr(t_{0i} \leq t) . \tag{14.17}$$

This formulation implies that on average, the area on a tray with random nucleation is determined by the area on the same tray under zero nucleation times, down-weighted by the proportion of islands, which have nucleated. This allows us to recover $E(S^0(t))$ under the specific case of zero nucleation times, since $Pr(t_{0i} \leq t) = 1$ for all t in that case. Substituting the expression for $E(S^0(t))$ from (14.12) into (14.17), and noting that $Pr(t_{0i} \leq t) = 1 - e^{-\alpha t}$, it follows that

$$E(S_.(t)) = S^*(1 - e^{-\alpha \omega})(1 - e^{-\alpha t})^2 . \tag{14.18}$$

By defining $S_i(t_{0i}, t) = 0$, whenever $t_{0i} > t$, and applying the law of iterated expectations on (14.5), we obtain

$$E(S_.(t)) = E(N(\infty)) \cdot E(S_i(t_{0i}, t)) .$$

Substituting (14.16) and (14.18) into the aforementioned expression, the final model for $E(S_i(t_{0i}, t))$ is given by

$$E(S_i(t_{0i}, t)) = \frac{\alpha(1 - e^{-\alpha \omega})}{\beta}(1 - e^{-\alpha t})^2 . \tag{14.19}$$

Expressions for $\xi(t)$, $\varphi(t)$ and $\varpi(t)$

Based on the models developed in Sects. 14.1–14.2, we can now derive expressions for $\xi(t)$, $\varphi(t)$ and $\varpi(t)$. By the law of total expectation, we know that

$$E(S_i(t_{0i}, t)) = E(S_i(t_{0i}, t)|t_{0i} \le t) \cdot Pr(t_{0i} \le t) + E(S_i(t_{0i}, t)|t_{0i} > t) \cdot Pr(t_{0i} > t).$$

Since we defined $S_i(t_{0i}, t) = 0$, whenever $t_{0i} > t$, we have that $E(S_i(t_{0i}, t)|t_{0i} > t) = 0$, and thus,

$$\xi(t) = E(S_i(t_{0i}, t)|t_{0i} \le t) = \frac{E(S_i(t_{0i}, t))}{Pr(t_{0i} \le t)} = \frac{\alpha(1 - e^{-\alpha\omega})}{\beta}(1 - e^{-\alpha t}).$$

$$\tag{14.20}$$

The last step follows from (14.19) and the fact that $Pr(t_{0i} \le t) = 1 - e^{-\alpha t}$. In order to obtain expressions for $\varphi(t)$ and $\varpi(t)$, we first recall from (14.13), after plugging in the model for $E(N(\infty))$, that

$$\tilde{S}_i(0, t) = E(S_i(0, t)) = \frac{\alpha(1 - e^{-\alpha\omega})}{\beta}(1 - e^{-\alpha t}).$$

Now,

$$E(\sqrt{\tilde{S}_i(0, t_{0i})}|t_{0i} \le t) = \sqrt{\frac{\alpha(1 - e^{-\alpha\omega})}{\beta}} \int_0^t \sqrt{(1 - e^{-\alpha t_{0i}})} \frac{f(t_{0i})}{Pr(t_{0i} < t)} dt_{0i}$$

$$= \sqrt{\frac{\alpha(1 - e^{-\alpha\omega})}{\beta}} \int_0^t \sqrt{(1 - e^{-\alpha t_{0i}})} \frac{\alpha e^{-\alpha t_{0i}}}{(1 - e^{-\alpha t})} dt_{0i}$$

$$= \sqrt{\frac{\alpha(1 - e^{-\alpha\omega})}{\beta}} \cdot \frac{2}{3} \sqrt{(1 - e^{-\alpha t})} \quad \text{after some manipulation,}$$

and

$$E(\tilde{S}_i(0, t_{0i})|t_{0i} \le t) = \frac{\alpha(1 - e^{-\alpha\omega})}{\beta} \int_0^t (1 - e^{-\alpha t_{0i}}) \frac{f(t_{0i})}{Pr(t_{0i} < t)} dt_{0i}$$

$$= \frac{\alpha(1 - e^{-\alpha\omega})}{\beta} \int_0^t (1 - e^{-\alpha t_{0i}}) \frac{\alpha e^{-\alpha t_{0i}}}{(1 - e^{-\alpha t})} dt_{0i}$$

$$= \frac{\alpha(1 - e^{-\alpha\omega})}{\beta} \cdot \frac{1}{2}(1 - e^{-\alpha t}) \quad \text{after some manipulation .}$$

We can now write

$$\varphi(t) = \frac{2}{3}\sqrt{\frac{\alpha(1 - e^{-\alpha\omega})}{\beta}}\sqrt{(1 - e^{-\alpha t})} , \qquad (14.21)$$

and

$$\varpi(t) = \frac{1}{2}\frac{\alpha(1 - e^{-\alpha\omega})}{\beta}(1 - e^{-\alpha t}) - \varphi^2(t) = \frac{\alpha(1 - e^{-\alpha\omega})}{18\beta}(1 - e^{-\alpha t}) . \qquad (14.22)$$

Modified Expression for the Kinetics

We are now in a position to obtain a closed-form expression of equation (14.7) for
the kinetics $q(t)$ governing the graphene growth on the tray under the models for the
shape, expected area, and distribution of nucleation times developed in Sects. 14.1–
14.2. Based on the expressions in (14.7), (14.20), (14.21) and (14.22), we obtain the
following form for $q(t)$:

$$q(t) = \frac{\left(\frac{1}{3} + \sqrt{\frac{17}{72}}\right)\sqrt{\alpha^3(1 - e^{-\alpha\omega})}e^{-\alpha t}}{\sqrt{A_{nuc}\beta\left(1 - e^{-\alpha t}\right)}} . \qquad (14.23)$$

The role each term plays in the kinetics can now be clearly seen. As A_{nuc}
becomes smaller, one should expect the kinetics driving the growth process to a fixed
maximum area to increase. An increase in β should naturally lead to a decrease in
the kinetics driving the growth of each graphene island when all else is fixed. This
is due to a decrease in the available space in which each island has to grow, as a
result of higher expected number of islands. An increase in ω, which signifies a
higher overall limiting area, should natually lead to an increase in the kinetics as
well. Increasing α will generally lead to an increase in the kinetics, especially in the
early stages of growth.

A Simple Example We now illustrate the effect of α on the proposed kinetics
model with a simple example. Assume we have three unique process conditions
corresponding to $\alpha_1 = 0.5$, $\alpha_2 = 1$ and $\alpha_3 = 1.5$. Suppose $S^* = 1$ cm^2,
$A_{nuc} = 6.02 \times 10^{-14}$ cm^2, $\omega = 4$ and $\beta = 50$. The plots of the intensity functions
and kinetics versus time are shown for the three different values of α_j (Fig. 14.4).

In Fig. 14.4a, the intensity starts at β and rapidly declines with time for each
value of α. Observe that higher values of α decay much faster, leading to a lower

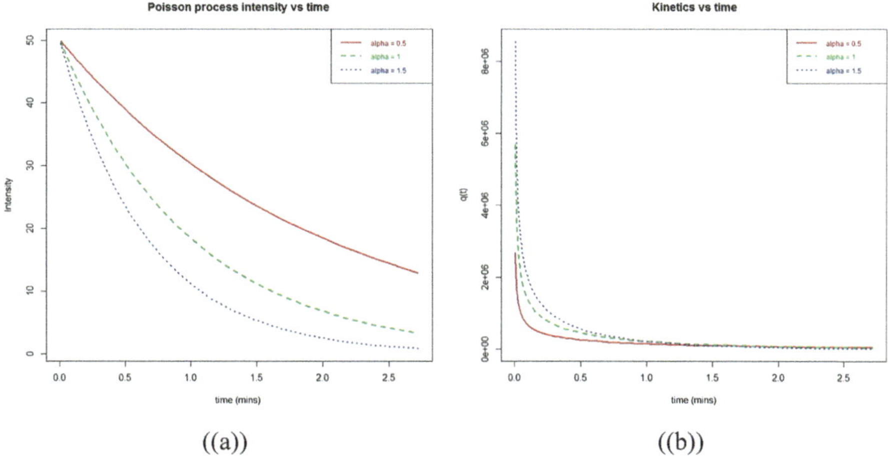

((a))   ((b))

Fig. 14.4 Intensity and Kinetic plots based on a simple example

final nucleation density overall. In Fig. 14.4b, note that higher values of α lead to higher growth kinetics especially in the beginning, when the copper trays are yet to be covered.

14.3   Observational Level Model for Experimental Data

In the previous sections, we determined an expression for the kinetics driving graphene growth under several parametric model assumptions. The next goal is to learn about those parameters from experimental data, and thus, we will need to propose a final observational level model for carrying out inference on the parameters. If we can observe $S_{.}(t)$ at different time points $t = t_1, \ldots, t_{n_T}$, then the inference on the parameters is somewhat straightforward, and in Sect. 14.5, we propose a method to estimate the kinetics under this scenario. However, in many situations, the data collection mechanism prohibits observation of the whole tray, and there is therefore a need to account for the extra noise generated at the observational level. In order to motivate the observational model, we first briefly describe the experimental data.

Multi-resolution Data

Graphene grown on copper substrate was monitored at a single process condition combination (temperature = 1050°C, methane concentration = 60 ppm). At four time points during the growth, 5–8 non-overlapping images are taken at random

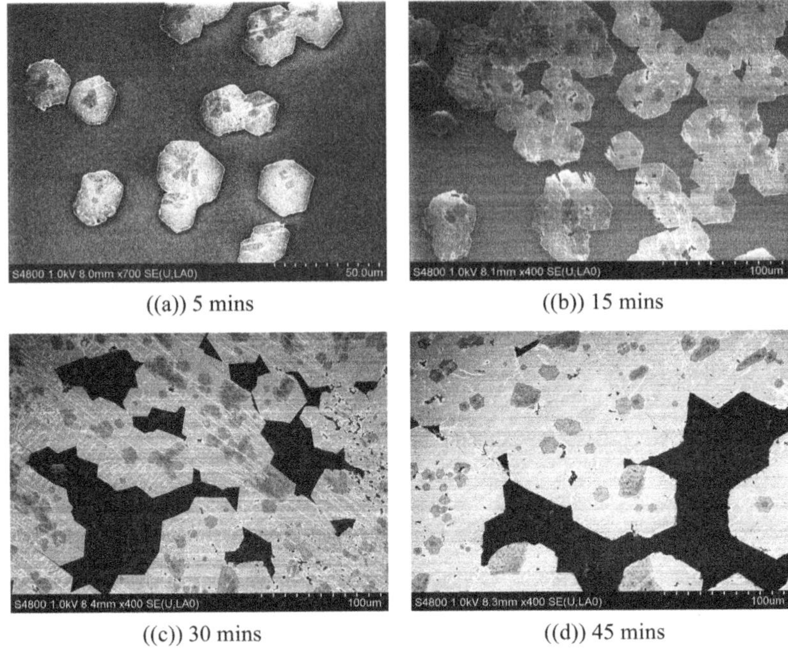

((a)) 5 mins   ((b)) 15 mins

((c)) 30 mins   ((d)) 45 mins

Fig. 14.5 A sample of the multi-resolution data

locations on the copper tray. The resolutions of the images differ, thus ensuring that the sizes of the images, at some baseline magnification, are different. Figure 14.5 shows a sample of the data, where each time point is represented.

Measuring the area of graphene in each image requires the use of image processing software, and we made use of ImageJ's Fiji software [15, 16] in the processing, and Fig. 14.6 shows the sample from Fig. 14.5 after processing.

A Noisy Observation Model

From the description of the data and Fig. 14.5, it is evident that due to the merging of islands, measurement of $N(t)$ will be difficult, if not impossible, particularly at later stages of the growth. Thus, the parameter β will not be estimable from the data. What that means is that we will know the kinetics, $q(t)$ only up to some proportionality constant. Since our eventual goal is to compare the kinetics under different process conditions, we do not lose anything if we make ratio-based comparisons.

In order to develop a model that incorporates information from multiple resolutions, consider the hypothetical state of a given tray, along with 5 images at different resolutions depicted in Fig. 14.7.

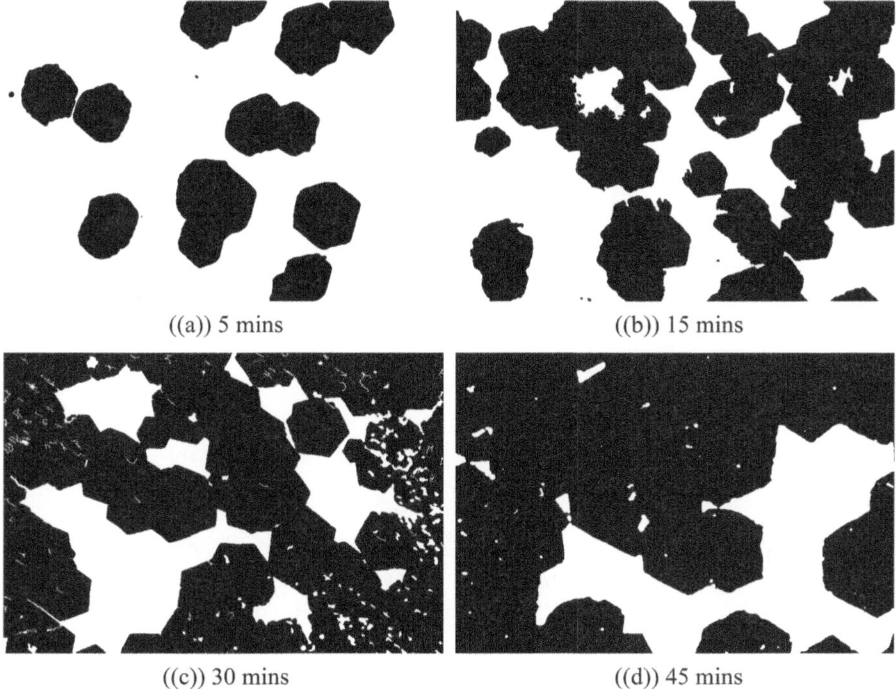

((a)) 5 mins   ((b)) 15 mins

((c)) 30 mins   ((d)) 45 mins

Fig. 14.6 A sample of the processed multi-resolution data

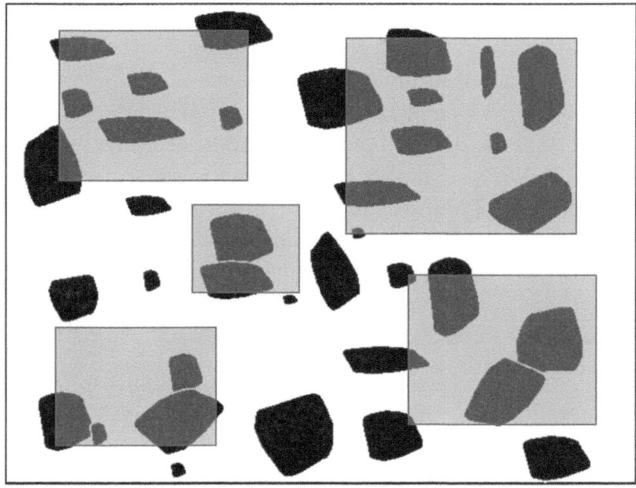

Fig. 14.7 A hypothetical copper tray at some time t. The gray rectangles represent sampled portions (images) of the tray, and the black objects represent graphene islands. Smaller rectangles correspond to higher resolutions/magnifications.

Let us denote the measured graphene area covered in lth image on the tray at time t by $u_{l,t}$, and the area of that lth image by $U_{l,t}$. In Fig. 14.7, the area of a gray rectangle is $U_{l,t}$, and the sum total of the black area within that rectangle is $u_{l,t}$. Also denote the total area of all images on the tray at time t as U_t^{obs}, and the corresponding non-observed area as U_t^{mis}. Further denote the total area covered by graphene in all images on the tray at time t as $S^{obs}(t)$, the unknown graphene area in the unobserved portion as $S^{mis}(t)$, and the number of images on the tray at time t as m_t. Recall that $S.(t)$ is the total area covered by graphene on the tray at time t, and S^* is the area of the copper tray. Then,

$$U_t^{obs} = \sum_{l=1}^{m_t} U_{l,t}, \quad U_t^{mis} = S^* - U_t^{obs},$$

$$S^{obs}(t) = \sum_{l=1}^{m_t} u_{l,t}, \quad \text{and} \quad S^{mis}(t) = S.(t) - S^{obs}(t) .$$

Note that $\{u_{1,t}, u_{2,t}, \ldots, u_{m_t,t}, S^{mis}(t)\}$, which we now denote by $\boldsymbol{u}_t$, defines a random partition of $S.(t)$, with $S^{mis}(t)$ always missing/unobserved. On the other hand, $\{U_{1,t}, U_{2,t}, \ldots, U_{m_t,t}, U_t^{mis}\}$, which we denote by $\boldsymbol{U}_t$, defines a known partition of S^*.

We can now build a model for the observations by making the following considerations:

(i) Each $u_{l,t}/U_{l,t}$ should be an unbiased estimator of $E(S.(t))/S^*$, based on the random location of the images on the tray.
(ii) As $U_{l,t} \to \infty$, the variance of $u_{l,t}/U_{l,t}$ should approach 0. On the other hand, as $U_{lt} \to 0$, u_{lt}/U_{lt} should increasingly exhibit bi-modality at 0 and 1, due to the clustering behavior of the graphene islands.
(iii) The variance of $u_{l,t}/U_{l,t}$ should approach zero when $E(S.(t)) \to S^*$ and when $E(S.(t)) \to 0$. These are the boundary conditions of the growth process.
(iv) Due to the clustering behavior of graphene islands, the correlation between any $u_{l,t}/U_{l,t}$ and $u_{k,t}/U_{k,t}$ should be positive and decay with the distance between the images. This distance can be defined in terms of the minimum distance between boundary points of the two images. In the absence of spatial information, this final consideration can be ignored based on a simplifying assumption of zero correlation, at the risk of underestimating the uncertainty in the model.

Based on considerations (i)–(iii), we propose the following:

$$\frac{u_{l,t}}{U_{l,t}} \sim Beta\left(vU_{l,t} \frac{E(S.(t))}{S^*}, \ vU_{l,t}\left(1 - \frac{E(S.(t))}{S^*}\right)\right) , \tag{14.24}$$

where v is an unknown parameter, which scales the influence of each $U_{l,t}$ on the variance, so that any rescaling of all $U_{l,t}$'s will have no effect, since $1/v$ will be rescaled as well. Relatively high values of v, for fixed $U_{l,t}$, result in $u_{l,t}$ very close to its expected value for all $\{l, t\}$ and are appropriate for situations where clustering behavior is not expected. Relatively low values, on the other hand, represent situations where clustering is the norm, as in the case of graphene growth. The parameter v thus provides some useful information about clustering behavior exhibited in the data.

By plugging in the model for $E(S.(t))$ from (14.10), (14.11), (14.15) and (14.17), we can now write the full hierarchical data generating model for the observed area as

$$\frac{u_{l,t}}{U_{l,t}} \sim Beta\Big(vU_{l,t}(1 - e^{-\alpha\omega})(1 - e^{-\alpha t})^2,$$

$$vU_{l,t}\Big(1 - (1 - e^{-\alpha\omega})(1 - e^{-\alpha t})^2\Big)\Big), \tag{14.25}$$

$$(v, \omega, \alpha) \sim \Pi(v, \omega, \alpha),$$

where $\Pi(\cdot)$ represents some specified distribution.

14.4 Model Estimation and Results

Bayesian Estimation

We denote the data by $\mathcal{D}$ and the parameters by $\boldsymbol{\phi}$, noting that $\mathcal{D} \equiv (\{u_{l,t}\}, \{U_{l,t}\})$ and $\boldsymbol{\phi} \equiv (v, \omega, \{\alpha\})$. Our goal is to draw posterior samples for $q(t)$ through posterior samples from $\Pi(\boldsymbol{\phi}|\mathcal{D})$. After obtaining posterior samples for $q(t)$ through posterior samples from $\Pi(\boldsymbol{\phi}|\mathcal{D})$, we can estimate $q(t)$ by its posterior mean $\hat{q}(t)$ and obtain the posterior intervals as a measure of the uncertainty of estimation (see [6] for more on Bayesian inference).

We first assume a flat hyper-prior distribution for the log-transformed parameters so that the posterior estimates correspond with the maximum likelihood estimates. i.e.,

$$\Pi(\log(v), \log(\omega), \log(\alpha)) \propto 1.$$

Table 14.1 shows the posterior estimates for the parameters, and Fig. 14.8 shows the fitted model and kinetics (where for the sake of plotting the shape, we have assumed

Table 14.1 Posterior statistics for the kinetics parameters

Parameters	Estimates	Lower	Upper
v	94.61	50.56	165.15
ω	8.82	5.87	12.90
α	0.16	0.13	0.21

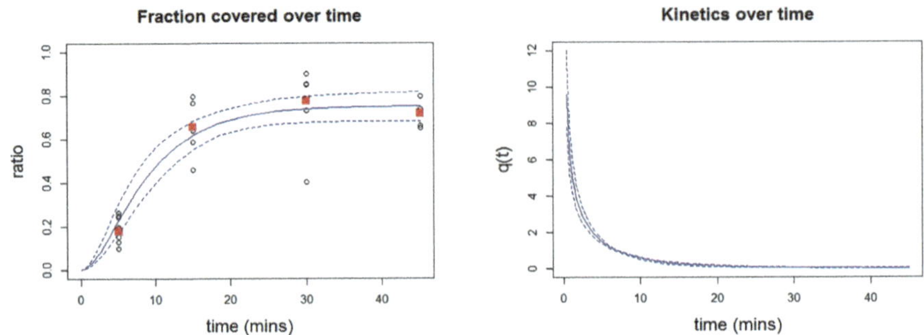

((a)) Proportion covered over time. The circles are $u_{l,t}/U_{l,t}$, the red squares are $S^{obs}(t)/U_t^{obs}$, the solid line is the estimated $E(S_.(t))/S^*$, and the broken lines are 95% posterior intervals.

((b)) The estimated kinetics along with 95% posterior intervals (assuming $\beta = 1$).

Fig. 14.8 Intensity and kinetic plots based on the Graphene data

$\beta = 1$). The posterior samples are obtained using the *MCMCpack* package [11] in R, and the diagnostic plots are shown in chapter Appendix A2.

Sensitivity to Sampling Design

Based on the estimated observation model, we are also interested in determining the sensitivity of the estimation procedure to the choice of sampling design. Specifically, we would like to know if and how the number of images, the similarity of resolutions across images, the number of time points chosen, and the percentage of missing data (unobserved tray portion) affect inference. This information will be used in designing optimum sampling plans for new experiments.

In determining the effect of the aforementioned sampling factors, we turn to simulations. In the simulations, we assume the data generating model is the same model given in (14.24) and is estimated using the Bayesian methodology as described in Sect. 14.4. We also assume the following sampling design features:

(a) The number of time points is n_T, and $n_T \in 4, 5, \ldots, 40$. Note that we select equi-spaced time points on the log scale, so as to finely capture the initial moments of the growth process.
(b) The fraction of missing data, U_t^{mis}/S^*, is ρ_{mis}, and $\rho_{mis} \in [0.1, 0.99]$.
(c) The number of images throughout a single growth process is n_P, and $n_P \in 4, 5, \ldots, 20$.
(d) The image sizes follow a Dirichlet distribution, i.e.,

$$(U_{1,t}, U_{2,t}, \ldots, U_{n_P,t}) \sim Dirichlet(\gamma \cdot \mathbf{1}_{n_P}),$$

where $\mathbf{1}_{np}$ denotes a vector of 1s, of length np. Note that increasing γ increases the similarity of resolutions (i.e., the balance of sizes across images at time t). We assume that $\log(\gamma) \in [0.5, 2.5]$.

A full-factorial design for the four factors listed earlier is computationally expensive due to the curse of dimensionality. For instance, a full-factorial design for four factors at 10 levels each requires 10^4 runs. In order to reduce the number of runs, while maintaining space filling properties for each factor, we make use of approximate Latin Hypercube Sampling (LHS) designs [12, 13], with the approximation due to integer rounding for the discrete factors. Specifically, we generate a 600-run LHS design for the four factors aforementioned, using the *SLHD* package [1] in R. The metric we use to determine sensitivity at each run is the posterior RMSE (root mean square error) defined as

$$RMSE = \sqrt{\frac{1}{n_T M} \sum_{k=1}^{n_T} \sum_{m=1}^{M} \left(\hat{q}_m(t_k) - q(t_k)\right)^2},$$

where $M(=5)$ denotes the number of replications at each factor combination. Since LHS designs generate irregularly spaced domain points, some smoothing is needed to interpolate unsampled domain points on a grid. In order to produce all two factor interactions, we apply a kriging interpolator on the calculated posterior RMSE for each two factor combination using the *geoR* package [14] in R, using the default Matérn covariance function. All the two factor interaction plots are shown in Fig. 14.9.

Remarks

(1) Figures 14.9a, b and c, which depict the two-factor joint effect of the number of time points ($n_T = 4, \ldots, 40$) with the other factors, suggest that a higher number of time points will in general lead to better inference.

(2) Figures 14.9a, d and e, which depict the two-factor joint effect of the fraction of missing data ($\rho_{mis} \in [0.1, 0.99]$) with the other factors, suggest that a lower fraction of missing data will also lead to better inference. In fact, judging from the plots, the fraction of missing data seems to be the most relevant factor in how good the inference procedure will be, and its effect changes depending on the values of the other factors.

(3) Figures 14.9c, e and f show the two-factor joint effect of the number of images ($n_P \in 4, 5, \ldots, 20$) with the other factors. The plots, especially Fig. 14.9e, suggest that a lower number of images generally leads to slightly improved inference, with the effect of the number of images being somewhat weak in comparison to the effect of the other factors.

(4) Figures 14.9b, d and f show the two-factor joint effect of the resolution balance ($\log(\gamma) \in [0.5, 2.5]$) with the other factors. Figure 14.9b does not yield much information, but the other two plots (especially Fig. 14.9d) imply that inference improves with more balanced resolutions across the images.

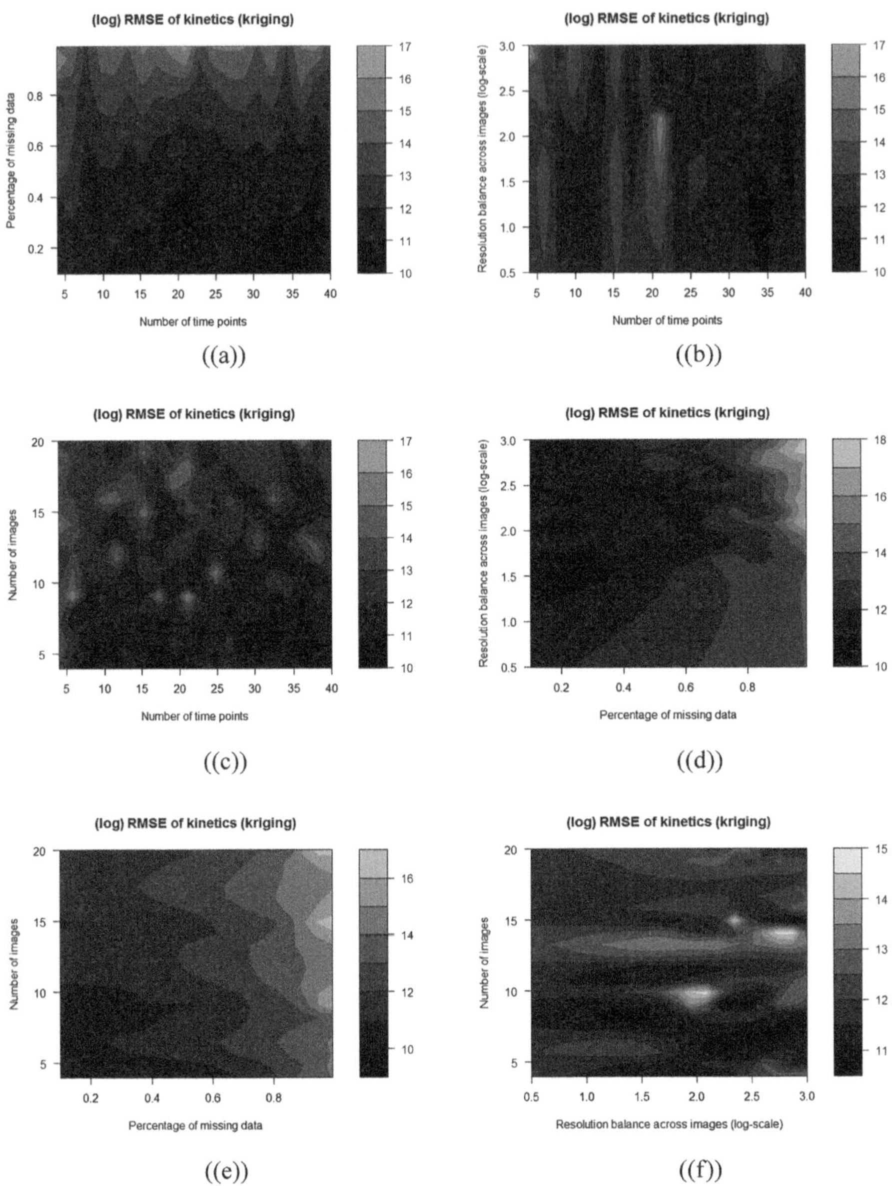

Fig. 14.9 All two factor interactions for $(n_T, \rho_{mis}, \gamma, n_P)$. Darker shades represent smaller RMSEs (i.e., lower estimation uncertainty) and are desirable

Judging from the plots in Fig. 14.9, the fraction of missing data seems to be the most important factor, followed by the number of time points, the number of images, and the resolution balance across the images. Combining this with the four interpretations aforementioned, we thus recommend observing the whole tray at as many time points as possible. If observation of the entire tray is not feasible, then the observable region should be partitioned into as few images as possible, while those images should be at approximately the same resolution.

14.5 Discussions

This chapter presents a stochastic framework for understanding the kinetics driving graphene growth by proposing models for the observed shape, time-varying area of each individual island, as well as the nucleation time distribution, while ensuring that each component model reflects practical considerations and expectations. The resulting inference and sensitivity analyses show the stability of the model and suggest some new sampling designs for graphene growth experiments.

It is worthwhile to point out that there are a lot of unsettled issues such as the coexistence of multiple growth mechanisms and multilayer growth, but we believe our work provides a starting point to address these issues. Specifically, the supersaturation assumption, considered debatable by some experts, has actually been relaxed in our work (Eqs. 14.10 and 14.11), in anticipation of the growth conditions affecting the final graphene coverage. Further, we have used the single growth mechanism only as a description of the average nucleation behavior $-$, the actual nucleation behavior at any site on the copper tray is assumed to be random (Eq. 14.4), thus ensuring a simple class of multiple growth mechanisms for each graphene island. More sophisticated classes of multiple growth mechanisms, based on more detailed information of the copper foil's roughness, are certainly possible. The graphene process can simultaneously have multilayer growth. The proposed modeling framework handles the growth of the first layer (right on top of the copper foil) and describes how the first layer covers the foil over time. Since only the data of the first layer are collected for modeling, the secondary and tertiary layers do not affect the modeling of the first layer. Thus, one limitation of our final model is that it can only handle monolayer (two-dimensional) graphene growth. Expanding the model to describe multilayer (three-dimensional) growth is an important research direction that we plan to explore.

We now propose a couple of model extensions/modifications for richer experimental datasets.

Observational Model for Entire Tray

If instead of $u_{l,t}$, we observe $S_.(t)$ as per our recommendation in Sect. 14.4, then we can update the observational level model based on the following considerations:

(i) $S.(t)$ should be non-decreasing over time.

(ii) The variance of $S.(t)$ should approach zero when $E(S.(t)) \rightarrow S^*$ and when $E(S.(t)) \rightarrow 0$. These are the boundary conditions of the growth process.

(iii) The models proposed in Sects. 14.1–14.2 remain valid.

We can thus propose

$$
\begin{aligned}
\frac{S.(t)}{S^*} | S.(\infty) &\sim \frac{S.(\infty)}{S^*} \cdot DP\left(v,\ (1 - e^{-\alpha t})^2\right), \\
\frac{S.(\infty)}{S^*} &\sim Beta\left(v(1 - e^{-\alpha \omega}),\ v e^{-\alpha \omega}\right),
\end{aligned}
\tag{14.26}
$$

where $DP(v, H(\cdot))$ denotes the Dirichlet process distribution with concentration parameter v and base function $H(\cdot)$. This observational level model captures the random behavior of carbon atom attachment to the edges of graphene islands, as well as ensuring that temporal dependence is accounted for. It is thus recommended for inference when the whole tray is observed.

Growth as a Function of Process Conditions

We can expand the component models to include covariate information, representing the different process conditions (e.g., growth temperature, methane flow rate, and ambient pressure). We can do this by first considering graphene growth on multiple trays and introduce an extra index, $j \in \{1, \ldots, J\}$, representing each tray. We then explicitly connect each tray specific α with X_j, the vector of covariates describing the process condition for tray j, i.e., we use $\alpha_j = \alpha(X_j)$, some positively valued function of the covariates X_j for tray j. We can then write $q_j(t)$ as $q(t|X_j)$ by simply replacing α_j with $\alpha(X_j)$, thus explicitly showing the dependence of the kinetics on the process covariates X_j.

Since $\alpha_j > 0$, one natural way of modeling $\alpha(X_j)$ is to use the log-linear model,

$$
\log(\alpha(X_j)) = \sum_{k=1}^{K} a_k \cdot b_k(X_j).
$$

where the basis functions $b_k(\cdot)$ can be determined through some exploratory data analysis.

It is our expectation that future experimental data will lead to improvements on the component models, especially the log-linear model for $\alpha(\cdot)$. Some future research directions include designing and evaluating experiments to obtain graphene data, along with process condition information.

Appendix

Appendix A0: Glossary of Notations

$g_i(\theta)$ $\quad\rightarrow\quad$ the random shape of island i on the copper tray.

$q(t)$ $\quad\rightarrow\quad$ the kinetics driving the growth on the tray.

$$\text{We denote } Q(t) = \int_0^t q(s)ds \text{ and define } Q(0) = 0.$$

t_{0i} $\quad\rightarrow\quad$ the nucleation time of island i on the tray.

$v_i(\theta, t)$ $\quad\rightarrow\quad$ the angular dependent growth velocity of island i.

$N(t)$ $\quad\rightarrow\quad$ the (random) number of islands on the tray at time t.

$S_i(t_{0i}, t)$ $\quad\rightarrow\quad$ the area covered by island i with nucleation time t_{0i}, by time t.

$S_{.j}(t)$ $\quad\rightarrow\quad$ the total area covered by the graphene islands on the tray by time t.

$$\text{Note that } S_.(t) = \sum_{i=1}^{\infty} S_i(t) \cdot 1\{t_{0i} \leq t\} = \sum_{i=1}^{N(\infty)} S_i(t)$$

$$\cdot 1\{t_{0i} \leq t\} = \sum_{i=1}^{N(t)} S_i(t)$$

where $1\{t_{0i} \leq t\}$ is the usual indicator function and

$$N(t) = \sum_{i=1}^{\infty} 1\{t_{0i} \leq t\}.$$

$S_.^0(t)$ $\quad\rightarrow\quad$ the total area covered by the graphene islands on the tray by time t, when $t_{0i} = 0$ for every i.

$r_i(\theta, t_{0i}, t)$ $\quad\rightarrow\quad$ the length/distance from the edge of the ith island with nucleation time t_{0i}, to its center, at a specific angle θ and time t (for a circular island shape, this is simply the radius at time t).
Also note that $r_i(\theta, t_{0i}, t) = 0$ if $t_{0i} > t$ (as implied by [22]).

$\tilde{v}_i(\theta, t)$ $\quad\rightarrow\quad$ the angular dependent growth velocity of island i on the tray for the average growth process (i.e., averaging out the random behavior of carbon atom attachment/detachment). Recall that we assumed $\tilde{v}_i(\theta, t) = g_i(\theta)q(t)$.

$\tilde{r}_i(\theta, t_{0i}, t)$ $\quad\rightarrow\quad$ the length/distance from the edge of the ith island with nucleation time t_{0i}, to its center, at a specific angle θ and time t, after averaging out the random carbon atom attachment/detachment.

$$\tilde{r}_i(\theta, t_{0i}, t) = \int_{t_{0i}}^t \tilde{v}_i(\theta, s)ds.$$

$\tilde{S}_i(t_{0i}, t)$ $\quad\rightarrow\quad$ the average area covered by island i with nucleation time t_{0i}, by time t i.e., averaging out the random behavior of carbon atom attachment/detachment.
Note: $\tilde{S}_i(t_{0i}, t) = E(S_i(t_{0i}, t)|t_{0i})$, and

$$\tilde{S}_i(t_{0i}, t) = \tfrac{1}{2} \int_0^{2\pi} \tilde{r}_i^2(\theta, t_{0i}, t).$$

Appendix A1: Proof of Theorem 1

Given $t_{0i} \le t$ for ith graphene island on the copper tray, we can write

$$
\begin{aligned}
\frac{d}{dt}\tilde{S}_i(t_{0i}, t) &= \frac{1}{2}\frac{d}{dt}\int_0^{2\pi}\tilde{r}_i^2(\theta, t_{0i}, t)\,d\theta \\
&= \int_0^{2\pi}\tilde{v}_i(\theta, t)\left(\int_{t_{0i}}^t \tilde{v}_i(\theta, s)\,ds\right)d\theta \\
&= \left(\int_0^{2\pi} g_i^2(\theta)\,d\theta\right) q(t)\int_{t_{0i}}^t q(s)\,ds \\
&= 2A_{nuc}\cdot q(t)\int_{t_{0i}}^t q(s)\,ds\;.
\end{aligned}
$$

Now integrating both sides with respect to time t, we obtain

$$
\tilde{S}_i(t_{0i}, t) + Const = 2A_{nuc}\left[\frac{1}{2}Q^2(t) - Q(t_{0i})Q(t)\right].
$$

When $t = t_{0ij}$, we have $\tilde{S}_i(t_{0i}, t) = 0$ and thus

$$
Const = -2A_{nuc}\left(\frac{1}{2}Q^2(t_{0i})\right),
$$

so that $\tilde{S}_i(t_{0i}, t) = A_{nuc}[Q(t) - Q(t_{0i})]^2$. Since $Q(0) = 0$, we also get $\tilde{S}_i(0, t_{0i}) = A_{nuc}Q^2(t_{0i})$, which implies that

$$
Q(t_{0i}) = \sqrt{\frac{\tilde{S}_i(0, t_{0i})}{A_{nuc}}}.
$$

The conditional mean for $\tilde{S}_i(t_{0i}, t)$ is given as

$$
\begin{aligned}
E(&\tilde{S}_i(t_{0i}, t)|t_{0i} \le t) \\
&= A_{nuc}\cdot E\left([Q(t) - Q(t_{0i})]^2\,|t_{0i} \le t\right) \\
&= A_{nuc}\cdot\left[(Q(t) - E(Q(t_{0i})|t_{0i} \le t))^2 + Var(Q_j(t_{0i})|t_{0i} \le t)\right].
\end{aligned}
$$

Denoting

$$
\xi(t) = E(\tilde{S}_i(t_{0i}, t)|t_{0i} \le t)\,,
$$

$$
\varphi(t) = \sqrt{A_{nuc}}\cdot E(Q(t_{0i})|t_{0i} \le t) = E(\sqrt{\tilde{S}_i(0, t_{0i})}|t_{0i} \le t)\,, \text{ and}
$$

$$\varpi(t) = A_{nuc} \cdot Var(Q(t_{0i})|t_{0i} \leq t) = Var(\sqrt{\tilde{S}_i(0, t_{0i})}|t_{0i} \leq t) ,$$

We can now write

$$Q(t) = \frac{1}{\sqrt{A_{nuc}}} \left[\sqrt{\xi(t) - \varpi(t)} + \varphi(t) \right] \quad \text{and}$$

$$q(t) = \frac{1}{\sqrt{A_{nuc}}} \left[\frac{\xi'(t) - \varpi'(t)}{2\sqrt{\xi(t) - \varpi(t)}} + \varphi'(t) \right] ,$$

completing the proof.

Appendix A2: Diagnostic Plots

See Figs. 14.10 and 14.11.

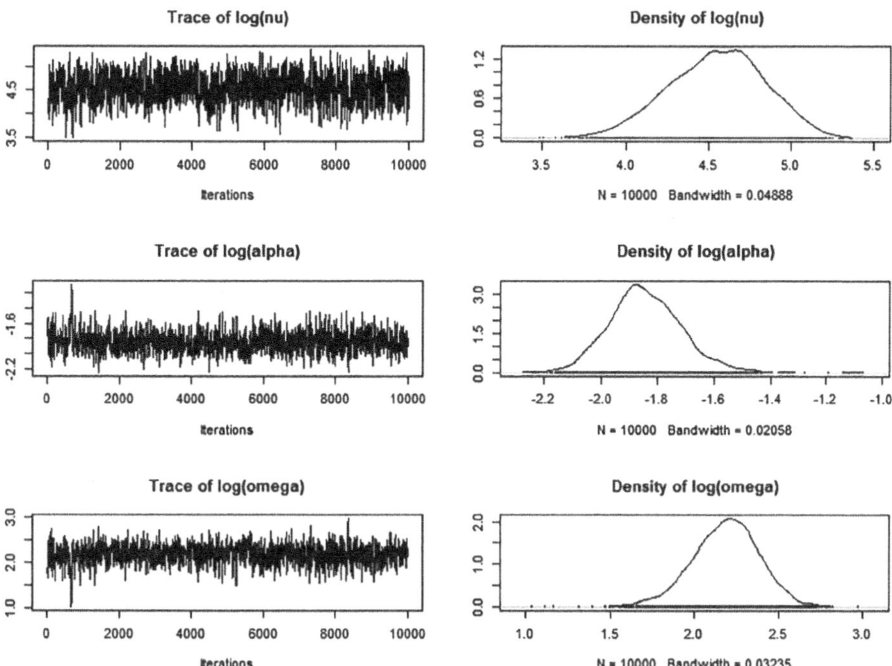

Fig. 14.10 10000 MCMC draws for each variable after discarding the burn-in. The plots indicate good mixing

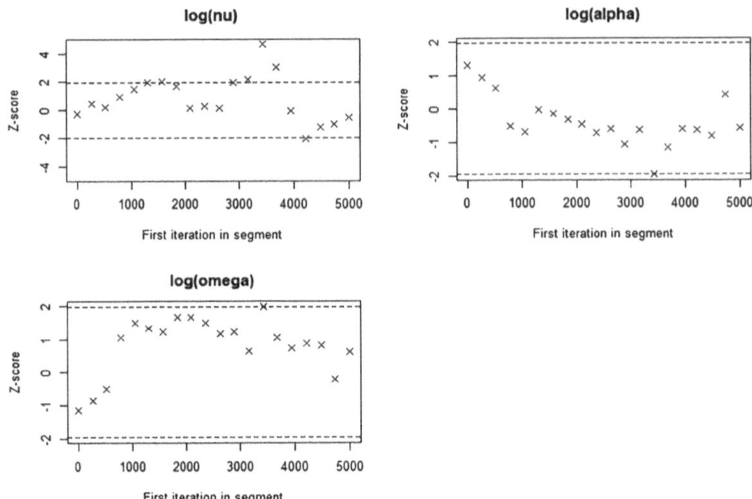

Fig. 14.11 The Geweke plots mostly appear to indicate that stationarity has been achieved

Acknowledgments The work in this chapter was supported by US National Science Foundation under CAREER grant number CMMI-1055394.

References

1. Ba, S.: SLHD: Maximin-Distance (Sliced) Latin Hypercube Designs. CRAN (2015). http:// CRAN.R-project.org/package=SLHD. R package version 2.1-1
2. Bhaviripudi, S., Jia, X., Dresselhaus, M.S., Kong, J.: Role of kinetic factors in chemical vapor deposition synthesis of uniform large area graphene using copper catalyst. Nano Lett. **10**(10), 4128–4133 (2010). http://doi.org/10.1021/nl102355e. PMID: 20812667
3. Bolotin, K., Sikes, K., Hone, J., Stormer, H., Kim, P.: Temperature-dependent transport in suspended graphene. Phys. Rev. Lett. **101**(9), 96802 (2008)
4. Celebi, K., Cole, M.T., Choi, J.W., Wyczisk, F., Legagneux, P., Rupesinghe, N., Robertson, J., Teo, K.B.K., Park, H.G.: Evolutionary kinetics of graphene formation on copper. Nano Lett. **13**(3), 967–974 (2013). http://doi.org/10.1021/nl303934v. PMID: 23339597
5. Geim, A.K., Novoselov, K.S.: The rise of graphene. Nature Mater. **6**(3), 183–191 (2007)
6. Gelman, A., Carlin, J.B., Stern, H.S., Rubin, D.B.: Bayesian Data Analysis, 2nd edn. Chapman and Hall/CRC, New York (2003)
7. Kim, H., Mattevi, C., Calvo, M.R., Oberg, J.C., Artiglia, L., Agnoli, S., Hirjibehedin, C.F., Chhowalla, M., Saiz, E.: Activation energy paths for graphene nucleation and growth on cu. ACS Nano **6**(4), 3614–3623 (2012)
8. Kim, H., Saiz, E., Chhowalla, M., Mattevi, C.: Modeling of the self-limited growth in catalytic chemical vapor deposition of graphene. New J. Phys. **15**(5), 053012 (2013). http://doi.org/10. 1088/1367-2630/15/5/053012
9. Kim, K., Zhao, Y., Jang, H., Lee, S., Kim, J., Kim, K., Ahn, J., Kim, P., Choi, J., Hong, B.: Large-scale pattern growth of graphene films for stretchable transparent electrodes. Nature **457**(7230), 706–710 (2009)

10. Li, X., Cai, W., An, J., Kim, S., Nah, J., Yang, D., Piner, R., Velamakanni, A., Jung, I., Tutuc, E., Banerjee, S.K., Colombo, L., Ruoff, R.S.: Large-area synthesis of high-quality and uniform graphene films on copper foils. Science **324**, 1312–1314 (2009)
11. Martin, A.D., Quinn, K.M., Park, J.H.: MCMCpack: Markov chain monte carlo in R. J. Stat. Software **42**(9), 22 (2011). http://www.jstatsoft.org/v42/i09/
12. McKay, M.D.: Latin hypercube sampling as a tool in uncertainty analysis of computer models. In: Proceedings of the 24th Conference on Winter Simulation, WSC '92, pp. 557–564. ACM, New York, NY, USA (1992). http://doi.org/10.1145/167293.167637. http://doi.acm.org.ezp-prod1.hul.harvard.edu/10.1145/167293.167637
13. McKay, M.D., Beckman, R.J., Conover, W.J.: A comparison of three methods for selecting values of input variables in the analysis of output from a computer code. Technometrics **21**(2), 239–245 (1979). http://www.jstor.org/stable/1268522
14. Ribeiro, P.J.J., Diggle, P.J.: geoR: Analysis of Geostatistical Data. CRAN (2015). http://CRAN.R-project.org/package=geoR. R package version 1.7-5.1
15. Schindelin, J., Arganda-Carreras, I., Frise, E., Kaynig, V., Longair, M., Pietzsch, T., Preibisch, S., Rueden, C., Saalfeld, S., Schmid, B., Tinevez, J.Y., White, D.J., Hartenstein, V., Eliceiri, K., Tomancak, P., Cardona, A.: Fiji - an open source platform for biological image analysis. Nat Methods **9**(7), (2012). http://doi.org/10.1038/nmeth.2019. http://www.ncbi.nlm.nih.gov/pmc/articles/PMC3855844/. 22743772[pmid]
16. Schneider, C.A., Rasband, W.S., Eliceiri, K.W.: Nih image to imagej: 25 years of image analysis. Nat Meth **9**(7), 671–675 (2012). http://doi.org/10.1038/nmeth.2089
17. Shafiei, S., Nourbakhsh, A., Ganjipour, B., Zahedifar, M., Vakili-Nezhaad, G.: Diameter optimization of VLS-synthesized ZnO nanowires, using statistical design of experiment. Nanontechnology **18**, 355708 (2007)
18. Stankovich, S., Dikin, D., Dommett, G., Kohlhaas, K., Zimney, E., Stach, E., Piner, R., Nguyen, S., Ruo, R.: Graphene-based composite materials. Nature **442**(7100), 282–286 (2006)
19. Stoller, M.D., Park, S., Zhu, Y., An, J., Ruoff, R.S.: Graphene-based ultracapacitors. Nano Lett. **8**(10), 3498–3502 (2008)
20. Vlassiouk, I., Regmi, M., Fulvio, P., Dai, S., Datskos, P., Eres, G., Smirnov, S.: Role of hydrogen in chemical vapor deposition growth of large single-crystal graphene. ACS Nano **5**(7), 6069–6076 (2011). http://doi.org/10.1021/nn201978y. PMID: 21707037
21. Vlassiouk, I., Smirnov, S., Regmi, M., Surwade, S.P., Srivastava, N., Feenstra, R., Eres, G., Parish, C., Lavrik, N., Datskos, P., Dai, S., Fulvio, P.: Graphene nucleation density on copper: Fundamental role of background pressure. J. Phys. Chem. C **117**(37), 18919–18926 (2013). http://doi.org/10.1021/jp4047648
22. Wu, J., Huang, Q.: Graphene growth process modeling: a physical-statistical approach. Appl. Phys. A Mater. Sci. & Process. **116**(4), 1747–1756 (2014)
23. Xing, S., Wu, W., Wang, Y., Bao, J., Pei, S.S.: Kinetic study of graphene growth: Temperature perspective on growth rate and film thickness by chemical vapor deposition. Chem. Phys. Lett. **580**, 62–66 (2013). http://doi.org/10.1016/j.cplett.2013.06.047. http://www.sciencedirect.com/science/article/pii/S0009261413008245

Part VI
Domain-Informed Nanostructure Characterization and Defection Detection in Nanomanufacturing

Chapter 15
Learning Interactions Among Nanostructures for Characterization and Defect Detection

Since properties of nanomaterials are determined by their structures, characterizing nanostructure feature variability and diagnosing structure defects are of great importance for quality control in scale-up nanomanufacturing. It is known that nanostructure interactions such as competing for source materials during growth contribute strongly to nanostructure uniformity and defect formation. However, there is a lack of rigorous formulation to describe nanostructure interactions and their effects on nanostructure variability. This chapter introduces a method to relate local nanostructure variability (quality measure) to nanostructure interactions under the framework of Gaussian Markov random field. With the developed modeling and estimation approaches, we are able to extract nanostructure interactions for any local region with or without defects based on its feature measurement. The established connection between nanostructure variability and interactions not only provides a metric for assessing nanostructure quality but also enables a method to automatically detect defects and identify their patterns based on the underlying interaction patterns. Both simulation and real case studies are conducted to demonstrate the developed methods. The insights obtained from real case study agree with physical understanding.

15.1 Nanostructure Interaction and Quality

Low-dimensional nanostructures such as carbon nanotubes and semiconductor nanowires are critical building blocks for nanodevices and nanosystems that have promising applications in fields like medicine and energy production [16, 18, 31]. For instance, high-density aligned nanotubes were fabricated to build submicron nanotube transistors with superior properties [28] and CMOS integrated circuits [27]. The improved performance is mainly due to the unique structures of nano-materials. Taking Si nanowire solar cells as an example, efficiencies of light absorption and surface recombination can be greatly enhanced by the geometry and arrangement of nanowires [11]. On the other hand, nonuniformity and defects of nanostructures can cause performance degradation in nanodevices. For scale-up nanomanufacturing, it is therefore critical to develop metrics and methods to quantify nanostructure spatial variability or nonuniformity and detect structure defects in each sample.

The formation of nanostructures and defects is sensitive to the interactions among neighboring structures such as competing for source materials during growth. Figure 15.1 shows ZnO nanowires grown via VLS (vapor–liquid–solid) mechanism. Large nanowire bundles can inhibit the growth of neighboring nanowires by absorbing majority of Zn vapor [23]. Hence, improved understanding of nanostructure interactions will facilitate the control of local morphology and diagnosis of structure defects.

To control nanostructure synthesis variations and achieve scale-up nanomanufacturing, research has been conducted on nanowire growth process modeling [8, 9, 14, 15], robust synthesis of nanostructures [24, 32], and nanoparticle dispersion quantification [6, 30]. The only work that quantifies the variability of continuous nanostructure features such as length is given by Huang [14]. Besides, only [14] and [6] have ever considered the interactions among nanostructures. Huang [14] attributed the local variability of nanowire length to the interacting effects among neighboring nanowires, and [6] associated the clustering of nanoparticles to their interacting forces. But the interaction analysis was not the focus for either of the papers.

This chapter characterizes interactions among nanostructures for quality assessment. For any local region of interest, we quantify the spatial variability of continuous quality features, e.g., length for nanostructures based on the extracted interaction patterns. With the established connection between nanostructure variability and nanostructure interactions, we can also detect the existence of defects within the region and identify detailed patterns for the defects.

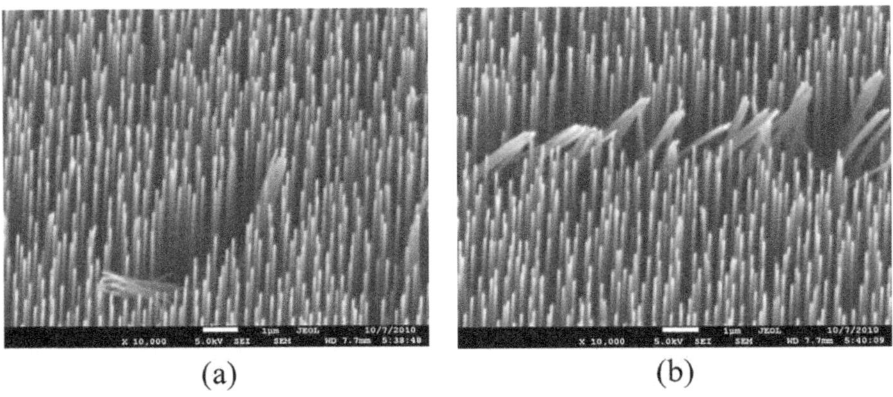

(a)   (b)

Fig. 15.1 Samples of ZnO nanowires with (trend of) bundles, grown by VLS

15.2   Problem Formulation

Our objective is to characterize nanostructure features in local regions of a substrate or "(nanostructure) local features" for short. Because of the important roles played by nanostructure interactions on local morphology, our characterization of local features will target on interaction patterns. Issues to be addressed include (1) modeling and estimating nanostructure interactions based on feature measurement and (2) quantifying local feature variability and identifying defects through interaction estimation.

We consider features in continuous scale like lengths of nanowires. With certain abstraction, nanostructures can be viewed to disperse in a real space with dimension 1, 2, or 3 depending on applications. We take a discrete sampling approach to measure nanostructures. Following steps are taken to measure nanostructure features in each local region of interest.

Step 1.   We divide the local region regularly into non-overlapping sites (grids) as illustrated in Fig. 15.2. Denote the sites as $s = \{s_1, s_2, \ldots, s_n\}$

Step 2.   Measure and summarize nanostructures in site s_i as $X(s_i)$. Altogether, we obtain a random field

$$X(s) = \{X(s_1), X(s_2), \ldots, X(s_n)\}. \tag{15.1}$$

In detail, if a site has only normally grown nanostructures, we measure a randomly selected nanostructure or average all nanostructures in that site. Here, "normally grown" means the nanostructure is a single crystal and has no structural defects. If a site has defective structures, we measure the defective structures approximately by following the same procedure as that for normally grown nanostructures. And if there is no growth in the site, we take the feature measurement as zero for that site.

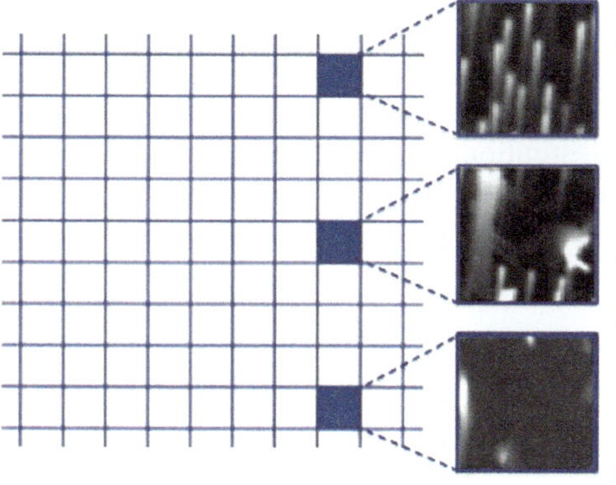

Fig. 15.2 Illustration of nanostructure feature measurement in a local region of interest

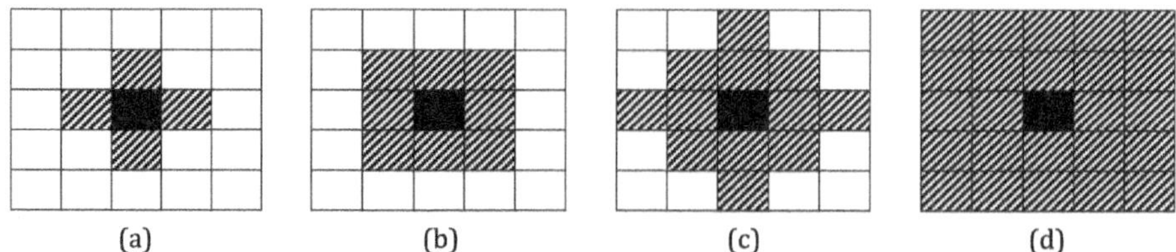

Fig. 15.3 Samples of candidate neighborhood structures: (**a**) first order (DS 1); (**b**) second order (SS 1); (**c**) DS 2; and (**d**) SS 2 (up diagonal filled grids are neighbors of the solid filled grid)

The grid size here establishes our sampling resolution and determines the length scale of corresponding interaction analysis. The length scale of interest will be determined by the structure property studied by nanoscientists. Our approach can be applied to any given scale.

Current techniques of measuring nanostructures are generally two-dimensional imaging such as SEM (Scanning Electron Microscopy) and TEM (Transmission Electron Microscopy). Therefore, it's usually difficult to characterize defective structures due to their complex shapes. We suggest measuring these defects in the same way as measuring normally grown nanostructures. This approximation permits automatic measurement processing. At the same time, we can still identify defect patterns based on extracted interactions. This argument is validated in real case study of ZnO nanowire bundles. We provide a method to calculate nanowire lengths from SEM images in the chapter appendix.

Based on the feature measurement $X(s)$, we model the feature field as a Gaussian Markov random field (GMRF) and estimate nanostructure interactions. More specifically, we will understand the interactions by finding for each site s_i: (1) its neighbors, i.e., those sites that have interactions with s_i and (2) the relation between site s_i and its neighbors. Candidate neighborhood structures are illustrated in Fig. 15.3. They are of two forms: diamond schemes (DS) and square schemes (SS) of different orders. These two forms can cover most commonly observed interaction patterns.

We first model local features as stationary GMRF by assuming all nanostructures are normally grown. We then develop normality tests to detect defects. If there is no defect, the estimated stationary GMRF characterizes features' local variability. Otherwise, we relax the stationarity assumption and estimate interaction patterns for every site in the field individually. Defect patterns are then identified by "abnormally" strong interactions. Details are given in following sections.

15.3 Modeling of Nanostructure Interactions for Local Feature Characterization

We first decompose local nanostructure feature field $X(s)$ into two parts: local trend or mean and local variability. Specifically, we take the following model

$$X(s) = Z\gamma + \Phi(s).$$ (15.2)

Here $Z\gamma$ describes the local mean by a linear combination of explanatory variables Z that can be spatial location functions or concomitant data with each site [7]. $\Phi(s)$ captures the local variability by describing interactions among neighboring nanostructures. This decomposition is different from Huang [14] because $Z\gamma$ here extracts the *local* trend for the region of interest and is not characterized by growth kinetics.

We adopt GMRF [1, 7] to model nanostructure interactions and associate the interactions with local feature variability. Specifically, for each site s_i, we define its neighbors and specify the conditional distribution of $\Phi(s_i)$:

$$\Phi(s_i)|\{\phi(s_j)\}_{j\neq i} \sim N\left(\sum_{s_j \sim s_i} \beta_{ij}\phi(s_j), \sigma^2(s_i)\right)$$ (15.3)

where $s_j \sim s_i$ means s_i and s_j are neighbors of each other, $\sigma^2(s_i)$ is the conditional variance and β_{ij} is a spatial coefficient that is zero unless $s_i \sim s_j$. We interpret interactions among nanostructures as their mutual contributions on each other's local variability. Our modeling states, the conditional expectation of local feature variability in site s_i is a "weighted" sum of the local variabilities in its neighbors. Intuitively, the larger the "weights" (β_{ij}), the stronger the interactions.

GMRF has been extensively used in areas including spatial statistics [21], image analysis [2, 12] and disease mapping [4] etc. By varying the neighborhood structures and model parameters therein, GMRF is capable of modeling complex and diverse interaction structures. This property will certainly assist us to identify patterns of nanostructure interactions and develop a proper characterization of local features.

Denote $D = \text{diag}\{\sigma^2(s_1), \ldots, \sigma^2(s_n)\}$ and $B = (\beta_{ij})$. From Brook's Lemma [5], when $(I - B)$ is invertible and $(I - B)^{-1}D$ is sysmetric and positive definite, $X(s)$ follows a joint normal distribution as following:

$$X(s) \sim N\left(Z\gamma, (I - B)^{-1}D\right)$$ (15.4)

which establishes the spatial distribution of nanostructure features. In particular, the covariance matrix $\Sigma = (I - B)^{-1}D$ or equivalently the precision matrix $Q = D^{-1}(I - B)$ clearly captures the local feature variability. Besides, it is easy to show the pairwise conditional correlation for any two sites s_i and s_j is [7]:

$$\text{corr}\left(X(s_i), X(s_j)|\{x(s_k)\}_{k\neq i,j}\right) = \frac{\beta_{ij}}{|\beta_{ij}|}\sqrt{\beta_{ij}\beta_{ji}}.$$ (15.5)

Quantitatively, the stronger the conditional correlations, the stronger the interactions among neighboring nanostructures.

For a local region that has no defect (i.e., normal/uniform growth of nanostructures), we may assume stationary interaction patterns for every site across the field. That is, $\sigma^2(s_i) = \sigma^2$ and $\beta_{ij} = \beta_q$ where $q = s_i - s_j$. The conditional characteristics are thus

$$\Phi(s_i)|\{\phi(s_j)\}_{j\neq i} \sim N\left(\sum_{s_j \sim s_i} \beta_q \phi(s_j), \sigma^2\right) \tag{15.6}$$

for any site s_i. Due to symmetry of the precision matrix, we have $\beta_{ij} = \beta_{ji}\ \forall i, j$ and $\beta_q = \beta_{-q}$ for any $q > 0$. Besides, $\mathrm{corr}\left(X(s_i), X(s_j)|\{x(s_k)\}_{k\neq i,j}\right) = \beta_q$. β_q directly explains interactions between site s_i and site s_j.

We call the modeling in Eq. (15.6) with stationarity assumptions "stationary GMRF modeling." Correspondingly, GMRF of general forms in Eq. (15.3) is called "nonstationary" or "varying coefficient GMRF." Because nonstationary GMRF involves much more unknown parameters, for model parsimony we always try stationary GMRF modeling first. Only for local regions that have been detected to have defects, we then turn to nonstationary GMRF to identify the detailed defect patterns.

15.4   Estimation of Nanostructure Interactions for Local Feature Characterization

We model nanostructure interactions in terms of conditional correlations between sites. Under the framework of GMRF, nanostructure interactions and local features are described by model parameters $\eta = \{\beta_{ij}, \sigma^2(s_i), \gamma\}_{i,j=1,...,n}$ in Eq. (15.4). Due to the computation of matrix determinant $|I - B|$, maximum likelihood estimation (MLE) approach is only feasible for stationary GMRF defined on regular fields. Therefore, we follow strategies in Fig. 15.4 to estimate nanostructure interactions. Particularly, we apply MLE to estimate stationary interactions in normal growth regions and local likelihood estimation (LLE) to estimate nonstationary interactions in regions with defects. Normality tests on transformed residuals detect defects initially based on stationary interaction estimation.

Maximum Likelihood Estimation for Stationary Nanostructure Interactions

If nanostructure interactions are modeled by stationary GMRF, the local features follow a normal distribution $N(Z\gamma, \sigma^2(I - B)^{-1})$. Since we divide regions of interest regularly, $I - B$ is a block circulant matrix under periodic boundary conditions. We can thus efficiently calculate $|I - B|$ through Fast Fourier Transformation (FFT).

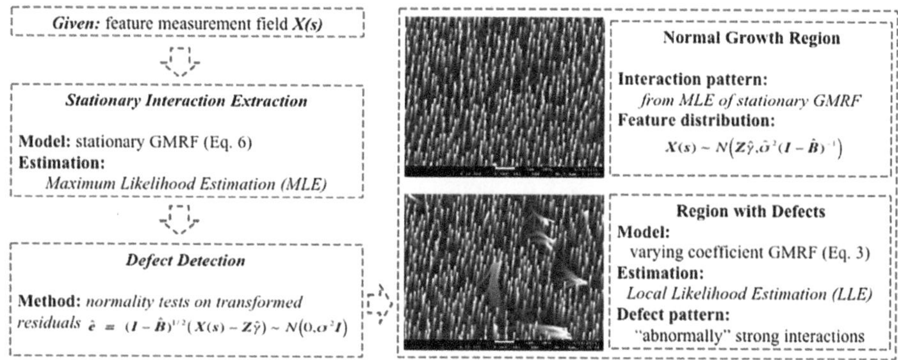

Fig. 15.4 Overall strategies of modeling and estimating nanostructure interactions for local feature characterization and automatic defect detection through feature measurement

Consequently, maximum likelihood estimation (MLE) is feasible and efficient in this case.

We summarize the MLE approach as following by referring to relevant literatures [1, 3, 7, 13, 22, 29]. First, we can write the negative log likelihood as

$$2L(\eta|x(s)) = -2 \log f(x(s)|\eta)$$

$$= \frac{1}{\sigma^2}(x(s) - Z\gamma)^T(I - B)(x(s) - Z\gamma)$$

$$+ n \log \sigma^2 - \log(|I - B|) + c \qquad (15.7)$$

where $\eta = \{\beta_q, \sigma^2, \gamma\}_{q>0}$ denotes the collection of unknown parameters, n is the total number of random variables (sites) in the field, and c is a known constant. Let $\hat{\eta} = \{\hat{\beta}_q, \hat{\sigma}^2, \hat{\gamma}\}$ be the MLE of η. Then setting $\partial L/\partial \sigma^2 = 0$ we obtain:

$$\hat{\sigma}^2 = n^{-1}(x(s) - Z\gamma)^T(I - B)(x(s) - Z\gamma). \qquad (15.8)$$

Similarly setting $\partial L/\partial \gamma = \mathbf{0}$ we can see $\hat{\gamma}$ is the generalized LSE for linear regression: $x(s) = Z\gamma + \epsilon$ with $\epsilon \sim MVN(\mathbf{0}, (I - B)^{-1})$, or equivalently,

$$\hat{\gamma} = (Z^T(I - B)Z)^{-1}Z^T(I - B)x(s) \qquad (15.9)$$

Therefore, $\hat{\beta} = \operatorname{argmin}\mathfrak{L}(\beta|x(s))$ with $\beta = \{\beta_q\}_{q>0}$ and

$$\mathfrak{L}(\beta|x(s)) = \log\{(x(s) - Z\hat{\gamma})^T(I - B)(x(s) - Z\hat{\gamma})\}$$

$$- n^{-1}\log(|I - B|) \qquad (15.10)$$

Based on $\hat{\boldsymbol{\beta}}$, we can obtain $\hat{\sigma}^2$ and $\hat{\boldsymbol{\gamma}}$ from Eqs. (15.8) and (15.9). In addition, if we denote τ^2 as the unconditional variance for $X(s_i)$, $\forall i$, then $\hat{\tau}^2 = \hat{\sigma}^2 (\boldsymbol{I} - \hat{\boldsymbol{B}})_{11}^{-1}$ (Eq. 15.4).

$\hat{\boldsymbol{\eta}}$ is a consistent estimator of $\boldsymbol{\eta}$ for stationary GMRF. Besides, $\hat{\boldsymbol{\eta}}$ follows a joint normal distribution with mean $\boldsymbol{\eta}$ and precision matrix $\boldsymbol{J}(\boldsymbol{\eta})$ (the fisher information matrix) asymptotically [7, 20]. Confidence intervals/elipsoid for $\hat{\boldsymbol{\eta}}$ can be obtained accordingly. Models of different neighborhood structures can also be compared through asymptotic likelihood ratio tests (LRTs) [29]. GMRF with the winning neighborhood structure will be selected to describe the stationary interactions.

Estimation of stationary nanostructure interactions can be viewed as an overall estimation of interactions for the local region. We use it as a benchmark to detect defects. For normal growth regions, it also specifies the spatial distribution of nanostructure features: $N(\boldsymbol{Z}\hat{\boldsymbol{\gamma}}, \hat{\sigma}^2 (\boldsymbol{I} - \hat{\boldsymbol{B}})^{-1})$ and determines the interaction patterns for each site.

Normality Tests on Transformed Residuals to Detect Defects

Since normally grown nanostructures have stationary interaction patterns across sites, deviation of our measurement data from the stationary interaction pattern indicates potential defects. Therefore, we can formulate the defect detection problem as a problem of testing following hypotheses.

H_0:   measured nanostructure feature field follows stationary GMRF;
H_1:   measured nanostructure feature field follows nonstationary GMRF.

We define the transformed residuals as

$$e = (\boldsymbol{I} - \boldsymbol{B})^{1/2}(X(s) - \boldsymbol{Z}\boldsymbol{\gamma}). \tag{15.11}$$

Under the null hypothesis H_0, we have $X(s) \sim N(\boldsymbol{Z}\boldsymbol{\gamma}, \sigma^2 (\boldsymbol{I} - \boldsymbol{B})^{-1})$ by referring to Eqs. (15.4) and (15.6). Therefore, the transformed residuals $e \sim N(0, \sigma^2 \boldsymbol{I})$. Based on stationary GMRF estimation, we obtain MLEs of transformed residuals as

$$\hat{e} = (\boldsymbol{I} - \hat{\boldsymbol{B}})^{1/2}(X(s) - \boldsymbol{Z}\hat{\boldsymbol{\gamma}}). \tag{15.12}$$

Taking into account the consistency property of $\hat{\boldsymbol{\eta}} = \{\hat{\beta}_q, \hat{\sigma}^2, \hat{\boldsymbol{\gamma}}\}_{q>0}$, MLEs of transformed residuals $\hat{e}_i$, $i = 1, \ldots, n$ are also i.i.d normal random variables asymptotically. That is,

$$\hat{e}_i \sim \text{i.i.d. } N(0, \sigma^2) \text{ for } i = 1, \ldots, n \tag{15.13}$$

asymptotically if stationary GMRF really captures the interaction patterns among measurement data. On the contrary, if there are some defects, due to the inaccurate interaction extraction for defective sites, transformed residuals of defective sites would be larger comparing to those of normal sites. Therefore, evaluating the deviation of $\hat{e}_i$, $i = 1, \ldots, n$ from i.i.d. normal distributions will help detecting defects.

We perform normality tests on transformed residuals $\hat{e}$ to detect defects. Normality tests detect "outliers" among $\hat{e}_i$, $i = 1, \ldots, n$. Comparing to other tests, e.g., χ^2 test, which test on the equal variance assumption, normality tests are more powerful defect detection method when the percentage of defects is small. Both graphical methods including normality plots and formal tests including Anderson–Darling test, Cramer–von-Mises test and Lilliefors test, etc., can be used. Thode [26] has a good summary of normality test methods. In this chapter, we recommend half-normal plot for visual judgment because it has a clearest indication of outliers. Anderson–Darling test is chosen for formal detection because its calculation on departure of empirical distributions from normality puts more weights on tails [25], and thus, it is more sensitive to defects.

If normality tests indicate the existence of defects in our measurement data, local likelihood estimation (LLE) technique is then used to estimate interactions across the field. Patterns of defects will be identified by patterns of interactions.

Local Likelihood Estimation for Regions with Defects

For regions detected to have defects, we use varying coefficient GMRF to model the nonstationary interactions among nanostructures. Since spatial parameters $\{\beta_{ij}\}_{i=1,\ldots,n, s_j \sim s_i}$ vary across sites, dimension of unknown parameters is larger than measurement data. Traditional estimation methods (e.g. MLE, pseudo-likelihood, coding method, etc. [1, 3, 7]), are not directly applicable here.

We adopt local likelihood estimation (LLE) to estimate nanostructure interactions. Parameters $\{\beta_{ij}\}_{s_j \sim s_i}$ and $\sigma^2(s_i)$ that define the interactions of any site s_i with its neighbors are estimated based on those sites physically near s_i. As illustrated in Fig. 15.5, given window size $\lambda = 4$ we use sites within the corresponding circles to estimate the interactions for colored target sites. Denote $R^\lambda(s_i)$ (including s_i) to be the collection of sites included for estimating s_i under λ. To estimate parameters $\{\beta_{ij}\}_{s_j \sim s_i}$ and $\sigma^2(s_i)$, we assume stationary interaction patterns among $R^\lambda(s_i)$. In addition, we assign different weight $W_k^\lambda(s_i)$ to each site $s_k \in R^\lambda(s_i)$. The weight

Fig. 15.5 Illustration of local likelihood estimation

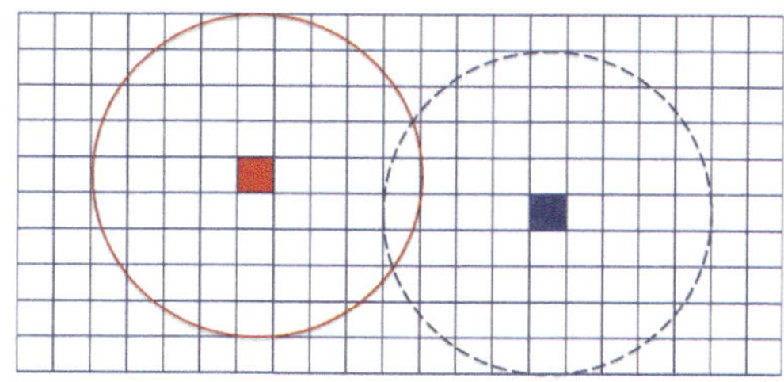

$W_k^\lambda(s_i)$ measures how much we believe the interaction pattern of s_k is close to that of s_i.

For each site s_i, pseudo likelihood is chosen to estimate its parameters because of the total computation complexity. In this way, $\{\beta_{ij}, \sigma^2(s_i)\}_{s_j \sim s_i}$ is the weighted least squares estimator for linear regression

$$X(s_k) = \beta_{i0} + \sum_{s_j \sim s_i} \beta_{ij} X(s_k + s_j - s_i) + \epsilon_{ik} \qquad (15.14)$$

where $s_k \in R^\lambda(s_i)$ for any $i = 1, \ldots, n$ under given λ. Here β_{i0} is the intercept, and ϵ_{ik} is the kth error term. We select the smoothing parameter λ (i.e., the window size) by cross validations [19]. Bias correction on raw estimation of $\{\beta_{ij}\}_{i=1,\ldots,n,s_j \sim s_i}$ from Eq. (15.14) is then performed by referring to Equation (15) in [17]. Interested readers please refer to [10, 17] and references therein for more details.

15.5 Simulation Case Studies

Simulation case studies are to illustrate and validate our proposed modeling and estimation procedures for extracting nanostructure interactions and characterizing nanostructure local features.

Setups of Simulation Experiments

We simulated 100 i.i.d. datasets defined on 50×100 regular lattice $s = \{(s, t)\}_{s=1,\ldots,50, t=1,\ldots,100}$. These datasets mimic (replicates of) feature measurement fields we may obtain from normal growth regions. Each dataset has structure $X(s) = Z\gamma + \Phi(s)$ as in Eq. (15.2). The local trend $eX(s) = Z\gamma$ is defined as:

$$eX(s, t) = \gamma_0 + \gamma_1 * s + \gamma_2 * t. \qquad (15.15)$$

The nanostructure interaction field $\Phi(s)$ follows stationary GMRF with second-order neighborhood structure. That is, for each site $s_i = (s, t) \in s$, we have

$$
\begin{aligned}
e\Phi(s, t) | &\{\phi(s', t')\}_{(s', t') \neq (s, t)} \\
&= \beta_1 \{\phi(s - 1, t - 1) + \phi(s + 1, t + 1)\} \\
&\quad + \beta_2 \{\phi(s - 1, t) + \phi(s + 1, t)\} \\
&\quad + \beta_3 \{\phi(s - 1, t + 1) + \phi(s + 1, t - 1)\} \\
&\quad + \beta_4 \{\phi(s, t - 1) + \phi(s, t + 1)\} \qquad (15.16)
\end{aligned}
$$

and $\mathrm{Var}(\Phi(s,t)|\{\phi(s',t')\}_{(s',t')\neq(s,t)}) = \sigma^2$. Values set for simulating the datasets are summarized in Table 15.1. We include both positive and negative values for β to explore the most general patterns of nanostructure interactions.

Interaction Extraction for Normal Growth Regions

For each previously simulated dataset (referred as *stationary datasets*), we model it as stationary GMRF with local trend $eX(s,t) = \gamma_0 + \gamma_1 * s + \gamma_2 * t$. We then estimate parameters and interaction schemes for GMRF models by MLE.

Different neighborhood structures are compared based on asymptotic likelihood ratio tests [29] at 95% confidence level. Among the 100 datasets, 95 datasets correctly select second-order scheme to describe the interaction structure while the other five datasets select diamond scheme (DS 2 in Fig. 15.3).

As to parameter estimation, we summarize MLEs of γ, β and σ^2 in Table 15.2 based on those 95 datasets that select second order scheme. From the results we can see MLEs of parameters are quite stable among different datasets and are very close to their corresponding true values. We also plot MLEs of $\beta_i, i = 1, \ldots, 4$ together with their 95% confidence bounds in Fig. 15.6. Vibrations of confidence bounds are due to the randomness among different datasets (i.i.d. samples from the same distribution). From the figure we can see the 95% confidence intervals of $\beta_i, i = 1, \ldots, 4$ contain corresponding true values almost all times.

Although we know there is no defect in our simulated data, we still do normality tests to detect defects as described in Section IV. Among the 100 simulated datasets, 97 of them have an Anderson–Darling p value greater than 5%. That is at 95% confidence level only 3 out of 100 datasets are falsely detected to have defects. Our method for detecting defects works well for normal growth regions.

The simulation study of stationary datasets shows that our method can correctly extract interaction patterns for normal growth regions. The estimated stationary

Table 15.1 Parameter values set in simulation studies

γ_0	γ_1	γ_2	β_1	β_2	β_3	β_4	σ^2
6.00	-0.03	-0.01	-0.1	0.25	0.05	0.15	1

Table 15.2 Summary of MLEs for γ, β and σ^2 based on stationary datasets (95 samples)

	True Val.	Est. Mean	Std. Dev.	5% Qt.	95% Qt.
γ_0	6.00	6.0020	0.0744	5.8740	6.1113
γ_1	-0.03	-0.0299	0.0016	-0.0328	-0.0276
γ_2	-0.01	-0.0101	0.0009	-0.0116	-0.0087
β_1	-0.10	-0.1013	0.0131	-0.1246	-0.0825
β_2	0.25	0.2512	0.0102	0.2324	0.2660
β_3	0.05	0.0477	0.0101	0.0333	0.0626
β_4	0.15	0.1506	0.0140	0.1300	0.1756
σ^2	1.00	0.9958	0.0219	0.9636	1.0355

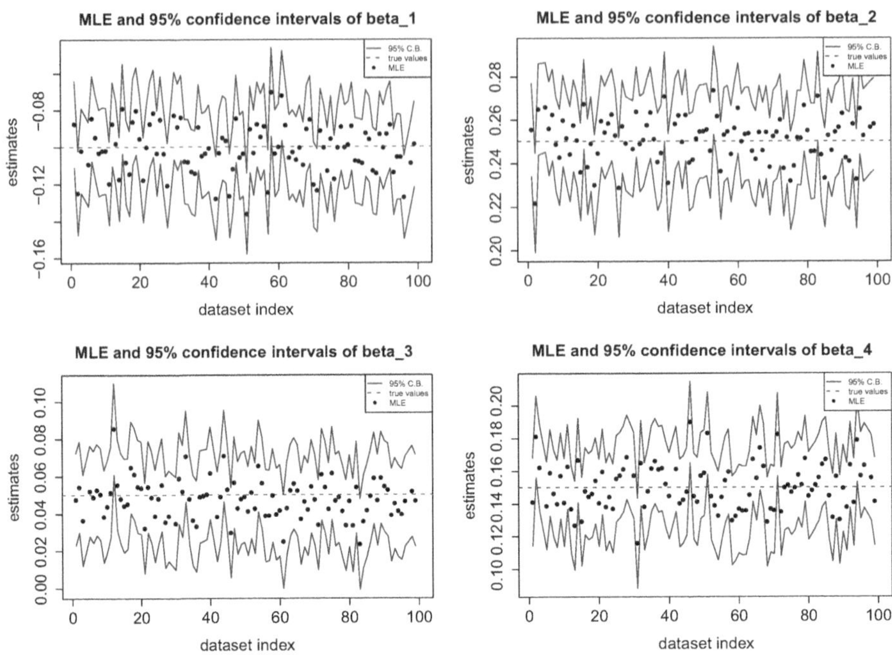

Fig. 15.6 MLEs and 95% confidence bounds of β_i, $i = 1, \ldots, 4$

GMRF accurately describes features' spatial variability. We also have high confidence that our suggested normality tests don't falsely indicate defects.

Interaction Extraction for Defects Detection

For each stationary dataset, we set randomly 50–250 sites in it as zeros (i.e. 1%–5% zero ratio). They mimic those no growth sites caused by underlying deformities of the substrate. By studying these *nonstationary datasets*, we hope our estimation of nanostructure interactions and proposed normality tests can detect them and identify their patterns.

Following the strategies summarized in Fig. 15.4, we first model each nonstationary dataset through stationary GMRF pretending that all nanostructures are normally grown. Then normality tests are performed to detect defects based on MLEs of the transformed residuals $(\boldsymbol{I} - \hat{\boldsymbol{B}})^{1/2}(\boldsymbol{x}(s) - \boldsymbol{Z}\hat{\boldsymbol{\gamma}})$. For the 100 nonstationary datasets, Anderson–Darling p values are all nearly zero ($< 10^{-32}$) indicating the existence of defects in all datasets. Our defect detection result here is 100% consistent with reality.

We then randomly select one dataset to illustrate the identification of detailed defect patterns by nonstationary interaction estimation. We model the selected dataset by varying coefficient GMRF (Eq. 15.3) with second-order neighborhood structure based on its overall interaction estimation. We then estimate the interactions by local likelihood technique under $\lambda = 12$ (Fig. 15.7). Level plots of estimated β_i, $i = 1, \ldots, 4$ are in Fig. 15.8. We also add locations of designed no growth sites to assist visual comparison. Here β_i, $\forall i$ has the same meaning as that explained in simulation setups. Therefore, values of β_i, $\forall i$ determine the intensity of interactions across the field. Directions of the interactions are indicated by corresponding red arrows beside the level plots. From the figure we can see that local peaks of nanostructure interactions are always related to those no growth sites. This observation is both consistent with the data: higher correlation between zeros and their neighboring values and the physics: defects have higher impact on neighboring sites due to competition of source materials, etc.

To summarize the simulation studies presented earlier, our proposed strategies to characterize local features and detect defects by extracting nanostructure interactions are feasible and effective. Although no growth may be an extreme form of "defects," our results suggest that it is feasible in general cases to identify defects by extracting interaction patterns from feature measurement. The real case study of large bundle defects supports this claim in the following section.

Fig. 15.7 λ selection by cross validation [10]

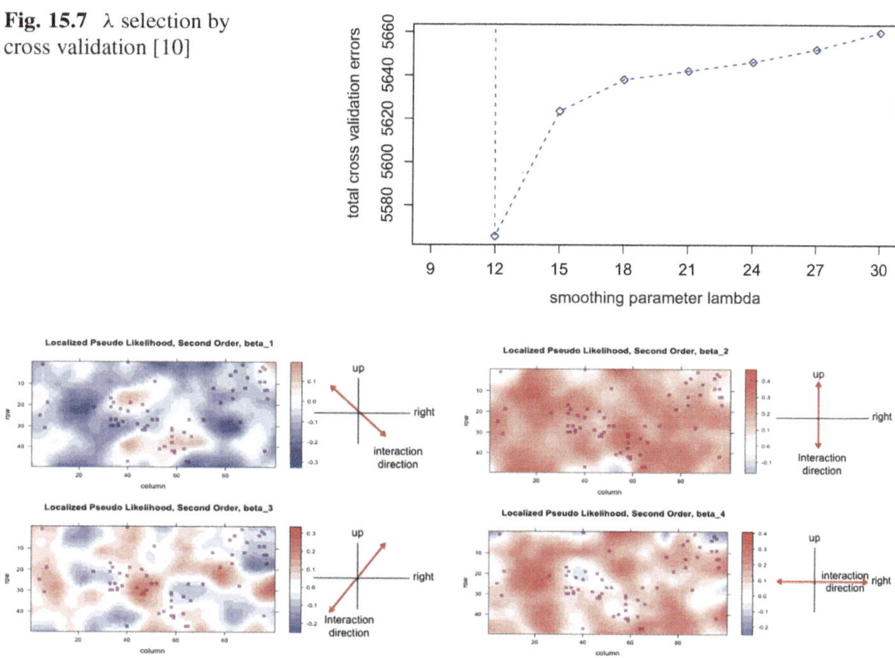

Fig. 15.8 Level plots of local likelihood estimated β together with simulated no growth sites

15.6   Real Case Study of ZnO Nanowire Data

In this section, we are going to study interactions among nanowires through length measurement to automatically detect nanowire bundles (insets of Fig. 15.9).

We measured a 27 um × 56.4 um local region of a ZnO nanowire sample, which was grown through VLS at 950°C, 1 atm. with 100 sccm Ar. flow for 25 minutes. Two SEM images are taken for the same region with one from the top and the other by tilting the sample for 8° (as shown in Fig. 15.9). By extracting tip locations of the same nanowire on the two images, we can calculate the length of the nanowire. Please refer to Appendix A for more details.

We then divide the region regularly into 45×94 grids (sites) each of size 0.6 um × 0.6 um. For each site, we measure a randomly selected nanowire if there is. If a site has no growth, we take the length measurement as zero. And if a site is occupied by a nanowire bundle, we measure the bundle approximately as measuring straight nanowires. Altogether, we obtain a 45×94 nanowire length field.

Preliminary Data Analysis

We first look at the data graphically. Figures 15.10 and 15.11 show level plot of the NW length field (with contour lines) and locations of nanowire bundles and no growth sites, respectively. From the figure we can see, (1) neglecting no growth sites, nanowire lengths generally decrease from top to bottom and from left to right;

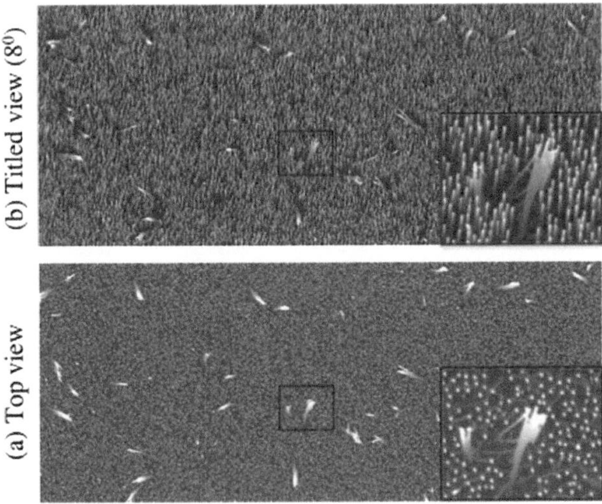

Fig. 15.9 SEM images of ZnO nanowire sample. Insets are enlarged images for the labeled area

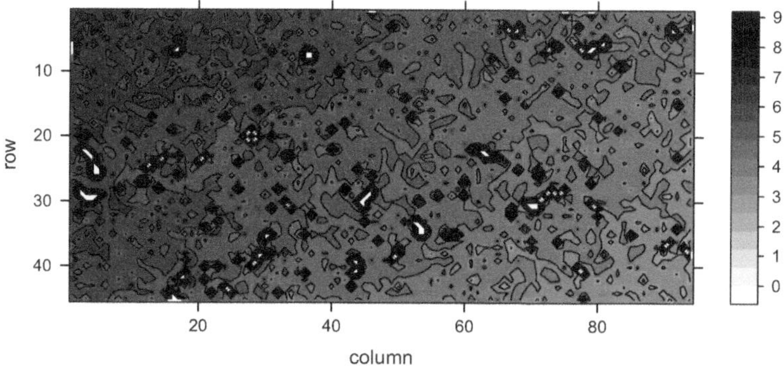

Fig. 15.10 Level plot of nanowire length field

Fig. 15.11 Locations of NW
bundles (◇) and no growth
sites (●)

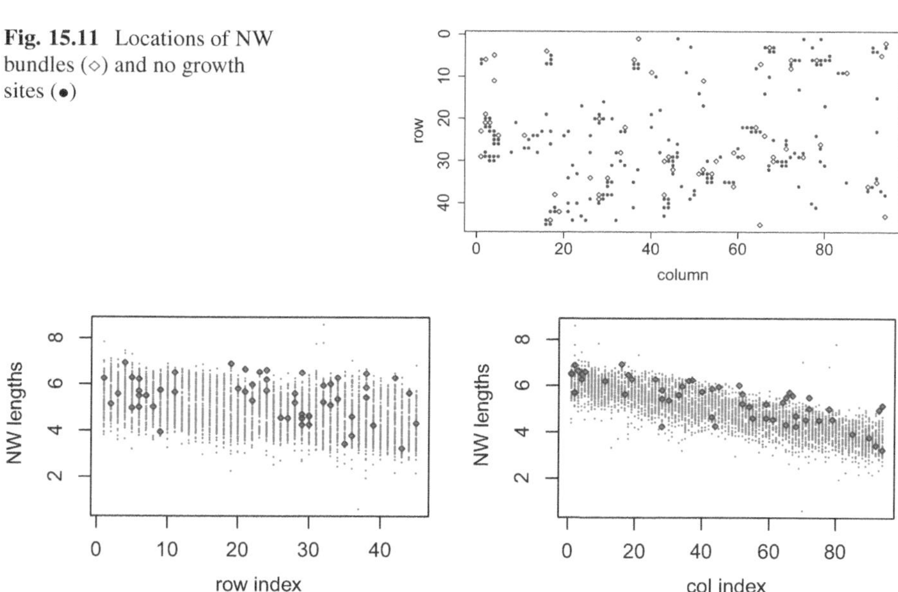

Fig. 15.12 NW lengths VS. row indexes and column indexes of the measured field

and (2) nanowire bundles are normally surrounded by no growth sites indicating
stronger impact of bundles on their neighboring sites.

We further plot nanowire lengths with corresponding row and column indexes
of the field in Fig. 15.12. Those no growth sites are neglected to give a clearer
view of the trend. Lengths of nanowire bundles (approximate) are indicated by blue
circles. Our findings are: (1) local trend of the lengths is generally linear with row
and column indexes; and (2) nanowire bundles cannot be identified directly by their
lengths.

Therefore, we propose to detect bundles by extracting interaction patterns. Besides, we take the local trend as $eX(s, t) = \gamma_0 + \gamma_1 * s + \gamma_2 * t$ (same as Eq. (15.15)) with $s = \{(s, t)\}_{s=1,2,...,45, t=1,2,...,94}$ in stationary GMRF modeling at first step.

Interaction Extraction for NW Bundle Detection

We first model the length field as a stationary GMRF with its local trend as in Eq. (15.15) pretending we don't know there are defects within the field. Our motivations are (1) to get an overall estimate of local trend and nanowire interactions and (2) to check whether our method can automatically detect defects without looking at the images.

By stationary GMRF modeling, we select second-order scheme at 95% confidence level. Table 15.3 shows MLEs and 95% confidence intervals of the parameters. Based on the estimation, we do normality tests on $\hat{e} = (I - \hat{B})^{1/2}(x(s) - Z\hat{\gamma})$ to see whether there is defect in the field. Anderson–Darling test gives a zero p value. Half-normal plot of $\hat{e}$ is also depicted in Fig. 15.13 to assist the detection. Both indexes strongly indicate the existence of defects. Our method indeed detects defects automatically without observing the images.

Then we model the NW length field by second-order varying coefficient GMRF to identify detailed defect patterns. Model parameters are estimated by local likelihood estimation (LLE) technique. Cross-validation errors under different λ values are shown in Fig. 15.14. So $\lambda = 12$ is chosen. We plot estimated β_i, $i = 1, \ldots, 4$ across the field on top of the top view SEM image in Fig. 15.15. As expected, local peaks of interactions always correspond to nanowires bundles.

Table 15.3 MLEs and 95% C.I.s of stationary GMRF parameters for NW length field

γ_1	γ_2	γ_3	σ^2
6.5648	−0.0295	−0.0254	1.0865
(6.424, 6.706)	(−0.034, −0.026)	(−0.027, −0.023)	(1.039, 1.134)
β_1	β_2	β_3	β_4
0.0605	0.0810	0.0585	0.1384
(0.032, 0.089)	(0.051, 0.111)	(0.030, 0.087)	(0.110, 0.167)

Fig. 15.13 Half-normal plot for transformed residuals $\hat{e}$, real case study

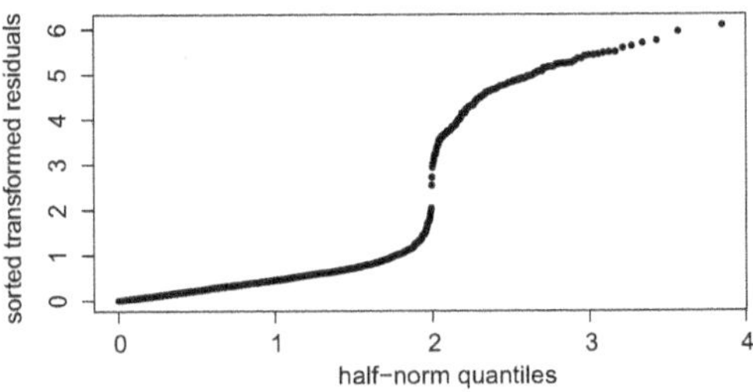

Fig. 15.14 λ selection by cross validation, real case study

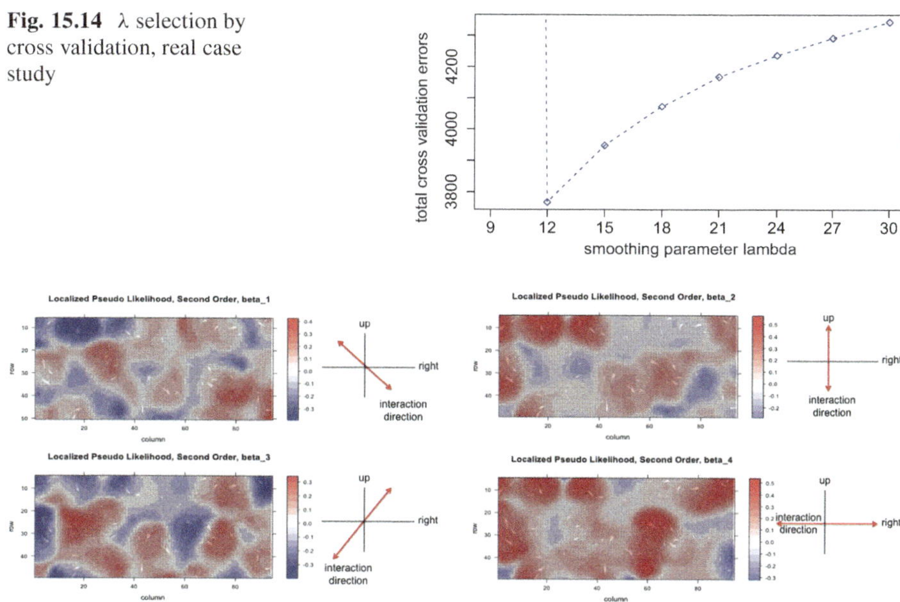

Fig. 15.15 Level plots of local likelihood estimated β together with top view SEM images

Although length measurements of nanowire bundles are approximate, our extracted patterns of nanowire interactions still identify the patterns of defects satisfactorily.

Nanowire bundles absorb more source materials during the growth process and occupy additional space due to their curved shapes (insets of Fig. 15.9). As a result, nanowire bundles inhibit the growth of neighboring sites and may even cause no growth areas around them (Fig. 15.11). Consequently, their estimated interactions from the length measurement data are stronger than other sites. That is why we may identify defects by "abnormally" strong interactions.

By studying real collected nanowire length data, we can confirm that our method to model and estimate nanostructure interactions does automatically detect defects and identify defect patterns. Our studies of nanostructure interactions are consistent with known physical understanding.

Remark 1 Our approach of modeling and estimating nanostructure interactions enables not only the characterization of nanostructure features but also the detection of defects based on "abnormal" interactions among nanostructures.

Remark 2 Simulation study and real case study show the identification of no growth sites caused by underlying deformities on the substrate and neighboring defective structures (large bundles), respectively. Further research can be done to distinguish these two failure mechanisms.

This chapter develops a learning technique to model and estimate interactions among neighboring nanostructures for any local region of interest to assist the characterization of local features and automatic diagnosis of structure defects.

We interpret interactions among nanostructures as the mutual contributions on neighbors' feature variability. Under the framework of Gaussian Markov random field, we incorporate nanostructure interactions into the modeling of local features and establish their connections with local feature variability by pairwise conditional correlations. Our strategies of interaction estimation consist of three sequential steps: estimating the stationary interactions by MLE (maximum likelihood estimation) assuming no defects, defect detection by normality tests on transformed residuals based on stationary interaction estimation, and estimating the nonstationary interactions by LLE (local likelihood estimation) if the region has defects. Estimated interaction patterns identify defect patterns.

Both simulation and real case studies of nanostructure interactions are consistent with existing physical knowledge. Our approach indeed quantifies local feature variability and diagnoses defects accurately and automatically. In addition, our established connection of nanostructure interactions and feature variability will advance both property analysis of nanostructures and physical investigation of structure formation. Further exploration of nanostructure interactions will assist to generate new understandstandings of process–structure–property relation.

Appendix: Nanowire Length Measurement

Denote $(.)_p$ and $(.)_d$ as location parameters on top view and tilted view images respectively where $(.)$ could be Y or α. Let D be the distance between NW tip and the reference point. As illustrated in Fig. 15.16, we then have the following relations:

$$K = \sqrt{D^2 - X^2} = \frac{Y_p}{\cos(\alpha_p)} = \frac{Y_d}{\cos(\alpha_d)} \tag{15.17}$$

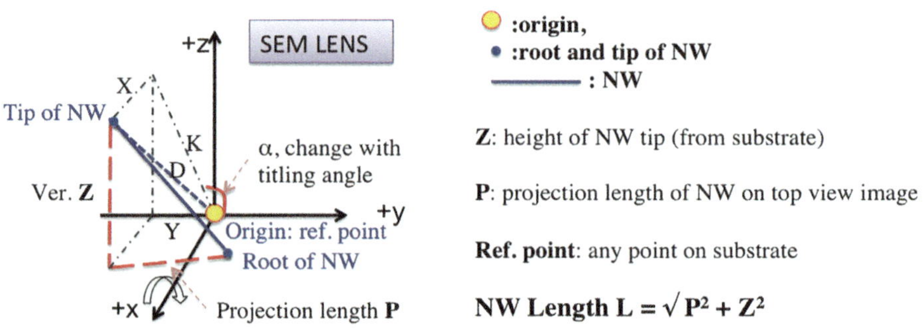

Fig. 15.16 Geometric model for measuring NW length

$$Z = Y_p \sin(\alpha_p) / \cos(\alpha_p) \tag{15.18}$$

$$\alpha_p = \alpha_d + \delta \tag{15.19}$$

where δ is the known tilting angle, which equals $8°$ in our case. Y_p and Y_d are measurable. Since we tilt around axis x, X and thus K are the same for both images. Therefore, the height of NW tip to the reference point is

$$Z = \frac{Y_p \cos(\delta) - Y_d}{\sin(\delta)} \tag{15.20}$$

Substituting Z as in Eq. (15.20), we obtain the NW length

$$L = \sqrt{P^2 + Z^2} = \sqrt{P^2 + \frac{(Y_p \cos(\delta) - Y_d)^2}{\sin^2(\delta)}} \tag{15.21}$$

where P is the projection length of NW on top view image, which is also measurable.

Acknowledgments The work in this chapter was supported by US National Science Foundation under grant CMMI-1000972.

References

1. Besag, J.: Spatial interaction and the statistical analysis of lattice systems. J. Roy. Stat. Soc. B (Methodol.) **36**(2), 192–236 (1974)
2. Besag, J.: On the statistical analysis of dirty pictures. J. Roy. Stat. Soc. B **48**(3), 259–302 (1986)
3. Besag, J., Moran, P.A.P.: On the estimation and testing of spatial interaction in Gaussian lattice processes. Biometrika **62**(3), 555–562 (1975)
4. Best, N., Ickstadt, K., Wolpert, R., Briggs, D.: Combiningmodels of Health and Exposure Data: The SAVIAH Study. Oxford University Press, Oxford (2000)
5. Brook, D.: On the distinction between the conditional probability and the joint probability approaches in the specification of nearest-neighbour systems. Biometrika **51**(3), 481–483 (1964)
6. Chang, C., Xu, L., Huang, Q., Shi, J.: Quantitative characterization and modeling strategy of nanoparticle dispersion in polymer composites. IIE Trans. **44**(7), 523–533 (2012). http://doi.org/10.1080/0740817X.2011.588995
7. Cressie, N.: Statistics for Spatial Data, 2nd edn. Wiley, New York (1993)
8. Dasgupta, T., Ma, C., Joseph, V., Wang, Z., Wu, C.: Statistical modeling and analysis for robust synthesis of nanostructures. J. Am. Stat. Assoc. **103**(482), 594–603 (2008)
9. Dasgupta, T., Weintraub, B., Joseph, V.: A physical-statistical model for density control of nanowires. IIE Trans. **43**(4), 233–241 (2011)
10. Dreesman, J., Tutz, G.: Non-stationary conditional models for spatial data based on varying coefficients. J. Roy. Stat. Soc. D (Statistician) **50**(1), 1–15 (2001)
11. Garnett, E., Yang, P.: Light trapping in silicon nanowire solar cells. Nano Lett. **10**(3), 1082–1087 (2010)

12. Geman, S., Geman, D.: Stochastic relaxation, Gibbs distributions and the Bayesian restoration of images. IEEE Trans. Pattern Anal. Mach. Intell. **6**, 721–741 (1984)
13. Guyon, X.: Parameter estimation for a stationary process on a d-dimensional lattice. Biometrika **69**(1), 95 (1982)
14. Huang, Q.: Physics-driven Bayesian hierarchical modeling of nanowire growth process at each scale. IIE Trans. Quality Reliab. **43**, 1–11 (2011)
15. Huang, Q., Wang, L., Dasgupta, T., Zhu, L., Sekhar, P.K., Bhansali, S., An, Y.: Statistical weight kinetics modeling and estimation for silica nanowire growth catalyzed by pd thin film. IEEE Trans. Autom. Sci. Eng. **8**, 303–310 (2011)
16. Johnson, J., Yan, H., Yang, P., Saykally, R.: Optical cavity effects in ZnO nanowire lasers and waveguides. J. Phys. Chem. B **107**(34), 8816–8828 (2003)
17. Kauermann, G., Tutz, G.: Local likelihood estimation in varying-coefficient models including additive bias correction. J. Nonparametric Stat. **12**(3), 343–371 (2000)
18. Law, M., Greene, L., Johnson, J., Saykally, R., Yang, P.: Nanowire dye-sensitized solar cells. Nature Mater. **4**(6), 455–459 (2005)
19. Van der Linde, A.: On cross-validation for smoothing splines in the case of dependent observations. Aust. New Zealand J. Stat. **36**(1), 67–73 (1994)
20. Mardia, K., Marshall, R.: Maximum likelihood estimation of models for residual covariance in spatial regression. Biometrika **71**(1), 135 (1984)
21. Mead, R.: A relationship between individual plant-spacing and yield. Ann. Botany **30**(118), 301–309 (1966)
22. Moran, P.: A Gaussian Markovian process on a square lattice. J. Appl. Probab. **10**(1), 54–62 (1973)
23. Song, J., Wang, X., Riedo, E., Wang, Z.L.: Systematic study on experimental conditions for large-scale growth of aligned ZnO nanowires on nitrides. J. Phys. Chem. B **109**(20), 9869–9872 (2005)
24. Song, J., Xie, H., Wu, W., Roshan Joseph, V., Jeff Wu, C., Wang, Z.: Robust optimization of the output voltage of nanogenerators by statistical design of experiments. Nano Res. **3**(9), 613–619 (2010)
25. Stephens, M.: Edf statistics for goodness of fit and some comparisons. J. Am. Stat. Assoc. **69**(347), 730–737 (1974)
26. Thode, H.: Testing for Normality, vol. 164. CRC Press, New York (2002)
27. Wang, C., Ryu, K., Badmaev, A., Zhang, J., Zhou, C.: Metal contact engineering and registration-free fabrication of complementary metal-oxide semiconductor integrated circuits using aligned carbon nanotubes. ACS Nano **5**(2), 1147–1153 (2011)
28. Wang, C., Ryu, K., De Arco, L., Badmaev, A., Zhang, J., Lin, X., Che, Y., Zhou, C.: Synthesis and device applications of high-density aligned carbon nanotubes using low-pressure chemical vapor deposition and stacked multiple transfer. Nano Res. **3**(12), 831–842 (2010)
29. Whittle, P.: On stationary processes in the plane. Biometrika **41**(3), 434–449 (1954)
30. Zeng, L., Zhou, Q., De Cicco, M., Li, X., Zhou, S.: Quantifying boundary effect of nanoparticles in metal matrix nanocomposite fabrication processes. IIE Trans. **44**(7), 551–567 (2012). http://doi.org/10.1080/0740817X.2011.635180
31. Zheng, G., Patolsky, F., Cui, Y., Wang, W., Lieber, C.: Multiplexed electrical detection of cancer markers with nanowire sensor arrays. Nature Biotechnol. **23**(10), 1294–1301 (2005)
32. Zhu, L., Dasgupta, T., Huang, Q.: A d-optimal design for estimation of parameters of an exponential-linear growth of nanostructures. Technometrics. Under review (2012)

Chapter 16
Characterization of Nanostructure Interactions with Incomplete Feature Measurement

Nanostructure interactions contribute strongly to structure uniformity and defect formation. Characterizing the interaction not only assists to understand the growth kinetics but also provides a metric to benchmark growth quality. Chapter 15 presented interaction modeling and estimation methods for one local region, which assume complete feature measurement. In order to map interaction patterns across the whole substrate, direct application of the methods would require complete information of a specimen. This implies a formidable metrology task using current inspection techniques such as scanning electron microscopy (SEM). In this chapter, we relax the metrology constraint and analyze nanostructure interactions with incomplete feature measurement. Specifically, we optimize Expectation-Maximization (EM) algorithm based on Markovian properties of interaction modeling and develop a tailored space filling design to select which sites to measure.

16.1   Nanostructure Characterization and Measurement Cost

Nanostructure interactions such as competition for source materials during growth process contribute strongly to structure uniformity and defect formation. Nonuniformity and defects of nanostructures can cause performance degradation in nanodevices. Therefore, characterizing the interaction not only assists to understand the growth kinetics but also provides a metric to benchmark growth quality. Chapter 15 introduced approaches of extracting interactions among neighboring nanostructures for single local region based on its feature measurement [25]. Estimated nanostructure interactions successfully quantify feature variation and enable the detection of structural defects within local regions.

Mapping interaction patterns across the whole substrate would demand complete information of all local regions based on our previous approach. However, characterization of nanostructure features is extremely labor-intensive and time-

Q. Huang, *Domain-informed Machine Learning for Smart Manufacturing*,
https://doi.org/10.1007/978-3-031-91631-1_16

consuming. On the one hand, current inspection techniques such as Scanning Electron Microscopy (SEM) and Transmission Electron Microscopy (TEM) capture a very tiny area in one image. We may need to carefully calibrate and take over 200 SEM images of neighboring areas to cover a 0.1 mm × 0.1 mm local region. To characterize dense quantities of nano elements under manufacturing-relevant time spans, it is necessary to have revolutionary instead of evolutionary advances but for which"no known solutions" are available for 5–10 years down the road [15]. On the other hand, extracting feature information from these two dimensional SEM/TEM images is also complicated and computation intensive. For example, to measure lengths of individual nanowires, we need to match SEM images taken from different angles and find the same nanowire on each image among the nanowire forest [22, 25].

To map interaction patterns across the substrate with affordable measurement of nanostructure features, this chapter introduces both a sampling strategy that selects which sites to measure over the substrate and an analysis technique that estimates nanostructure interactions based on corresponding "incomplete" feature measurement. With the developed approaches, we will be able to quantify nanostructure feature variation and detect structural defects with greatly reduced metrology efforts.

16.2   Strategy of Characterization with Measurement Reduction

The difficulties of characterizing nanostructures include (i) taking and tediously calibrating SEM images for seemless connection and (ii) extracting and matching feature information from SEM images. Correspondingly, to reduce metrology efforts but maintain the desired interaction analysis resolution, we take following strategy to sample and measure nanostructures: (i) we selectively measure multiple separated tiny regions spreading on the substrate; and (ii) for each selected region, we divide it regularly into non-overlapping sites (grids) and only measure nanostructures in selected sites. The aforementioned approach is illustrated in Fig. 16.1.

To better explore the whole substrate, we use *maximin distance Latin hypercube designs* to sample measurement regions (i.e., those regions selected to measure). And for each selected region, we will develop a *tailored space filling design* to choose which sites to measure. By selectively measuring partial of the sites (e.g., white grids in Fig. 16.1b) we can not only lighten the burden of feature extraction but also maintain an appropriate interaction analysis resolution. We call those sites to be measured "measurement sites" and those not measured "non-measurement sites" for convenience. The resulting feature measurement is denoted as "partial" feature measurement or "incomplete" feature measurement interchangeably in this work.

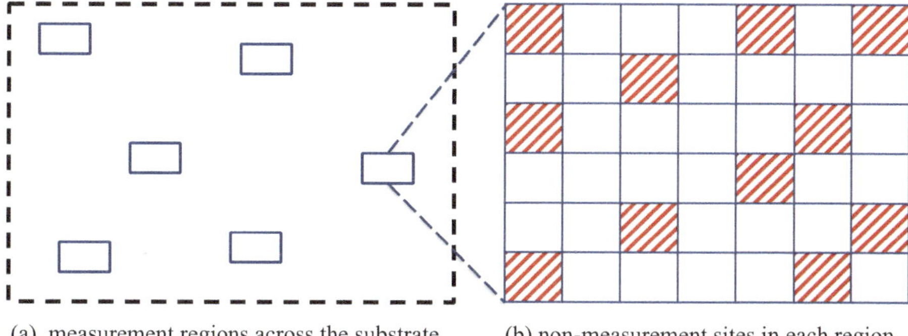

(a) measurement regions across the substrate   (b) non-measurement sites in each region

Fig. 16.1 Strategies of feature measurement (**a**) measurement regions (blue squares) across the substrate (**b**) non-measurement sites (patterned grids) in each measurement region

Although we take a two-step procedure to sample nanostructures, our strategy is different from the sequential or nested designs in literature [8, 16–19, 21] for two reasons. First, regions and sites are all determined before the real measurement takes place. There is no sequential evaluation. Second, our dataset consists only feature measurement in the sites. There is no multiple levels of sampling or approximation.

In addition, Latin hypercube-based space filling designs [13] are not applicable to select non-measurement sites for selected regions, because we have both pre-established division of sites driven by desired resolution of interaction analysis and predetermined number of sites to measure due to metrology constraints.

Besides sampling of regions and sites, other issues to address in this chapter are the modeling and estimation of nanostructure interactions with incomplete feature measurement. Details are discussed in following sections.

16.3 Learning of Nanostructure Interactions with Incomplete Measurements

Notation for measurement regions, sites, and nanostructure features is as following.

- $R = \{r_1, r_2, \ldots, r_R\}$ represents the collection of R measurement regions sampled from the substrate. Standard space filling designs such as maximin distance Latin hypercube designs are applied to select the measurement regions.
- $r_r = \{s_1^{(r)}, s_2^{(r)}, \ldots, s_n^{(r)}\}$ denotes the complete division of sites in region r.
- m_r and $o_r = r_r \backslash m_r$ are respectively measurement sites and non-measurement (omitted) sites in region r. We will develop a tailored space filling design in Sect. 16.4 to select which sites to measure (i.e. m_r) for each region.
- $X(s_i^{(r)})$ represents the nanostructure feature in site $s_i^{(r)}$.

Although here we assume that different measurement regions are of the same size (i.e. $n_r = n, \forall r$), the modeling and estimation techniques developed in this chapter apply in general cases. In addition, the equal size assumption is not hard to satisfy in practice since we normally determine the region sizes before doing the real measurement.

Extension of Stationary Nanostructure Interaction Modeling to the Whole Substrate

In Chap. 15 [25], we have studied interaction modeling for single local region. Specifically, we adopted Gaussian Markov random field (GMRF) to describe the conditional dependence among physically close (i.e., neighboring) nanostructures. We started from stationary interaction modeling, which assumes homogeneous interaction patterns among nanostructures, and then turn to non-stationary interaction modeling for detailed pattern recognition if defects are detected.

We extend the stationary interaction modeling for single local region in [25] to the whole substrate by assembling interaction relations from multiple measured regions and analyzing nanostructure interactions with incomplete feature measurement. We will also briefly mention the detection of defects in real case study but leave non-stationary interaction analysis to interested readers. The extension of stationary interaction analysis to non-stationary interaction analysis will be easy by using local likelihood methods [10, 25].

To separate different sources of uncertainties, we decompose the feature field $X(r_r)$ for each region r into local trend $Z(r_r)\gamma$ and local variability $\Phi(r_r)$ similarly as in [25]. That is:

$$X(r_r) = Z(r_r)\gamma + \Phi(r_r) \tag{16.1}$$

Here the local trend $Z(r_r)\gamma$ is a linear combination of explanatory variables Z that can be spatial location functions or concomitant data with each site [3]. Before we have further information, we assume that the local trend for each measurement region has the same functional form and shares the same set of regression parameters γ. It implies that the global trend for the whole substrate is a smooth surface described by $Z\gamma$. If measurement data object this assumption, it means we have a more complex trend profile due to large process variations among different regions. In this case, one possible extension of the trend modeling here is to take different regression parameters γ_r for different regions and connect γ_r with the process conditions such as temperature T. Interested readers may refer to Huang [7] for the connection.

We model local variabilities $\Phi(r_r)$ for regions $r = 1, 2, \ldots, R$ as *independent* GMRF [1, 3] to capture interactions among neighboring nanostructures, since the measurement regions are purposely selected to be tiny and fairly separated. For each site $s_i^{(r)}$, $i = 1, 2, \ldots, n$ in region $r = 1, 2, \ldots, R$, we assume:

$$\Phi(s_i^{(r)})|\{\phi(s_j^{(r)})\}_{j \neq i} \sim N \left(\sum_{s_j^{(r)} \sim s_i^{(r)}} \beta_q \phi(s_j^{(r)}), \sigma^2 \right) \tag{16.2}$$

Here $s_j^{(r)} \sim s_i^{(r)}$ means $s_j^{(r)}$ and $s_i^{(r)}$ are neighbors of each other. $\{\beta_q : q = s_i^{(r)} - s_j^{(r)}\}$ are spatial coefficients that are zero unless $s_j^{(r)} \sim s_i^{(r)}$. In this model, we express interactions among neighboring sites as their "weighted" contributions to each others' local variability. Intuitively, the larger the weights β_q, the stronger the interactions between $s_i^{(r)}$ and $s_j^{(r)}$. Since our stationary interaction modeling assumes that the weight parameter β_q between any two sites only depends on their relative locations, it assumes homogeneous and nonisotropic interaction patterns across the substrate. When the local trend is correctly captured, violation of the stationary interaction assumption indicates the existence of defects within measured regions.

Denote $B = (\beta_{ij})_{i,j=1,\ldots,n}$ where $\beta_{ij} = \beta_{s_i^{(r)} - s_j^{(r)}}$. We can easily obtain the distribution of nanostructure features for each region r as $X(r_r) \sim N\left(Z(r_r)\gamma, \sigma^2(I - B)^{-1}\right)$ [2], where B is block circulant under periodic boundary conditions. Denote A_r to be the matrix that maps the incomplete collection of sites $m_r \subset r_r$ to the complete collection of sites r_r, i.e. $m_r = A_r r_r$. We get the distribution of measured feature field $X(m_r)$ as

$$X(m_r) \sim N\left(A_r Z(r_r)\gamma, \sigma^2 A_r(I - B)^{-1}A_r^T\right) \tag{16.3}$$

for any region $r = 1, 2, \ldots, R$. Parameters $\eta = \{\gamma, \beta, \sigma^2\}$ with $\beta = \{\beta_q : q > 0, s_i^{(r)} + q \sim s_i^{(r)}, \forall i, r\}$ specify the distribution of nanostructure features.

EM Estimation of Nanostructure Interactions with Incomplete Feature Measurement

Since nanostructure interactions are specified by parameters $\eta = \{\gamma, \beta, \sigma^2\}$ in GMRF modeling, estimating nanostructure interactions is to estimate values of η based on the R independent pieces of incomplete feature measurement $x(m) = \{x(m_r)\}_{r=1,2,\ldots,R}$. The major estimation difficulty comes from the "incompleteness." The non-square mapping matrix A_r in Eq. (16.3) makes MLE (Maximum Likelihood Estimation) [1, 3, 6, 23, 25] computationally infeasible based on measured feature field $x(m)$.

We use EM (Expectation-Maximization) algorithm to estimate parameters η. Those omitted sites (i.e., non-measurement sites) are treated as "missing at random" observations, though here we omit them purposely. Comparing to other techniques dealing with missing values such as imputation, interpolation, and Bayesian data analysis [12], EM algorithm takes advantages of the Markovian property in GMRF modeling of nanostructure interactions (Eq. 16.2) and the analytical property of likelihood function for complete feature field.

EM algorithm is an iterative method for finding the maximum likelihood estimates of η. Instead of maximizing $\log f\left(\{x(m_r)\}_{r=1}^{R}\right)$ directly (what MLE does), EM algorithm iteratively calculates (**E-Step**) and maximizes (**M-Step**)

$$Q(\eta|\eta^p) = \mathbb{E}\left\{\log f\left(\{X(r_r)\}_{r=1}^{R}|\eta\right) |\{x(m_r)\}_{r=1}^{R}, \eta^p\right\} \qquad (16.4)$$

based on the current fit η^p of parameters η at step p [4, 24]. In our case, because different regions are assumed independent, we may further simplify $Q(\eta|\eta^p)$ into a sum of R similar pieces, one for each region:

$$Q(\eta|\eta^p) = \sum_{r=1}^{R} \mathbb{E}\left\{\log f(X(r_r)|\eta) |x(m_r), \eta^p\right\}. \qquad (16.5)$$

Since the calculation is similar for each region r, we only demonstrate EM approach for *one single* measurement region. To keep notation simpler, we denote Z for $Z(r_r)$ and omit any subscript r in the notation for following discussions.

Expectation Step of EM Estimation For complete feature field $X(r)$ in a single regular region, we have its distribution as $X(r)|\eta \sim N\left(Z\gamma, \sigma^2(I - B)^{-1}\right)$. Given partial feature measurement $x(m)$, we can thus evaluate the expectation of its conditional log likelihood function as following:

$$-2Q(\eta|\eta^p) = -2\,\mathbb{E}\left\{\log f(X(r)|\eta)|x(m), \eta^p\right\}$$
$$= 1/\sigma^2 e_{cond}\{(X(r) - Z\gamma)^T(I - B)(X(r) - Z\gamma)\}$$
$$+ n\log(2\pi\sigma^2) - \log(|I - B|) \qquad (16.6)$$

where $e_{cond}(.) = e\{.|x(m), \eta^p\}$ denotes the conditional expectation. Let $\mathrm{Var}_{cond}(.) = \mathrm{Var}\{.|x(m), \eta^p\}$ as the conditional variance. We can expand the conditional expectation in Eq. (16.6) as following.

$$e_{cond}\left\{(X(r) - Z\gamma)^T(I - B)(X(r) - Z\gamma)\right\}$$
$$= (e_{cond}X(r) - Z\gamma)^T(I - B)(e_{cond}X(r) - Z\gamma)$$
$$+ \mathrm{Tr}\left\{(I - B)\mathrm{Var}_{cond}X(r)\right\} \qquad (16.7)$$

Therefore, to evaluate $Q(\eta|\eta^P)$ for given η and η^P, keys are to compute the *conditional expectation* $\mathrm{e}_{cond}\{X(s_i)\}$ for any non-measurement site s_i (i.e. $s_i \in o$, $\forall i$) and the *conditional covariance* $\mathrm{Cov}_{cond}(X(s_i), X(s_j))$ between any two non-measurement sites s_i and s_j (i.e. $\forall i, j$ s.t. $s_i \in o$ and $s_j \in o$).

We take advantages of the Markovian property in our interaction modeling (Eq. 16.2) to simplify the calculation. Specifically, we classify non-measurement sites o into conditionally independent groups g_l. So that we have

$$f(X(o)|x(m), \eta^P) = \Pi_l f(X(g_l)|x(m), \eta^P)$$

$$= \Pi_l f(X(g_l)|x(n(g_l)), \eta^P) \qquad (16.8)$$

where $n(g_l)$ denotes the neighboring measurement sites for g_l i.e. $n(g_l) = \{s_i : s_i \in m,$ and $\exists\, s_j \in g_l$ s.t. $s_i \sim s_j\}$. Based on the Markovian property of GMRF modeling, $X(g_l)$ is conditionally independent to any site in the field other than $n(g_l)$ for given $x(m)$. Consequently, we may obtain the conditional expectations and covariances of non-measurement sites o by evaluating the conditional distribution $f(X(g_l)|x(n(g_l)), \eta^P)$ for each group g_l individually.

The algorithm for classifying non-measurement sites o is as the following. If a non-measurement site has all measured neighbors, e.g., site A in Fig. 16.2, we call it *isolated non-measurement site*. Given $x(m)$, each "isolated non-measurement site" is conditionally independent from any other non-measurement sites. If a non-measurement site also has non-measurement sites as neighbors, e.g., site B in Fig. 16.2, we call it *clustered non-measurement site*. We further group "clustered non-measurement sites" into smallest *clusters* such that a site in one cluster is not the neighbor of any site in other clusters. For instance, in Fig. 16.2, sites B,C, and D form a cluster and sites E–J form another cluster under second order neighborhood structure (neighboring sites are the eight nearest sites, e.g., sites 1–8 for site A). It is easy to prove that clusters constructed this way are conditionally independent. Their

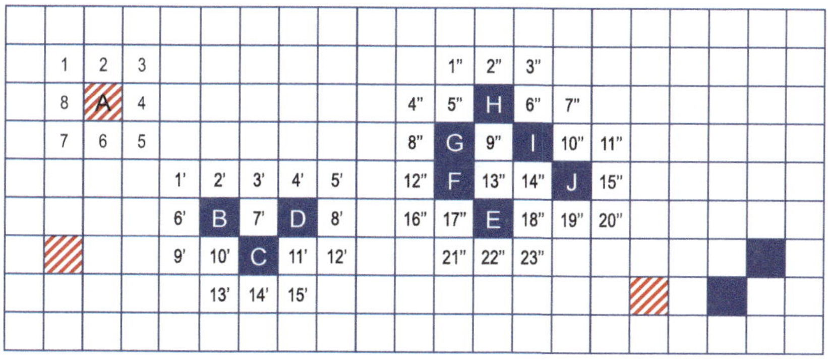

Fig. 16.2 Schematic illustration of non-measurement site classification

conditional distributions will only be dependent on their neighboring measurement sites, e.g., sites 1"–23" for the cluster of E–J in Fig. 16.2.

For *isolated non-measurement sites*, their conditional distributions can be directly obtained from our interaction modeling (Eq. 16.2). That is $\mathbb{E}_{cond}\{X(s_i)\} = (\mu_i)^p + \sum \beta^p_{s_j - s_i}(x(s_j) - (\mu_j)^p)$ and $\text{Var}_{cond}\{X(s_i)\} = (\sigma^2)^p$ with $(\mu_i)^p = E(X(s_i)|\eta^p)$, for each isolated non-measurement site s_i.

For each *cluster* of non-measurement sites c_k, we need to calculate its conditional distribution given neighboring sites $n(c_k)$. That is calculating $f(X(c_k)|x(n(c_k)), \eta^p)$ to extract $e_{cond}\{X(c_k)\}$ and $\text{Var}_{cond}\{X(c_k)\}$. Since $X_k = (X(c_k), X(n(c_k)))$ is part of the whole field $X(r)$, which is a multivariate normal random vector, conditional distribution of $X(c_k)$ given $x(n(c_k))$ can be routinely obtained.

It is clear that EM estimation is mostly affected by those large clusters of non-measurement sites. Therefore, to efficiently estimate nanostructure interactions, it is desirable to have non-measurement sites grouped into smaller clusters. This is exactly the motivation to develop tailored space filling design in Sect. 16.4.

Maximization Step of EM Estimation Maximizing $Q(\eta|\eta^p)$ based on its evaluation in Sect. 16.3 is a quite standard procedure. Denote $\eta^{p+1} = \{\beta^{p+1}, (\sigma^2)^{p+1}, \gamma^{p+1}\}$ to be the EM estimates at step $p + 1$. We obtain η^{p+1} by setting $\partial Q(\eta|\eta^p)/\partial \eta = 0$.

We iteratively perform the Expectation and Maximization steps described earlier until EM estimates η^p converges with certain tolerance. The converged values $\tilde{\eta}$ of η^p will be our final parameter estimation.

16.4 Tailored Space Filling Design for Site Selection in Each Region

This section aims to develop a sampling approach of choosing non-measurement sites o_r in each measurement region r with given non-measurement ratio (number of non-measurement sites to total number of sites). Our objectives are to optimize both the accuracy and the efficiency of nanostructure interaction estimation.

A problem with simple random sampling is that with moderately large non-measurement ratios, it often produces clustering and poorly covered area. The associated consequences include: (1) nanostructures measured may not be representative over the whole region, as a result of which the sampling brings extra variations to interaction estimation; and (2) EM estimation of interactions is not computation-efficient due to those large clusters of dependent non-measurement sites (Eq. 16.8). While classical space-filling designs including distance-based designs [9, 14] and uniform designs [5] may well solve problem (1), they have no control in clustering of dependent non-measurement sites. Figure 16.3 shows a sample maximin distance design with nine points on a 5 × 5 griding. All non-measurement sites

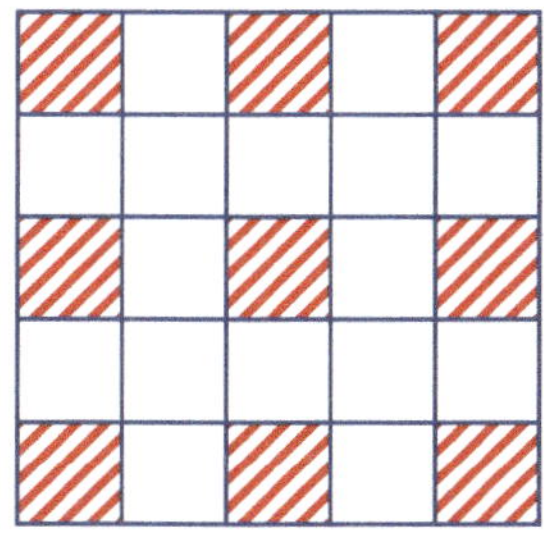

(a) maximin space filling design

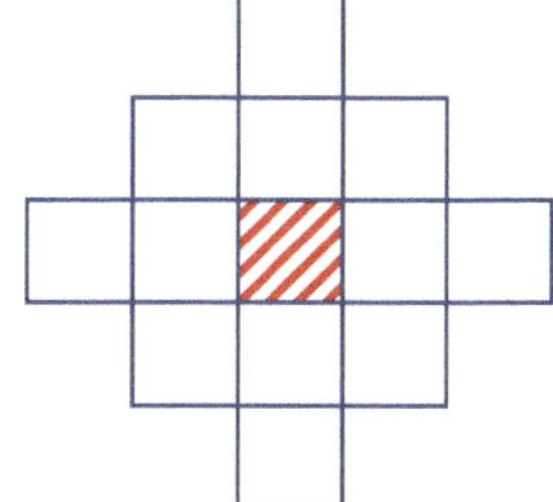

(b) diamond neighborhood structure

Fig. 16.3 Sample maximin distance design that produces *single* conditionally dependent group of non-measurement sites under diamond neighborhood structure

(patterned sites) are conditionally dependent under diamond neighborhood structure (Fig. 16.3b), which is usually evaluated for interaction estimation. Consequently, all non-measurement sites form a single cluster making EM algorithm inefficient (Eq. 16.8).

To optimize EM estimation efficiency, we develop a tailored space filling design of non-measurement sites. **Design criteria** are as the following with descending priorities.

C.1 size of the largest cluster of dependent non-measurement sites, *smaller the better*
C.2 number of isolated non-measurement sites, *larger the better*
C.3 number of conditionally independent groups, *larger the better*

It is intuitive that with smaller clusters, more isolated non-measurement sites, and more conditionally independent groups, we have more "space filling" non-measurement sites.

We use simulated annealing [11, 14], which is an iterative optimization method to search the best space filling design with given non-measurement ratio. Four components are contained in the algorithm, which are (1) generation of initial designs, (2) perturbation scheme, (3) updating criteria, and (4) termination conditions.

Initial Design Generation
When non-measurement ratio is low (e.g. $< 15\%$), we use *simple random sampling* to generate the initial design. When non-measurement ratio is high, we either inherit *optimized designs of lower non-measurement ratios* or use *stratified sampling* to select sites from conditionally independent strata.

Perturbation Scheme
At each step, we perturb the current design $\mathbb{D}_{current}$ by randomly replacing a non-measurement site in the largest cluster with a measurement site. Denote the new design as $\mathbb{D}_{try}$, the newly added non-measurement site as *moveToSite* and the replaced design point as *moveFromSite*. Instead of summarizing the new design

$\mathbb{D}_{try}$ from scratch, we re-classify only those non-measurement sites affected by the movement. The re-classification algorithm is in Algorithm 1.

Updating Criteria

For each perturbation, we will either keep $\mathbb{D}_{current}$ unchanged or update it with $\mathbb{D}_{try}$. The probability of updating is

$$p(\mathbb{D}_{try}, \mathbb{D}_{current}) = \min\{1, \exp(\frac{g(\mathbb{D}_{try}) - g(\mathbb{D}_{current})}{T})\} \qquad (16.9)$$

where $g(\mathbb{D}) = -a \times \text{largestClusterSize} + b \times \text{numIsoSites} + c \times \text{numClusters}$ with $a > b > c$ being positive constants. T is the temperature of annealing that decreases along iterations. It is clear a better design based on criteria **C.** 1–3 has larger chances to be accepted. Besides, we put most efforts to control the largest cluster size.

Termination Conditions

The optimization process stops when either of the following two is met: (i) the largest cluster size in $\mathbb{D}_{current}$ is smaller than a pre-defined value; or (ii) there is no design update in certain number of consequent iterations.

Algorithm 1 Re-classification of non-measurement sites

---

if moveToSite is an isolated non-measurement site **then**
    isoSites in $\mathbb{D}_{try}$ = isoSites in $\mathbb{D}_{current}$ + moveToSite
    The origCluster may decompose after the replacement
    classiSites = otherSitesInOrigCluster in $\mathbb{D}_{current}$
else
    Find non-measurement neighbors of moveToSite
    nonMNeighs = non-measurement neighbors of
              moveToSite in $\mathbb{D}_{try}$
    if ALL nonMNeighs belong to origCluster in $\mathbb{D}_{current}$ **then**
        classiSites = otherSitesInOrigCluster + moveToSite
    else if NONE of nonMNeighs belong to origCluster   **then**
        nonNeigGroups = groups nonMNeighs belong to in
                $\mathbb{D}_{current}$
        moveToSite joins nonNeigGroups together
        aNewCluster = moveToSite + sitesInNonNeigGroups
        classiSites = otherSitesInOrigCluster
    else
        Some belong to origCluster, some don't
        nonNeigGroups = groups nonMNeighs belong to in
                $\mathbb{D}_{current}$
        otherSites = sites in nonNeigGroups but not in
             origCluster
        classiSites = otherSitesInOrigCluster + moveToSite
             + otherSites
Classify classiSites
Update $\mathbb{D}_{try}$ accordingly

Non-measurement sites form large clusters more easily when non-measurement ratio increases. Therefore, more efforts are needed to obtain the space filling design with increasing non-measurement ratios. However, as a design can be used for any dataset with compatible dimensions, calculation time here is not a big concern.

16.5 Simulation Case Studies

In this section, we will demonstrate and validate by simulation case studies the effectiveness of our proposed approaches.

Multiple Tiny Regions Versus One Large Region

For the same amount of data collection, if we distribute it in multiple separated regions instead of a single large region, we can reduce SEM calibration since calibration is only needed for neighboring images. In this subsection, we will validate that the interaction estimation accuracy is maintained by this multiple region approach as well.

We simulated 100 i.i.d. datasets of size 50×100 to mimic replicates of feature measurement from large continuous regions. At the same time, we simulated 100 i.i.d. data collections each of which contains eight independent 25×25 datasets to represent replicates of feature measurement obtained in eight separated small regions. Structure of simulated datasets is depicted in Fig. 16.4. All datasets are GMRF with linear trend and second-order interaction patterns. Specifically for any site $s_i = (s, t)$ where s and t are row and column indexes in the datasets, we assume

$$e\{X(s, t)\} = \gamma_0 + \gamma_1 * s + \gamma_2 * t \tag{16.10}$$

$$e\,\Phi(s, t)|\{\phi(s', t')\}_{(s',t')\neq(s,t)}$$
$$= \beta_1\{\phi(s - 1, t - 1) + \phi(s + 1, t + 1)\}$$
$$+ \beta_2\{\phi(s - 1, t) + \phi(s + 1, t)\}$$
$$+ \beta_3\{\phi(s - 1, t + 1) + \phi(s + 1, t - 1)\}$$
$$+ \beta_4\{\phi(s, t - 1) + \phi(s, t + 1)\} \tag{16.11}$$

and $\mathrm{Var}\,\Phi(s, t)|\{\phi(s', t')\}_{(s',t')\neq(s,t)} = \sigma^2$. Values set for simulating the datasets are summarized in Table 16.1 under the column True Val.

Parameter estimation results for multiple region approach and single large region approach are summarized and compared in Table 16.1. Results show that both approaches give excellent quantification of feature variability. Sampling multiple regions indeed reduces SEM calibration and achieves as good interaction estimation accuracy.

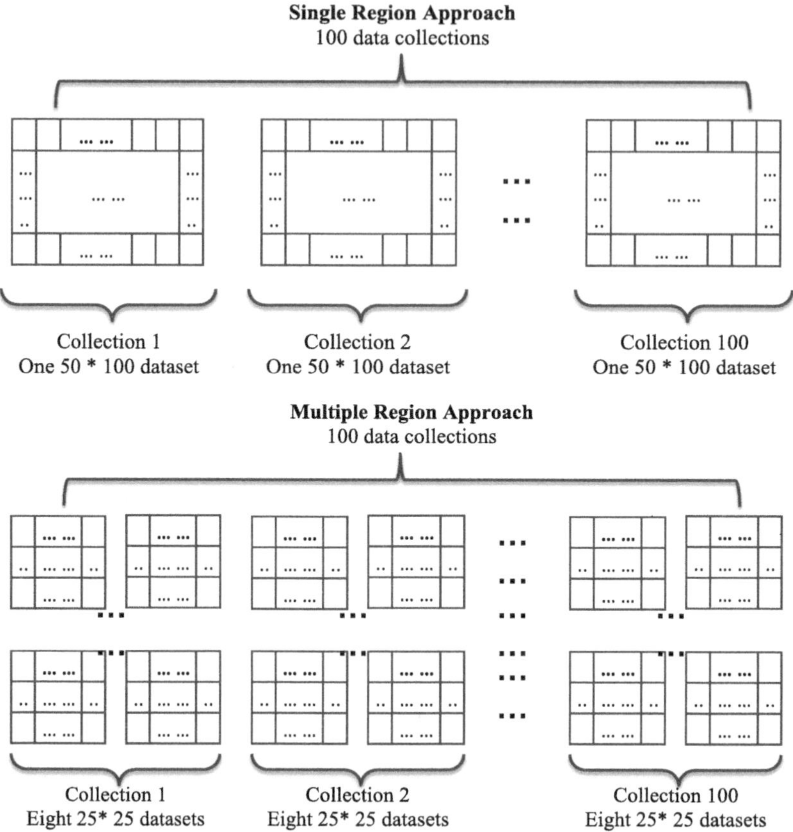

Fig. 16.4 Schematic illustration of simulated data structures

Table 16.1 MLE estimation of nanostructure interactions based on simulated datasets

	True Val.	Multiple Region		Single Region	
		Est. Mean	95% C.I.	Est. Mean	95% C.I.
γ_0	6.00	6.010	[5.888, 6.158]	6.001	[5.838, 6.142]
γ_1	−0.03	−0.031	[−0.037, −0.026]	−0.030	[−0.033, −0.027]
γ_2	−0.01	−0.010	[−0.018, −0.003]	−0.010	[−0.012, −0.008]
β_1	−0.10	−0.099	[−0.129, −0.078]	−0.101	[−0.127, −0.080]
β_2	0.25	0.250	[0.223, 0.269]	0.251	[0.230, 0.270]
β_3	0.05	0.049	[0.027, 0.074]	0.048	[0.030, 0.069]
β_4	0.15	0.150	[0.125, 0.175]	0.150	[0.129, 0.182]
σ^2	1.00	0.997	[0.956, 1.037]	0.996	[0.952, 1.038]

EM Estimation of Nanostructure Interactions with Incomplete Measurement

We use the 100 i.i.d 50×100 datasets simulated in Sect. 16.5 as baseline and study the EM estimation of interactions with different non-measurement ratios in this subsection. In particular, we simulate 10 extra data collections by adding 3%–30% missing values in the complete datasets. Non-measurement sites in these data collections are generated by *simple random sampling* approach.

We summarize EM estimated means and 95% empirical confidence intervals for interaction parameters $\beta_1, \ldots, \beta_4$ in Fig. 16.5. Results for each non-measurement ratio are calculated based on 100 i.i.d datasets with corresponding amount of "missed" measurement. MLEs of β for complete datasets are also depicted for comparison. It shows that we can reduce more than 30% nanostructure feature measurement but still achieve comparable quantification of nanostructure variations with our proposed strategies.

Besides estimation accuracy, we also investigate the dependence of EM estimation speed on non-measurement ratios. From Fig. 16.6, we can see both averages and standard deviations of estimation time increase monotonically with non-

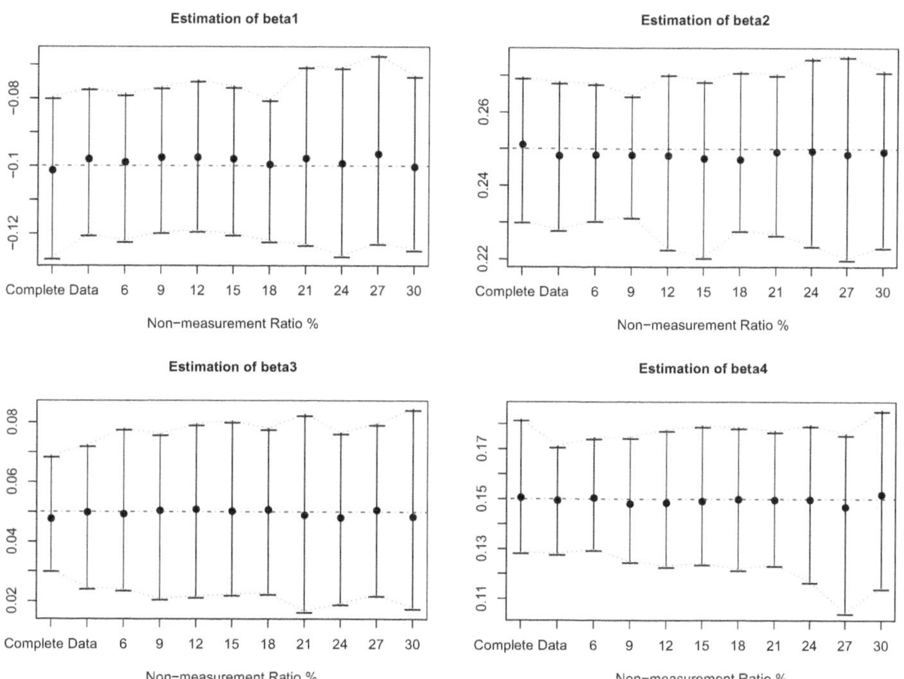

Fig. 16.5 EM estimation of interactions for simulated datasets under different non-measurement ratios. − −: True Val., •: Est. Mean, |: 95% empirical C.I. based on 100 datasets

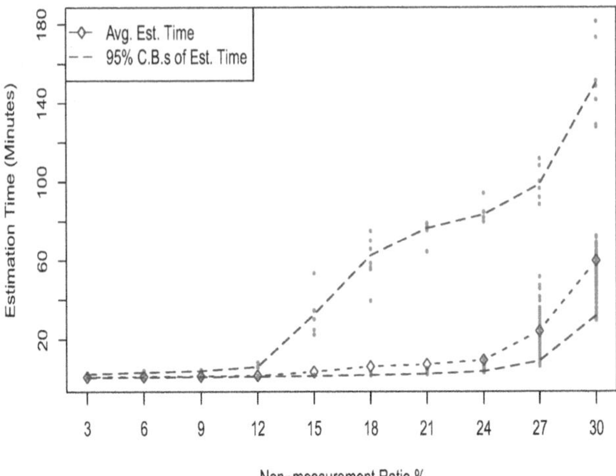

Fig. 16.6 EM estimation time with non-measurement ratios (*simple random sampling*)

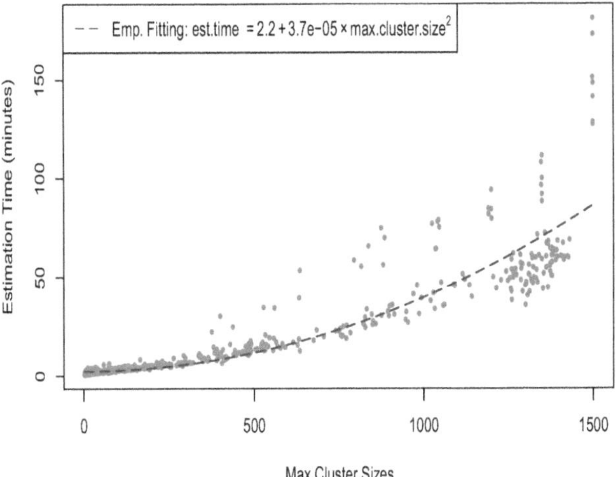

Fig. 16.7 EM estimation time with maximum cluster sizes (*simple random sampling*)

measurement ratios. This rapid increase roots in those large clusters of dependent non-measurement sites generated by *simple random sampling*. Figure 16.7 validates our claim by depicting EM estimation time with the largest cluster size under its highest evaluated interaction structure. Empirically, estimation time increases at least quadratically with maximum cluster sizes since we have to evaluate the covariance matrix for each cluster.

With *simple random sampling*, there are even cases where all non-measurement sites are clustered together, e.g., the case with maximum cluster size $50 \times 100 \times 30\% = 1500$. Therefore, to enhance EM estimation efficiency so as to further reduce metrology efforts, the key is to develop a new sampling approach that controls the largest cluster size. This is exactly what we do in Sect. 16.4. We will examine the effectiveness of our tailored space filling design in Sect. 16.5.

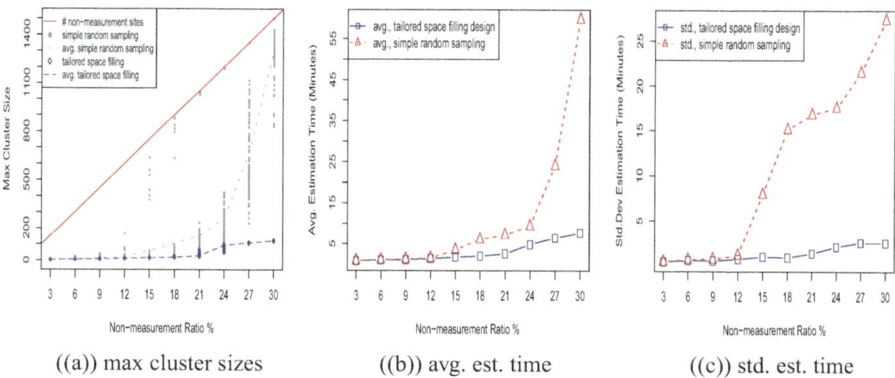

((a)) max cluster sizes   ((b)) avg. est. time   ((c)) std. est. time

Fig. 16.8 Comparison of *tailored space filling sampling* and simple random sampling of non-measurement sites

Tailored Space Filling Design to Select Non-measurement Sites in Each Region

We simulate 10 data collections with 3% to 30% non-measurement sites similar to those in Sect. 16.5. The only difference is that non-measurement sites here are selected by our *tailored space filling design* developed in Sect. 16.4 instead of *simple random sampling*. With our new approach, we obtain the largest cluster sizes in Fig. 16.8a for each dataset under their highest evaluated interaction structures. Comparing to simple random sampling, our tailored space filling design success-fully precludes large clusters. As a result, both averages and standard deviations of EM estimation time are greatly reduced (Figs. 16.8b and c).

Summarizing simulation case studies, we conclude that our approaches of measuring nanostructures and analyzing nanostructure interactions indeed greatly reduce the metrology efforts and achieve excellent interaction estimation accuracy and efficiency.

16.6   Real Case Studies

In this section, we will study a 45×94 real collected ZnO nanowire length dataset that includes measurement for both normally grown nanowires and nanowire bundles (one type of structural defects). Interested readers may find more details in [25] about the dataset. Our primary interest is to investigate whether and how much incomplete length measurement affects interaction analysis and nanowire bundle detection. Similar to simulation studies, we generate 10 data collections of 3%–30% "missing observations" based on the original 45×94 length measurement data. Each data collection consists of 100 datasets with different space filling designs of

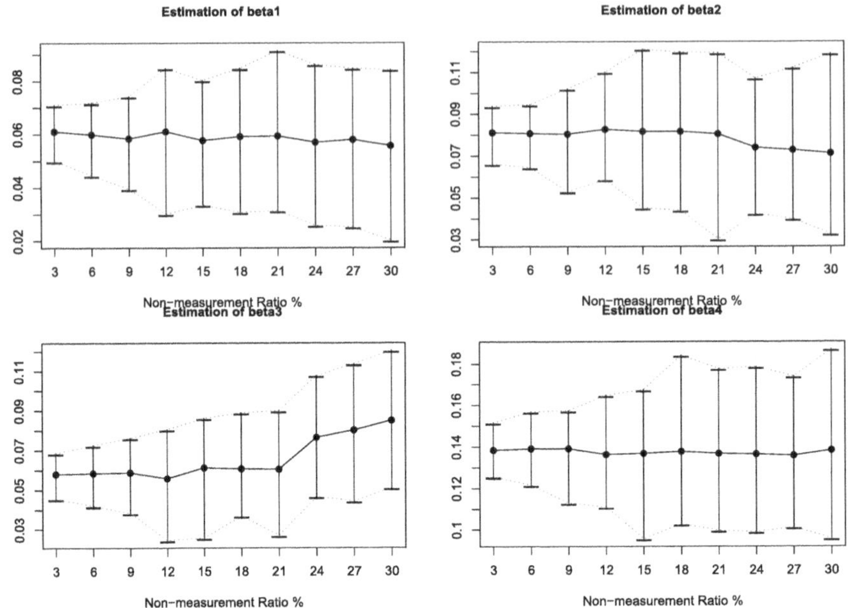

Fig. 16.9 EM estimation of interactions for datasets generated from real nanowire length data. •: Est. Mean, |: 95% empirical C.I. based on 100 datasets.

"missing observations." The "missing observations" are used to mimic the sites we neglect during real measurement.

We use optimized EM algorithm (Sect. 16.3) to estimate interactions among nanowires. Since most of the datasets identify second-order interaction structure (Eq. 16.11), we summarize estimation results for interaction parameters $\beta_1, \ldots, \beta_4$ in Fig. 16.9. It shows that mean estimation of interaction parameters doesn't change much with different levels of incompleteness except for β_3 when non-measurement ratio exceeds 21%. However, empirical confidence intervals are generally wider with larger non-measurement ratios for all β_i, $i = 1, \ldots, 4$. Comparing to simulated stationary datasets (Fig. 16.5), unstable confidence intervals here indicate larger variations among the 100 datasets with different missing observations, which qualitatively implies the existence of defects.

To quantitatively detect defects, we do normality tests on transformed residuals of measured length field similarly as complete data case [25]. Specifically, under the null hypothesis that there are no defects within the field, interaction patterns are stationary across the field and thus measured feature field $X(\boldsymbol{m}_r) \sim N(\boldsymbol{A}_r\boldsymbol{Z}(\boldsymbol{r}_r)\boldsymbol{\gamma}, \sigma^2\boldsymbol{A}_r(\boldsymbol{I} - \boldsymbol{B})\boldsymbol{A}_r^T)$ (Eq. 16.3). The mapping matrix $\boldsymbol{A}_r$ is known for each measurement region $\boldsymbol{r}_r$ and $\boldsymbol{\gamma}$, σ^2 and $\boldsymbol{B}$ are estimated as $\tilde{\boldsymbol{\gamma}}, \tilde{\sigma}^2, \tilde{\boldsymbol{B}}$ by EM algorithm. Based on the consistency of EM estimates, we have asymptotically:

$$\left\{ \tilde{e}_r = \left[A_r (I - \tilde{B}) A_r^T \right]^{-\frac{1}{2}} \left[X(m_r) - A_r Z(r_r)\tilde{\gamma}) \right] \right\}_{r=1}^{R}$$

$$\sim \text{ i.i.d. } N(0, \sigma^2) \tag{16.12}$$

Therefore, we may perform Anderson–Darling tests on transformed residuals $\{\tilde{e}_r\}_{r=1,2,...,R}$ to determine the existence of structural defects [20, 25]. With this strategy, all datasets here containing up to 30% missing observations are detected to have defects. The defect detection result is 100% correct.

Real case studies in this section show that our new method of interaction analysis guarantees to reduce metrology efforts more than 30% and at the same time provides comparable interaction estimation accuracy and competitive defect detection precision. It thus provides a supporting tool for scale-up nanomanufacturing.

In this chapter, we introduce both sampling strategies and analysis techniques to map nanostructure interactions across the whole substrate with minimum metrology efforts and best interaction estimation efficiency.

To minimize metrology efforts, we sample multiple tiny regions over the substrate and develop tailored space filling designs to selectively measure partial of the sites in each measurement region. At the same time, to enhance interaction estimation efficiency, we take advantages of the Markovian property of our interaction modeling and optimize the Expectation-Maximization algorithm by classifying non-measurement sites into conditionally independent groups. We focus on stationary interaction analysis in this chapter. Similar to interaction analysis with complete feature measurement, this stationary interaction estimation quantifies feature variation for normally grown nanostructures and provides initial defect detection for fields with defective structures.

Both simulation and real case studies show that our methods achieve comparable interaction estimation efficiency with significantly reduced SEM measurement. Our approach of interaction analysis with incomplete feature measurement enables further investigations of nanostructure growth kinetics. It thus serves as a supporting tool for nanomanufacturing.

Acknowledgments The work in this chapter was supported by US National Science Foundation under grant number CMMI-1000972.

References

1. Besag, J.: Spatial interaction and the statistical analysis of lattice systems. J. Roy. Stat. Soc. B (Methodol.) **36**(2), 192–236 (1974)
2. Brook, D.: On the distinction between the conditional probability and the joint probability approaches in the specification of nearest-neighbour systems. Biometrika **51**(3), 481–483 (1964)
3. Cressie, N.: Statistics for Spatial Data, 2nd edn. Wiley, New York (1993)

4. Dempster, A., Laird, N., Rubin, D.: Maximum likelihood from incomplete data via the em algorithm. J. Roy. Stat. Soc. B (Methodol.) **39**(1), 1–38 (1977)
5. Fang, K., Lin, D., Winker, P., Zhang, Y.: Uniform design: theory and application. Technometrics, 237–248 (2000)
6. Guyon, X.: Parameter estimation for a stationary process on a d-dimensional lattice. Biometrika **69**(1), 95 (1982)
7. Huang, Q.: Physics-driven bayesian hierarchical modeling of the nanowire growth process at each scale. IIE Trans. **43**(1), 1–11 (2011)
8. Husslage, B., Van Dam, E., Den Hertog, D.: Nested maximin latin hypercube designs in two dimensions. CentER Discussion Paper No. 2005–79 (2005)
9. Johnson, M., Moore, L., Ylvisaker, D.: Minimax and maximin distance designs. J. Stat. Plann. Infer. **26**(2), 131–148 (1990)
10. Kauermann, G., Tutz, G.: Local likelihood estimation in varying-coefficient models including additive bias correction. J. Nonparametric Stat. **12**(3), 343–371 (2000)
11. Laarhoven, P., Aarts, E.: Simulated Annealing: Theory and Applications, vol. 37. Springer (1987)
12. Little, R., Rubin, D.: Statistical Analysis with Missing Data. Wiley, New York (1987)
13. McKay, M., Beckman, R., Conover, W.: A comparison of three methods for selecting values of input variables in the analysis of output from a computer code. Technometrics, 239–245 (1979)
14. Morris, M., Mitchell, T.: Exploratory designs for computational experiments. J. Stat. Plann. Infer. **43**(3), 381–402 (1995)
15. Postek, M., Hocken, R.: Instrumentation and metrology for nanotechnology: report of the national nanotechnology initiative workshop, January 27–29, 2004. Monograph (technical report), National Nanotechnology Coordination Office (2006)
16. Qian, P.: Nested latin hypercube designs. Biometrika **96**(4), 957–970 (2009)
17. Qian, P., Ai, M., Wu, C.: Construction of nested space-filling designs. Ann. Stat. **37**(6A), 3616–3643 (2009)
18. Qian, P., Tang, B., Jeff Wu, C.: Nested space-filling designs for computer experiments with two levels of accuracy. Statistica Sinica **19**(1) (2009)
19. Rennen, G., Husslage, B., Van Dam, E., Den Hertog, D.: Nested maximin latin hypercube designs. Struct. Multidiscip. Optim. **41**(3), 371–395 (2010)
20. Thode, H.: Testing for Normality. CRC, New York (2002)
21. van Dam, E., Husslage, B., den Hertog, D.: One-dimensional nested maximin designs. J. Global Optim. **46**(2), 287–306 (2010)
22. Wang, F., Hwang, Y., Qian, P., Wang, X.: A statistics-guided approach to precise characterization of nanowire morphology. ACS Nano **4**(2), 855–862 (2010)
23. Whittle, P.: On stationary processes in the plane. Biometrika **41**(3), 434–449 (1954)
24. Wu, C.: On the convergence properties of the em algorithm. Ann. Stat., 95–103 (1983)
25. Xu, L., Huang, Q.: Modeling the interactions among neighboring nanostructures for local feature characterization and defects detection. IEEE Trans. Autom. Sci. Eng. **Conditionally Accepted** (2012)

Index